D1240949

T
351
.5
J46
2006

INTERPRETING
ENGINEERING DRAWINGS

Fifth Canadian Edition

LIBRARY
NSCC, WATERFRONT CAMPUS
80 MAWIO'MI PLACE
DARTMOUTH, NS B2Y 0A5 CANADA

INTERPRETING ENGINEERING DRAWINGS

Fifth Canadian Edition

Cecil H. Jensen and Jay D. Helsel

LIBRARY
NSCC, WATERFRONT CAMPUS
80 MAWIO'MI PLACE
DARTMOUTH, NS B2Y 0A5 CANADA

THOMSON

NELSON

Australia • Canada • Mexico • Singapore • Spain • United Kingdom • United States

THOMSON
NELSON

Interpreting Engineering Drawings
Fifth Canadian Edition

by Cecil H. Jensen and Jay D. Helsel

Associate Vice President, Editorial Director:
Evelyn Veitch

Publisher:
Veronica Visentin

Senior Acquisitions Editor:
Kevin Smulan

Marketing Manager:
Nigel Corish

Developmental Editors:
Elaine Freedman/Heather Parker

Permissions Coordinator:
Kristiina Bowering

Production Service:
Lachina Publishing Services

Copy Editor:
Elaine Freedman

Proofreader:
Diane Kimmel

Indexer:
Katherine Stimson

Manufacturing Coordinator:
Charmaine Lee-Wah

Design Director:
Ken Phipps

Cover Design:
Pam Connell

Compositor:
Carol Kurila

Printer:
Edwards Brothers

COPYRIGHT © 2007, 2002 by Nelson, a division of Thomson Canada Limited.

Adapted from *Interpreting Engineering Drawings*, Seventh Edition, by Cecil H. Jensen and Jay D. Helsel, published by Delmar Learning. Copyright © 2007 by Thomson Delmar Learning.

Printed and bound in the United States
1 2 3 4 09 08 07 06

For more information contact Nelson, 1120 Birchmount Road, Toronto, Ontario, M1K 5G4. Or you can visit our Internet site at http://www.nelson.com

ALL RIGHTS RESERVED. No part of this work covered by the copyright herein may be reproduced, transcribed, or used in any form or by any means—graphic, electronic, or mechanical, including photocopying, recording, taping, Web distribution, or information storage and retrieval systems—without the written permission of the publisher.

For permission to use material from this text or product, submit a request online at www.thomsonrights.com

Every effort has been made to trace ownership of all copyrighted material and to secure permission from copyright holders. In the event of any question arising as to the use of any material, we will be pleased to make the necessary corrections in future printings.

Library and Archives Canada Cataloguing in Publication Data

Jensen, C. H. (Cecil Howard), 1925-
  Interpreting engineering drawings / Cecil H. Jensen, Jay D. Helsel. -- 5th Canadian ed.

Includes bibliographical references and index.

ISBN 13: 978-0-17-641609-6
ISBN 10: 0-17-641609-9

1. Engineering drawings. I. Helsel, Jay D. II. Title.

T351.5.J46 2006   604.2'5
C2006-906235-8

CONTENTS

Preface xv
About the Authors xvii
Acknowledgments xviii

Unit 1   1

Bases for Interpreting Drawings 1
   Commonly Used Descriptive Terms 1
   The Need for Standardization 1
Engineering Drawings 2
Line Styles and Lettering 2
   Line Styles 2
   Lettering 3
   Symbols and Abbreviations 3
Sketching 3
Materials for Sketching 4
   Sketching Paper 4
   Pencils and Erasers 5
   Trigonometry Set 5
   Templates 5
Sketching Techniques 5
Information Shown on Assignment
   Drawings 6
References 6
Internet Resources 6
Assignments
   A-1M . . .Sketching Lines, Circles,
            and Arcs 7
   A-2 . . . .Inlay Designs 7

Unit 2   8

Lines Used to Describe the Shape of a Part . .8
   Visible Lines 8
   Hidden Lines .8
   Break Lines 8
Title Blocks and Title Strips 9
Drawing to Scale 9

Reference 10
Internet Resources 10
Assignments
   A-3 . . . .Garden Gate 11
   A-4 . . . .Roof Truss 12

Unit 3   13

Circular Features 13
Centre Lines 13
Sketching Circles and Arcs 13
   Using a Circle Template 14
   Using a Compass 14
   Using Freehand Sketching Techniques . .16
   Sketching a Complete View
      Containing Circles and Arcs 16
References 17
Internet Resources 17
Assignments
   A-5 . . . .Sketching Circles
            and Arcs–1 18
   A-6M . . .Sketching Circles
            and Arcs–2 19

Unit 4   20

Working Drawings 20
Arrangement of Views 21
ISO Projection Symbol 22
Third-Angle Projection 22
   Objects with Circular Figures 22
Sketching Views in Third-Angle
   Projection 24
   Using a Mitre Line to Construct the
      Right-Side View 24
   Using a Mitre Line to Construct the
      Top View 24
References 26

Internet Resources 26
Assignments
    A-7 . . . .Matching Drawings–1 28
    A-8 . . . .Matching Drawings–2 29
    A-9 . . . .Orthographic Sketching–
              Visible and Hidden Lines 30
    A-10 . . .Orthographic Sketching
              of Parts with Circular
              Features 31
    A-11 . . .Orthographic Sketching
              of Parts Having Flat
              Surfaces–Decimal-Inch
              Dimensioning 32
    A-12M . .Orthographic Sketching
              of Parts with Flat
              Surfaces–Millimetre
              Dimensioning 33
    A-13 . . .Orthographic Sketching of
              Parts with Circular
              Features–Decimal-Inch
              Dimensioning 34

Unit 5   35

Dimensioning 35
  Reading Direction 35
Dimensioning Flat Surfaces 35
  Dimension Lines 35
  Extension Lines 35
  Leaders 35
Linear Units of Measurement 35
  Inch Units of Measurement 36
  SI (Metric) Units of Measurement 36
Choice of Dimensions 38
Basic Rules for Dimensioning 38
Dimensional Cylindrical Features 38
Dimensioning Cylindrical Holes 39
Drilling, Reaming, and Boring 39
Dimensioning Rounds and Fillets 41
Dimensioning Repetitive Features 41
Identifying Similarly Sized Features 41
References 42
Internet Resources 42
Assignments
    A-14 . . .Feed Hopper 43
    A-15 . . .Coupling 44
    A-16 . . .Third-Angle Projection and
              Dimensioning 45
    A-17M . .Third-Angle Projection and
              Dimensioning 46

Unit 6   47

Inclined Surfaces 47

Measurement of Angles 47
Symmetrical Outlines 47
Machine Slots 49
Reference Dimensions 50
References 50
Internet Resources 50
Assignments
    A-18 . . .Base Plate 51
    A-19 . . .Compound Rest Slide 52
    A-20 . . .Orthographic Sketch of Objects
              Having Sloped Surfaces Using
              Grid Lines 54
    A-21 . . .Orthographic Sketching of
              Parts Having Sloped Surfaces
              Using Decimal-Inch
              Dimensioning 55
    A-22M . .Orthographic Sketching
              of Parts Having Sloped
              Surfaces Using Millimetre
              Dimensioning 56

Unit 7   57

Pictorial Sketching 57
  Viewing Direction 57
Isometric Sketching 58
  Isometric Grid Sheets 58
  Inclined Surfaces 58
  Circles and Arcs 58
  Basic Steps for Isometric Sketching
    (Figure 7–6) 59
Oblique Sketching 59
  Oblique Grid Sheets 60
  Inclined Surfaces 61
  Circles and Arcs 61
  Basic Steps for Oblique Sketching
    (Figure 7–13) 62
References 62
Internet Resource 62
Assignments
    A-23 . . .Pictorial Sketching of Parts
              with
              Flat Surfaces Using Decimal-
              Inch Dimensioning 64
    A-24M . .Pictorial Sketching of Parts
              with Flat Surfaces Using
              Metric Dimensioning 65
    A-25 . . .Pictorial Sketching of Parts
              with Circular Features Using
              Decimal-Inch Dimensioning . .66
    A-26M . .Pictorial Sketching of Parts
              with Circular Features Using
              Metric Dimensioning 67

Unit 8   **68**

Machining Symbols .68
   Indicating Machining Allowance 69
   Removal of Material Prohibited 70
Not-to-Scale Dimensions 70
Drawing Revisions .71
Break Lines 71
References 71
Internet Resources .73
Assignments
   A-27M . .Offset Bracket 74
   A-28   . . .Guide Bar 76

Unit 9   **78**

Sectional Views 78
   The Cutting-Plane Line 78
   Section Lining 79
Types of Sections .80
   Full Sections .80
   Half Sections .80
   Offset Sections 80
   Other Types of Sections 80
Countersinks, Counterbores,
and Spotfaces 82
Intersection of Unfinished Surfaces 83
Reference 83
Internet Resources .83
Assignments
   A-29   . . .Sketching Full Sections 85
   A-30   . . .Slide Bracket 86
   A-31M . .Base Plate 88
   A-32   . . .Sketching Half Sections 90

Unit 10   **91**

Chamfers 91
Undercuts 91
Tapers 91
   Conical Tapers 91
   Flat Tapers 91
Knurls 91
Reference 92
Internet Resources .92
Assignments
   A-33   . . .Handle 94
   A-34M . .Indicator Rod 95

Unit 11   **96**

Selection of Views 96
One- and Two-View Drawings 96
Multiple-Detail Drawings 96
Functional Drafting 97

References 97
Internet Resources .99
Assignments
   A-35   . . .Centering Connector
             Details 100
   A-36M . .Link 102

Unit 12   **103**

Surface Texture 103
   Surface Texture Definitions 103
Surface Texture Symbol 106
Surface Texture Ratings 106
   Notes 107
Control Requirements 107
References 111
Internet Resources 111
Assignments
   A-37M . .Caster Details 112
   Λ-38   . . .Hanger Details 114

Unit 13

Tolerances and Allowances 116
Definitions 116
Tolerancing Methods 117
   Limit Dimensioning 117
   Plus and Minus Tolerancing 117
   Inch Tolerances 118
   Millimetre Tolerances 118
References 119
Internet Resource 119
Assignments
   A-39   . . .Inch Tolerances and
             Allowances .120
   Λ-40M . .Millimetre Tolerances and
             Allowances .121

Unit 14   **122**

Inch Fits 122
   Clearance Fits 122
   Interference Fits 122
   Transition Fits 122
Description of Fits 122
   Running and Sliding Fits 122
   Location Fits .122
   Drive and Force Fits 124
Standard Inch Fits 124
   Running and Sliding Fits 124
   Locational Clearance Fits 125
   Locational Transition Fits 125
   Locational Interference Fits 125
   Force or Shrink Fits 125
   Basic Hole System 126

Basic Shaft System 126
References 126
Internet Resource 126
Assignments
   A-41 . . .Inch Fits–Basic Hole
         System 127
   A-42 . . .Inch Fits 128

Unit 15   129

Metric Fits 129
   Metric Tolerance Symbol 129
   Fit Symbol 130
   Hole Basis Fits System 130
   Shaft Basis Fits System 130
   Drawing Callout 131
Reference 132
Internet Resource 132
Assignments
   A-43M . .Metric Fits–Basic Hole
         System 136
   A-44M . .Metric Fits 137
   A-45M . .Bracket 138
   A-46M . .Swivel 140

Unit 16   141

Threaded Fasteners 141
   Thread Representation 141
Threaded Assemblies 142
   Thread Standards 142
Threaded Holes 142
   Inch Threads 142
Inch Thread Designation 144
Right- and Left- Handed Threads 144
Metric Threads 145
Metric Thread Designation 145
References 149
Internet Resources 149
Assignments
   A-47M . .Drive Support Details 150
   A-48 . . .Housing Details 152
   A-49M . .V-Block Assembly 154

Unit 17   155

Revolved and Removed Sections 155
   Resolved Sections 155
   Removed Sections 157
References 157
Internet Resources 157
Assignments
   A-50 . . .Shaft Intermediate
         Support 159
   A-51 . . .Shaft Supports 160

Unit 18   162

Keys 162
   Dimensioning of Keyways and
      Keyseats 163
Set Screws 163
Flats 164
Bosses and Pads 164
Dimension Origin Symbol 164
Rectangular Coordinate Dimensioning
   without Dimension Lines 165
Rectangular Coordinate Dimensioning
   in Tabular Form 165
References 166
Internet Resources 166
Assignments
   A-52 . . .Terminal Block 168
   A-53M . .Terminal Stud 169
   A-54 . . .Rack Details 170
   A-55 . . .Support Bracket 172

Unit 19   174

Oblique Surfaces 174
References 175
Internet Resource 175
Assignments
   A-56 . . .Identifying Oblique
         Surfaces 176
   A-57 . . .Completing Oblique
         Surfaces 176

Unit 20   178

Primary Auxiliary Views 178
References 179
Internet Resource 179
Assignments
   A-58 . . .Gear Box 181
   A-59 . . .Inclined Stop 182

Unit 21   183

Secondary Auxiliary Views 183
References 184
Internet Resource 184
Assignments
   A-60 . . .Hexagon Bar Support 186
   A-61 . . .Control Block 188

Unit 22   190

Development Drawings 190
Joints, Seams, and Edges 190
Sheet Metal Sizes 190

Straight Line Development 190
Stampings 192
Internet Resources 194
Assignments
A-62 . . .Letter Box 195
A-63M . .Bracket 195

Unit 23   196

Arrangement of Views 196
Reference 198
Internet Resource 198
Assignments
A-64 . . .Mounting Plate 199
A-65 . . .Index Pedestal 200

Unit 24   202

Piping 202
Kinds of Pipe 202
Pipe Joints and Fittings 203
Valves 204
Piping Drawings 204
Single-Line Drawings 204
Drawing Projection 206
Pipe Drawing Symbols 206
Dimensioning 206
Orthographic Piping Symbols 206
References 209
Internet Resources 209
Assignments
A-66 . . .Engine Starting Air System . .210
A-67 . . .Boiler Room 212

Unit 25   214

Bearings 214
Plain Bearings 214
Antifriction Bearings 215
Premounted Bearings 216
References 216
Internet Resources 216
Assignments
A-68M . .Adjustable Shaft Support . . .217
A-69 . . .Corner Bracket 218

Unit 26   220

Manufacturing Materials 220
Cast Irons 220
Types of Cast Iron 220
Steel 221
SAE and AISI Systems of Steel
Identification 221
Effect of Alloys on Steel 221

Structural Steel 223
Plastics 223
Thermoplastics 223
Thermosetting Plastics 223
Rubber 223
Mechanical Rubber 223
Cellular Rubber 223
Internet Resources 224
Assignments
A-70M . .Crossbar 225
A-71 . . .Oil Chute 226
A-72M . .Parallel Clamp Details 228
A-73M . .Caster Assembly 229

Unit 27   230

Casting Processes 230
Sand Mould Casting 230
Full Mould Casting 232
Casting Design 232
Simplicity of Moulding from Flat
Back Patterns 232
Irregular or Odd-Shaped Castings 232
Set Cores 233
Coping Down 234
Split Patterns 234
Core Castings 234
Machining Lugs 235
Surface Coatings 235
References 236
Internet Resources 236
Assignments
A-74 . . .Offset Bracket 237
A-75 . . .Trip Box 238
A-76 . . .Auxiliary Pump Base 240
A-77M . .Slide Valve 242

Unit 28   243

Chain Dimensioning 243
Base Line (Datum) Dimensioning 243
References 243
Internet Resource 243
Assignments
A-78 . . .Interlock Base 246
A-79M . .Contact Arm 248
A-80 . . .Contactor 249

Unit 29   250

Alignment of Parts and Holes 250
Foreshortened Projection 251
Holes Resolved to Show True Centre
Distance 251
Partial Views 251

Naming of Views for Spark Adjuster 252
Drill Sizes 252
References 252
Internet Resources 253
Assignments
   A-81 . . .Spark Adjuster 254
   A-82 . . .Control Bracket 256

Unit 30   258

Broken-Out and Partial Sections 258
Webs in Section 258
Ribs in Section 258
Spokes in Section .261
References 261
Internet Resource .261
Assignments
   A-83M . .Raise Block 262
   A-84   . . .Coil Frame 264

Unit 31   266

Pin Fasteners 266
   Machine Pins .266
   Radial-Locking Pins 266
Section through Shafts, Pins, and Keys . . .270
Arrangement of Views of Assignment
   A-85M 270
References 271
Internet Resources 271
Assignments
   A-85M . .Spider 272
   A-86   . . .Hood 274

Unit 32   276

Drawings for Numerical Control 276
Dimensioning for Numerical Control 276
Dimensioning for Two-Axis Coordinate
   System 276
   Origin (Zero Point) 277
   Setup Point 277
   Relative Coordinate (Point-to-Point)
     Programming 278
   Absolute Coordinate Programming 278
Internet Resources 278
Assignments
   A-87   . . .Cover Plate 279
   A-88M . .Terminal Board 280

Unit 33   281

Assembly Drawings 281

Subassembly Drawings 281
   Identifying Parts of an Assembly
     Drawing 281
Bill of Material (Items List) 283
Helical Springs 283
References 285
Internet Resources 285
Assignments
   A-89   . . .Fluid Pressure Valve 286
   A-90M . .Parallel Clamp Assembly . . .288

Unit 34   289

Structural Steel Shapes 289
   Abbreviations .289
Phantom Outlines 290
Conical Washers .291
References 291
Internet Resources 291
Assignments
   A-91   . . .Four-Wheel Trolley 292

Unit 35   294

Welding Drawings 294
Welding Symbols .294
   Tail of Welding Symbol 296
   Multiple Reference Lines 297
   Weld Locations on Symbol 297
Fillet Welds 297
References 299
Internet Resources 299
Assignments
   A-92   . . .Fillet Welds 301
   A-93   . . .haft Support 302

Unit 36   303

Groove Welds 303
Supplementary Symbols 305
   Back and Backing Welds 305
   Melt-Through Symbol 307
References 307
Internet Resources 307
Assignments
   A-94   . .Base Skid 308
   A-95   . . .Groove Welds 310

Unit 37   311

Other Basic Welds 311
   Plug and Slot Welds 311
   Plug Welds (Figure 37-1) 311

Slot Welds (Figure 37-2) 312
Spot Welds (Figure 37-3) 313
Seam Welds (Figure 37-4) 314
Flange Welds (Figure 37-5) 316
References 317
Internet Resources 317
Assignment
   A-96 . . .Base Assembly 318
   A-97 . . .Plug, Slot, and Spot
            Welds 320
   A-98 . . .Seam and Flange Welds 321

Unit 38   322

Gears 322
Spur Gears 323
   Gear Terms 323
   Chordal Thickness and Corrected
      Addendum 324
   Working Drawings of Spur Gears 325
   Examples of Spur Gear Calculations . . .326
Reference 326
Internet Resources 326
Assignments
   A-99 . . .Spur Gear 328
   A-100 . .Spur Gear Calculations 330

Unit 39   331

Bevel Gears 331
Reference 331
Internet Resources 332
Assignment
   A-101 . .Mitre Gear 334

Unit 40   336

Gear Trains 336
   Centre Distance 336
   Motor Drive .337
Internet Resources 337
Assignments
   A-102 . .Motor Drive Assembly 338
   A-103 . .Gear Train Calculations . . .340

Unit 41   341

Cams 341
   Types of Cams 341
   Cam Displacement Diagrams 341
Internet Resources 342
Assignments
   A-104 . .Cylindrical Feeder Cam . . .344
   A-105 . .Plate Cam 346

Unit 42   347

Antifriction Bearings 347
   Ball Bearings 347
   Roller Bearings 349
Retaining Rings 349
O-Ring Seals 349
Clutches 349
Belt Drives 349
   V-Belt Sizes .352
   Sheaves and Bushings 353
References 353
Internet Resources 353
Assignment
   A-106 . .Power Drive 354

Unit 43   356

Ratchet Wheels 356
   Mechanical Advantage 356
Internet Resources 357
Assignment
   A-107 . .Winch 358

Unit 44   360

Modern Engineering Tolerancing 360
Geometric Tolerancing 362
   Points, Lines, and Surfaces 362
Feature Control Frame 362
   Application to Surfaces
      (Figure 44–6A) 363
   Application to Features of Size
      (Figure 44–6B) 364
Form Tolerances 365
Straightness 365
Straightness Controlling Surface Elements 365
References 368
Internet Resources 368
Assignment
   A-108 . .Straightness Tolerance
            Controlling Surface
            Elements 370

Unit 45   372

Straightness of a Feature of Size 372
   Features of Size 372
   Circular Tolerance Zones 372
Feature of Size Definitions 372
Material Condition Symbols 374
   Applicability of RFS, MMC,
      and LMC .374

Examples 374
Maximum Material Condition (MMC) . . .375
   Application with Maximum Value 376
Regardless of Feature Size (RFS) 376
Least Material Condition (LMC) 376
Straightness of a Feature of Size 376
   Straightness—RFS 377
   Straightness—MMC 377
   Straightness—Zero MMC 377
   Straightness with a Maximum Value . . .377
   Straightness per Unit Length 377
References 377
Internet Resources 378
Assignment
   A-109 . . Straightness of a Feature
            of Size 382

Unit 46   384

Form Tolerances .384
Flatness 384
   Flatness per Unit Area 384
Circularity 385
   Circularity of Noncylindrical Parts 386
Cylindricity 386
References 387
Internet Resources 388
Assignment
   A-110 . . Form Tolerances 390

Unit 47   392

Datums and the Three-Plane Concept 392
Datums for Geometric Tolerancing 392
Three-Plane System 392
   Primary Datum 392
   Secondary Datum 393
   Tertiary Datum 393
Uneven Surfaces .393
Datum Feature Symbol 395
   For Datum Features Not Subject
      to Size Variation 395
   For Datum Features Subject to Size
      Variation 396
   Former ANSI Datum Feature
      Symbol 397
   Association with Geometric
      Tolerances .397
   Multiple Datum Features 398
   Partial Surfaces as Datums 398
References 399
Internet Resources 399
Assignments

A-111 . . Datums 400
A-112M . Axle 402

Unit 48   403

Orientation Tolerances 403
   Reference to a Datum 403
   Angularity Tolerance 403
   Perpendicularity Tolerance 404
   Parallelism Tolerance 404
Orientation Tolerancing for Flat
   Surfaces 405
   Control in Two Directions 405
   Applying Form and Orientation
      Tolerances to a Single Feature 405
References 406
Internet Resources 406
Assignments
   A-113 . . Stand 407
   A-114M . Cut-Off Stand 408

Unit 49   409

Orientation Tolerancing for Features
   of Size 409
   Angularity Tolerance 409
   Perpendicularity Tolerance 409
   Parallelism Tolerance 411
   Control in Two Directions 412
   Control on an MMC Basis 412
Internal Cylindrical Features 412
   Specifying Parallelism for an Axis 413
Perpendicularity for a Median Plane 413
Perpendicularity for an Axis (Both
   Feature and Feature and Datum RFS) . .413
Perpendicularity for an Axis (Tolerance
   at MMC) 414
Perpendicularity for an Axis (Zero
   Tolerance at MMC) 414
Perpendicularity with a Maximum
   Tolerance Specified 414
External Cylindrical Features 415
   Perpendicularity for an Axis
      (Pin or Boss RFS) 415
   Perpendicularity for an Axis (Pin or
      Boss at MMC) 415
References 417
Internet Resources 418
Assignment
   A-115 . . Orientation Tolerancing for
            Features of Size 420

Unit 50   421

Datum Targets 421
Datum-Target Symbol 421
  Identification of Targets 422
  Targets Not in the Same Plane 424
  Dimensioning for Target Location 425
References 425
Internet Resources 426
Assignment
  A-116 . .Bearing Housing 427

Unit 51   428

Tolerancing of Features by Position 428
Tolerancing Methods 428
Coordinating Tolerancing 429
  Maximum Permissible Error 429
  Use of Chart .431
Advantages of Coordinate Tolerancing . . .432
Disadvantages of Coordinate
  Tolerancing .432
Positional Tolerancing 432
Material Condition Basis 432
Positional Tolerancing for Circular
  Features 434
  Positional Tolerancing—MMC 434
  Positional Tolerancing at Zero
    MMC 436
  Positional Tolerancing—RFS 438
  Positional Tolerancing—LMC 438
Advantages of Positional Tolerancing 438
References 439
Internet Resources 440
Assignment
  A-117 . .Positional Tolerancing 442

Unit 52   444

Selection of Datum Features for
  Positional Tolerancing 444
Long Holes 444
Circular Datums 445
Multiple Hole Datums 445
References 445
Internet Resources 446
Assignment
  A-118 . .Datum Selection for
    Positional Tolerancing 450

Unit 53   451

Profile Tolerances 451

Profile Symbols 451
Profile of a Line 451
  Bilateral and Unilateral Tolerances 452
  Specifying All-Around Profile
    Tolerancing 452
  Method of Dimensioning 452
  Extent of Controlled Profile 453
Surface Profile (Profile of a Surface) 454
References 456
Internet Resources 456
Assignments
  A-119 . .Profile Tolerancing 457

Unit 54   458

Runout Tolerances 458
Circular Runout 458
Total Runout 460
Establishing Datums 461
  Measuring Principles 461
References 463
Internet Resources 463
Assignments
  A-120 . .Runout Tolerances 464
  A-121 . .Housing 466
  A-122M .End Plate 468

APPENDIX   469

Table 1 Chart for Converting Inch
  Dimensions to Millimetres 469
Table 2 Abbreviations and Symbols Used
  on Technical Drawings 470
Table 3 Number and Letter Size Drills . . .471
Table 4 Metric Twist Drill Sizes 472
Table 5 Unified and American (Inch)
  Threads .473
Table 6 Metric Threads 474
Table 7 Common Cap Screws 475
Table 8 Hexagon-Head Bolts and Cap
  Screws 476
Table 9 Set Screws 477
Table 10 Hexagon-Head Nuts 478
Table 11 Hex Flange Nuts 479
Table 12 Common Washer Sizes 480
Table 13 Square and Flat Stock Keys 482
Table 14 Woodruff Keys 482
Table 15 American Standard Wrought
  Steel Pipe 483
Table 16 Sheet Metal Gauges and
  Thicknesses 484
Table 17 Running and Sliding Fits (Values
  in Thousandths of an Inch) . . .485

Table 18 Locational Clearance Fits
(Values in Thousandths
of an Inch) 486
Table 19 Locational Transition Fits
(Values in Thousandths
of an Inch) 487
Table 20 Locational Interference Fits
(Values in Thousandths
of an Inch) 488
Table 21 Force and Shrink Fits
(Values in Thousandths
of an Inch) 489
Table 22 Preferred Hole Basis Metric Fits
Description 490
Table 23 Preferred Shaft Basis Metric Fits
Description 491
Table 24 Preferred Hole Basis Metric Fits
(Values in Millimetres) 492
Table 25 Preferred Shaft Basis Metric Fits
(Values in Millimetres) 494
Table 26 Metric Conversion Tables 496

INDEX   *497*

PREFACE

We are proud to present the Fifth Canadian Edition of *Interpreting Engineering Drawings*, the most comprehensive and up-to-date text of its kind. The authors have worked diligently to provide a text that will best prepare students to enter twenty-first-century technology-intensive industries. It is also useful for people working in technology-based industries who feel the need to enhance their understanding of key aspects of twenty-first-century technology. To that end, the text offers the flexibility needed to provide instruction in as broad a customized program of study as is required or desired. Clearly, it provides the theory and practical application for individuals to develop the intellectual skills needed to communicate technical concepts used throughout the international marketplace.

Flexibility is the key to developing a program of studies to meet the needs of every student. *Interpreting Engineering Drawings* is designed to allow instructors and students to choose specific units of instruction based on individual needs and interests. While all students should cover the core material in Units 1 to 20, advanced topics in the remaining 34 units provide opportunities for them to learn selected advanced subjects or a broad range of subjects spread over nearly all aspects of modern industry. Ancillary materials on the Instructor's CD-ROM and *Internet Resources* listed at the end of each unit provide for a more in-depth coverage of the material. Through the use of these materials, the depth of understanding achieved is limited only by the students' time constraints and their desire to master the material provided.

The entire text is developed around the latest drafting standards accepted throughout industry. This includes both decimal-inch and metric (millimetre) sizes and related concepts. Both systems are introduced early in the text and are reinforced in both theory and practical application through the assignments at the end of each unit and the Appendix. Tables in the Appendix are given in both systems of measure.

Features that made *Interpreting Engineering Drawings* highly successful in previous editions continue to be used in the Fifth Canadian Edition. For example, the text carefully examines the basic concepts needed to understand technical drawings and methodically takes the student through progressively more complex issues. Plenty of carefully developed illustrations, most using a second colour, provide a clear understanding of material covered in the written text. Assignments at the end of each unit are designed to measure students' understanding of the material covered as well as to reinforce the theoretical concepts. Only after the students develop a clear understanding of basic concepts are they introduced to more advanced concepts, such as modern engineering tolerancing (geometric tolerancing and true positioning), manufacturing materials and processes, welding drawings, and piping.

While *Interpreting Engineering Drawings* has always used sketching practices as a means of reinforcing students' understanding of technical information, this edition greatly expands this important technique. Not only does sketching enhance the students' understanding of technical concepts, it also enhances their ability to communicate technical concepts more effectively.

In keeping with the dynamic changes in the field of engineering graphics, various new features have been added to this edition.

NEW FEATURES OF THE FIFTH CANADIAN EDITION

■ **Additional assignments.** Significantly more assignments designed to broaden and reinforce basic concepts have been added to Units 1 to 20, the core of all print-reading concepts for mechanical programs.

■ **Internet resources.** Considerably more information is provided by listing important Internet resources at the end of each unit.

■ **Test generator.** For evaluating the progress of each student, a bank of test questions and answers (test generator) has been provided for each unit in the text.

■ **Instructor's CD-ROM.** The complete Instructor's Manual is now packaged on CD-ROM for faster and more convenient access to review questions, answers, problem solutions, and tests.

■ **Standards update.** All drawings in the text have been updated to conform to the latest CSA and ASME drawing standards.

■ **Expanded sketching practice.** As computer-aided drafting continues to replace board drafting, the use of sketching techniques as a means of better understanding and communicating technical information has been greatly expanded throughout this text.

■ **Co-author.** A co-author has been added to offer his expertise and experience to this edition.

The authors and the publisher hope you find the Fifth Canadian Edition of *Interpreting Engineering Drawings* to be as practical and as useful as you have the previous editions. If you have questions or comments about the book, please feel free to contact the publisher.

ABOUT THE AUTHORS

Cecil H. Jensen took an early retirement from teaching to devote his time to technical writing. He held the position of Technical Director at the McLaughlin Collegiate and Vocational Institute, Oshawa, Ontario and had 27 years of teaching experience in mechanical drafting. He was an active member of the Canadian Standards Association (CSA) Committee on Technical Drawings. Mr. Jensen represented Canada at ISO conferences on engineering drawing standards, which took place in Oslo, Norway and Paris, France. He also represented Canada on the ANSI Y14.5M Committee on Dimensioning and Tolerancing. He was the successful author of numerous texts, including *Engineering Drawing and Design*, *Fundamentals of Engineering Drawing*, *Geometric Dimensioning and Tolerancing for Engineering and Manufacturing Technology*, *Computer-Aided Engineering Drawing*, and *Home Planning and Design*. Before he began teaching, Mr. Jensen spent several years in industrial design. He also supervised the evening courses in Oshawa and was responsible for teaching selected courses for General Motors Corporation apprentices. Cecil Jensen passed away in April, 2005.

Jay D. Helsel has worked more than 35 years in education, having served as a professor of applied engineering and technology courses, chairperson of the Department of Applied Engineering and Technology, and Vice-President for Administration and Finance at California University of Pennsylvania. Dr. Helsel has had extensive experience teaching mechanical drafting at both the secondary and post-secondary levels and has worked in industry as well. He holds an undergraduate degree from California University of Pennsylvania, a master's degree from the Pennsylvania State University, and a doctoral degree from the University of Pittsburgh. Dr. Helsel is now a full-time writer and has authored such publications as *Engineering Drawing and Design*, *Fundamentals of Engineering Drawing*, *Programmed Blueprint Reading*, *Computer-Aided Engineering Drawing*, and *Mechanical Drawing: Board and CAD Techniques*, as well as various workbooks and other ancillary products associated with the above publications.

ACKNOWLEDGMENTS

The authors would like to thank and acknowledge the many professionals who reviewed the Fourth Canadian Edition in order to direct the revision to create the Fifth Canadian Edition.

A special thank you to the following for their detailed comments that shaped this new edition:

Robert Deeks, Mohawk College

Ed Espin, Humber College

Charles John Holroyd, Algonquin College

Arthur J. Houghton, Fleming College

Dragan Jovanov, Sheridan College

BASES FOR INTERPRETING DRAWINGS

Commonly Used Descriptive Terms

When looking at objects, we normally see them as three-dimensional: having width, depth, and height; or length, width, and height. The choice of terms used depends on the shape and proportions of the object.

Spherical shapes, such as a basketball, would be described as having a certain *diameter* (one term).

Cylindrical shapes, such as a baseball bat, would have *diameter* and *length*. However, a hockey puck would have *diameter* and *thickness* (two terms).

Objects that are not spherical or cylindrical require three terms to describe their overall shape. The terms used for a car would probably be *length, width,* and *height;* for a filing cabinet—*width, height,* and *depth,* even though the longest measurement (length) could be the width, height, or depth; for a sheet of drawing paper—*length, width,* and *thickness.* The terms used are interchangeable according to the *proportions* of the object being described, and the *position* it is in when being viewed. For example, a telephone pole lying on the ground would be described as having *diameter* and *length,* but when placed in a vertical position, its dimensions would be *diameter* and *height.*

To avoid confusion, distances from left to right are referred to as width, distances from front to back as depth, and vertical distances (except when very small in proportion to the others) as height.

The Need for Standardization

Engineering drawings are more complicated and require a set of rules, terms, and symbols that everyone can understand and use. A drawing showing a part may be drawn in Toronto, the part made in Vancouver, and then the part sent to Hamilton for assembly. If this is to be successfully accomplished, the drawing must have only one interpretation.

Most countries set up standards committees to accomplish this feat. These committees must decide on such factors as the best methods of representation, dimensioning, and tolerancing, and the adopting of drawing symbols. Different styles of lines must be established to represent visible or hidden lines, or to indicate the centre of a feature. If only one interpretation of a drawing is to be made, then the rules must be followed and interpreted correctly.

In Canada, drawing standards are established by the *Canadian Standards Association* (CSA); in the United States by the *American Society of Mechanical Engineers* (ASME). Members of these committees are part of the worldwide committee on standardization, known as the *International Organization for Standardization* (ISO).

The drawings and information shown throughout this text are based on CSA drawing standards. In some areas of drawing practice, such as simplified drafting, national standards have not yet been established. The authors have, in such cases, adopted the practices used by leading industries in Canada and the United States.

Engineering, or *technical, drawings* furnish a description of the shape and size of an object. Other information necessary for constructing the object is given in a way that renders it readily recognizable to anyone familiar with engineering drawings.

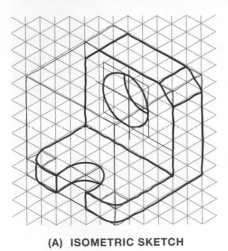

(A) ISOMETRIC SKETCH

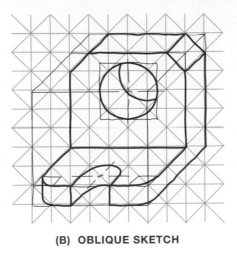

(B) OBLIQUE SKETCH

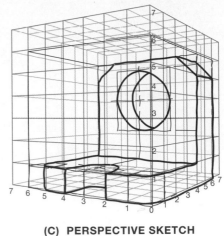

(C) PERSPECTIVE SKETCH

FIGURE 1–1 ■ Pictorial sketches

Pictorial drawings are similar to photographs because they show objects as they would appear to the observer, Figure 1–1. Such drawings, however, are not often used for technical designs because interior features and complicated detail are easier to understand and dimension on orthographic drawings. The drawings used in industry must clearly show the exact shape of objects. This usually cannot be accomplished in just one pictorial view, because many details of the object may be hidden or not clearly shown when the object is viewed from only one side.

For this reason, the drafter must show a number of views of the object as seen from different directions. These views, referred to as front view, top view, right-side view, and so forth, are systematically arranged on the drawing sheet and projected from one another, Figure 1–2. This type of projection is called *orthographic projection* and is explained in Unit 4. The ability to understand and visualize an object from these views is essential in the interpretation of engineering drawings.

ENGINEERING DRAWINGS

Throughout the history of engineering drawings, many drafting conventions, terms, abbreviations, and practices have come into common use. If drafting and sketching are to serve as a reliable means of communicating technical theory and applications, it is essential that all drafters, designers, and engineers use the same practices.

An engineering drawing consists of a variety of lines styles, symbols, and lettering. When positioned correctly on the drawing paper, they convey precise information to the reader.

LINE STYLES AND LETTERING

Line Styles

A line is the fundamental, and perhaps the most important single, entity on an engineering drawing. Lines are used to illustrate and describe the shape and size of objects that will later become real parts. The various types of lines used on engineering drawings form the alphabet of the drafting language. Like letters of the alphabet, they have different appearances. Two widths of lines are used.

Visible Lines

Thick continuous lines are used to indicate all visible edges of an object. These are known as *visible*, or *object*, lines. They should stand out clearly in contrast to other lines, so that the shape of an object is quickly apparent to the eye. Other types of lines are normally drawn as thin lines.

Construction Lines

In the initial sketch, light thin solid lines are used to develop the shape and location of features. These *construction* lines are normally left on the sketch. The applications of the other types of lines are explained in detail throughout this text.

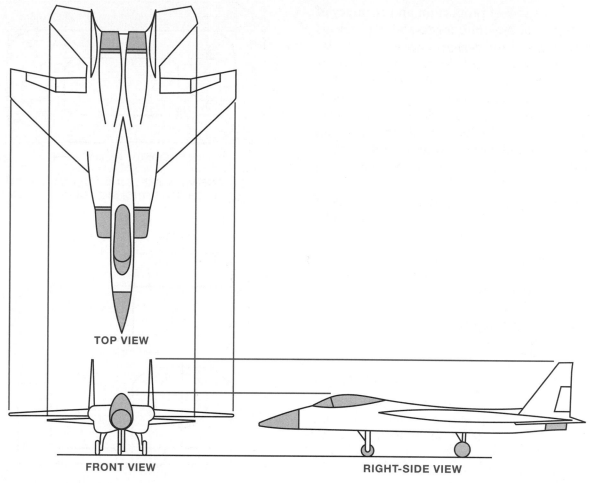

TOP VIEW

FRONT VIEW

RIGHT-SIDE VIEW

FIGURE 1–2 ■ Systematic arrangement of views

Lettering

The most important requirements for lettering in engineering sketches are legibility and reproducibility, best met by the style of lettering known as standard uppercase Gothic, shown in Figure 1–3. Suitable lettering size for notes and dimensions is .10 inch (in.) for decimal-inch drawings and 3 millimetres (mm) for metric drawings. Larger characters are used for titles and numbers, and where necessary to emphasize some part of the drawing.

Symbols and Abbreviations

Symbols and abbreviations are extensively used on engineering drawings. They reduce drawing time and save valuable drawing space. The symbols are truly a universal language, as their meanings are understood in all countries. The first abbreviations and symbols you will see on the drawings in this text are:

IN. *meaning* inch
mm *meaning* millimetre

ABCDEFGHIJKLMNOP
QRSTUVWXYZ
1234567890

FIGURE 1–3 ■ Recommended lettering for use on engineering drawings

FT *meaning* foot
Ø *meaning* diameter
R *meaning* radius

SKETCHING

Sketching is the simplest form of drawing, a quick way to express ideas. The drafter, technician, or engineer may use sketches to help simplify and explain (communicate) thoughts and concepts to other people.

Sketching is a part of drafting and design because the drafter frequently sketches ideas and designs prior to making the final drawing using CAD (computer-aided drawing). Practice in sketching helps develop a good sense of proportion and accuracy of observation. It is also helpful for resolving problems in the early stages of the design process.

CAD is replacing board drafting because of its speed, versatility, and economy. Sketching, like drafting, is also changing, and cost-saving methods are being used to produce a sketch. For example, using sketching paper reduces sketching time and produces a neater and more accurate sketch, with its built-in ruler for measuring distance and lines acting as a straightedge for drawing lines.

The entire sketch need not be drawn freehand if faster and easier methods can be used. For example, long lines can be drawn faster and more accurately with a straightedge. Large circles and arcs can be drawn or positioned with a compass; small circles and arcs with a circular template.

MATERIALS FOR SKETCHING

Sketching has two main advantages over formal drawing. First, only a few materials and instruments are required to produce a sketch. Second, you can produce a sketch anywhere. If many sketches are to be made, such as in working from this text, the following materials for sketching should be considered:

Sketching Paper

This type of paper has light lines. Various grid sizes (spacings) and formats are available to suit most drawing requirements.

Two-dimensional sketching paper is designed for single-view and multiview (orthographic) sketching, which are covered in this unit and in Unit 4. The paper has uniformly spaced horizontal and vertical lines forming squares. These are available in a number of grid sizes, Figure 1–4. The most commonly used spaces or grids are the decimal-inch, fractional-inch, and centimetre. These spaces are further subdivided into smaller spaces, such as eighths or tenths of an inch or millimetres. Since the units of measure are not shown on these sheets, the spaces can represent any desired unit of length.

Three-dimensional sketching paper is designed for sketching pictorial drawings. There are three basic types: isometric, oblique, and perspective, Figure 1–5.

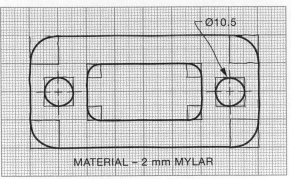

(A) ONE-VIEW SKETCH ON DECIMAL-INCH (.01 INCH DIVISIONS) SKETCHING PAPER

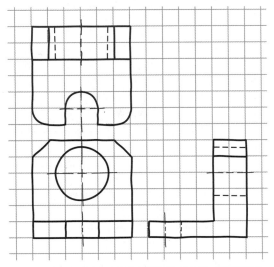

(B) ORTHOGRAPHIC SKETCH ON .25-INCH DIVISIONS SKETCHING PAPER

FIGURE 1–4 ■ Two-dimensional sketching paper

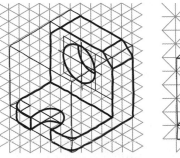 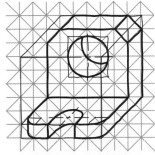

(A) ISOMETRIC SKETCH   **(B) OBLIQUE SKETCH**

FIGURE 1–5 ■ Three-dimensional sketching paper

Isometric sketching paper has evenly spaced lines running in three directions. Isometric sketching is covered in Unit 7.

Oblique sketching paper is similar to the two-dimensional sketching paper except that 45° lines, which pass through the intersecting horizontal and vertical lines, are added in one or both directions. Oblique sketching is covered in Unit 7.

One-, two-, and three-point perspective sketching papers are designed with worm's- and bird's-eye views. The spaces on the receding axis are proportionately shortened to create a perspective illusion. The sketches made on this type of paper provide a more realistic view than the sketches made on the isometric and oblique sketching papers.

Pencils and Erasers

Soft lead pencils (grades F, H, or HB), properly sharpened, are the best for sketching. Plastic or kneaded-rubber erasers, good for soft leads, are most commonly used.

Trigonometry Set

This small compact math set, which includes a compass, plastic ruler, and triangles, is very useful for sketching.

Templates

A circle template improves the quality of sketches by making circles and arcs neat and

uniform. It also reduces sketching time. Elliptical circle templates, used for pictorial sketching, are normally made available in the drafting classroom for use by students.

SKETCHING TECHNIQUES

In Figure 1–6, the following sketching techniques were used:

■ A 1-inch grid subdivided into tenths was selected for the part to be sketched. It required decimal-inch dimensioning. The part was sketched to half scale (half size). This type of sketching paper simplified the measuring of sizes and spacing and ensured accuracy when parallel and vertical lines were drawn. The grid lines also acted as guidelines for lettering notes and helped produce neat, legible lettering.

■ A straightedge was used for drawing long lines. This method was faster and more accurate than if the lines were drawn freehand.

■ A circle template was used for drawing the circular holes. Freehand sketching of round holes is time consuming, and not accurate or pleasing to the eye.

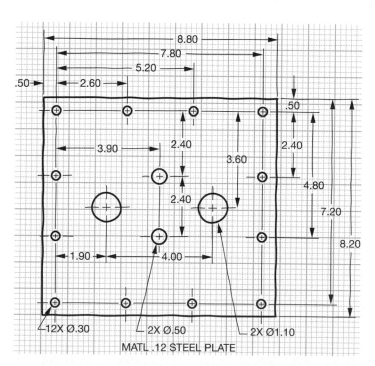

FIGURE 1–6 ■ Sketch of a cover plate

INFORMATION SHOWN ON ASSIGNMENT DRAWINGS

Assignment problems are either in inch units of measurement or millimetres (metric). Metric assignments are distinguishable by the letter M shown after the assignment number located at the bottom right-hand corner of the assignment sheet. Circled numbers and letters shown in colour are used to identify lines, distances, and surfaces, so that questions may be asked about these features, as shown on Assignment A-14. For clarity, the actual working drawing is shown in black. The information shown in colour is for instructional purposes only and would not appear on working drawings found in industry.

REFERENCES

CAN3-B78.1-M83 Technical Drawings—General Principles

ASME Y14.2M-1992 (R2003) Line Conventions and Lettering

ASME Y14.38-1999 Abbreviations and Acronyms

INTERNET RESOURCES

Canadian Standards Association For information on Canadian drafting standards, see: http://www.csa.ca

IDS Development–Nebraska Education For information on the various line types used on engineering drawings, see: http://idsdev.mccneb.edu/djackson/lineintro.htm

Incompetech For information on grid sheets, see: http://www.incompetech.com/beta/plainGraph Paper/

technologystudent.com For information on third-angle projection and related subjects, see: http://www.technologystudent.com/designpro/ortho2.htm

Wikipedia, the Free Encyclopedia For information on engineering drawings and various line types, see: http://en.wikipedia.org/wiki/Engineering_drawing

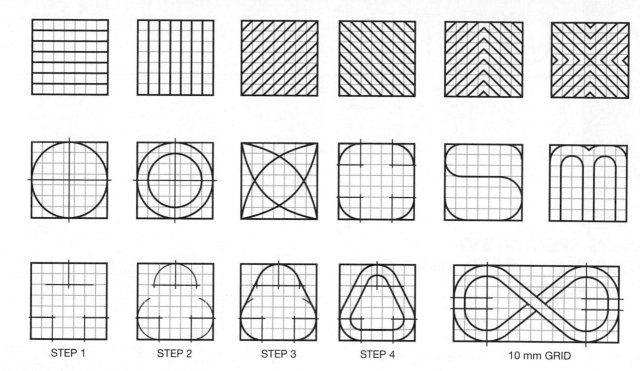

STEP 1   STEP 2   STEP 3   STEP 4   10 mm GRID

ASSIGNMENT:
ON A CENTIMETRE GRID (1 mm SQUARES),
SKETCH THE SHAPES SHOWN ABOVE. ALLOW
5 mm BETWEEN BLOCKS. THICK OBJECT LINES
ARE TO BE USED FOR THE SQUARES AND LINE
FEATURES. THIN LINES ARE TO BE USED FOR THE
CONSTRUCTION LINES.

| **SKETCHING LINES, CIRCLES, AND ARCS** | **A-1M** |

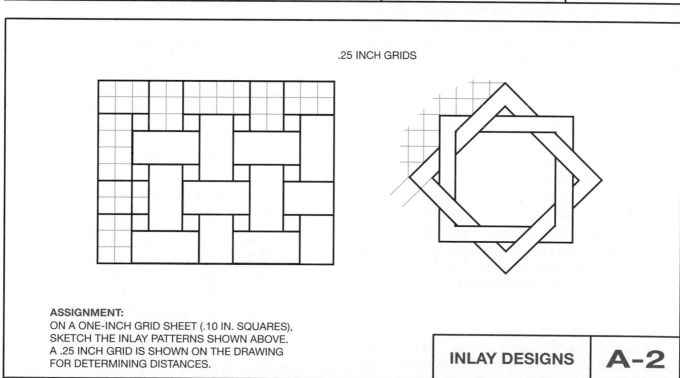

.25 INCH GRIDS

ASSIGNMENT:
ON A ONE-INCH GRID SHEET (.10 IN. SQUARES),
SKETCH THE INLAY PATTERNS SHOWN ABOVE.
A .25 INCH GRID IS SHOWN ON THE DRAWING
FOR DETERMINING DISTANCES.

| **INLAY DESIGNS** | **A-2** |

2 UNIT

LINES USED TO DESCRIBE THE SHAPE OF A PART

Visible Lines

Most objects drawn in engineering offices are complicated and contain many surfaces and edges. In Unit 1, thick solid lines, called *visible lines*, were introduced. Visible lines clearly stand out on the drawing and define the exterior shape of the object. However, many interior features, such as lines and holes, cannot be seen when viewed from the outside of the part. To show these hidden features, a special *hidden line* is used. Hidden and visible lines are used on engineering drawings to fully describe both the exterior and interior of the part.

Hidden Lines

Hidden lines, used to describe features that cannot be seen, are positioned on the view in the same manner as visible lines. These lines consist of short, evenly spaced thin dashes and spaces. The dashes are three to four times as long as the space. These lines should begin and end with a dash in contact with the line in which they start and end, except when such a dash would form a continuation of a visible line. Dashes should join at corners. Figure 2–1 shows some hidden line applications.

Break Lines

Break lines serve many purposes. They are used to shorten the view of long, uniform sections, which saves valuable drawing space, Figure 2–2(A). They are used to remove a segment of a part, which serves

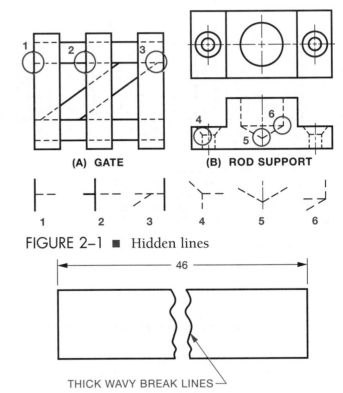

(A) GATE **(B) ROD SUPPORT**

FIGURE 2–1 ■ Hidden lines

THICK WAVY BREAK LINES

(A) SHORTENING LENGTH

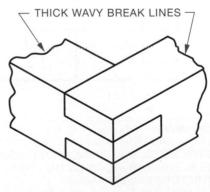

THICK WAVY BREAK LINES

(B) NOT SHOWING UNNECESSARY DETAIL

FIGURE 2–2 ■ The use of break lines

no useful purpose on the drawing, which saves valuable drawing or sketching time, Figure 2–2(B). The break line shown here is one of several break line styles used on engineering drawings. Other break lines are explained in Unit 8.

This particular type of break line is shown as a thick solid line since it forms part of the outline of the object being drawn. It is the third line style used to show the outline of a part.

TITLE BLOCKS AND TITLE STRIPS

Title blocks vary greatly and are usually preprinted. They are located in the lower right-hand corner of the drawing. The arrangement and size of the title block are optional, but the following four items must be shown:

■ Drawing number

■ Name of firm or organization

NORDALE MACHINE COMPANY TORONTO, ONTARIO		
PHONE 1-800-564-7832	E-MAIL NORDALE@att.net	
COVER PLATE		
MATL- SAE 1020 STL	NO. REQD- 4	
SCALE- 1:5	DN BY *D Scott*	**C2694**
DATE- 04/07/04	CH BY *E Johnson*	

FIGURE 2–3 ■ A typical title block

■ Title or description

■ Scale

The title block may also contain other pertinent information, such as the date of issue, signatures, approvals, and tolerance notes. A typical title block is shown in Figure 2–3.

In classrooms, where smaller sheet sizes are used, a title strip is commonly used. A typical title strip is shown in Figure 2–4(A). Unless otherwise designated by your instructor, the title strip shown in Figure 2–4(B) will be used on your sketching assignments.

DRAWING TO SCALE

When objects are drawn their actual size, the drawing is called *full scale* or *scale 1:1*. Many objects, however, including buildings, ships, and airplanes, are too large to be drawn full or *scale*. Therefore, they must be drawn to a *reduced scale*. An example would be the drawing of a house to a scale of 1:48 (1/4″ = 1 foot) in the inch-foot scale.

Frequently, small objects, such as watch parts, are drawn larger than their actual size to clearly define their shapes. This is drawing to an *enlarged scale*. The minute hand of a wrist watch, for example, could be drawn to scale 5:1 or 10:1.

Many mechanical parts are drawn to half scale, 1:2, and fifth scale, 1:5. The scale of the drawing is

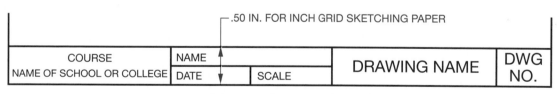

.50 IN. FOR INCH GRID SKETCHING PAPER

COURSE NAME OF SCHOOL OR COLLEGE	NAME		DRAWING NAME	DWG NO.
	DATE	SCALE		

(A) TYPICAL TITLE STRIP LAYOUT

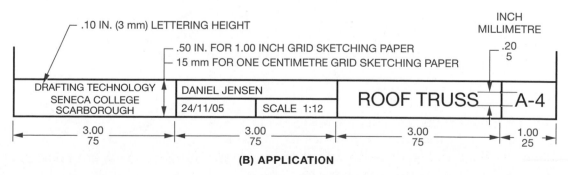

.10 IN. (3 mm) LETTERING HEIGHT

.50 IN. FOR 1.00 INCH GRID SKETCHING PAPER
15 mm FOR ONE CENTIMETRE GRID SKETCHING PAPER

INCH MILLIMETRE
.20
5

DRAFTING TECHNOLOGY SENECA COLLEGE SCARBOROUGH	DANIEL JENSEN		ROOF TRUSS	A-4
	24/11/05	SCALE 1:12		

3.00 / 75   3.00 / 75   3.00 / 75   1.00 / 25

(B) APPLICATION

FIGURE 2–4 ■ Title strips

expressed as a ratio: The left side of the ratio represents a unit of measurement of the size drawn; the right side represents the measurement of the actual object. Thus, 1 unit of measurement on the drawing equals 5 units of measurement on the actual object.

REFERENCE

ASME Y14.2M-1992 (R2003) Line Conventions and
Lettering

INTERNET RESOURCES

IDS Development–Nebraska Education For information on the various line types used on engineering drawings, see: http://idsdev.mccneb .edu/djackson/lineintro.htm

Metrication.com For information on metrication in drafting and engineering, see: http://www .metrication.com

Integrated Publishing For information on title blocks, see: http://www.tpub.com/content/ draftsman

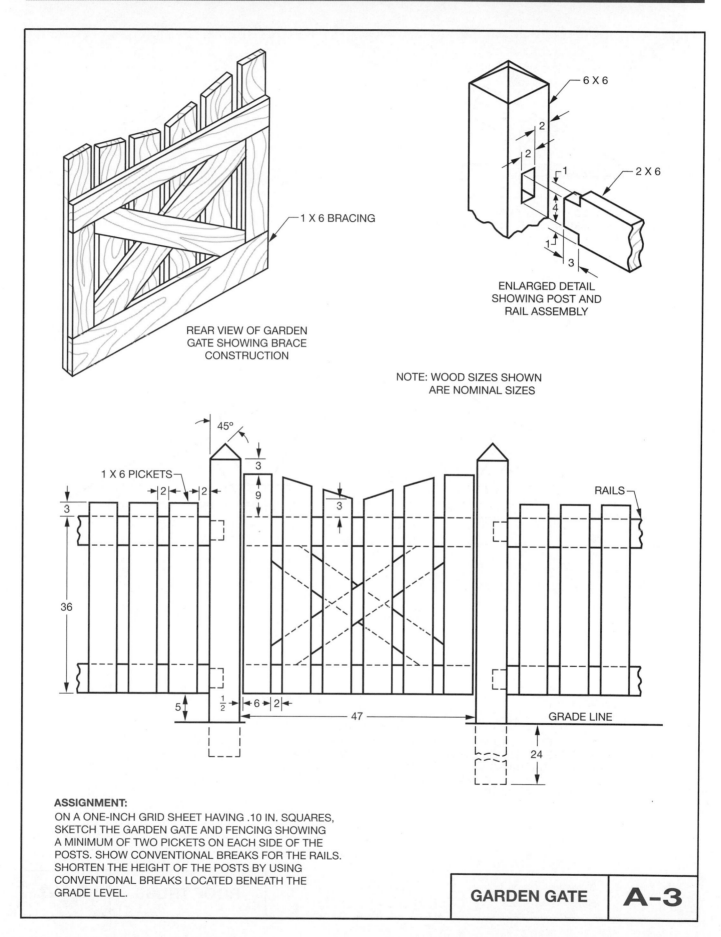

1 X 6 BRACING

REAR VIEW OF GARDEN
GATE SHOWING BRACE
CONSTRUCTION

6 X 6

2

2

1

4

1

3

2 X 6

ENLARGED DETAIL
SHOWING POST AND
RAIL ASSEMBLY

NOTE: WOOD SIZES SHOWN
ARE NOMINAL SIZES

45°

3

1 X 6 PICKETS

2

2

9

3

3

RAILS

36

5

½

6

2

47

GRADE LINE

24

ASSIGNMENT:
ON A ONE-INCH GRID SHEET HAVING .10 IN. SQUARES,
SKETCH THE GARDEN GATE AND FENCING SHOWING
A MINIMUM OF TWO PICKETS ON EACH SIDE OF THE
POSTS. SHOW CONVENTIONAL BREAKS FOR THE RAILS.
SHORTEN THE HEIGHT OF THE POSTS BY USING
CONVENTIONAL BREAKS LOCATED BENEATH THE
GRADE LEVEL.

GARDEN GATE | **A-3**

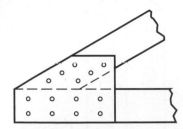

ENLARGED VIEW SHOWING NAILING
ARRANGEMENT OF .50 IN. THICK GUSSETS

NOTE: LUMBER SIZES SHOWN
ARE NOMINAL INCH SIZES

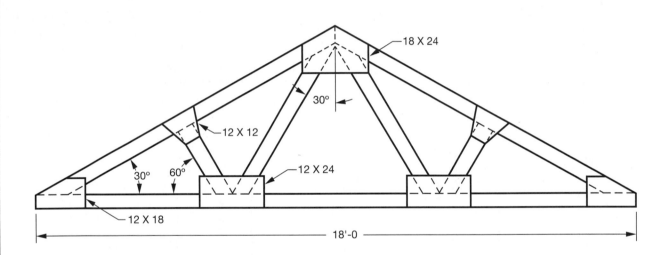

18 X 24

30°

12 X 12

12 X 24

30°   60°

12 X 18

18'-0

ASSIGNMENT:
ON A DECIMAL-INCH GRID SHEET HAVING .10 IN. DIVISIONS,
SKETCH THE LEFT HALF OF THE ROOF TRUSS TO THE SCALE
OF 1 IN. = 1 FT. EXTEND THE TRUSS A SHORT DISTANCE
BEYOND THE CENTRE OF THE TRUSS AND USE CONVENTIONAL
BREAKS ON THE TRUSS MEMBERS. INCLUDE AN ENLARGED
VIEW (2 IN. = 1 FT) OF THE END GUSSET ASSEMBLY SHOWING
THE NAILING REQUIREMENTS.

ROOF TRUSS	A-4

UNIT 3

CIRCULAR FEATURES

Circular features consist of full circles and arcs (parts of circles). Typical drawings with circular features are illustrated in Figure 3–1. Example 1 shows centre lines and two circles having the same centre point (*concentric circles*). Example 2 shows four small circles, two half circles, and four quarter circles (rounded corners). The half and quarter circles are called *arcs*. A point where a straight line joins a curved line is called a *point of tangency*, shown in Example 3.

CENTRE LINES

Due to tooling and manufacturing requirements, circular, cylindrical, and symmetrical parts, including holes, must have their centres located. A special line, referred to as a *centre line,* is used to locate these features.

A centre line is drawn as a thin broken line of alternating long and short dashes, Figure 3–2. The long and short dashes may vary in length, depending on the size of the drawing. Centre lines can indicate centre points, axes (singular, *axis*) of cylindrical parts, and axes of symmetrically shaped surfaces or parts. Solid centre lines are often used on small holes (Figure 3–1, Example 2), but the broken line is preferred. Centre lines should project for a short distance beyond the outline of the part or feature to which they refer. They may be lengthened (extended) for use as extension lines for dimensioning. In this case, the extended portion is not broken, Example 1, Figure 3–2.

SKETCHING CIRCLES AND ARCS

Circular features can be drawn with a circle template, a compass, or freehand. Since speed and accuracy of detail are important in preparing sketches to communicate technical ideas, basic drafting tools such as a circle template or compass are often used. The method chosen depends on available instruments and personal preference.

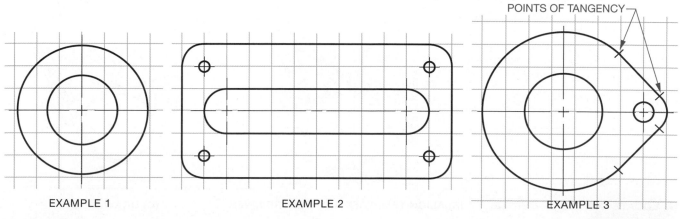

EXAMPLE 1   EXAMPLE 2   POINTS OF TANGENCY   EXAMPLE 3

FIGURE 3–1 ■ Simple objects with circular features

Using a Circle Template

Circle templates are often used to draw circles and arcs on sketches to improve quality and to speed up the process. Circle templates are made of thin plastic sheets with multiple holes in a range of diameters up to 1.50 inches (about 38 mm). The holes are labelled with their sizes in decimal-inches or millimetres, and each hole has register marks for quick and accurate alignment with vertical and horizontal centre lines, Figure 3–3.

To construct a circle or arc using a circle template:

■ Locate the centre of the circle or arc by drawing its centre lines, Figure 3–3(A).

■ Using the appropriate hole size, place the circle template over the centre lines and align the register marks with the centre lines, Figure 3–3(B).

■ With a pencil, trace around the hole in the template to draw the circle or arc, Figure 3–3(C).

Constructing an arc (rounds or fillets) using a circle template requires a somewhat different technique, Figure 3–4.

■ Sketch construction lines outlining the part that requires arcs, Figure 3–4(A).

■ Using the appropriate hole size, place the circle template over the sketching area, align the circumference of the circle with the two construction lines, and draw the arc, Figure 3–4(B). The arc should be a thick solid line.

■ Repeat this procedure for the remaining arcs, Figure 3–4(C).

■ Join these arcs with straight object lines, Figure 3–4(D).

Using a Compass

While a circle template is recommended for sketching circles up to its largest hole size (generally 1.50 inches in diameter), a *compass* can be used for larger circles and arcs. When used for sketching,

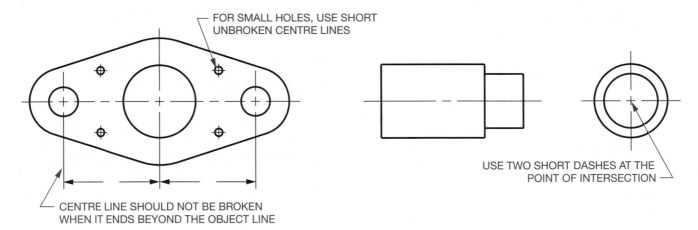

FIGURE 3–2 ■ Centre line application

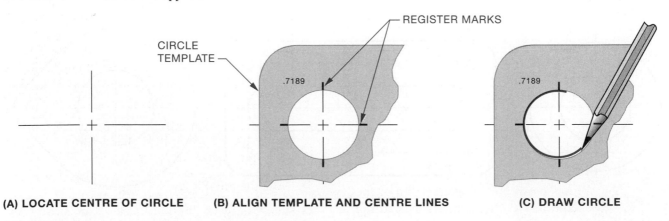

(A) LOCATE CENTRE OF CIRCLE   **(B) ALIGN TEMPLATE AND CENTRE LINES**   **(C) DRAW CIRCLE**

FIGURE 3–3 ■ Drawing a circle using a circle template

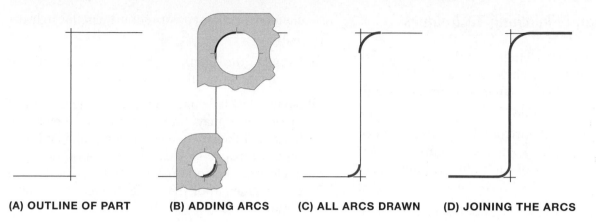

| (A) OUTLINE OF PART | (B) ADDING ARCS | (C) ALL ARCS DRAWN | (D) JOINING THE ARCS |

FIGURE 3–4 ■ Constructing arcs using a circle template

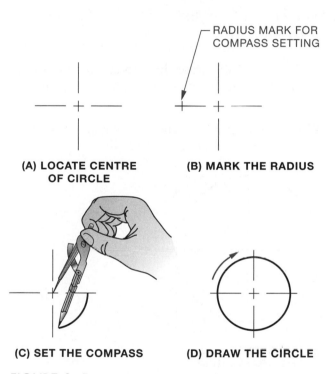

(A) LOCATE CENTRE OF CIRCLE   **(B) MARK THE RADIUS**

RADIUS MARK FOR COMPASS SETTING

(C) SET THE COMPASS   **(D) DRAW THE CIRCLE**

FIGURE 3–5 ■ Drawing a circle using a compass

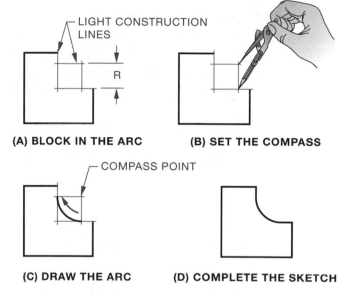

LIGHT CONSTRUCTION LINES

R

(A) BLOCK IN THE ARC   **(B) SET THE COMPASS**

COMPASS POINT

(C) DRAW THE ARC   **(D) COMPLETE THE SKETCH**

FIGURE 3–6 ■ Drawing an arc using a compass

almost any size and type of compass is adequate. The compass found in the instrument set described in Unit 1 generally holds a common pencil, sharpened with a standard classroom pencil sharpener.

Laying out and drawing circles with a compass is illustrated in Figure 3–5.

■ Locate the centre of the circle by drawing centre lines, Figure 3–5(A).

■ Estimate the length of the radius and mark it off on the centre lines, Figure 3–5(B).

■ Set the compass point on the intersection of the centre lines and adjust the compass lead to the radius mark.

■ Proceed to draw the circle by starting the arc in the lower right quadrant, Figure 3–5(C).

■ Complete the circle by rotating the compass in a clockwise direction. Left-handed people may find it easier to reverse the direction of compass rotation, Figure 3–5(D).

Laying out and drawing arcs is illustrated in Figure 3–6.

■ Use construction lines to locate and block in the extent of the arc. The radius of the arc is used to locate its centre, Figure 3–6(A).

■ Set the compass point on the intersection of the centre lines and adjust the compass lead to the radius mark, Figure 3–6(B).

■ Draw the arc, Figure 3–6(C).

■ Sketch tangent lines and other details as necessary, Figure 3–6(D).

Using Freehand Sketching Techniques

While the use of a circle template or compass is preferred for drawing circles and arcs, these may not be available. To sketch circles and arcs freehand, follow the directions in Figure 3–7.

■ Sketch vertical and horizontal construction lines to locate the circle or arc, Figure 3–7(A). Estimate the length of the radius (plural, *radii*) and mark it off on the centre lines.

■ With the radius marks as guides, sketch a square using construction lines into which you will sketch the circle or arc, Figure 3–7(B).

It is generally good practice to first sketch the circle or arc using construction lines and then darken the line when you are satisfied with the size and shape. Sketching on a grid sheet makes drawing circles and arcs easier.

Another common method is shown in Figure 3–8.

■ Locate the centre and construct vertical and horizontal centre lines, Figure 3–8(A). Next, sketch bisecting construction lines through the centre.

■ Estimate the length of the radius and mark off this distance on all the lines, Figure 3–8(B).

■ Sketch the circle or arc by connecting the radius marks. You may find it easier to sketch the bottom of the curve. If so, sketch it first and then turn the paper so that another portion of the circle is on the bottom and sketch it. Continue in this manner until the circle or arc is complete, Figure 3–8(C).

Sketching a Complete View Containing Circles and Arcs

Figure 3–9 illustrates how to lay out and sketch a complete view containing straight lines, circles, and arcs.

■ Lay out centre lines and radius marks for all circles and arcs, Figure 3–9(A).

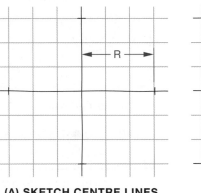

(A) SKETCH CENTRE LINES AND MARK RADIUS

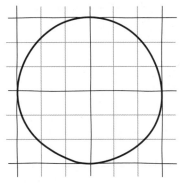

(B) CONSTRUCT SQUARE AND DRAW CIRCLE

FIGURE 3–7 ■ Sketching a circle within a square

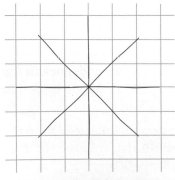

(A) LOCATE CENTRE AND SKETCH BISECTING LINES

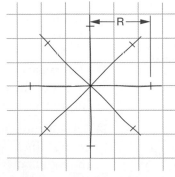

(B) MARK RADIUS

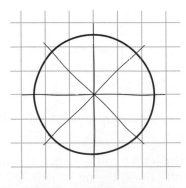

(C) SKETCH CIRCLE THROUGH RADIUS MARKS

FIGURE 3–8 ■ Alternate method for sketching a circle

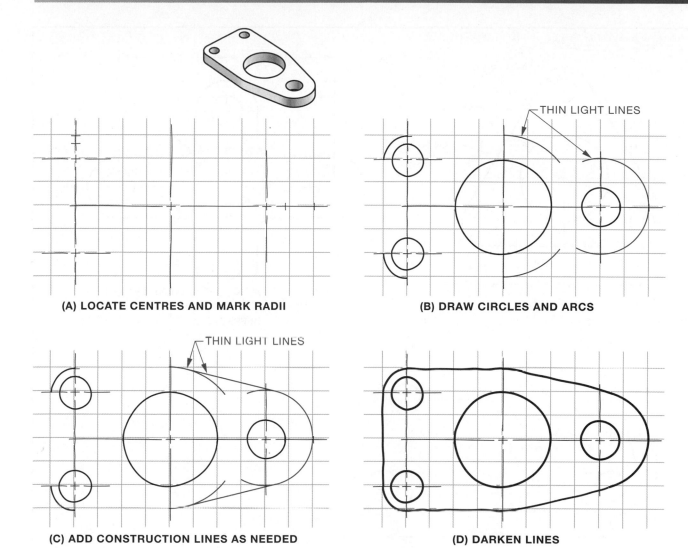

(A) LOCATE CENTRES AND MARK RADII

(B) DRAW CIRCLES AND ARCS

(C) ADD CONSTRUCTION LINES AS NEEDED

(D) DARKEN LINES

FIGURE 3–9 ■ Sketching a complete view containing straight lines, circles, and arcs

- Use a circle template, compass, or freehand sketching technique to draw circles and arcs, Figure 3–9(B).

- Sketch construction lines to lay out straight tangent lines that do not follow grid lines, Figure 3–9(C).

- Darken all lines as appropriate, Figure 3–9(D).

REFERENCES

CAN3-B78.1-M83 Technical Drawing—General Principles

ASME Y14.2M-1992 (R2003) Line Conventions and Lettering

INTERNET RESOURCES

The Mayline Company For information on furniture and equipment for the drafting office, see: http://www.mayline.com/

Goldengrovehs For additional information on sketching circular features, see: http://www.goldengrovehs.sa.edu.au/home/tech/yr8drawassignments/Drawingassignments.htm

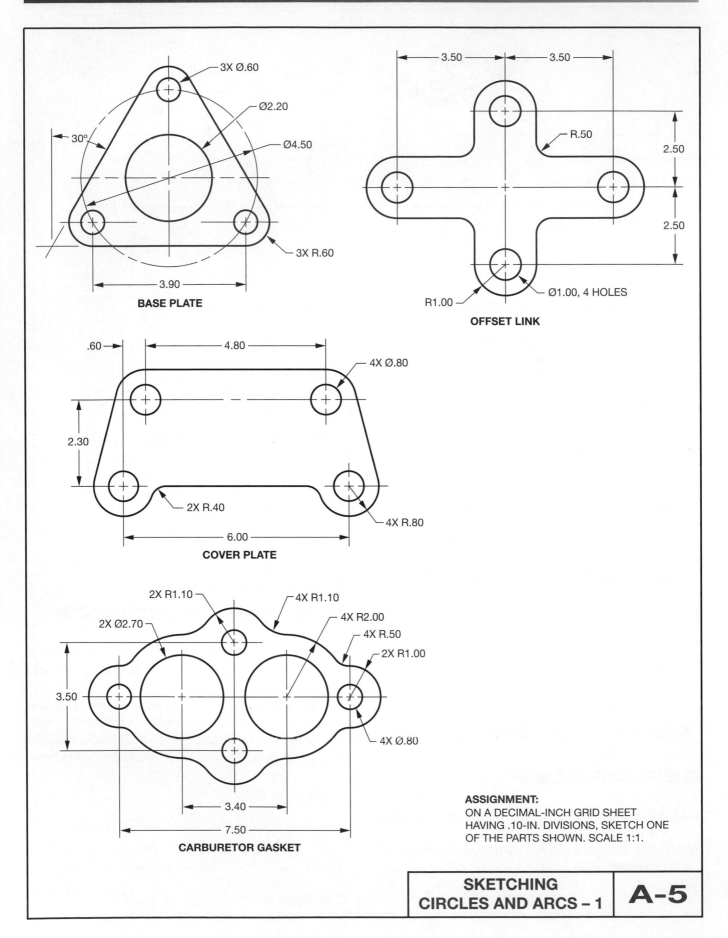

3X Ø.60

Ø2.20

30°

Ø4.50

3X R.60

3.90

BASE PLATE

3.50 3.50

R.50

2.50

2.50

Ø1.00, 4 HOLES

R1.00

OFFSET LINK

.60

4.80

4X Ø.80

2.30

2X R.40

4X R.80

6.00

COVER PLATE

2X R1.10   4X R1.10

2X Ø2.70   4X R2.00

4X R.50

2X R1.00

3.50

4X Ø.80

3.40

7.50

CARBURETOR GASKET

ASSIGNMENT:
ON A DECIMAL-INCH GRID SHEET
HAVING .10-IN. DIVISIONS, SKETCH ONE
OF THE PARTS SHOWN. SCALE 1:1.

SKETCHING CIRCLES AND ARCS – 1	A-5

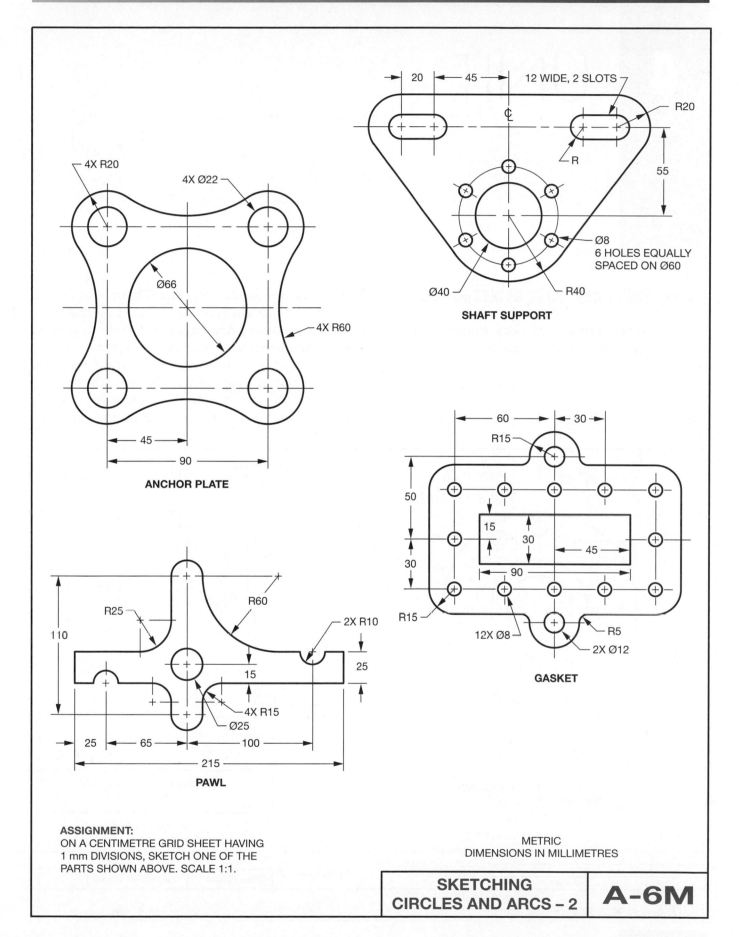

ANCHOR PLATE

SHAFT SUPPORT

GASKET

PAWL

ASSIGNMENT:
ON A CENTIMETRE GRID SHEET HAVING
1 mm DIVISIONS, SKETCH ONE OF THE
PARTS SHOWN ABOVE. SCALE 1:1.

METRIC
DIMENSIONS IN MILLIMETRES

**SKETCHING
CIRCLES AND ARCS – 2**

A-6M

4 UNIT

WORKING DRAWINGS

A *working drawing* supplies information and instructions for the manufacture or construction of machines or structures. Generally, working drawings are classified as either: *detail drawings* (Figure 4–1), which provide the necessary information for the manufacture of the parts for a specific product or structure; or *assembly drawings* (Figure 4–2), which supply information necessary for their assembly.

Since working drawings may be sent to another plant, another company, or even to another country to manufacture, construct, or assemble the final product, the drawing should conform to the drawing standards of that country. For example, drawing standards approved and adopted by the American

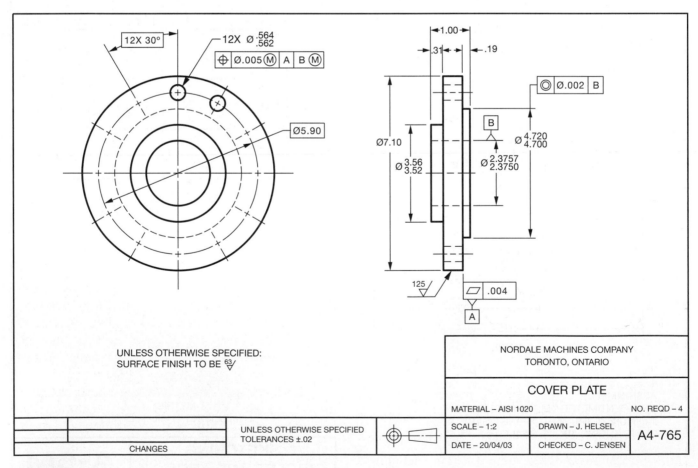

FIGURE 4–1 ■ A simple detail drawing

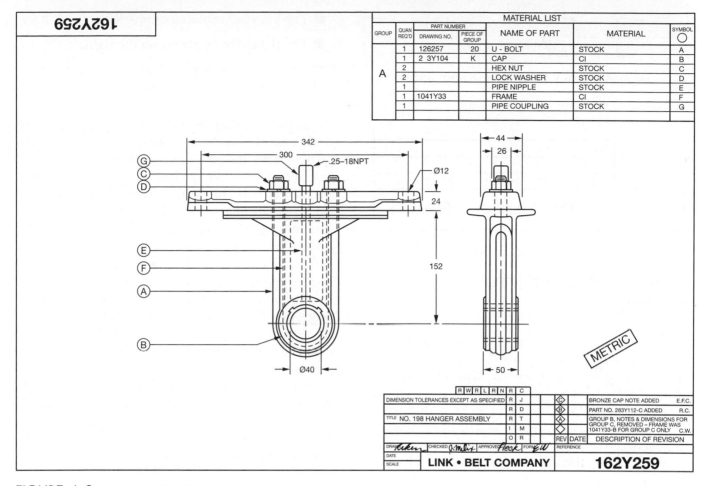

MATERIAL LIST					
GROUP	QUAN REQ'D	PART NUMBER		NAME OF PART	MATERIAL	SYMBOL ◯

GROUP	QUAN REQ'D	DRAWING NO.	PIECE OF GROUP	NAME OF PART	MATERIAL	SYMBOL
A	1	126257	20	U - BOLT	STOCK	A
	1	2 3Y104	K	CAP	CI	B
	2			HEX NUT	STOCK	C
	2			LOCK WASHER	STOCK	D
	1			PIPE NIPPLE	STOCK	E
	1	1041Y33		FRAME	CI	F
	1			PIPE COUPLING	STOCK	G

162Y259

.25–18NPT

342
300
Ø12
24
152
44
26
Ø40
50

METRIC

DIMENSION TOLERANCES EXCEPT AS SPECIFIED	R	W	R	L	R	N	R	C			BRONZE CAP NOTE ADDED	E.F.C.
		R		J							PART NO. 283Y112-C ADDED	R.C.
TITLE NO. 198 HANGER ASSEMBLY		R		D							GROUP B, NOTES & DIMENSIONS FOR GROUP C, REMOVED – FRAME WAS 1041Y33-B FOR GROUP C ONLY C.W.	
		R		T								
		I		M								
		O		R				REV	DATE		DESCRIPTION OF REVISION	

DRAWN	CHECKED	APPROVED	FORM	REFERENCE
DATE				
SCALE	**LINK • BELT COMPANY**			**162Y259**

FIGURE 4–2 ■ An assembly drawing

Society of Mechanical Engineers (ASME) have been adopted by most industries throughout the United States. Similarly, the Canadian Standards Association sets the drawing standards for industries throughout Canada. Fortunately, these two sets of standards are similar.

The information on working drawings can be classified under three headings:

- **Shape**, or **shape description**. This refers to the selection and number of views and other details used to show or describe the shape of the part. Multiview drawings are generally used as working drawings; pictorial drawings are also sometimes used.

- **Dimensions**, or **size description**. Approved dimensioning methods for engineering drawings are explained throughout this text starting in Unit 5. The units of measurement recommended are the decimal-inch and the millimetre.

- **Specifications**. Additional information, including general notes, type or material, heat treatment, surface texture finish, and other similar data needed to manufacture the part, are included on the drawing or in the title block.

ARRANGEMENT OF VIEWS

Since several views of a part are normally required to describe its shape, how they are positioned on the drawing must be clearly understood and have only one interpretation. Two systems of arranging or positioning of views are used on engineering drawings: *first-angle and third-angle orthographic projection.*

This text presents third-angle orthographic projection, used by many countries, including the United States and Canada. Most European and Asian countries have adopted first-angle projection. The shapes and sizes of views are identical in both systems; only the positioning of views differs.

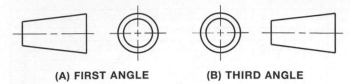

(A) FIRST ANGLE   **(B) THIRD ANGLE**

FIGURE 4–3 ■ ISO projection standards

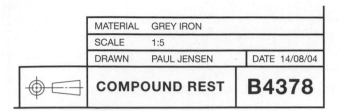

MATERIAL	GREY IRON	
SCALE	1:5	
DRAWN	PAUL JENSEN	DATE 14/08/04

COMPOUND REST   **B4378**

FIGURE 4–4 ■ Location of ISO projection symbol

ISO PROJECTION SYMBOL

Because these two types of arrangements or views are used on engineering drawings, it is necessary to be able to identify the type of projection used. Therefore, the International Organization for Standardization (ISO) has recommended that one of the symbols shown in Figure 4–3 be marked on all engineering drawings, preferably in the lower right-hand corner of the drawing, adjacent to the title block, Figure 4–4.

THIRD-ANGLE PROJECTION

The third-angle system of projection, used almost exclusively on mechanical engineering drawings in North America, permits each feature of the object to be drawn in true proportion and without distortion along all dimensions.

Three views are usually sufficient to describe the shape of an object. The views most commonly used are the front, top, and right side, Figure 4–5(A). In third-angle projection, the object can be assumed to be enclosed in a glass box, Figure 4–5(B). A view of the object drawn on each side of the box represents what is seen when looking perpendicularly at each face of the box. If the box were unfolded as if hinged around the front face, the desired orthographic projection would result, Figure 4–5(C) and (D). These views are identified by names as shown. With reference to the front view:

■ The top view is placed above.

■ The bottom view is placed underneath.

■ The left view is placed on the left.

■ The right view is placed on the right.

■ The rear view is placed at the extreme left or right, whichever is convenient.

Before making a drawing, the drafter must decide on the number of views necessary to adequately show the part, and which of the six sides of the part would make the best principal (front) view. Factors such as the most informative view and the avoidance of hidden lines help influence the decision making. The front view of the drawing need not be the "front" of the finished part.

The front view on the drawing shows the width and height of the object. The term *length* should be avoided when describing the views in orthographic projection, because it normally refers to the longest dimension. When describing the width, height, or depth of a part, any one of these may be the longest measurement.

Once the front view is selected, the next step is to decide what other views are required to adequately show the shape and features of the part. Seldom are more than three views necessary to completely describe the part. Therefore, the simple object in Figure 4–6 can be used to illustrate the position of these principal dimensions.

In Figure 4–6, the object is shown in pictorial form (A) and orthographic projection (B). The orthographic drawing uses each view to represent the exact shape and size of the object and the relationship of the three views to one another. This principle of projection is used in all mechanical drawings. The isometric drawing shows the relationship of the front, top, and right-side surfaces in a single view. Typical parts with flat surfaces are shown in both pictorial and third-angle projection in Figure 4–7.

Objects with Circular Features

Typical parts with circular features are illustrated in Figure 4–8. The circular feature appears circular in one view only, and no line is used to indicate where a curved surface joins a flat surface. Hidden circles, like hidden edges of flat surfaces, are represented on drawings by a hidden line. Objects having circular features are shown in both pictorial and third-angle projection in Figure 4–8. Often only two views are required to show the shape of the part, Figure 4–8 (E) and (F).

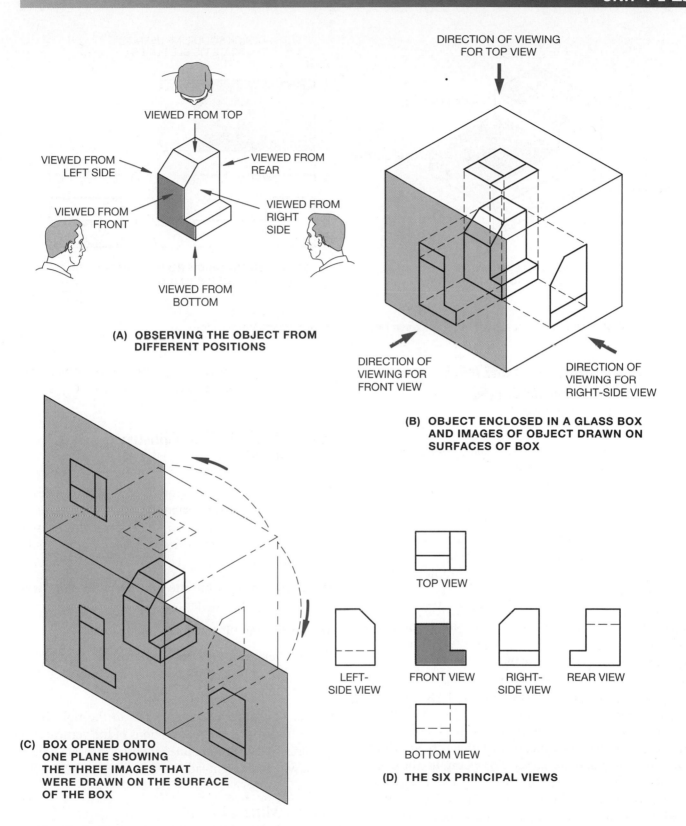

(A) **OBSERVING THE OBJECT FROM DIFFERENT POSITIONS**

(B) **OBJECT ENCLOSED IN A GLASS BOX AND IMAGES OF OBJECT DRAWN ON SURFACES OF BOX**

(C) **BOX OPENED ONTO ONE PLANE SHOWING THE THREE IMAGES THAT WERE DRAWN ON THE SURFACE OF THE BOX**

(D) **THE SIX PRINCIPAL VIEWS**

TOP VIEW

LEFT-SIDE VIEW   FRONT VIEW   RIGHT-SIDE VIEW   REAR VIEW

BOTTOM VIEW

FIGURE 4–5 ■ Third-angle orthographic projection

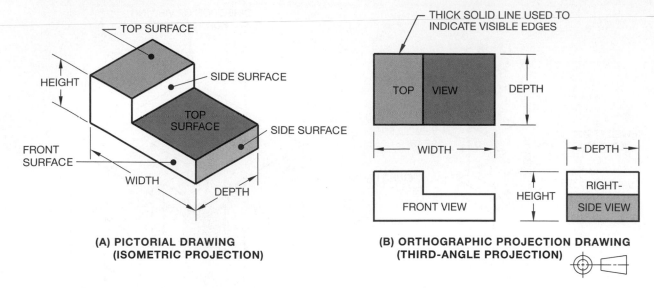

FIGURE 4–6 ■ A simple object shown in (A) pictorial form and (B) orthographic projection

SKETCHING VIEWS IN THIRD-ANGLE PROJECTION

Objects drawn using *orthographic projection* are represented with one or more views, depending on the shape and complexity of the part. For example, a part made from a flat sheet of material having uniform thickness, such as a gasket, can easily be represented with one view and a note describing the material and its thickness. More complex objects may require two or more views for complete shape description.

Figure 4–9 is a pictorial drawing of an object (latch) that would best be described using front, top, and right-side views. The arrow points to the view that best describes the shape of the object and, therefore, will become the front view.

Step 1 in Figure 4–10 shows a front view of the latch and a top view that has been projected from it. In some cases, features are projected back and forth from view to view as a means of efficiently arriving at finished details in all views. For example, the outline of the front view is first drawn and projected upward to develop the top view. The edges of the small holes are then projected back down to the front view so that the hidden edges can be accurately located. Adequate space must be provided between views for dimensions. You will learn about dimensioning practices in Unit 5.

The next step is to develop the third view. The use of a mitre line provides a fast and accurate method of constructing the third view once two views have been developed.

Using a Mitre Line to Construct the Right-Side View

■ Given the top and front views (Step 1, Figure 4–10), project horizontal lines to the right of the top view (Step 2). The projection lines should be thin light lines that later can be easily erased or ignored.

■ Decide how far from the front view the side view is to be drawn (distance D).

■ Construct a 45° mitre line as shown.

■ Drop vertical projection lines from points where the horizontal projection lines intersect the mitre line.

■ Run horizontal projection lines to the right from the front view. The intersections of the vertical and horizontal projection lines are used to locate points on which the right-side view is developed (Step 3).

Using a Mitre Line to Construct the Top View

■ Given the front and right-side views (Step 1, Figure 4–11), project vertical lines up from the right-side view (Step 2). Decide how far from the front view the top view is to be drawn (distance D).

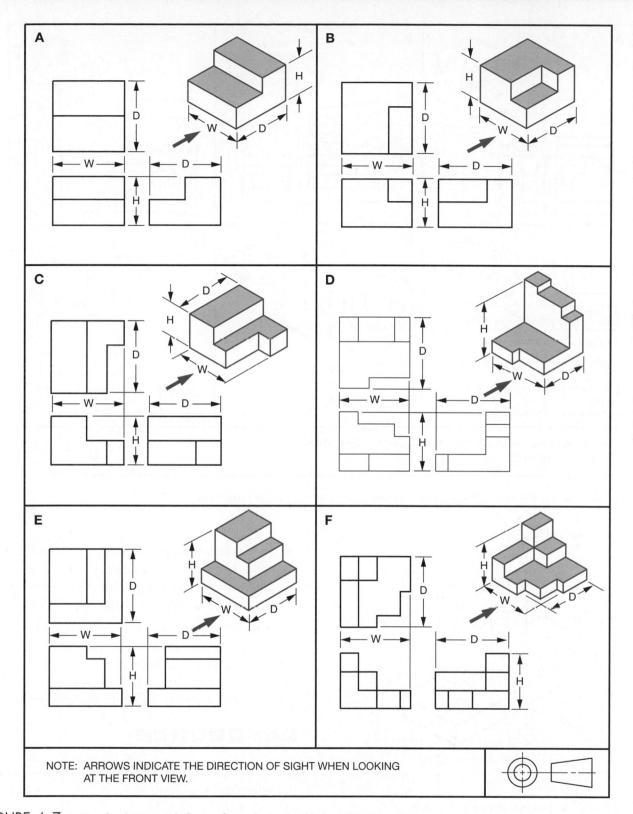

NOTE: ARROWS INDICATE THE DIRECTION OF SIGHT WHEN LOOKING AT THE FRONT VIEW.

FIGURE 4–7 ■ Simple objects with flat surfaces drawn in third-angle projection

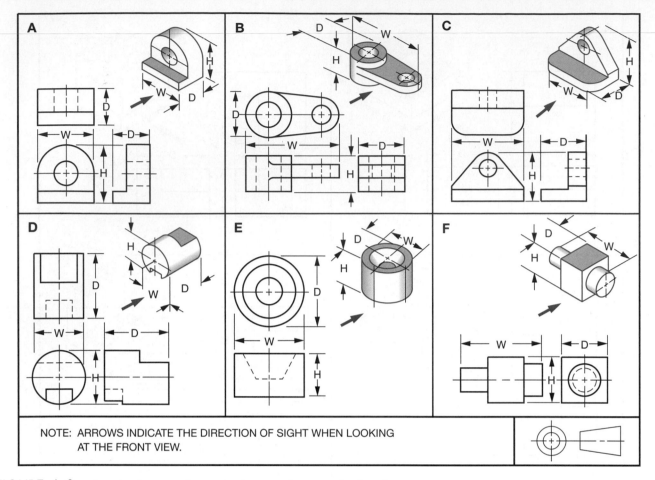

NOTE: ARROWS INDICATE THE DIRECTION OF SIGHT WHEN LOOKING AT THE FRONT VIEW.

FIGURE 4–8 ■ Simple objects with circular features drawn in third-angle projection

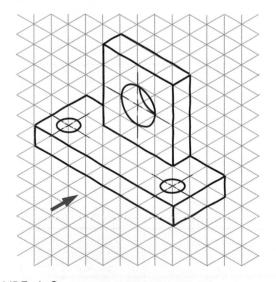

FIGURE 4–9 ■ Pictorial drawing of a latch

■ Construct a 45° mitre line as shown.

■ Run horizontal projection lines from points where the vertical projection lines intersect the mitre line.

■ Run vertical projection lines up from the front view. The intersections of the vertical and horizontal projection lines are used to locate points on which the top view is developed (Step 3).

After some practice using the mitre-line method, you should be able to construct the various views by projecting details visually from one view to another by following the grid lines. With practice, your sketching efficiency and accuracy will improve rapidly.

REFERENCE

CAN3-B78.1-M83, Technical Drawings—General Principles

INTERNET RESOURCES

Animated Worksheets For information on third-angle projection and mitre lines, see: http://www.animatedworksheets.co.uk (third-angle projection)

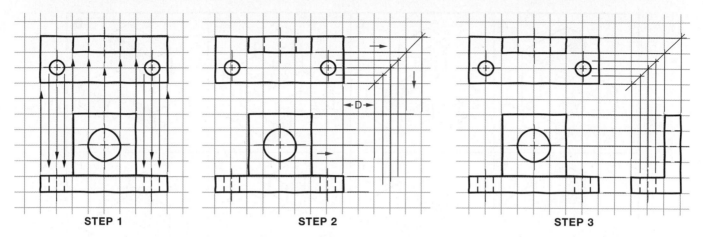

FIGURE 4–10 ■ Using a mitre line to construct the right-side view

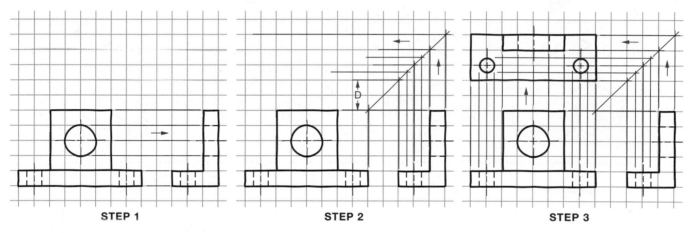

FIGURE 4–11 ■ Using a mitre line to construct the top view

Drafting Zone For information on dimensioning angles, see: http://www.draftingzone.com

technologystudent.com For information on third-angle projection and related subjects, see: http://www.technologystudent.com/designpro/ortho2.htm

Wikipedia, the Free Encyclopedia For information on third-angle angle projection, see: http://en.wikipedia.org/wiki/Orthographic_projection

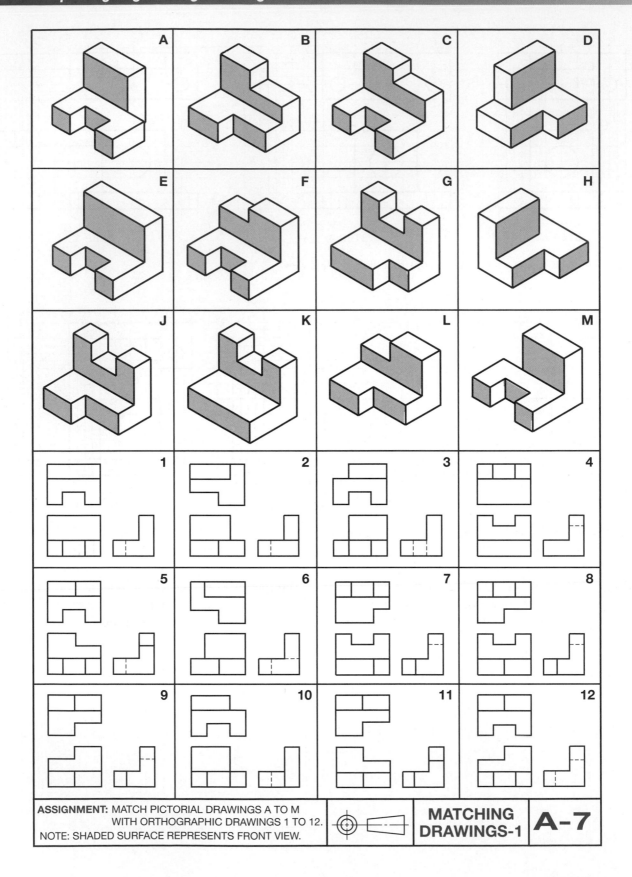

ASSIGNMENT: MATCH PICTORIAL DRAWINGS A TO M
WITH ORTHOGRAPHIC DRAWINGS 1 TO 12.
NOTE: SHADED SURFACE REPRESENTS FRONT VIEW.

MATCHING DRAWINGS-1

A-7

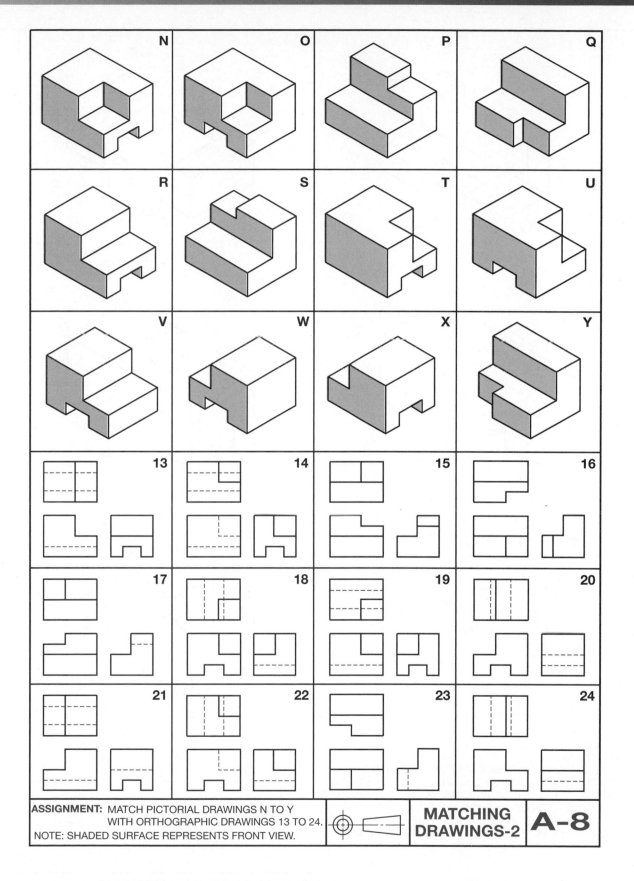

ASSIGNMENT: MATCH PICTORIAL DRAWINGS N TO Y
WITH ORTHOGRAPHIC DRAWINGS 13 TO 24.
NOTE: SHADED SURFACE REPRESENTS FRONT VIEW.

MATCHING DRAWINGS-2

A-8

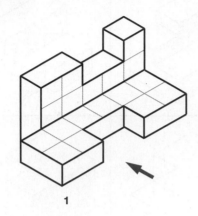

1

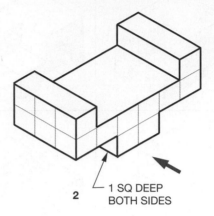

2   1 SQ DEEP
BOTH SIDES

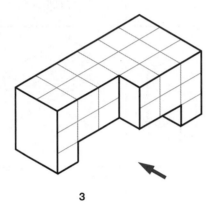

3

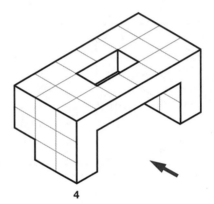

4

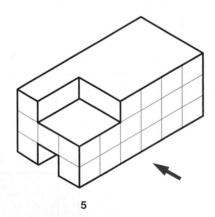

5

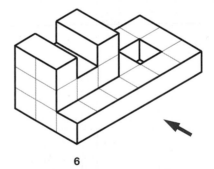

6

ASSIGNMENT:
ON A .25-INCH GRID SHEET, SKETCH THE TOP, FRONT,
AND RIGHT-SIDE VIEWS OF THE OBJECTS SHOWN USING
THIRD-ANGLE ORTHOGRAPHIC PROJECTION. ONE SQUARE
ON THE OBJECT REPRESENTS ONE SQUARE ON THE
SKETCHING PAPER. ALLOW ONE SQUARE BETWEEN VIEWS
AND A MINIMUM OF TWO SQUARES BETWEEN OBJECTS.

NOTE: ARROW INDICATES DIRECTION OF FRONT VIEW.

| ORTHOGRAPHIC SKETCHING–VISIBLE AND HIDDEN LINES | A-9 |

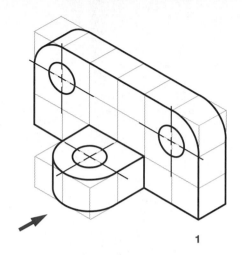

1

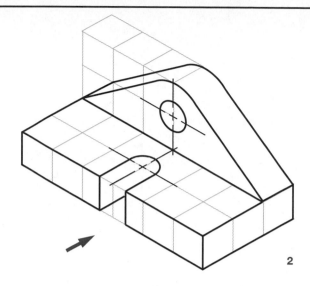

2

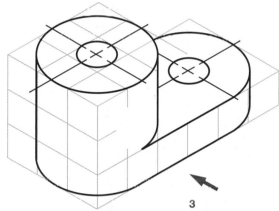

3

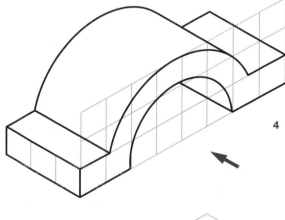

4

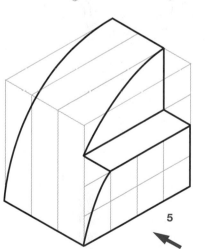

5

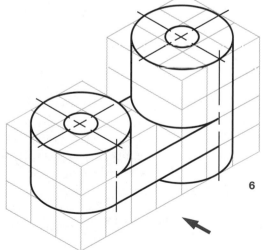

6

ASSIGNMENT:
ON A .25 INCH GRID SHEET, SKETCH THE TOP, FRONT, AND RIGHT-SIDE VIEWS OF THE PARTS SHOWN ABOVE IN ORTHOGRAPHIC PROJECTION. EACH SQUARE SHOWN ON THE OBJECT REPRESENTS ONE SQUARE ON THE GRID SHEET. ALLOW ONE GRID SPACE BETWEEN VIEWS AND A MINIMUM OF TWO GRID SPACES BETWEEN THE OBJECTS.

NOTE: ARROW INDICATES DIRECTION OF FRONT VIEW.

ORTHOGRAPHIC SKETCHING OF PARTS WITH CIRCULAR FEATURES

A-10

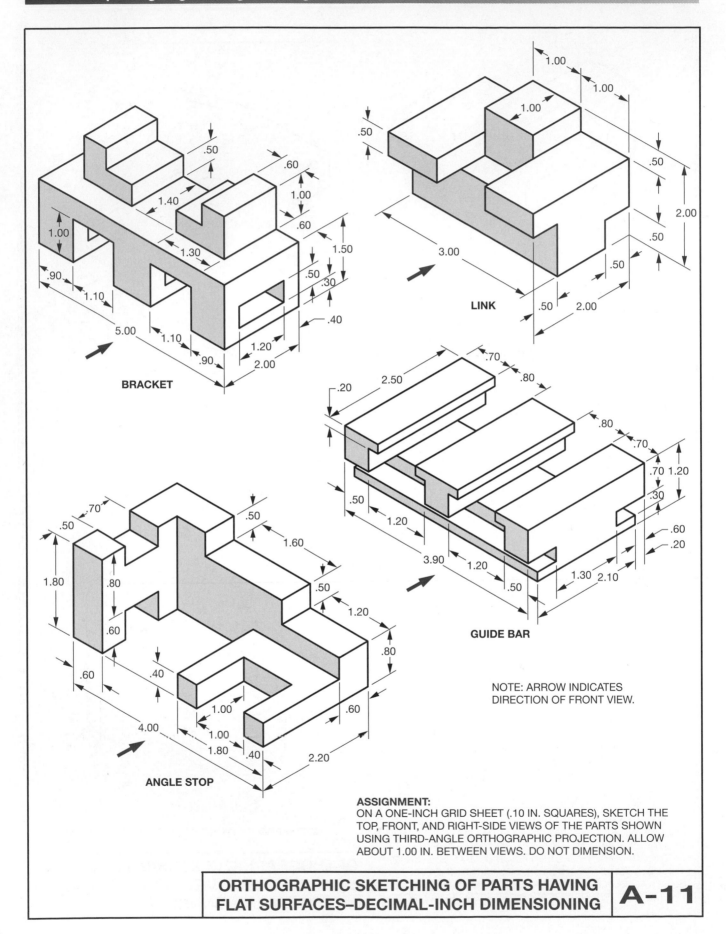

BRACKET

LINK

GUIDE BAR

ANGLE STOP

NOTE: ARROW INDICATES
DIRECTION OF FRONT VIEW.

ASSIGNMENT:
ON A ONE-INCH GRID SHEET (.10 IN. SQUARES), SKETCH THE
TOP, FRONT, AND RIGHT-SIDE VIEWS OF THE PARTS SHOWN
USING THIRD-ANGLE ORTHOGRAPHIC PROJECTION. ALLOW
ABOUT 1.00 IN. BETWEEN VIEWS. DO NOT DIMENSION.

ORTHOGRAPHIC SKETCHING OF PARTS HAVING
FLAT SURFACES–DECIMAL-INCH DIMENSIONING

A-11

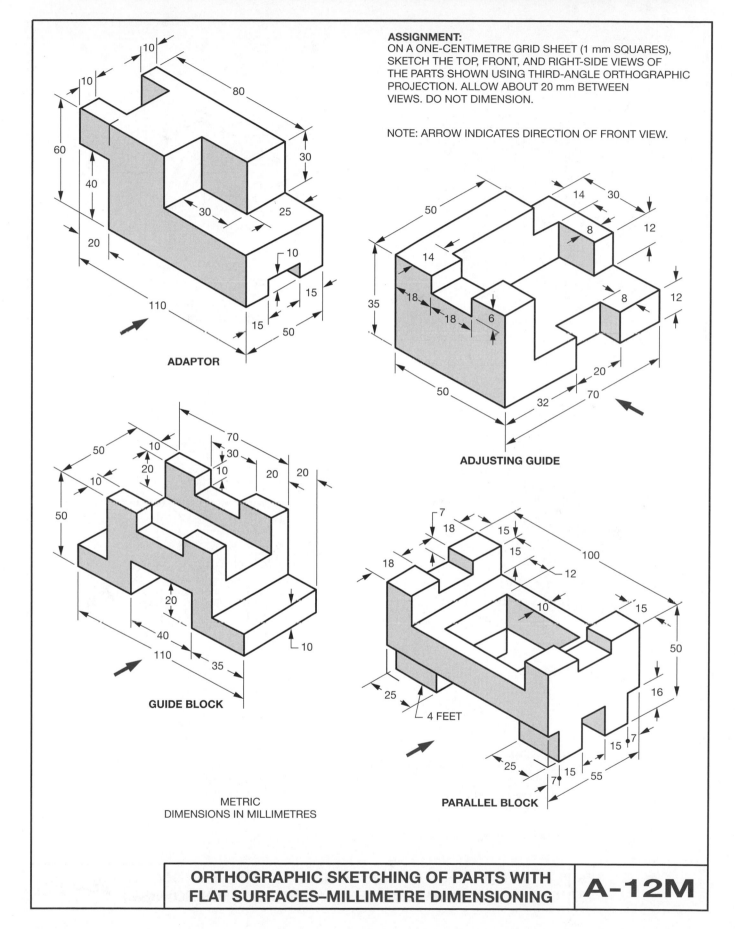

ASSIGNMENT:
ON A ONE-CENTIMETRE GRID SHEET (1 mm SQUARES),
SKETCH THE TOP, FRONT, AND RIGHT-SIDE VIEWS OF
THE PARTS SHOWN USING THIRD-ANGLE ORTHOGRAPHIC
PROJECTION. ALLOW ABOUT 20 mm BETWEEN
VIEWS. DO NOT DIMENSION.

NOTE: ARROW INDICATES DIRECTION OF FRONT VIEW.

ADAPTOR

ADJUSTING GUIDE

GUIDE BLOCK

PARALLEL BLOCK

METRIC
DIMENSIONS IN MILLIMETRES

**ORTHOGRAPHIC SKETCHING OF PARTS WITH
FLAT SURFACES–MILLIMETRE DIMENSIONING**

A-12M

NEL

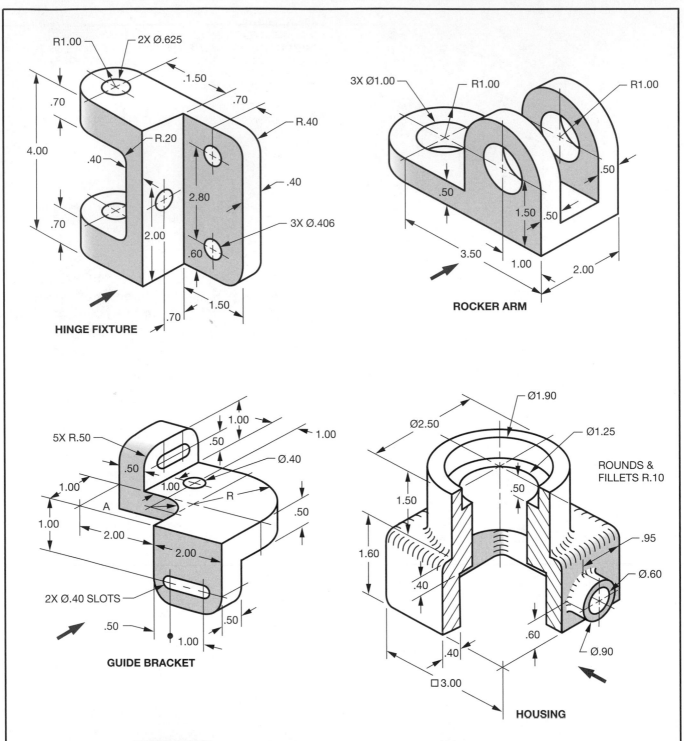

R1.00 2X Ø.625
.1.50
.70
.70
R.40
4.00
R.20
.40
.40
2.80
2.00
3X Ø.406
.70
.70
1.50
.70

HINGE FIXTURE

3X Ø1.00 R1.00 R1.00
.50
.50
1.50
.50
3.50
1.00
2.00

ROCKER ARM

5X R.50
1.00
.50
1.00
Ø.40
.50
.50
A
1.00
R
1.00
2.00
.50
2.00
2X Ø.40 SLOTS
.50
.50
1.00

GUIDE BRACKET

Ø1.90
Ø2.50
Ø1.25
ROUNDS &
FILLETS R.10
1.50
.50
1.60
.95
.40
Ø.60
.60
Ø.90
□3.00

HOUSING

ASSIGNMENT:
ON A ONE-INCH GRID SHEET (.10 IN. SQUARES), SKETCH THE TOP,
FRONT, AND RIGHT-SIDE VIEWS OF ONE OF THE PARTS SHOWN USING
THIRD-ANGLE ORTHOGRAPHIC PROJECTION. ALLOW 1.00 IN. BETWEEN
VIEWS. DO NOT DIMENSION.

NOTE: ARROW INDICATES DIRECTION OF FRONT VIEW.

ORTHOGRAPHIC SKETCHING OF PARTS WITH CIRCULAR FEATURES–DECIMAL-INCH DIMENSIONING

A-13

DIMENSIONING

Dimensions are indicated on drawings by extension lines, dimension lines, leaders, arrowheads, figures, notes, and symbols. These lines and dimensions define such geometrical characteristics as distances, diameters, angles, and locations, Figure 5–1. The lines used in dimensioning are thinner than the outline of the object. The dimension must be clear and concise, permitting only one interpretation. In general, each surface, line, or point is located by only one set of dimensions. Exceptions to these rules are the two types of rectangular coordinate dimensioning discussed in Unit 18.

Reading Direction

Dimensions and notes on engineering drawings should be placed to read from the bottom of the drawing, Figure 5–1.

DIMENSIONING FLAT SURFACES

Dimension Lines

Dimension lines denote particular sections of the object. They should be drawn parallel to the section they define. Dimension lines terminate in arrowheads, which touch an extension line and are broken to allow the insertion of the dimension, Figure 5–1. Where space does not permit the insertion of the dimension line and the dimension between the extension lines, the dimension line may be placed outside the extension line. The dimension can also be placed outside the extension line if the space

between the extension lines is limited. In restricted areas, a common industrial practice is to replace the two arrowheads with a circular dot. These methods are shown in Figure 5–2.

Extension Lines

Extension lines denote the points or surfaces between which a dimension applies. They extend from object lines and are drawn perpendicular to the dimension lines, Figure 5–1. A small gap (.03 to .06 in.) is left between the extension line and the outline to which it refers.

Where extension lines cross arrowheads, a break in the extension line is permitted, Figure 5–3.

Leaders

Leaders direct dimensions or notes to the surface or points to which they apply, Figure 5–4. A leader consists of a line with or without a short horizontal bar adjacent to the note or dimension, and an inclined portion terminating in an arrowhead touching the line or point to which it applies. A leader may terminate with a dot when it refers to a surface within the outline of a part.

NOTE: If a dimension is omitted on a drawing, contact the drafting department. *Never* scale a drawing for a missing dimension.

LINEAR UNITS OF MEASUREMENT

Although the metric system of dimensioning is expected to become the official standard of

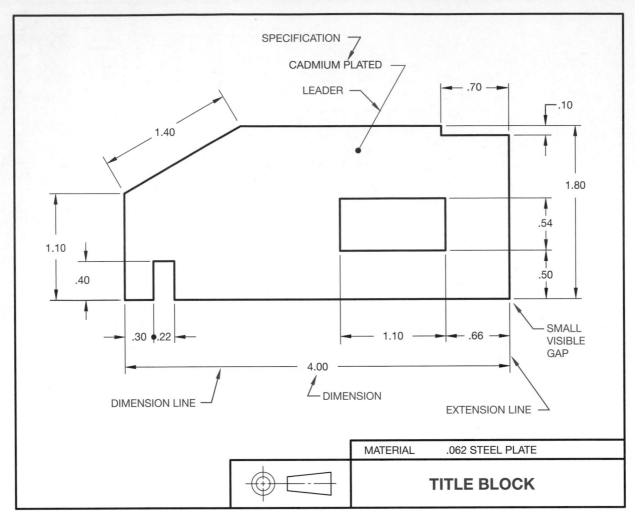

FIGURE 5–1 ■ Basic dimensioning elements

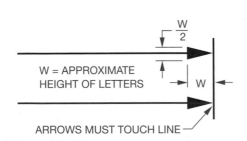

(A) SIZE OF ARROWHEADS

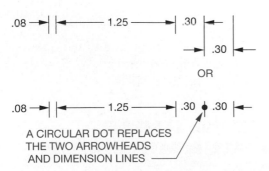

(B) APPLICATION OF ARROWHEADS AND DIMENSIONS

FIGURE 5–2 ■ Arrowheads

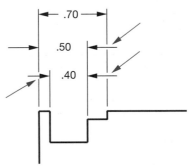

FIGURE 5–3 ■ Breaks in extension lines

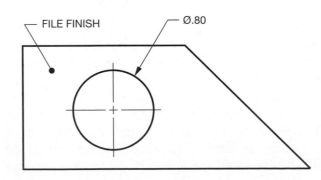

FIGURE 5–4 ■ Using leaders for dimensions and notes

measurement for engineering drawing, most drawings in North America are dimensioned in decimal-inches, or feet and inches. For this reason, drafters and people involved in reading engineering drawings should be familiar with all the dimensioning systems they may encounter.

The dimensions used in this book are primarily decimal-inch. However, metric dimensions are often used. When metric units of measurement are used, the drawing prominently displays the word METRIC and a note stating the dimensions are in millimetres.

Inch Units of Measurement

The Decimal-Inch System (Imperial)

In the decimal-inch system, parts are designed in basic decimal increments, preferably .02 inch, and are expressed as two-place decimal numbers. Using the .02 module, the second decimal place (hundredths) is an even number or zero, Figure 5–5. Sizes other than these, such as .25, are used when they are essential to meet design requirements. When greater accuracy is required, sizes are expressed as three- or four-place decimal numbers, such as 1.875 or 4.5625.

The Fractional-Inch System

This system was replaced by the decimal-inch system of dimensioning of engineering drawings, over 50 years ago. Due to existing tools (drills, reamers, etc.) and pipe sizes that were, and still are, made to fractional-inch sizes, you should be aware of this system of dimensioning. In this system, sizes are expressed in common fractions, the smallest divisions being 64ths, Figure 5–6. Sizes other than common fraction are expressed as decimals.

SI (Metric) Units of Measurement

The standard metric units on engineering drawings are the millimetre for linear measure and the micrometre for surface roughness. For architectural drawings metre and millimetre units are used, Figure 5–7.

In metric dimensioning, as in decimal-inch dimensioning, numerals to the right of the decimal point indicate the degree of precision.

Whole dimensions do not require a zero to the right of the decimal point.

2 not 2.0
10 not 10.0

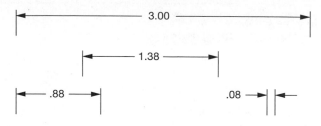

FIGURE 5–5 ■ Decimal-inch dimensioning

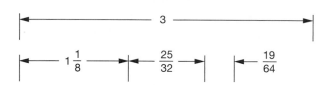

FIGURE 5–6 ■ Fractional-inch dimensioning

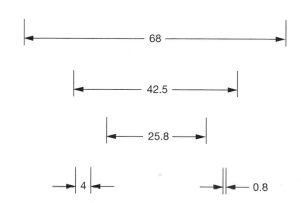

FIGURE 5–7 ■ Metric (millimetre) dimensioning

A millimetre value of less than 1 is shown with a zero to the left of the decimal point.

0.2 not .2
0.26 not .26

Commas should not be used to separate groups of three numbers in metric values. A space should be used in place of the comma.

32 541 not 32,541
2.562 826 6 not 2.5628266

Identification

A metric drawing should include a general note, such as UNLESS OTHERWISE SPECIFIED, DIMENSIONS ARE IN MILLIMETRES. In addition, a metric drawing should be identified by the word METRIC prominently displayed near the title block.

CHOICE OF DIMENSIONS

The choice of the most suitable dimensions and dimensioning methods often depends on whether the drawings are intended for unit production or mass production.

Unit production refers to applications in which each part is to be made separately, using general-purpose tools and machines. Details on custom-built machines, jigs, fixtures, and gauges required for the manufacture of production parts are made in this way. Frequently, only one of each part is required.

Mass production refers to parts produced in quantity, for which special tooling is usually provided. Most part drawings for manufactured products are considered to be for mass-produced parts.

Functional dimensioning should be expressed directly on the drawing, especially for mass-produced parts. This will result in the selection of datum features on the basis of function and assembly. For unit-produced parts, it is generally preferable to select datum features on the basis of manufacture and machining, Figure 5–8.

BASIC RULES FOR DIMENSIONING

■ Place dimensions between the views when possible, Figure 5–9(A).

■ Place the dimension line for the shortest width, height, and depth nearest the outline of the object, Figure 5–9(B). Parallel dimension lines are placed in order of size, making the longest dimension line the outermost line.

■ Place dimensions near the view that best shows the characteristic contour or shape of the object, Figure 5–9(C). In following this rule, dimensions will not always be between views.

■ When several dimension lines are directly above or next to one another, stagger the dimensions to improve the clarity of the drawing. The spacing suitable for most drawings between parallel dimension lines is .30 in. (8 mm), and the spacing between the outline of the object and the nearest dimension line should be about .40 in. (10 mm), Figure 5–10. Other rules for

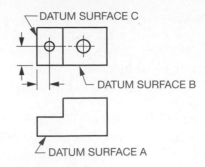

(A) THE PART WITH DATUM SURFACES SELECTED

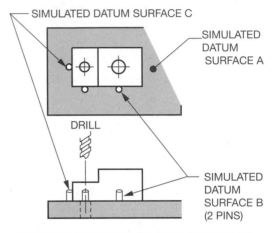

(B) POSITIONING PART FOR DRILLING HOLES

FIGURE 5–8 ■ Selection of datum surfaces for dimensioning

the placement of dimensions will appear in later units and assignments of this book.

■ On large drawings, dimensions can be placed on the view to improve clarity.

DIMENSIONING CYLINDRICAL FEATURES

Features shown as circles are normally dimensioned by one of the methods in Figure 5–11. Where the diameters of a number of concentric cylinders are to be given, it may be more convenient to show them on the side view. The diameter symbol Ø should always precede the diametral dimension.

The radius of the arc is used in dimensioning a circular arc. The letter R is shown before the radius dimension to indicate that it is a radius. Approved methods for dimensioning arcs are shown in Figure 5–12.

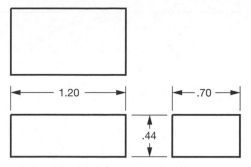

(A) PLACE DIMENSIONS BETWEEN VIEWS

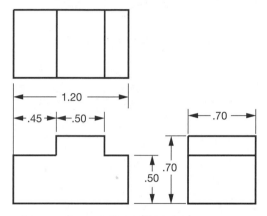

(B) PLACE SMALLEST DIMENSION NEAREST THE VIEW BEING DIMENSIONED

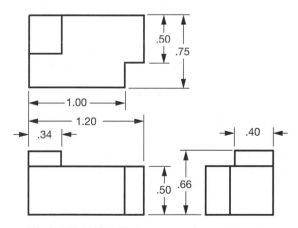

(C) DIMENSION THE VIEW THAT BEST SHOWS THE SHAPE

FIGURE 5–9 ■ Basic dimensioning rules

DIMENSIONING CYLINDRICAL HOLES

Specification of the diameter with a leader, Figure 5–13, is the preferred method for designating the size of small holes. For larger diameters, use one of the methods illustrated in Figure 5–11. When the leader is used, the symbol Ø precedes

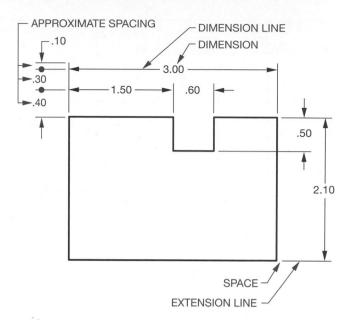

FIGURE 5–10 ■ Placement of dimensions

the size of the hole. The note end of the leader terminates in a short horizontal bar, which is adjacent to the beginning or the end of the note. When two or more holes of the same size are required, the number of holes is specified. If a blind hole is required, the depth of the hole is included in the dimensioning note; otherwise, it is assumed that all holes shown are through holes.

DRILLING, REAMING, AND BORING

Drilling is the process of using a drill to cut a hole through a solid, or to enlarge an existing hole. For some types of work, holes must be drilled smooth and straight and to an exact size. In other work, accuracy of location and size of the hole are not as important.

When accurate holes of uniform diameter are required, they are first drilled slightly undersize and then reamed. *Reaming* is the process of sizing a hole to a given diameter with a reamer to produce a hole that is round, smooth, straight, and accurate.

Boring is one of the more dependable methods of producing holes that are round and concentric. The term *boring* refers to enlarging a hole by means of a boring tool. The use of reaming is limited to the sizes of available reamers. However, holes may be bored to any size desired.

The degree of accuracy to which a hole is to be machined is specified on the drawing.

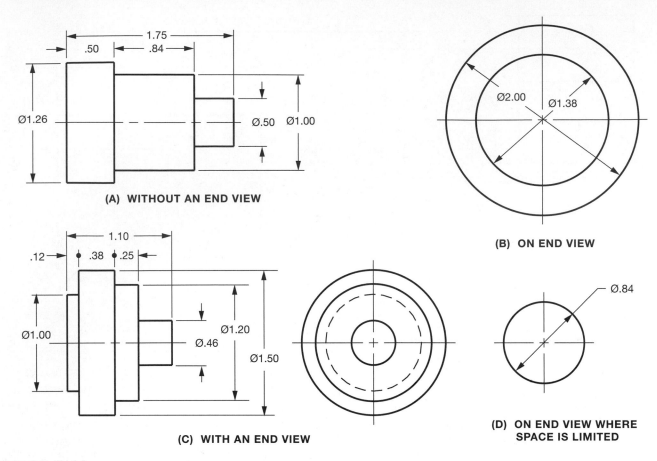

(A) WITHOUT AN END VIEW

(B) ON END VIEW

(C) WITH AN END VIEW

(D) ON END VIEW WHERE SPACE IS LIMITED

FIGURE 5–11 ■ Dimensioning diameters

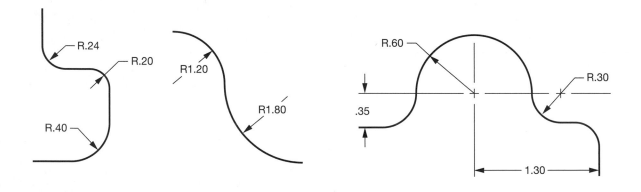

FIGURE 5–12 ■ Dimensioning radii

The use of operational names, such as *turn*, *bore*, *grind*, *ream*, *tap*, and *thread*, with dimensions should be avoided. Although the drafter should be aware of the methods by which a part can be produced, the method of manufacture is best left to the shop. If the part is adequately dimensioned and has the surface texture symbols showing the finish quality desired, it remains a shop problem to meet the drawing specifications.

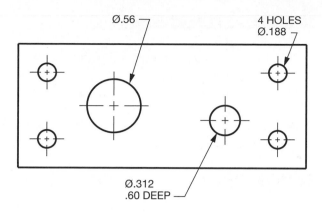

FIGURE 5–13 ■ Dimensioning circular holes

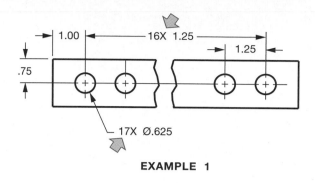

EXAMPLE 1

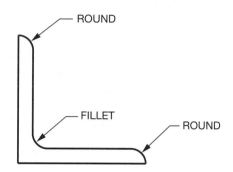

FIGURE 5–14 ■ Fillets and rounds

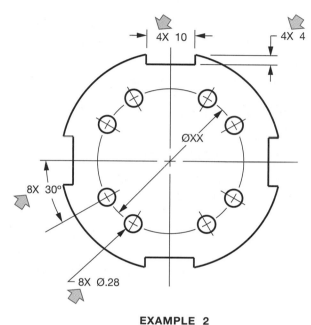

EXAMPLE 2

FIGURE 5–15 ■ Dimensioning repetitive features

DIMENSIONING ROUNDS AND FILLETS

A round, or radius, or chamfer, is put on the outside of a piece to improve its appearance and to avoid forming a sharp edge that might chip off under a sharp blow or cause interference. It is also a safety feature. A fillet is additional metal allowed in the inner intersection of two surfaces, Figure 5–14. This increases the strength of the object. A general note, such as ROUNDS AND FILLETS R10 or ROUNDS AND FILLETS R10 UNLESS OTHERWISE SHOWN, is normally used on the drawing instead of individual dimensions.

DIMENSIONING REPETITIVE FEATURES

Repetitive features and dimensions may be specified on a drawing by the use of an × in conjunction with the numeral to indicate the "number of times" or "places" they are required. A space is inserted between the × and the dimension, as shown in Figure 5–15.

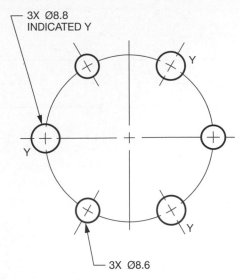

FIGURE 5–16 ■ Identifying similarly sized holes

IDENTIFYING SIMILARLY SIZED FEATURES

Where many similarly sized holes or features appear on a part, some form of identification may be desirable to ensure the legibility of the drawing, Figure 5–16.

REFERENCES

CAN/CSA-B78.2-M91 Dimensioning and Tolerancing of Technical Drawings

ASME Y14.5M-1994 (R2004) Dimensioning and Tolerancing

ASME Y14.2M-1992 (R2003) Line Conventions and Lettering

INTERNET RESOURCES

Drafting Zone For information on dimensioning circular features, see: http://www.draftingzone .com

Integrated Publishing For information on dimensioning, see: www.tpub.com/content/draftsman

IDS Development–Nebraska Education For information on the various line types used on engineering drawings, see: http://idsdev.mccneb .edu/djackson/lineintro.htm

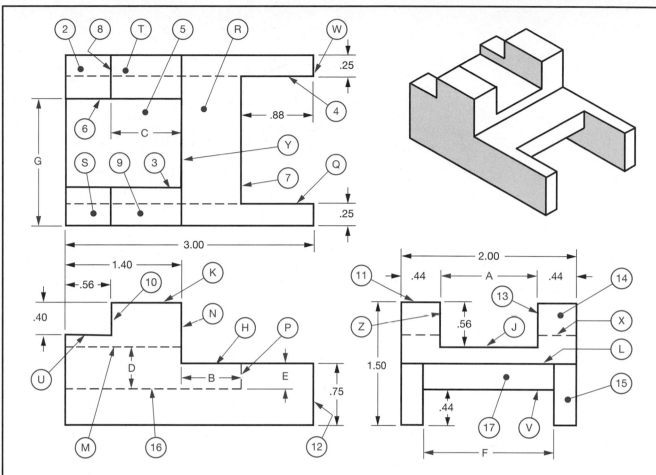

QUESTIONS:

1. What is the name of the object?
2. What is the drawing number?
3. How many castings are required?
4. What material is the part made of?
5. What is the overall width?
6. What is the overall height?
7. What is the overall depth?
8. Calculate distances A through G.
9. Which line in the top view represents surface Ⓟ ?
10. Which line in the side view represents surface ⑤ ?
11. Which line in the side view represents surface Ⓡ ?
12. Which surface in the top view does line Ⓚ in the front view represent?
13. Which surface in the top view does line Ⓜ in the front view represent?
14. Which line in the side view represents the same surface represented by line Ⓜ of the front view?
15. What kind, or type of line, is line Ⓜ ?
16. Which front view line does line Ⓧ in the side view represent?

17. Which front view line does line Ⓨ in the top view represent?
18. Which line in the front view does the surface ⑮ in the side view represent?
19. Which front view line represents surface Ⓡ in the top view?
20. Which surface in the side view represents line Ⓝ of the front view?
21. Which line in the side view represents surface ② ?
22. Which surface in the side view does line Ⓟ represent?
23. Which surface in the top view does line ⑪ represent?
24. Which line in the side view does line ③ in the top view represent?
25. Which line in the side view does line ⑯ in the front view represent?
26. Which surface in the side view does line Ⓦ represent?

QUANTITY	875	
MATERIAL	MALLEABLE IRON	
SCALE	NOT TO SCALE	
DRAWN		DATE

 FEED HOPPER | A-14

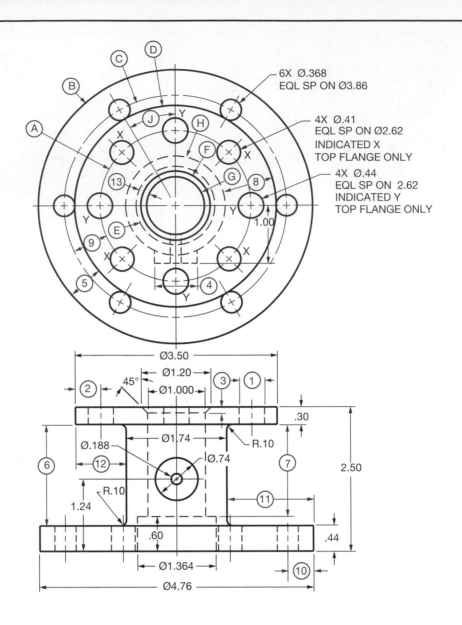

6X Ø.368
EQL SP ON Ø3.86

4X Ø.41
EQL SP ON Ø2.62
INDICATED X
TOP FLANGE ONLY

4X Ø.44
EQL SP ON 2.62
INDICATED Y
TOP FLANGE ONLY

QUESTIONS:

1. What are the diameters of circles Ⓐ to Ⓗ?

2. How many holes are in the bottom surface?

3. How many holes are in the top surface?

4. How deep is the Ø1.000 hole from the top of the coupling?

5. What is angle Ⓙ?

6. How thick is the largest flange?

7. What size bolts would be used for the Y holes located on the top flange?
   Allow about .06 total clearance. (Refer to the bolt sizes in the Appendix.)

8. What size bolts would be used for the bottom flange? Allow about
   .06 total clearance. (Refer to the bolt sizes in the Appendix.)

9. Calculate distances ① to ⑬.

MATERIAL	GREY IRON	
SCALE	NOT TO SCALE	
DRAWN		DATE

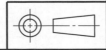

COUPLING

A-15

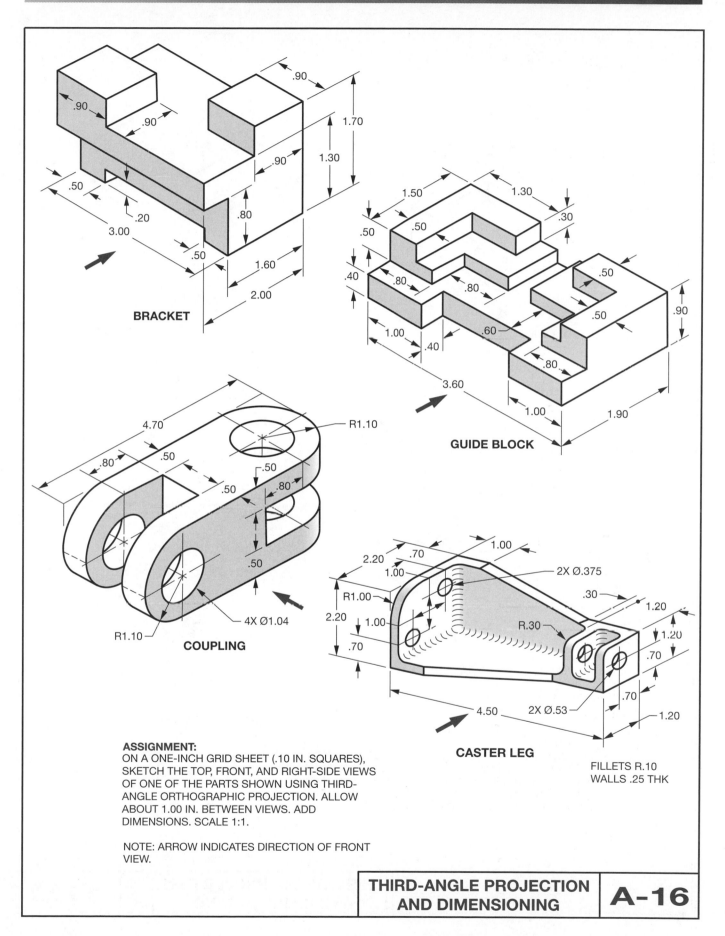

BRACKET

GUIDE BLOCK

COUPLING

R1.10

4X Ø1.04

R1.10

CASTER LEG

2X Ø.375

R.30

2X Ø.53

FILLETS R.10
WALLS .25 THK

ASSIGNMENT:
ON A ONE-INCH GRID SHEET (.10 IN. SQUARES),
SKETCH THE TOP, FRONT, AND RIGHT-SIDE VIEWS
OF ONE OF THE PARTS SHOWN USING THIRD-
ANGLE ORTHOGRAPHIC PROJECTION. ALLOW
ABOUT 1.00 IN. BETWEEN VIEWS. ADD
DIMENSIONS. SCALE 1:1.

NOTE: ARROW INDICATES DIRECTION OF FRONT
VIEW.

**THIRD-ANGLE PROJECTION
AND DIMENSIONING** | **A-16**

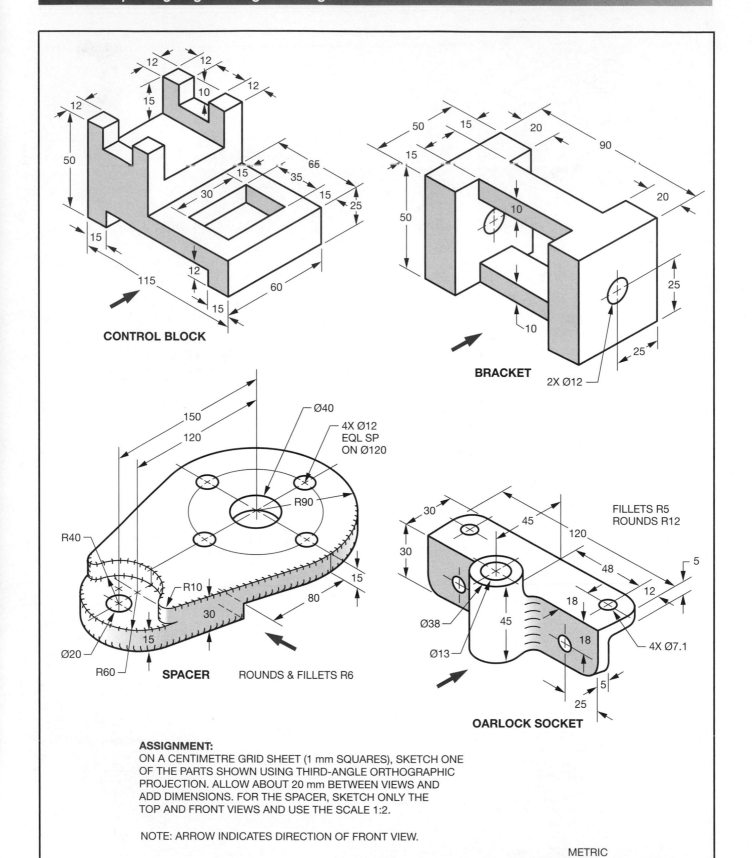

CONTROL BLOCK

BRACKET

2X Ø12

Ø40

4X Ø12
EQL SP
ON Ø120

R90

R40

R10

R60

Ø20

SPACER   ROUNDS & FILLETS R6

FILLETS R5
ROUNDS R12

Ø38

Ø13

4X Ø7.1

OARLOCK SOCKET

ASSIGNMENT:
ON A CENTIMETRE GRID SHEET (1 mm SQUARES), SKETCH ONE
OF THE PARTS SHOWN USING THIRD-ANGLE ORTHOGRAPHIC
PROJECTION. ALLOW ABOUT 20 mm BETWEEN VIEWS AND
ADD DIMENSIONS. FOR THE SPACER, SKETCH ONLY THE
TOP AND FRONT VIEWS AND USE THE SCALE 1:2.

NOTE: ARROW INDICATES DIRECTION OF FRONT VIEW.

METRIC
DIMENSIONS IN MILLIMETRES

**THIRD-ANGLE PROJECTION
AND DIMENSIONING**   **A-17M**

INCLINED SURFACES

If the surfaces of an object lie in either a horizontal or vertical position, they appear in their true shape in one of the three views and as a line in the other two.

When a surface is sloped or inclined in only one direction, it is not seen in its true shape in the top, front, or side views. It is, however, seen in two views as a distorted surface; on the third view, it appears as a line.

The true length of surfaces A and B in Figure 6–1 is seen in the front view only. In the top and side views, only the width of surfaces A and B appears in its true size. The length of these surfaces is foreshortened.

Where an inclined surface has important features that must be shown clearly and without distortion, an auxiliary or helper view must be used. These views are discussed in detail later in the book.

Illustrations of simple objects having inclined surfaces appear in Figure 6–2.

MEASUREMENT OF ANGLES

Some objects do not have all their features positioned so that all surfaces can be in the horizontal and vertical planes at the same time. The design of the part may require that some lines in the drawing be drawn in a direction other than horizontal or vertical, at an angle.

The amount of this divergence, or obliqueness, of lines may be indicated by either an offset dimension or an angle dimension, Figure 6–3.

Angle dimensions may be expressed in degrees and decimal parts of a degree. They may also be expressed in degrees, minutes, and seconds. The former method is now preferred.

The symbols for degrees (°), minutes (′), and seconds (″) are included with the appropriate values. For example, 2°; 30°; 28°10′; 0°15′; 27°13′15″; 0°0′30″; 0.25°; 30°0′0″; ±0°2′30″; and 2°±0.5° are all correct forms.

SYMMETRICAL OUTLINES

Symmetrical outlines or features may be indicated on a drawing with the symmetry symbol, Figure 6–4.

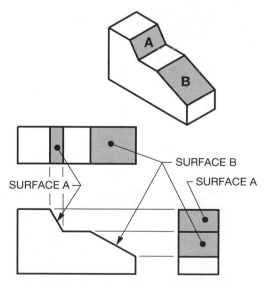

NOTE: THE TRUE SHAPES OF SURFACES A AND B DO NOT APPEAR ON THE TOP OR SIDE VIEWS.

FIGURE 6–1 ■ Inclined surfaces

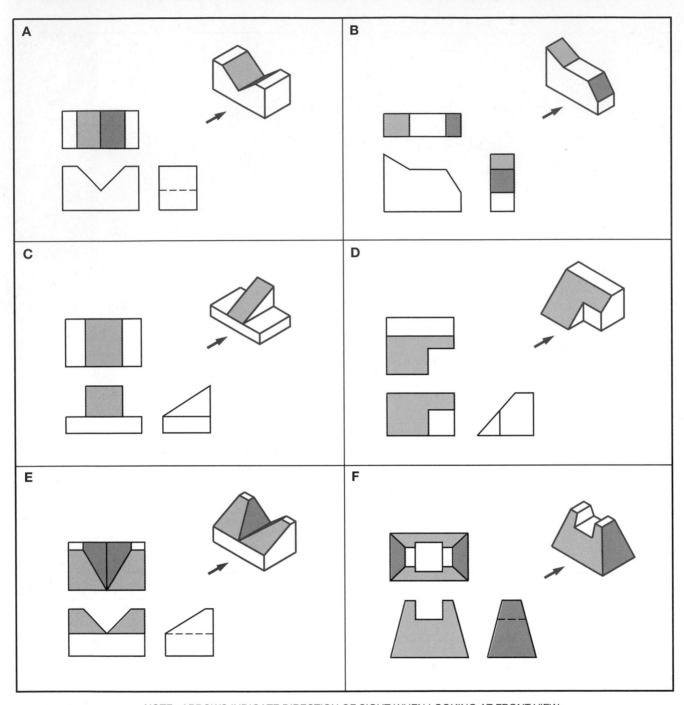

NOTE: ARROWS INDICATE DIRECTION OF SIGHT WHEN LOOKING AT FRONT VIEW.

FIGURE 6–2 ■ Simple objects with inclined surfaces

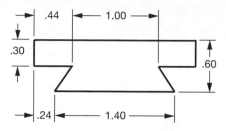

(A) LINEAR MEASUREMENTS

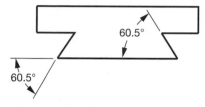

(B) ANGLE MEASUREMENTS USING DECIMAL DEGREES

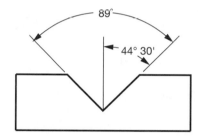

(C) ANGLE MEASUREMENTS USING DEGREES AND MINUTES

FIGURE 6–3 ■ Dimensioning angles

Two thick parallel lines are placed on the centre lines above and below the feature. The use of this symbol means the part or feature is symmetrical about the centre line or feature.

MACHINE SLOTS

Slots are used chiefly in machines to hold parts together. The two principal types are *T slots* and *dovetails*, Figure 6–4.

A dovetail is a groove or slide whose sides are cut on an angle. This forms an interlocking joint between two pieces, enabling the slot to resist pulling apart in any direction other than along the lines of the dovetail slide itself.

The dovetail is commonly used in the design of slides, including lathe cross slides and milling machine table slides. The two parts of a dovetail slide are shown in Figure 6–4.

When dovetail parts are to be machined to a given width, they may be gauged by using accurately sized cylindrical rods or wires.

Dovetails are usually dimensioned, Figure 6–5. The dimensions limit the boundaries within which the machinist works.

The edges of a dovetail are usually broken to remove the sharp corners. On large dovetails the external and internal corners are often machined, Figure 6–6(A) or (B).

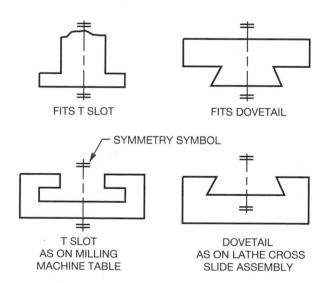

FIGURE 6–4 ■ Using the symmetry symbol to indicate symmetry or machining slots

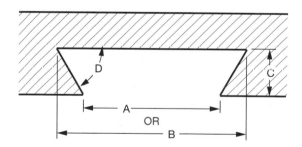

FIGURE 6–5 ■ Dimensions for dovetail

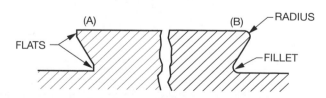

FIGURE 6–6 ■ Corners used on large dovetails

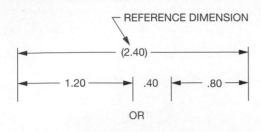

OR

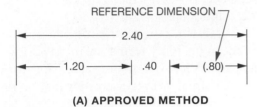

(A) APPROVED METHOD

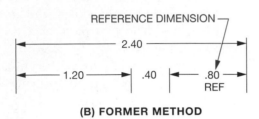

(B) FORMER METHOD

FIGURE 6–7 ■ Reference dimensions

REFERENCE DIMENSIONS

If a reference dimension is shown on a drawing for information only and is not used for the manufacture of the part, it must be clearly labelled. The approved method for indicating reference dimensions on a drawing is enclosing them inside parentheses, Figure 6–7. Previously, reference dimensions were indicated by placing the abbreviation REF after or below the dimension.

REFERENCE

ASME Y14.5M-1994 (R2004) Dimensioning and Tolerancing

INTERNET RESOURCES

Drafting Zone For information on drafting symbols and reference dimensioning, see: http://www.draftingzone.com

Metrication.com For metrication in drafting and engineering, see: http://www.metrication.com

QUESTIONS:

1. Calculate distances A to G.
2. At what angle is line ⑥ to the vertical?
3. At what angle is line ⑦ to the horizontal?
4. Locate surface ⑥ in the side view.
5. Locate surface ① in the side view.
6. Locate surface ⑥ in the top view.
7. Which lines in the side view are represented by line ② in the front view?
8. Locate ⑱ in the top view.
9. Locate surface ⑨ in the side view.
10. Locate surface ⑫ in the front view.
11. Locate surface ③ in the top view.
12. Which lines in the side view are represented by point ④ in the front view?
13. Which line in the side view is line ⑯ in the top view?
14. Locate surface ⑩ in the side view.

15. Locate surface ⑩ in the front view.
16. Locate surface ⑫ in the side view.
17. Which line does point ④ represent in the top view?
18. Locate line ㉔ in the top view.
19. Locate line ㉘ in the top view.
20. Locate line ㉕ in the top view.
21. Which line in the front view is surface ⑨ in the top view?

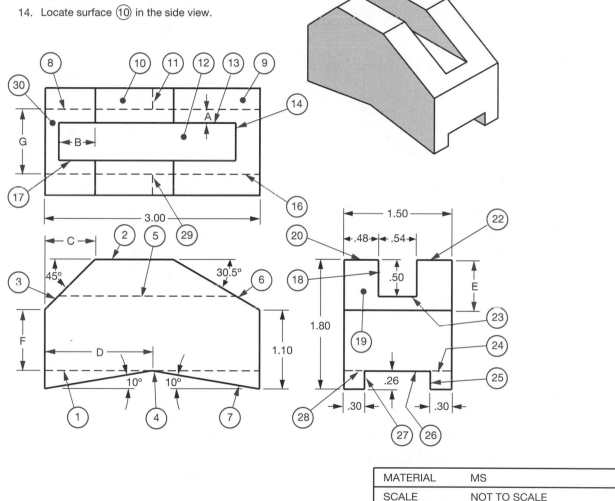

MATERIAL	MS	
SCALE	NOT TO SCALE	
DRAWN		DATE

BASE PLATE — **A-18**

QUESTIONS:

1. In which view is the shape of the dovetail shown?
2. In which view is the shape of the T slot shown?
3. How many rounds are shown in the top view?
4. In which view is a fillet shown?
5. Which line in the top view represents surface Ⓡ of the side view?
6. Which line in the front view represents surface Ⓡ?
7. Which line of the top view represents surface Ⓛ of the side view?
8. Which line in the front view represents surface Ⓛ?
9. Which line in the side view represents surface Ⓐ on the top view?
10. Which dimension in the front view represents the width of surface Ⓐ?
11. What type of lines are Ⓑ, Ⓙ, and Ⓚ?
12. How far apart are the two hidden edge lines on the side view?
13. What dimension indicates how far line Ⓙ is from the base of the slide?
14. How wide is the opening in the dovetail?
15. Which two lines in the top view indicate the opening of the dovetail?
16. At what angle to the horizontal is the dovetail cut?
17. In the side view, how far is the lower left edge of the dovetail from the left side of the piece?
18. What are the lengths of dimensions Y, V and X?
19. What is the height of the dovetail?

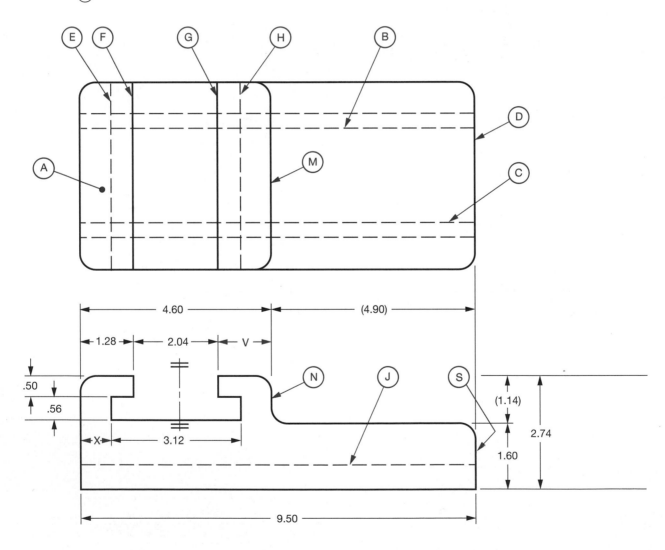

20. How much material remains between the surface represented by line Ⓠ and the top of the dovetail after the cut has been taken?

21. What is the vertical distance from the surface represented by line Ⓠ to that represented by line Ⓣ?

22. Which dimension represents the distance between lines Ⓕ and Ⓖ?

23. What is the overall height of the T slot?

24. What is the distance between the bottom of the T slot and the top of the dovetail?

25. What is the width of the bottom of the T slot?

26. What is the height of the opening of the bottom of the T slot?

27. What is the horizontal distance from line Ⓝ to line Ⓢ?

28. What is the unit of measurement for the angles shown?

29. How many reference dimensions are shown on the drawing?

30. What is the size of the largest reference dimension?

ROUNDS AND FILLETS R.38

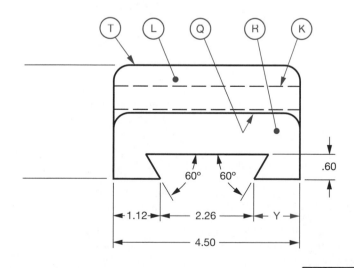

MATERIAL	GREY IRON	
SCALE	NOT TO SCALE	
DRAWN		DATE

COMPOUND REST SLIDE

A-19

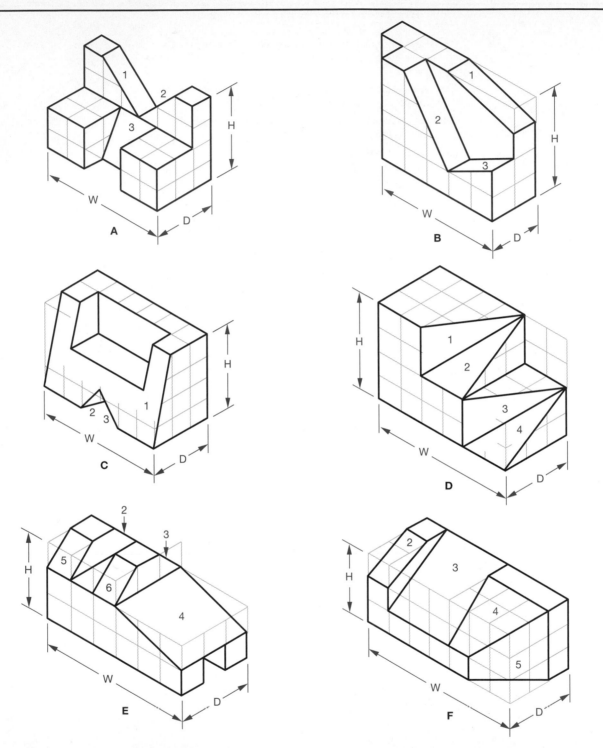

ASSIGNMENT:
ON A .25 INCH GRID SHEET, SKETCH THE TOP, FRONT, AND RIGHT-SIDE VIEWS OF
THE SIX PARTS SHOWN. EACH SQUARE SHOWN ON THE OBJECT REPRESENTS ONE
SQUARE ON THE GRID SHEET. ALLOW ONE GRID SPACE BETWEEN THE VIEWS AND A
MINIMUM OF TWO GRID SPACES BETWEEN THE OBJECTS. IDENTIFY THE SLOPED
SURFACES ON THE THREE VIEWS WITH THE CORRESPONDING NUMBERS SHOWN
ON THE PICTORIAL DRAWING.

ORTHOGRAPHIC SKETCHING OF OBJECTS HAVING SLOPED SURFACES USING GRID LINES	A-20

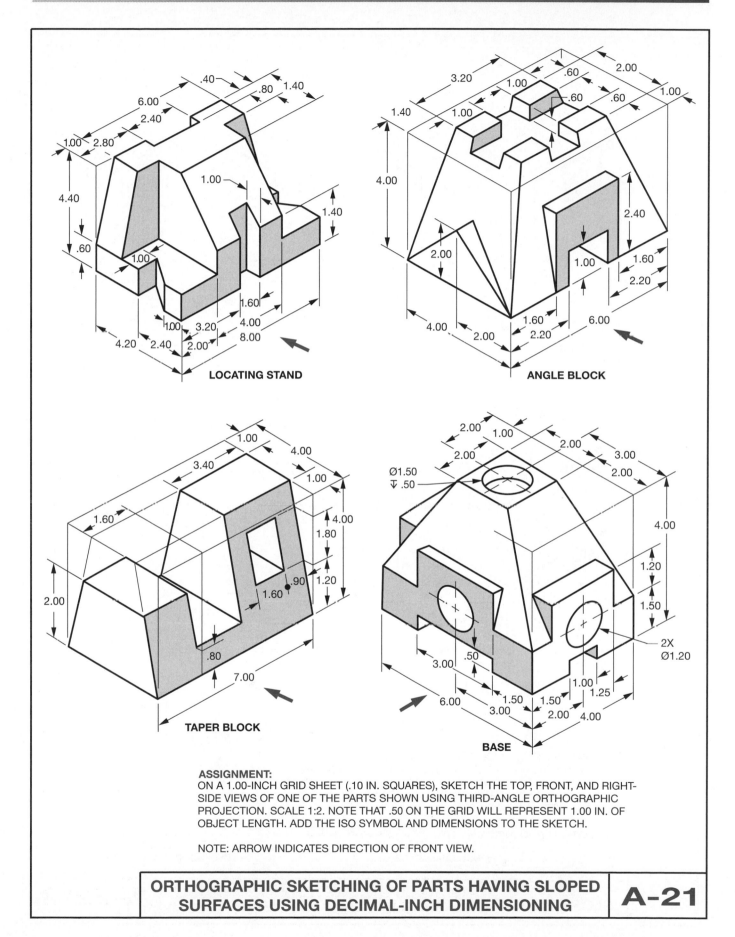

LOCATING STAND

ANGLE BLOCK

TAPER BLOCK

BASE

ASSIGNMENT:
ON A 1.00-INCH GRID SHEET (.10 IN. SQUARES), SKETCH THE TOP, FRONT, AND RIGHT-SIDE VIEWS OF ONE OF THE PARTS SHOWN USING THIRD-ANGLE ORTHOGRAPHIC PROJECTION. SCALE 1:2. NOTE THAT .50 ON THE GRID WILL REPRESENT 1.00 IN. OF OBJECT LENGTH. ADD THE ISO SYMBOL AND DIMENSIONS TO THE SKETCH.

NOTE: ARROW INDICATES DIRECTION OF FRONT VIEW.

ORTHOGRAPHIC SKETCHING OF PARTS HAVING SLOPED SURFACES USING DECIMAL-INCH DIMENSIONING	A-21

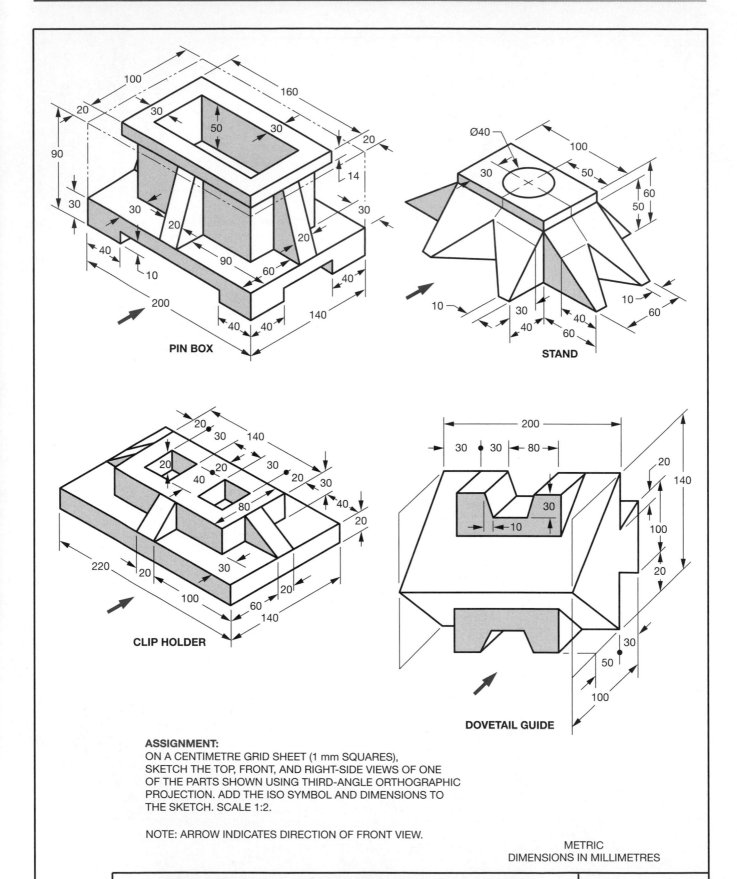

PIN BOX

STAND

CLIP HOLDER

DOVETAIL GUIDE

ASSIGNMENT:
ON A CENTIMETRE GRID SHEET (1 mm SQUARES),
SKETCH THE TOP, FRONT, AND RIGHT-SIDE VIEWS OF ONE
OF THE PARTS SHOWN USING THIRD-ANGLE ORTHOGRAPHIC
PROJECTION. ADD THE ISO SYMBOL AND DIMENSIONS TO
THE SKETCH. SCALE 1:2.

NOTE: ARROW INDICATES DIRECTION OF FRONT VIEW.

METRIC
DIMENSIONS IN MILLIMETRES

ORTHOGRAPHIC SKETCHING OF PARTS HAVING
SLOPED SURFACES USING MILLIMETRE DIMENSIONING

A-22M

PICTORIAL SKETCHING

Pictorial sketching is widely used in industry because it is easy to read and understand, Figure 7–1. It is also a quick and easy means of communicating technical ideas. Isometric sketching, one of several types of pictorial drawing, is the most frequently used. With the use of pictorial grid sheets and ellipse templates, pictorial drawings can be sketched quickly and accurately.

Viewing Direction

The pictorial sketch can be drawn so the part is viewed from above (bird's-eye view) or from below (worm's-eye view), Figure 7–2. The part features you wish to show normally govern the viewing direction selected.

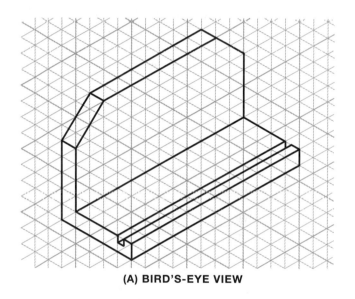

(A) BIRD'S-EYE VIEW

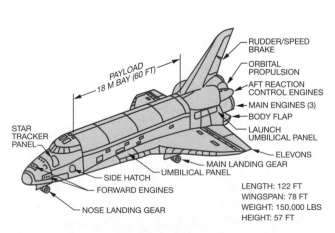

PAYLOAD
18 M BAY (60 FT)

RUDDER/SPEED BRAKE
ORBITAL PROPULSION
AFT REACTION CONTROL ENGINES
MAIN ENGINES (3)
BODY FLAP
LAUNCH UMBILICAL PANEL
ELEVONS
MAIN LANDING GEAR
UMBILICAL PANEL
SIDE HATCH
FORWARD ENGINES
NOSE LANDING GEAR
STAR TRACKER PANEL

LENGTH: 122 FT
WINGSPAN: 78 FT
WEIGHT: 150,000 LBS
HEIGHT: 57 FT

SPACE SHUTTLE ORBITER

FIGURE 7–1 ■ Application of a pictorial sketch

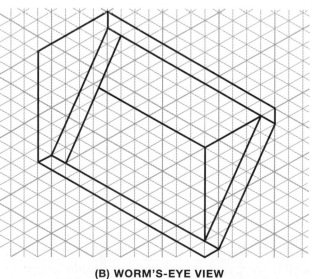

(B) WORM'S-EYE VIEW

FIGURE 7–2 ■ Viewing direction

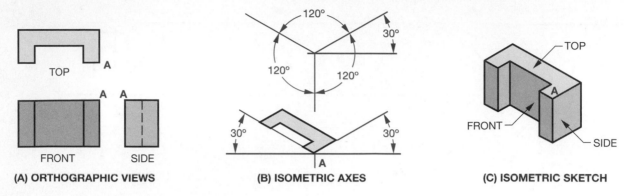

FIGURE 7–3 ■ Isometric axes and projection

ISOMETRIC SKETCHING

All isometric sketches are started by constructing the isometric axes, which are a vertical line for height and isometric lines to the left and right, at an angle of 30° from the horizon, for width and depth. The three faces seen in the isometric view are the same as those seen in the normal orthographic views: top, front, and side, Figure 7–3(A). Figure 7–3(B) shows the selection of the front corner A and the construction of the isometric axes. Figure 7–3(C) shows the completed isometric view. All lines are drawn to their true length, measured along the isometric axes; hidden lines are usually omitted.

Isometric Grid Sheets

Isometric sketching paper has evenly spaced lines running in three directions. Two sets of lines are sloped in the direction of the isometric axes; the third set is vertical, passing through the intersection of the sloping lines, Figure 7–2. The most commonly used grids are the inch, which is further subdivided into either 10 or 4 equal grids, and the centimetre, which is further subdivided into 10 equal grids of 1 mm. No units of measure are shown on these sheets; therefore, the spaces can represent any convenient unit of size.

Inclined Surfaces

Many objects have inclined surfaces, which are represented by sloping lines in orthographic views. In isometric drawings, sloping surfaces appear as **nonisometric lines**. To create them, their endpoints, which are found on the ends of isometric lines, are joined with a straight line, Figure 7–4.

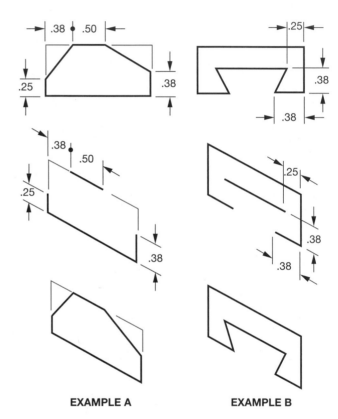

EXAMPLE A   **EXAMPLE B**

FIGURE 7–4 ■ Construction of non-isometric lines

Circles and Arcs

A circle on the three faces of an object drawn in isometric has the shape of an ellipse, Figure 7–5. Practically all circles and arcs shown on isometric sketches are drawn with an isometric ellipse template. The template, Figure 7–5, combines ellipses, scales, and angles. Markings on the ellipse coincide with the centre lines of the holes, speeding up the drawing of circles and arcs.

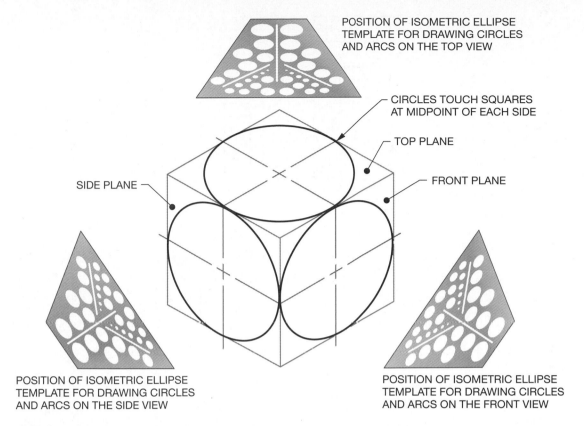

POSITION OF ISOMETRIC ELLIPSE
TEMPLATE FOR DRAWING CIRCLES
AND ARCS ON THE TOP VIEW

CIRCLES TOUCH SQUARES
AT MIDPOINT OF EACH SIDE

TOP PLANE

SIDE PLANE

FRONT PLANE

POSITION OF ISOMETRIC ELLIPSE
TEMPLATE FOR DRAWING CIRCLES
AND ARCS ON THE SIDE VIEW

POSITION OF ISOMETRIC ELLIPSE
TEMPLATE FOR DRAWING CIRCLES
AND ARCS ON THE FRONT VIEW

FIGURE 7–5 ■ Using the isometric ellipse template to draw circles and arcs

Basic Steps for Isometric Sketching (Figure 7–6)

To save time and make an accurate and neat sketch, use an isometric ellipse template for drawing arcs and circles and a straightedge for drawing long lines. A common technique is to sketch a box having the maximum height, width, and depth of the object, and then remove the parts of the box that are not part of the object, leaving the parts that form the total object.

■ **Build a frame.** The frame (or box) is the overall size of the part to be drawn. It is drawn with construction lines.

■ **Block in the overall sizes for each detail.** These sub blocks, or frames, enclose each detail. They are drawn with construction lines.

■ **Add the details.** Lightly sketch the shapes of the details using construction lines. For circles, draw squares equal to the size of the diameter. Also sketch in the lines to represent the centre lines of the circle.

■ **Darken the lines.** Using a soft lead pencil, darken the visible object lines.

OBLIQUE SKETCHING

This method of pictorial drawing requires placing the object with one face parallel to the frontal plane and the other two faces on oblique (or receding) planes, to left or right, top or bottom, at any convenient angle. The three axes of projection are vertical, horizontal, and receding. Figure 7–7 illustrates a cube drawn in typical positions with the receding axes at 60°, 45°, and 30°. This form of projection has the advantage of showing one face of the object without distortion. The face with the greatest irregularity of outline or contour, with the greatest number of circular features, or with the longest dimension faces the front, Figure 7–8.

Two types of oblique projection are used extensively. In *cavalier oblique*, all lines are made to their true length, measured on the axes of the projection. In *cabinet oblique*, the lines on the receding axis are shortened by one-half their true length to compensate for distortion and to approximate more closely what the eye would see. For this reason, and because of the simplicity of projection, cabinet oblique is often used especially when circles and arcs are to be drawn. Figure 7–9 shows a comparison of cavalier and cabinet oblique. Hidden lines are omitted unless

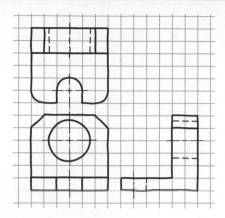

(A) THE PART

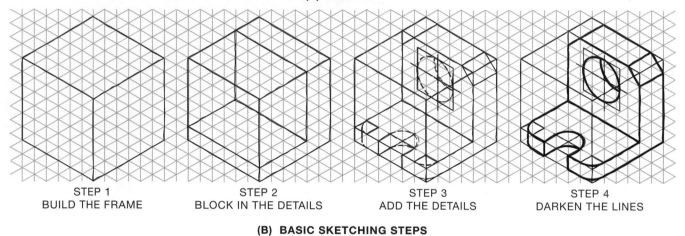

STEP 1
BUILD THE FRAME

STEP 2
BLOCK IN THE DETAILS

STEP 3
ADD THE DETAILS

STEP 4
DARKEN THE LINES

(B) BASIC SKETCHING STEPS

FIGURE 7–6 ■ Basic steps for isometric sketching

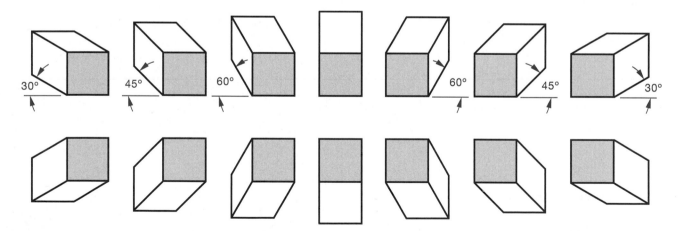

FIGURE 7–7 ■ Typical positions of receding axes for oblique projection

required for clarity. Most of the drawing techniques for isometric projection apply to oblique projection.

Oblique Grid Sheets

This type of sketching paper is similar to the two-dimensional sketching paper except that 45° lines, which pass through the intersecting horizontal and vertical lines, are added in one or both directions. The most commonly used grids are the inch, which is subdivided into smaller evenly spaced grids, and the centimetre. As no units of measurement are shown on these sheets, the spaces can represent any convenient unit of length, Figure 7–10.

Inclined Surfaces

Angles that are parallel to the picture plane are drawn as their true size. Other angles can be laid off by locating the ends of the inclined line.

A part with notched corners is shown in Figure 7–11(A). An oblique drawing with the angles parallel to the picture plane is shown in Figure 7–11(B). In Figure 7–11(C), the angles are parallel to the profile plane. In each case, the angle is laid off by measurement parallel to the oblique axes, shown by the construction lines. Since the part, in each case, is drawn in cabinet oblique, the receding lines are shortened by one-half their true length.

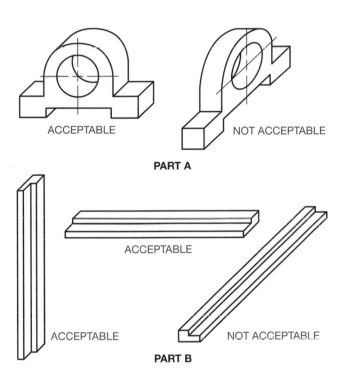

FIGURE 7–8 ■ Two general rules for oblique projection

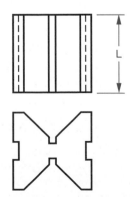

FIGURE 7–9 ■ Types of oblique projection

Circles and Arcs

Whenever possible, the face of the object having circles or arcs should be selected as the front face, so that such circles or arcs can be easily drawn in their true shape, Figure 7–12.

When circles or arcs must be drawn on one of the oblique faces, the following method is recommended, Figure 7–12(B):

■ Block off an oblique square with centre lines equal to the diameter of the circle required. Block in the circle first to get the proper size and shape of the ellipse. If an ellipse template is available, select an ellipse that fits within the square and touches the sides of the square at its midpoints. Using thick dark lines (object lines), draw the oblique circle (ellipse), Figure 7–12(C).

■ If an ellipse template is not available, lightly sketch an ellipse within this square with the circumference making contact with the square at its midpoints, Figure 7–12(B).

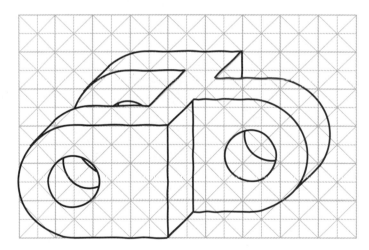

FIGURE 7–10 ■ Oblique sketching paper

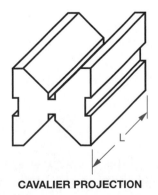

CAVALIER PROJECTION

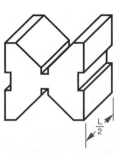

CABINET PROJECTION

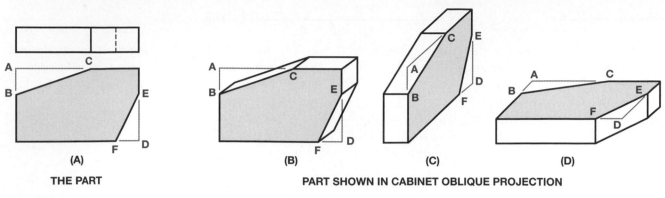

(A)
THE PART

(B)   (C)   (D)
PART SHOWN IN CABINET OBLIQUE PROJECTION

FIGURE 7–11 ■ Drawing inclined surfaces

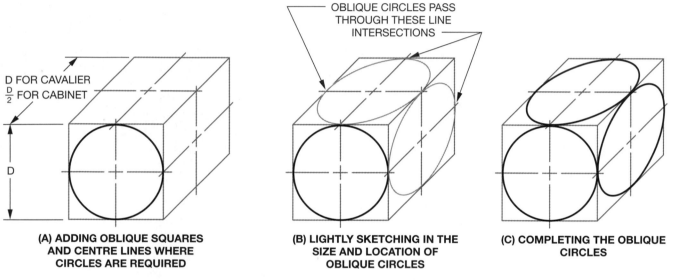

(A) ADDING OBLIQUE SQUARES AND CENTRE LINES WHERE CIRCLES ARE REQUIRED

(B) LIGHTLY SKETCHING IN THE SIZE AND LOCATION OF OBLIQUE CIRCLES

(C) COMPLETING THE OBLIQUE CIRCLES

FIGURE 7–12 ■ Sketching oblique circles

■ Using object lines, darken the oblique circle, Figure 7–12(C).

Basic Steps for Oblique Sketching (Figure 7–13)

■ **Build a frame.** The frame or box is the overall size of the part to be drawn. It is drawn with light thin lines.

■ **Block in the overall size of each detail.** These sub blocks or frames enclose each detail. For circles, draw squares equal to the diameter size. Sketch the centre lines, using light thin lines.

■ **Add the details.** Lightly sketch the shape of the details in each of their frames, using light thin lines. If an oblique circle (ellipse) template is available, draw the arcs and circles using thick dark (visible object) lines.

■ **Darken the lines.** Use a soft lead pencil to darken the lines.

REFERENCES

CAN 3-B78.1 M83 Technical Drawings–General Principles
ASME Y14.4M-1989 (R2004) Pictorial Drawings

INTERNET RESOURCE

Animated Worksheets For information on isometric and perspective drawings, see: http://www .animatedworksheets.co.uk (perspectives)

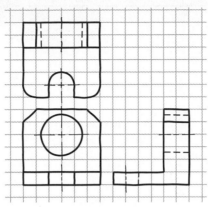

(A) THE PART

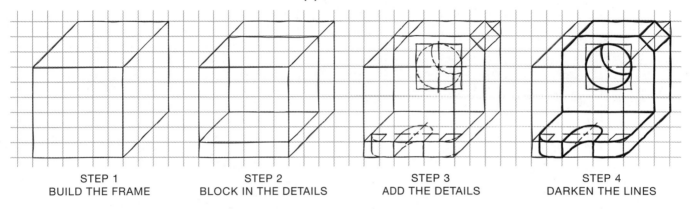

STEP 1
BUILD THE FRAME

STEP 2
BLOCK IN THE DETAILS

STEP 3
ADD THE DETAILS

STEP 4
DARKEN THE LINES

(B) BASIC SKETCHING STEPS

FIGURE 7–13 ■ Basic steps for oblique sketching

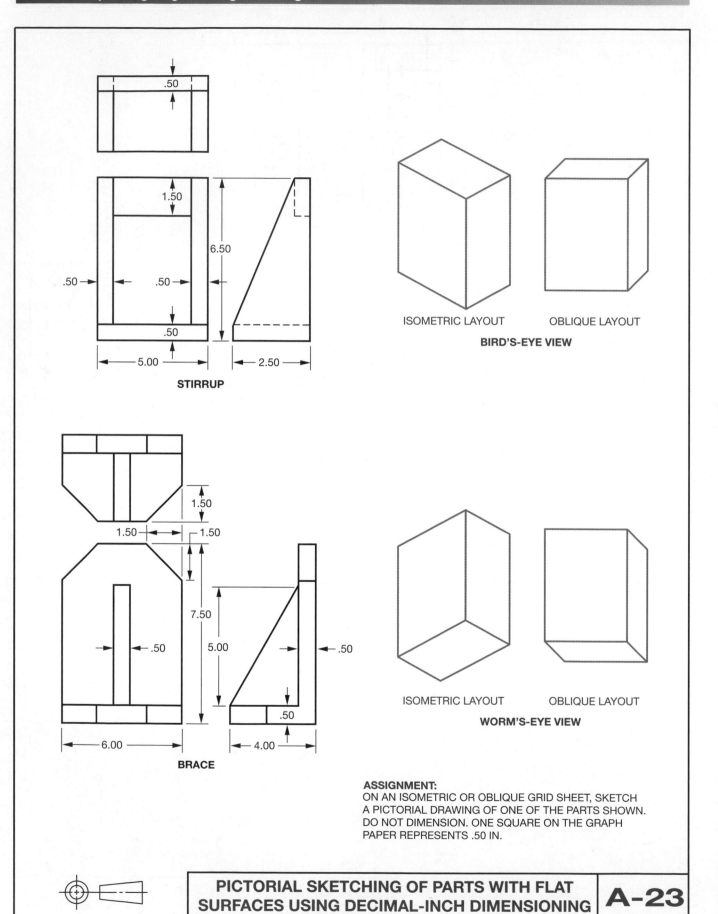

.50

1.50

6.50

.50 → ← .50 → ←

.50

← 5.00 →

STIRRUP

← 2.50 →

ISOMETRIC LAYOUT   **OBLIQUE LAYOUT**

BIRD'S-EYE VIEW

1.50

1.50 ← 1.50

7.50

5.00

.50

← 6.00 →

BRACE

← 4.00 →

.50

.50

ISOMETRIC LAYOUT   **OBLIQUE LAYOUT**

WORM'S-EYE VIEW

ASSIGNMENT:
ON AN ISOMETRIC OR OBLIQUE GRID SHEET, SKETCH
A PICTORIAL DRAWING OF ONE OF THE PARTS SHOWN.
DO NOT DIMENSION. ONE SQUARE ON THE GRAPH
PAPER REPRESENTS .50 IN.

PICTORIAL SKETCHING OF PARTS WITH FLAT
SURFACES USING DECIMAL-INCH DIMENSIONING

A-23

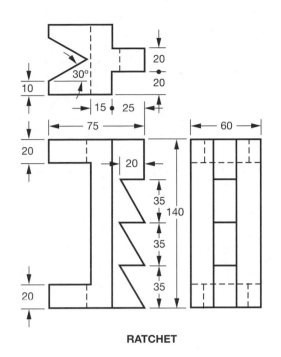

20
20
30°
10
15 ● 25
75
60
20
20
35
140
35
35
20

RATCHET

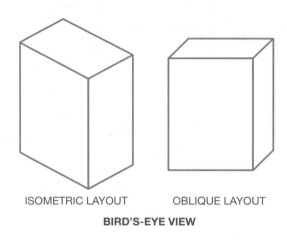

ISOMETRIC LAYOUT   OBLIQUE LAYOUT

BIRD'S-EYE VIEW

90
60
15
20
10 ● 15
25 ●
15
60
140
40
8
30
20

TABLET

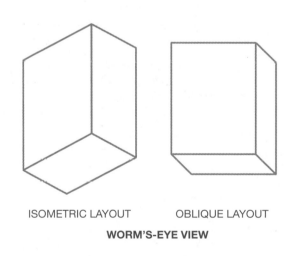

ISOMETRIC LAYOUT   OBLIQUE LAYOUT

WORM'S-EYE VIEW

ASSIGNMENT:
ON AN ISOMETRIC OR OBLIQUE GRID SHEET, SKETCH
A PICTORIAL DRAWING OF ONE OF THE PARTS SHOWN.
DO NOT DIMENSION. ONE SQUARE ON THE GRAPH
PAPER REPRESENTS 10 mm.

METRIC
DIMENSIONS IN MILLIMETRES

**PICTORIAL SKETCHING OF PARTS WITH FLAT
SURFACES USING METRIC DIMENSIONING**

A-24M

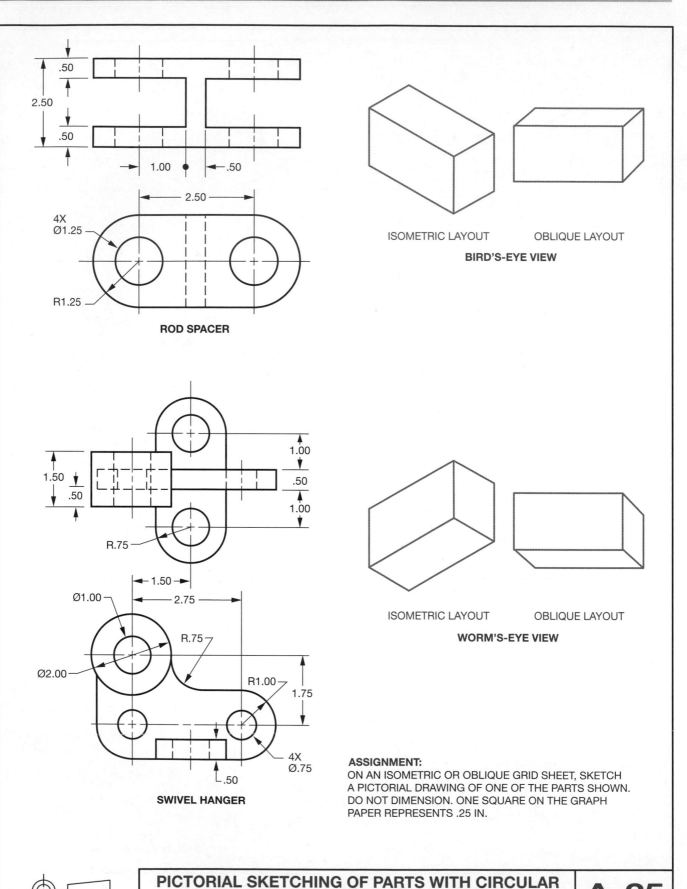

.50

2.50

.50

1.00

.50

2.50

4X
Ø1.25

R1.25

ROD SPACER

1.50

.50

1.00

.50

1.00

R.75

1.50

2.75

Ø1.00

R.75

Ø2.00

R1.00

1.75

4X
Ø.75

.50

SWIVEL HANGER

ISOMETRIC LAYOUT   OBLIQUE LAYOUT

BIRD'S-EYE VIEW

ISOMETRIC LAYOUT   OBLIQUE LAYOUT

WORM'S-EYE VIEW

ASSIGNMENT:
ON AN ISOMETRIC OR OBLIQUE GRID SHEET, SKETCH
A PICTORIAL DRAWING OF ONE OF THE PARTS SHOWN.
DO NOT DIMENSION. ONE SQUARE ON THE GRAPH
PAPER REPRESENTS .25 IN.

PICTORIAL SKETCHING OF PARTS WITH CIRCULAR FEATURES USING DECIMAL-INCH DIMENSIONING

A-25

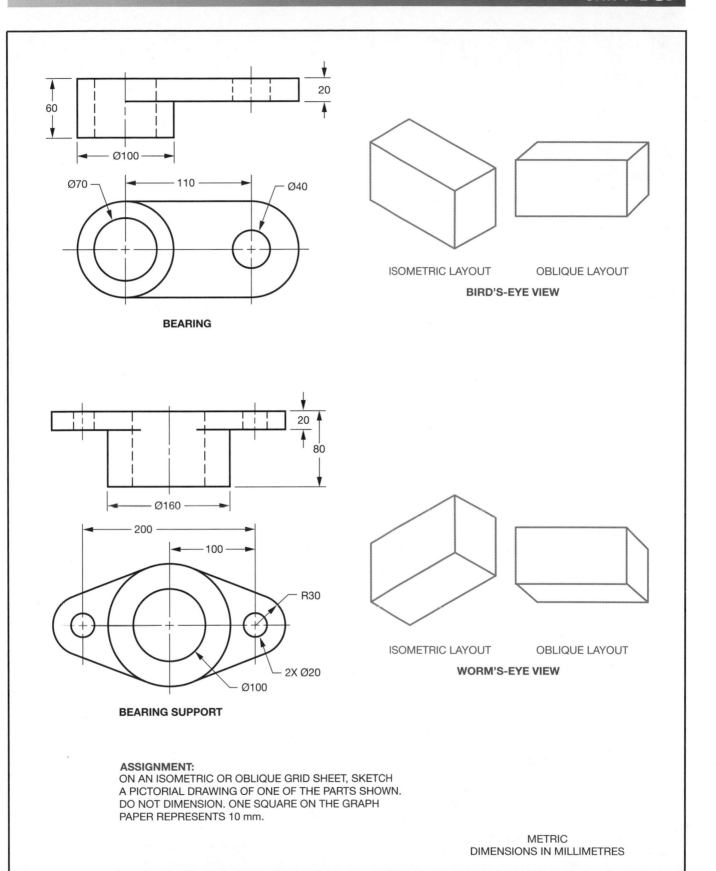

60

20

Ø100

Ø70   110   Ø40

BEARING

ISOMETRIC LAYOUT   OBLIQUE LAYOUT

BIRD'S-EYE VIEW

20

80

Ø160

200

100

R30

2X Ø20

Ø100

BEARING SUPPORT

ISOMETRIC LAYOUT   OBLIQUE LAYOUT

WORM'S-EYE VIEW

ASSIGNMENT:
ON AN ISOMETRIC OR OBLIQUE GRID SHEET, SKETCH
A PICTORIAL DRAWING OF ONE OF THE PARTS SHOWN.
DO NOT DIMENSION. ONE SQUARE ON THE GRAPH
PAPER REPRESENTS 10 mm.

METRIC
DIMENSIONS IN MILLIMETRES

**PICTORIAL SKETCHING OF PARTS WITH
CIRCULAR FEATURES USING METRIC DIMENSIONING**

A–26M

8 UNIT

MACHINING SYMBOLS

When preparing working drawings of parts to be cast or forged, the drafter must identify part surfaces that require machining or finishing. This is done by adding a machining allowance symbol to the surface or surfaces that must be finished by the removal of material, Figure 8–1. Figure 8–2 shows the current machining symbol and those that were formerly used on drawings. This information is essential to alert the patternmaker and diemaker to provide extra metal on the casting or forging to allow for the finishing process. Depending on the material to be cast or forged, between .04 and .10 inch is usually allowed on small castings and forgings for each surface that requires finishing, Figure 8–3.

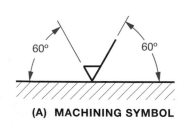

(A) MACHINING SYMBOL

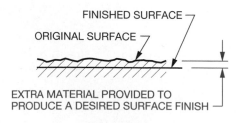

(B) MEANING

FIGURE 8–1 ■ Machining symbol

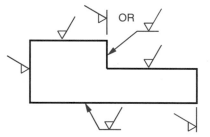

(A) RECOMMENDED SYMBOL

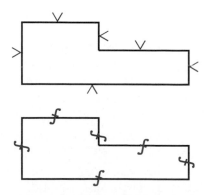

(B) FORMER MACHINING SYMBOLS

FIGURE 8–2 ■ Application of machining symbols

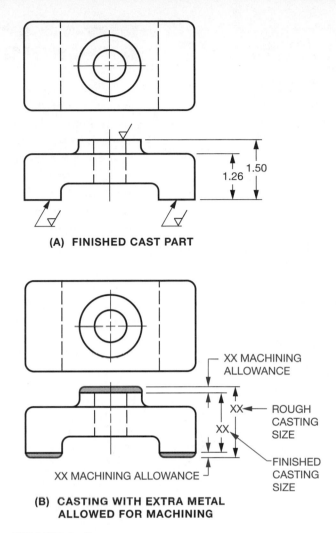

(A) FINISHED CAST PART

(B) CASTING WITH EXTRA METAL
ALLOWED FOR MACHINING

FIGURE 8–3 ■ Allowances for machining

Like dimensions, machining symbols are not duplicated on the drawing. They should be used on the same view as the dimensions that give the size or location of the surfaces. The symbol is placed on the line representing the surface or on a leader or an extension line locating the surface. The symbol and the inscription should be oriented so they can be read from the bottom or right-hand side of the drawing.

Where all the surfaces are to be machined, a general note, such as FINISH ALL OVER (FAO), may be used and the symbols on the drawings omitted.

The outdated machining symbols shown in Figure 8–2(B) are found on many older drawings still in use. When called upon to make changes or revisions on a drawing already in existence, a drafter must adhere to the drawing conventions shown on that drawing.

The machining symbol does not indicate surface-finish quality. The surface texture symbol, which is defined and discussed in Unit 12, is used to signify the desired surface-finish quality.

Indicating Machining Allowance

When the value of the machining allowance must be specified, it is indicated to the left of the symbol, Figures 8–4 and 8–5. This value is expressed in inches or millimetres, depending on which units of measurement are used on the drawing.

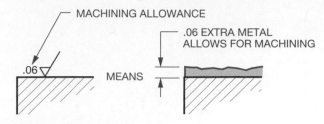

FIGURE 8–4 ■ Indicating the value of the machining allowance

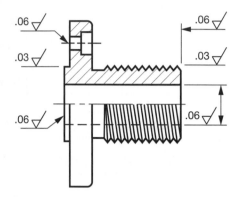

FIGURE 8–5 ■ The value of the machining allowance shown on the drawing

Removal of Material Prohibited

A surface from which the removal of material is prohibited is indicated by the symbol shown in Figure 8–6. This symbol indicates that a surface must be left as it was when it left the preceding manufacturing process, regardless of the removal of material or other changes. This machining symbol is part of the surface texture symbol described in Unit 12.

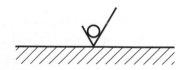

FIGURE 8–6 ■ Symbol for removal of material not permitted

NOT-TO-SCALE DIMENSIONS

When a dimension on a drawing is altered, making it not to scale, a straight freehand line is drawn below the dimension to indicate that the dimension is not drawn to scale, Figure 8–7(A).

This is a change from earlier methods of indicating not-to-scale dimensions. Formerly, a wavy line below the dimension or the letters NTS beside the dimension were used to indicate not-to-scale dimensions, Figure 8–7(B).

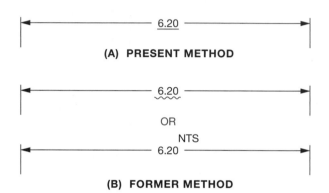

FIGURE 8–7 ■ Indicating dimensions that are not to scale

There is, however, an exception to the general use of the not-to-scale symbol. In cases where the dimension refers to an element of a part shortened by either long-break lines or short-break lines, the not-to-scale symbol may be omitted. Refer to Assignments A-27M and A-28. In both cases, dropping the not-to-scale symbol would have no effect on the final size of the manufactured part. In this case, it is used simply because the dimension has been revised from its original design size.

The not-to-scale dimension symbol found in Assignment A-35 is used in a somewhat different manner. In this case, the length of an element is not foreshortened by break lines. The not-to-scale symbol is used in this case simply because the length of a particular element of the part has been altered due to a design change (revision) and the dimension no longer represents the size measured on the original drawing.

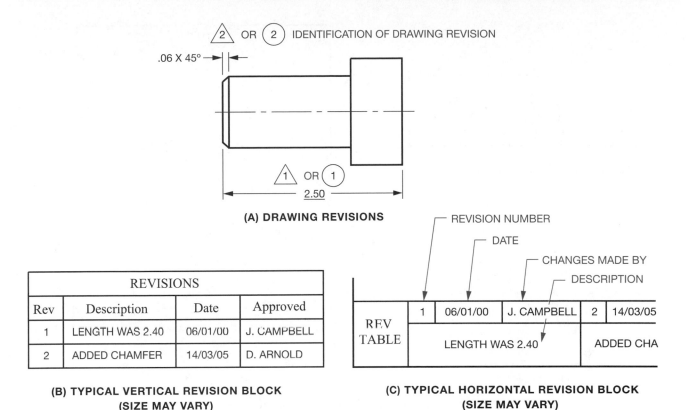

FIGURE 8–8 ■ Drawing revisions

DRAWING REVISIONS

Drawing revisions are made to accommodate improved manufacturing methods, reduce costs, correct errors, and improve design. A clear record of these revisions must be registered on the drawing.

All drawings should carry a change or revision table located at the bottom or down the top right-hand side of the drawing. The revision number, enclosed in a circle or triangle, should be located near the revised dimension for easy identification, Figure 8–8(A). The revision block should include a revision number or symbol, the date, the drafter's name or initials, and approval of the change. Should the drawing revision cause a dimension or dimensions to be different from the scale indicated, the dimensions that are not to scale should be indicated. Typical revision tables are shown in Figures 8–8(B) and (C).

When many revisions are needed, a new drawing is often made. The words REDRAWN AND REVISED should appear in the revision column of the new drawing when this is done.

After a revision to a drawing is made, a new set of prints of that drawing is distributed to the appropriate departments and the prints of the original drawing destroyed.

BREAK LINES

Break lines, Figure 8–9, are used to shorten the view of long uniform sections. They are also used when only a partial view is required. Such lines are used on both detail and assembly drawings. The thin line with freehand zigzags is recommended for long breaks, and the jagged line for wood parts. The special breaks shown for cylindrical and tubular parts are useful when an end view is not shown; otherwise, the thick break line is adequate.

REFERENCES

CAN3-B95-1962 (R1996) Surface Texture
ASME Y14.36M-1996 (R2002) Surface Texture
   Symbols

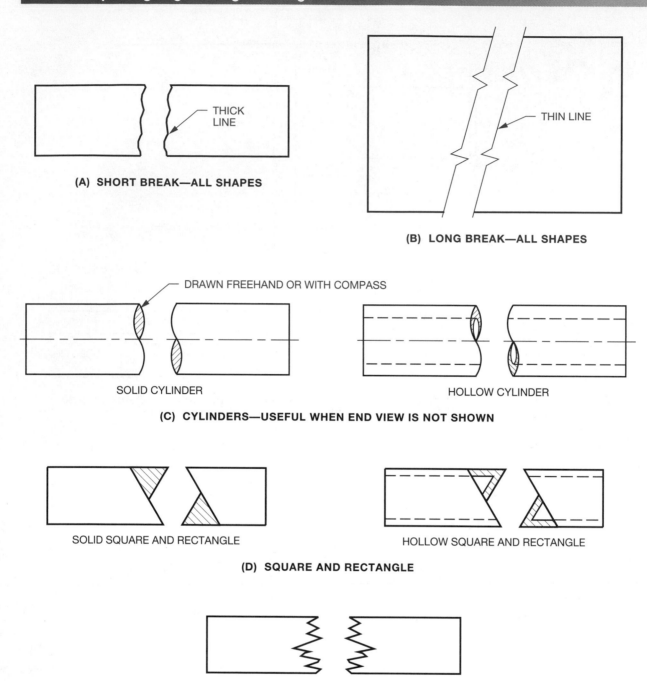

(A) SHORT BREAK—ALL SHAPES

THICK LINE

(B) LONG BREAK—ALL SHAPES

THIN LINE

DRAWN FREEHAND OR WITH COMPASS

SOLID CYLINDER

HOLLOW CYLINDER

(C) CYLINDERS—USEFUL WHEN END VIEW IS NOT SHOWN

SOLID SQUARE AND RECTANGLE

HOLLOW SQUARE AND RECTANGLE

(D) SQUARE AND RECTANGLE

(E) WOOD

FIGURE 8–9 ■ Conventional break lines

INTERNET RESOURCES

American Society of Mechanical Engineers For information on surface texture, machining symbols, and conventional breaks, refer to ASME Y14.36M-1996 (R2002) (*Surface Texture Symbols*) at: http://www.asme.org

For information on drawing revisions, refer to ASME Y14.35M-1997 (R2003) (*Revision of Engineering Drawings and Associated Documents*) at: http://www.asme.org

Drafting Zone For information on conventional breaks and drawing revisions, see: http://www.draftingzone.com

Integrated Publishing For information on break lines, see: http://www.tpub.com/content/draftsman

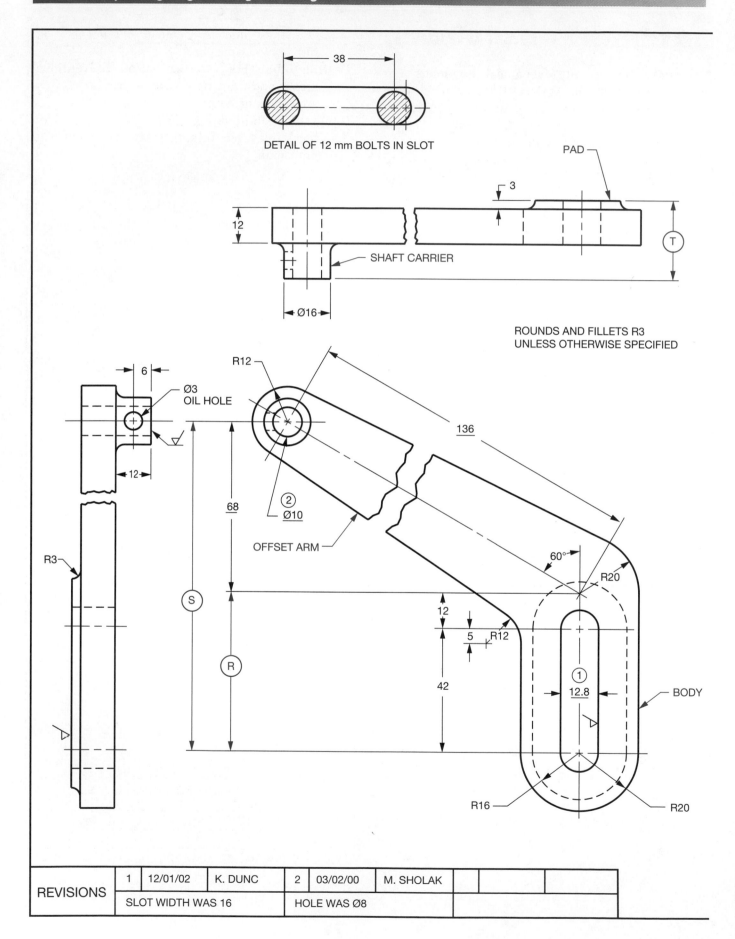

38

DETAIL OF 12 mm BOLTS IN SLOT

PAD

3

12

SHAFT CARRIER

Ø16

T

ROUNDS AND FILLETS R3
UNLESS OTHERWISE SPECIFIED

6

Ø3
OIL HOLE

R12

136

12

② Ø10

OFFSET ARM

68

60°

R20

12

S

5 R12

R3

R

① 12.8

BODY

42

R16

R20

REVISIONS	1	12/01/02	K. DUNC	2	03/02/00	M. SHOLAK			
	SLOT WIDTH WAS 16			HOLE WAS Ø8					

QUESTIONS:

1. At what angle is the offset arm to the body of the piece?

2. What is the centre-to-centre measurement of the length of the offset arm?

3. Which radius forms the upper end of the offset arm?

4. Which radii form the lower end of the offset arm where it joins the body?

5. What is the width of the bolt slot in the body of the bracket?

6. What is the centre-to-centre length of this slot?

7. What was the slot width before revision?

8. Which radii forms the ends of the pad?

9. What is the overall length of this pad?

10. What is the overall width of this pad?

11. What is the radius of the fillet between the pad and the body?

12. What is the diameter of the shaft carrier body?

13. What is the diameter of the shaft carrier hole?

14. What is the distance from the face of the shaft carrier to the face of the pad?

15. What is the radius of the inside fillet between the arm and the body of the piece?

16. If M12 bolts are used in holding the bracket to the machine base, what is the clearance on each side of the slot?

17. If the centre-to-centre distance of the two M12 bolts that fit the slot is 38 mm, how much play is there lengthwise in the slot?

18. What size oil hole is in the shaft carrier?

19. How far is the centre of the oil hole from the face of the shaft carrier?

20. How thick is the combined body and pad?

21. Calculate distance (R), (S), and (T).

22. The hole in the shaft carrier was revised. What is the difference in size between the new and old hole?

23. How many dimensions indicate that they are not drawn to scale?

24. If 2 mm is allowed for each surface to be machined, what would be the overall thickness of the original casting?

25. How many conventional breaks are shown?

26. When was the last drawing revision made?

METRIC
DIMENSIONS IN MILLIMETRES

MATERIAL	MI	
SCALE	1:1	
DRAWN	T. LOGAN	DATE 15/12/99

OFFSET BRACKET

A-27M

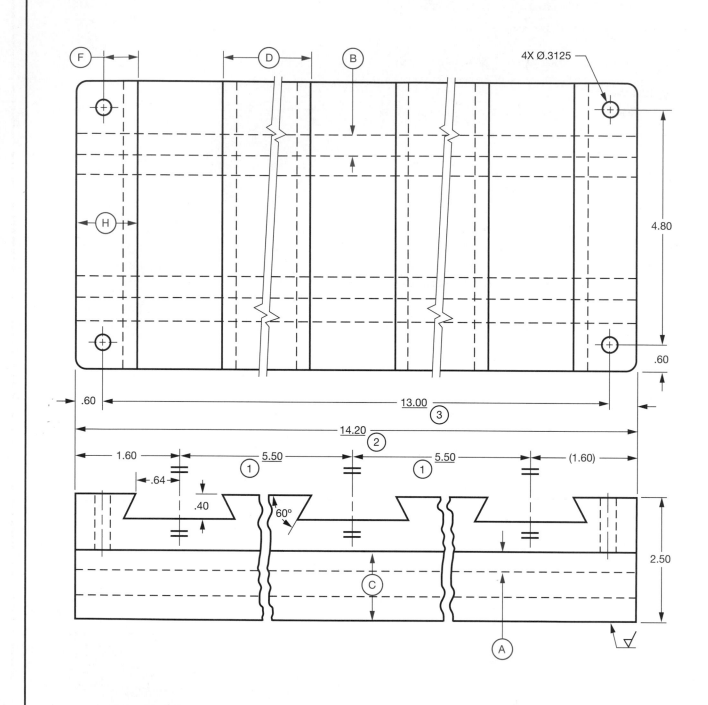

4X Ø.3125

NOTE:
ROUNDS AND FILLETS R.10
▽SHOWN TO BE .10▽

REV TABLE	1	18/07/05	R. HINES	2	18/07/05	R. HINES	3	18/07/05	R. HINES
	5.50 WAS 6.25			14.20 WAS 15.70			13.00 WAS 14.50		

QUESTIONS:

1. What is the drawing scale?
2. How many machine slots are shown?
3. How many reference dimensions are shown?
4. How many conventional breaks are shown?
5. How many not-to-scale dimensions are shown?
6. What was the total number of dimensions altered due to the drawing revisions?
7. How many symmetrical outlines are indicated by means of the symmetry symbol?
8. What machining allowance is called for on the machined surfaces?
9. How many fillets are shown?
10. How many rounds are shown?
11. What size bolts would be used in the holes?
12. What was the original length of the casting?
13. What was the overall height of the casting before the top and bottom surfaces were machined?
14. What is the overall height of the T slot?
15. What is the change in distance between the centres of dovetail slots from the present and the original drawing?
16. Determine distances (A) to (H).

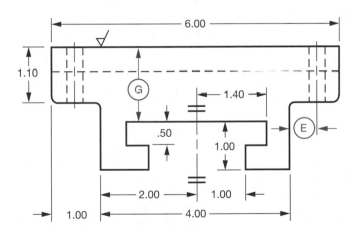

MATERIAL	GREY IRON		
SCALE	1:2		
DRAWN	P. JENSEN	DATE	30/11/04

GUIDE BAR A-28

SECTIONAL VIEWS

Sectional views, commonly called *sections,* are used to show interior detail too complicated to be shown clearly and dimensioned by outside views and hidden lines. A sectional view is obtained by supposing the nearest part of the object has been cut or broken away on an imaginary cutting plane. The exposed or cut surfaces are identified by section lining or crosshatching. Hidden lines and details behind the cutting-plane line are usually omitted unless required for clarity. Only in the sectional view is any part of the object shown as having been removed.

A sectional view frequently replaces one of the regular views. For example, a regular front view is replaced by a front view in section in Figure 9–1.

The Cutting-Plane Line

A *cutting-plane line* indicates where the imaginary cutting takes place. The position of the cutting plane is indicated, when necessary, on a view of the object or assembly by a cutting-plane line, Figure 9–2. The ends of the cutting-plane line are bent at 90° and terminated by arrowheads to indicate the direction of sight for viewing the section. Cutting planes are not shown on sectional views. The cutting-plane line may be omitted when it corresponds to the centre line of the part or when only one sectional view appears on a drawing.

If two or more sections appear on the same drawing, the cutting-plane lines are identified by two identical large, single-stroke Gothic letters. One letter is placed at each end of the line near the arrowhead. Sectional view subtitles are given when identification letters are used and appear directly below the view, incorporating the letters at each end of the cutting-plane line, for example, SECTION A-A, or abbreviated, SECT A-A. See Assignment A-31M.

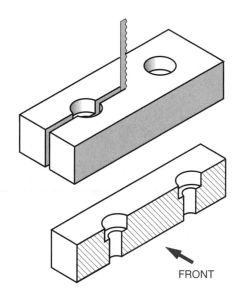

FRONT

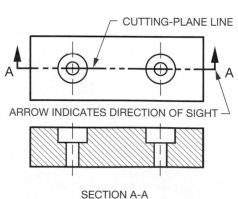

CUTTING-PLANE LINE

A   A

ARROW INDICATES DIRECTION OF SIGHT

SECTION A-A

FIGURE 9–1 ■ Section drawing

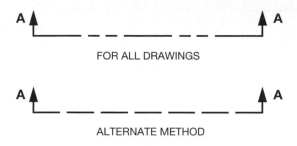

FOR ALL DRAWINGS

ALTERNATE METHOD

NOTE: LETTERS PLACED BESIDE ARROWS

FIGURE 9–2 ■ Cutting-plane lines

Section Lining

Section lining identifies the surface that has been cut and makes it stand out clearly. Section lines usually consist of thin parallel lines, Figure 9–3, drawn at an angle of about 45° to the principal edges or axis of the part.

Because the exact material specifications for a part are given elsewhere, the general-use section lining, Figure 9–4, is recommended for general use.

When it is desirable to indicate differences in materials, other symbolic section lines are used, Figure 9–4. If the part shape would cause section lines to be parallel or nearly parallel to one of the sides or features of the part, an angle other than 45° is chosen.

The spacing of the hatching lines is uniform to give a good appearance to the drawing. The pitch, or distance, between lines varies from .06 to .18 inch, depending on the size of the area to be sectioned. Section lining is uniform in direction and spacing in all sections of a single component.

Wood and concrete are the only two materials usually shown symbolically. When wood symbols are used, the direction of the grain is shown.

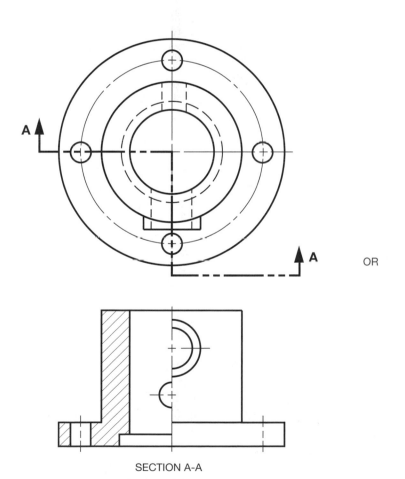

SECTION A-A

LETTERS, SUBTITLE, AND CUTTING-PLANE LINE USED WHEN MORE THAN ONE SECTION APPEARS ON A DRAWING OR WHEN THEY MAKE THE DRAWING CLEARER.

OR

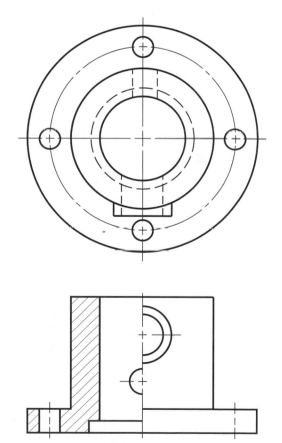

LETTERS, SUBTITLE, AND CUTTING-PLANE LINE MAY BE OMITTED WHEN THEY CORRESPOND WITH THE CENTRE LINE OF THE PART AND WHEN THERE IS ONLY ONE SECTION VIEW ON THE DRAWING.

FIGURE 9–3 ■ Identification of cutting plane and sectional views

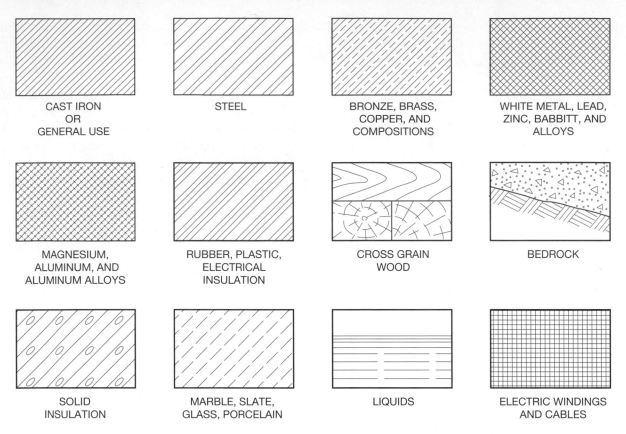

FIGURE 9–4 ■ Symbolic section lining

TYPES OF SECTIONS

Full Sections

When the cutting plane extends entirely through the object in a straight line and the front half of the object is theoretically removed, a *full section* is obtained, Figure 9–5(B). This type of section is used for both detail and assembly drawings. When the cutting plane divides the object into two identical parts, it is not necessary to indicate its location. However, the cutting plane may be drawn and labelled in the usual manner to increase clarity.

Half Sections

A symmetrical object or assembly may be drawn as a *half section*, Figure 9–5(C), showing one half in section and the other half in full view. A normal centre line is listed on the section view.

The half section drawing is not normally used where the dimensioning of internal diameters is required, because many hidden lines would have to be added to the portion showing the external features. This type of section is used mostly for assembly drawings where internal and external features are clearly shown and only overall and centre-to-centre dimensions are required.

Offset Sections

To include features that are not in a straight line, the cutting-plane line may be offset or bent to include several planes or curved surfaces, Figures 9–6 and 9–7.

An offset section is similar to a full section in that the cutting plane extends through the object from one side to the other. The change in direction of the cutting-plane line is not shown on the sectional view.

Other Types of Sections

Other types of sections, such as *revolved, removed, partial* or *broken-out,* and *sectioned assembly drawings,* are explained in detail in other units in this text, as are sectioning techniques for showing ribs, spokes, shafts, rivets, and keys.

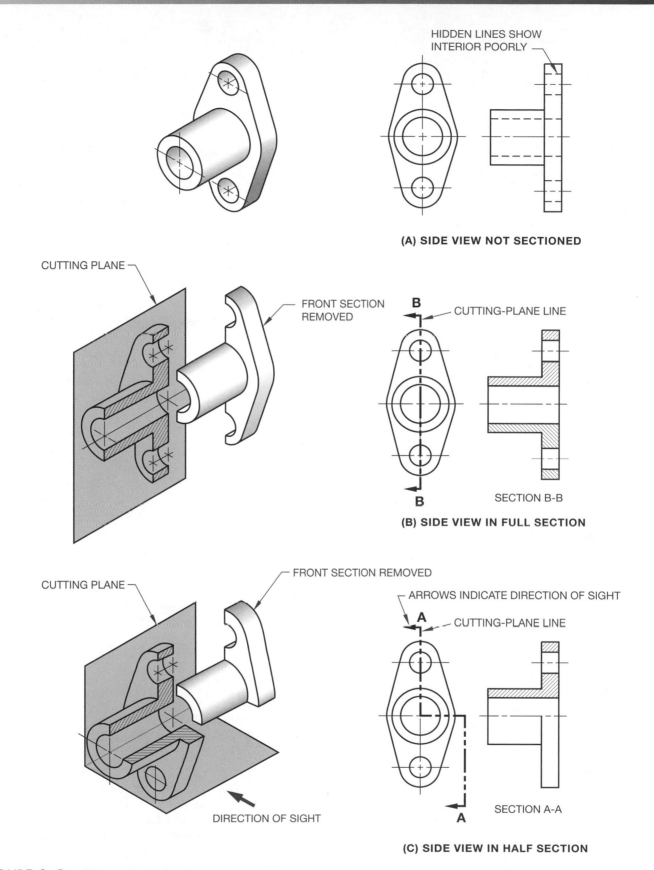

(A) SIDE VIEW NOT SECTIONED

HIDDEN LINES SHOW
INTERIOR POORLY

CUTTING PLANE

FRONT SECTION
REMOVED

B

CUTTING-PLANE LINE

B

SECTION B-B

(B) SIDE VIEW IN FULL SECTION

CUTTING PLANE

FRONT SECTION REMOVED

ARROWS INDICATE DIRECTION OF SIGHT

A

CUTTING-PLANE LINE

A

SECTION A-A

DIRECTION OF SIGHT

(C) SIDE VIEW IN HALF SECTION

FIGURE 9–5 ■ Full and half sections

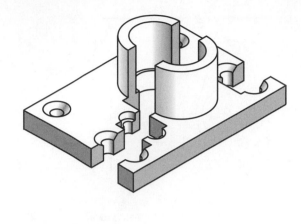

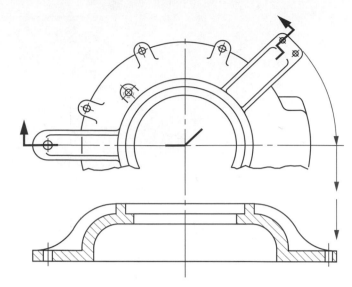

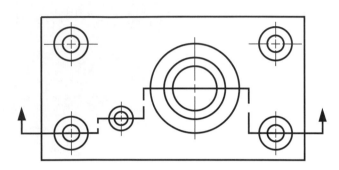

FIGURE 9–7 ■ Bent cutting-plane line for offset sections

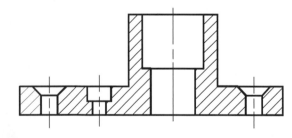

FIGURE 9–6 ■ An offset section

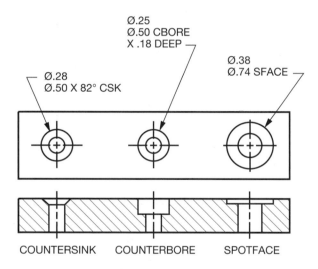

COUNTERSINK   COUNTERBORE   SPOTFACE

FIGURE 9–8 ■ Using words to dimension countersinks, counterbores, and spotfaces

COUNTERSINKS, COUNTERBORES, AND SPOTFACES

A *countersunk* hole is a conical depression cut in a piece to receive a countersunk type of flathead screw or rivet, Figure 9–8. The size is usually shown by a note listing the diameter of the hole first, followed by the diameter of the countersink, the abbreviation CSK, and the angle. A *counterbored hole* is one that has been machined larger to a given depth to receive a fillister, hex-head, or similar type of bolt head. Counterbores are specified by a note giving the diameter of the hole first, followed by the counterbore

diameter, the abbreviation CBORE, and the depth of the counterbore. The counterbore and depth may also be indicated by direct dimensioning. A *spotface* is an area where the surface is machined just enough to provide a level seating surface for a bolt head, nut, or washer. A spotface is specified by a note listing the diameter of the hole first, followed by the spotface diameter, and the abbreviation SFACE. The depth of the spotface is not usually given.

The symbolic means of specifying a countersink, counterbore, or spotface, and the depth of a feature are shown in Figure 9–9. In each case, the symbol precedes the dimension.

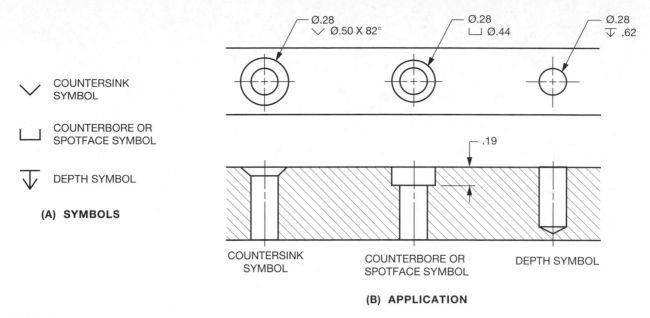

(A) SYMBOLS

COUNTERSINK SYMBOL

COUNTERBORE OR SPOTFACE SYMBOL

DEPTH SYMBOL

Ø.28
⌵ Ø.50 X 82°

Ø.28
⌴ Ø.44

Ø.28
�downarrow .62

.19

COUNTERSINK SYMBOL

COUNTERBORE OR SPOTFACE SYMBOL

DEPTH SYMBOL

(B) APPLICATION

FIGURE 9–9 ■ Using symbols to indicate shape and depth of holes

INTERSECTION OF UNFINISHED SURFACES

The intersection of unfinished surfaces that are rounded or filleted at the point of theoretical intersection is indicated by a line coinciding with the theoretical point of intersection. The need for this convention is shown by the examples in Figure 9–10. For a large radius, Figure 9–10(C), no line is drawn.

Members such as ribs and arms that blend into other features end in curves called *runouts*.

REFERENCE

CAN3-B78.1-M83 Technical Drawings—General Principles

INTERNET RESOURCES

American Society of Mechanical Engineers For information on sectional views and related topics, refer to ASME Y14.3M-1994 (R2003) (*Multi- and Sectional-View Drawings*) at: http://www.asme.org

Drafting Zone For information on sectioning, see: http://www.draftingzone.com

IDS Development–Nebraska Education For information on the various line types used on engineering drawings, see: http://idsdev.mccneb.edu/djackson/lineintro.htm

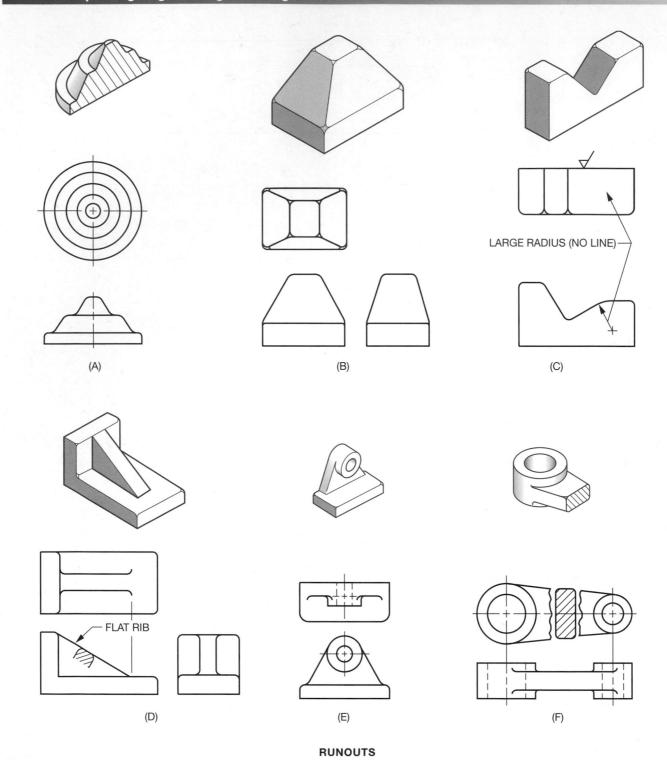

LARGE RADIUS (NO LINE)

(A)   (B)   (C)

FLAT RIB

(D)   (E)   (F)

RUNOUTS

FIGURE 9–10 ■ Rounded and filleted intersections

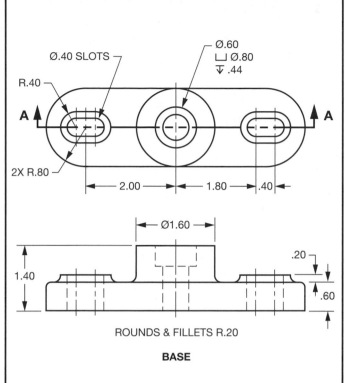

Ø.40 SLOTS

Ø.60
⌴ Ø.80
▽ .44

R.40

A

2X R.80

2.00 1.80 .40

Ø1.60

1.40 .20 .60

ROUNDS & FILLETS R.20

BASE

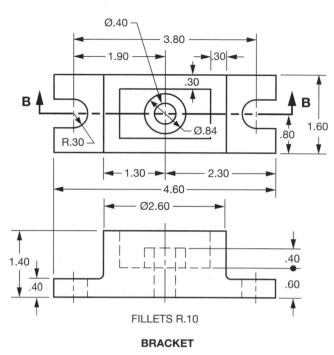

Ø.40

3.80

1.90 30

.30

B B

R.30 Ø.84

.80 1.60

1.30 2.30

4.60

Ø2.60

1.40 .40 .40 .60

FILLETS R.10

BRACKET

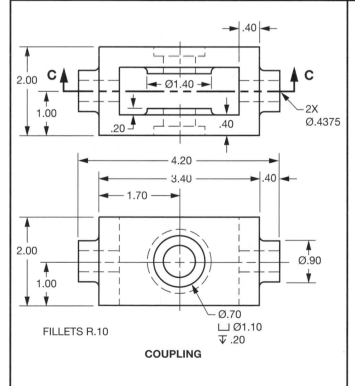

.40

2.00

C C

1.00

.20 .40 2X Ø.4375

4.20

3.40 .40

1.70

2.00

1.00 Ø.90

Ø.70
⌴ Ø1.10
▽ .20

FILLETS R.10

COUPLING

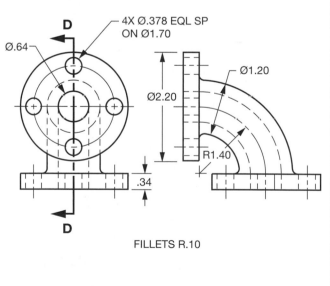

4X Ø.378 EQL SP
ON Ø1.70

Ø.64

D

Ø1.20

Ø2.20

R1.40

.34

D

FILLETS R.10

FLANGED ELBOW

ASSIGNMENT:
USE 1.00-INCH GRID SHEETS (.10 IN. SQUARES).
SKETCH THE FOUR FULL-SECTION VIEWS. SCALE 1:1.

**SKETCHING
FULL SECTIONS**

A-29

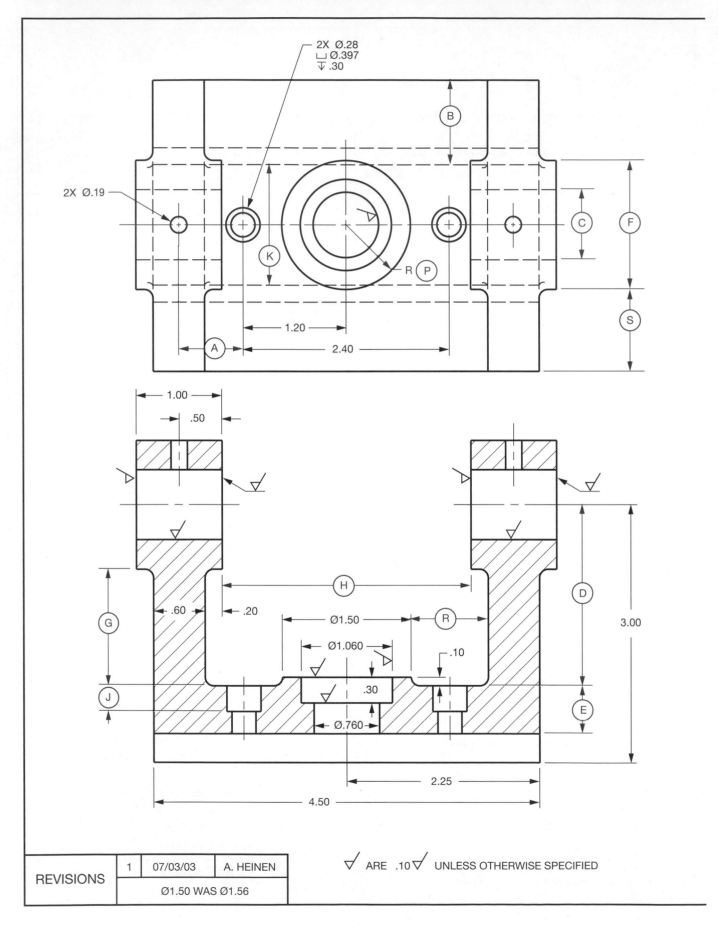

2X Ø.28
⌴ Ø.397
▼ .30

2X Ø.19

B

C

F

K

R P

1.20

A

2.40

S

1.00

.50

G

J

.60

.20

H

Ø1.50

Ø1.060

.30

Ø.760

R

.10

D

E

3.00

2.25

4.50

REVISIONS

1	07/03/03	A. HEINEN
Ø1.50 WAS Ø1.56		

✓ ARE .10 ✓ UNLESS OTHERWISE SPECIFIED

QUESTIONS:

1. What is the overall width?

2. What is the overall height?

3. What is the centre-to-centre distance of the Ø.19 holes?

4. If .03 is allowed for each surface requiring finishing, calculate the width of the casting before machining.

5. At what angle to the vertical is the dovetail slot?

6. How many different surfaces require finishing?

7. Which type of lines in the top view represent the dovetail?

8. What was the original size of the Ø1.50?

9. How wide is the opening in the dovetail?

10. How high is the dovetail?

11. When was the Ø1.50 dimension altered?

12. How many reference dimensions are shown?

13. Calculate the distances Ⓐ to Ⓢ.

14. What is the (A) size and (B) type of cap screw required to fasten the slide bracket to its mating part?

15. How many not-to-scale dimensions are shown?

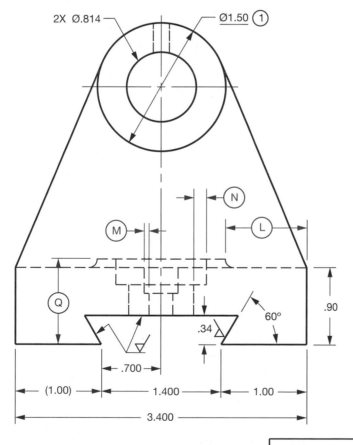

2X Ø.814   Ø1.50 ①

N
M
L
Q
60°
.90
.34
.700
(1.00)   1.400   1.00
3.400

ROUNDS AND FILLETS R.10

MATERIAL	GREY IRON	
SCALE	NOT TO SCALE	
DRAWN	A. HEINEN	DATE 10/01/02

SLIDE BRACKET

A-30

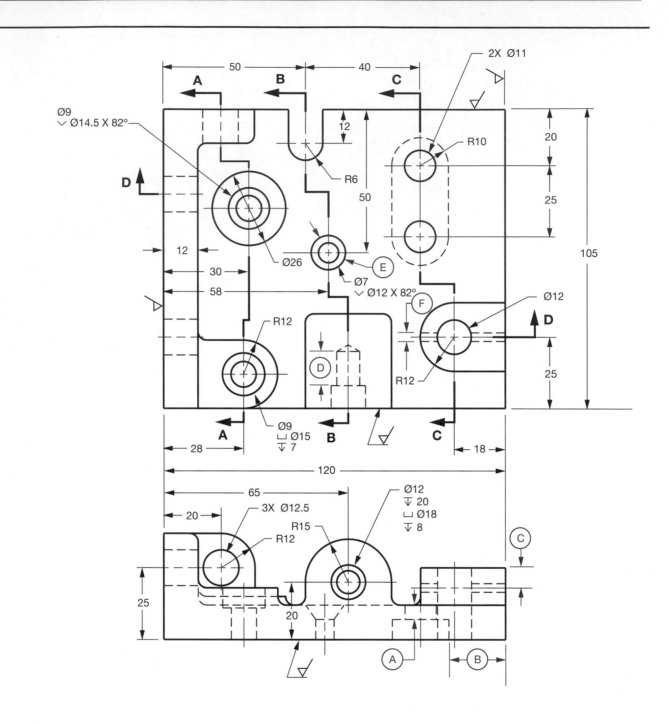

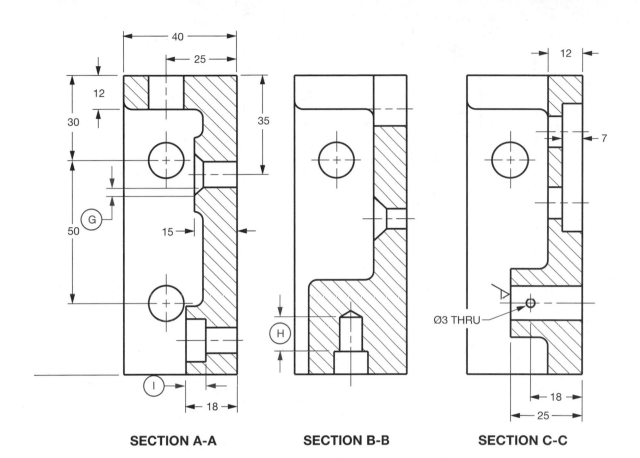

SECTION A-A   **SECTION B-B**   **SECTION C-C**

ROUNDS AND FILLETS R3

∀ SHOWN TO BE   2 ∀

QUESTIONS:

1. What type of sectional view is used?

2. How many surfaces require finishing?

3. What is the total number of holes in the part?
    Note: A THRU hole is considered one hole.

4. How many countersunk holes are there?

5. What was the (A) width, (B) height, and (C) depth of the casting
    before finishing?

6. With reference to the counterbored hole, what is the (A) size and
    (B) type of cap screw required if the head of the cap screw
    must not protrude above the surface of the part? Refer to Table
    7 of the Appendix.

7. What is the total number of offsets used on the cutting-plane
    lines?

8. Calculate distances A through I.

METRIC
NOTE: DIMENSIONS IN MILLIMETRES

ASSIGNMENT:
ON A CENTIMETRE GRID SHEET (1 mm SQUARES),
SKETCH OFFSET SECTION D-D HAVING THE CUTTING-
PLANE PASS THROUGH THE CENTRES OF THE
COUNTERSUNK HOLES.

MATERIAL	MALLEABLE IRON	
SCALE	NOT TO SCALE	
DRAWN	S. LOGAN	DATE 12/04/04

BASE PLATE   **A-31M**

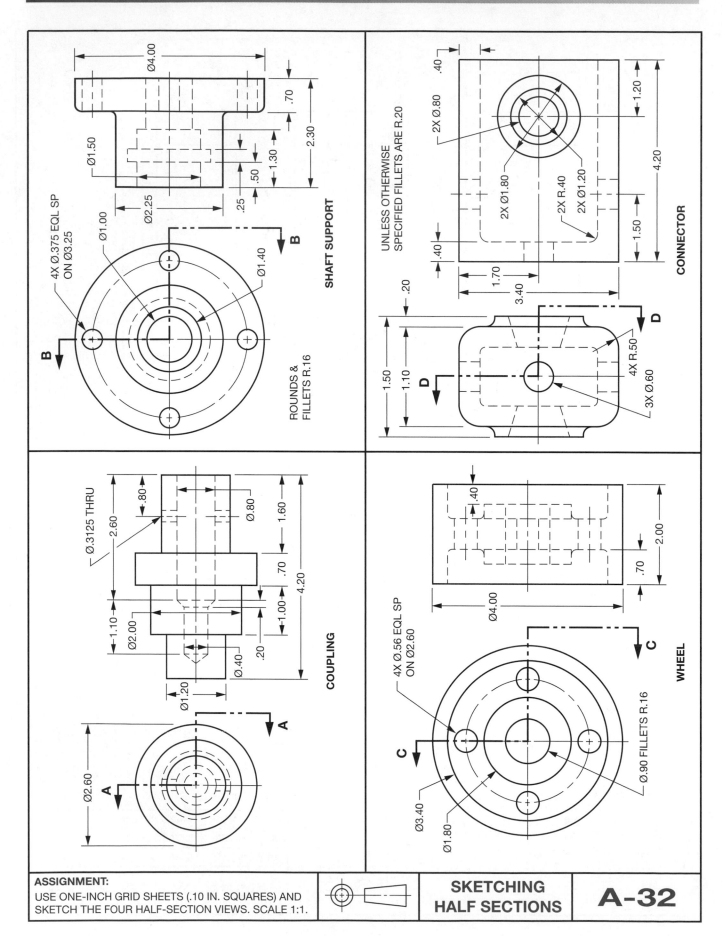

SHAFT SUPPORT

Ø4.00
.70
2.30
Ø1.50
1.30
.50
.25
Ø2.25
4X Ø.375 EQL SP ON Ø3.25
Ø1.00
Ø1.40
B
B
ROUNDS & FILLETS R.16

CONNECTOR

UNLESS OTHERWISE SPECIFIED FILLETS ARE R.20
2X Ø.80
.40
1.20
4.20
1.50
2X Ø1.80
2X R.40
2X Ø1.20
.40
1.70
3.40
.20
1.50
1.10
D
D
4X R.50
3X Ø.60

COUPLING

Ø.3125 THRU
.80
Ø.80
2.60
1.60
.70
4.20
1.10
2.00
1.00
.20
Ø.40
Ø1.20
Ø2.60
A
A

WHEEL

.40
2.00
.70
Ø4.00
4X Ø.56 EQL SP ON Ø2.60
C
C
Ø.90 FILLETS R.16
Ø3.40
Ø1.80

ASSIGNMENT:
USE ONE-INCH GRID SHEETS (.10 IN. SQUARES) AND SKETCH THE FOUR HALF-SECTION VIEWS. SCALE 1:1.

SKETCHING HALF SECTIONS

A-32

CHAMFERS

The process of *chamfering,* or cutting away the inside or outside corner of an object, is done to facilitate assembly, Figure 10–1. The recommended method of dimensioning a chamfer is to give an angle and the linear length, or an angle and a diameter. For angles of 45° only, a note form may be used. This method is permissible only with 45° angles because the size may apply to either the longitudinal or radial dimension. Chamfers are never measured along the angular surface.

UNDERCUTS

The operation of *undercutting,* or *necking,* is the cutting of a recess in a cylinder. Undercuts permit two parts to join, Figure 10–1. Undercutting is indicated on a drawing by a note listing the width first and then the diameter. If the radius is shown at the bottom of the undercut, it is assumed that the radius will be equal to half the width unless specified differently, and the diameter will apply to the centre of the undercut. Where the size of the neck is unimportant, the dimension may be omitted from the drawing.

TAPERS

A taper is the ratio of the difference in diameters of two sections along a conical part (perpendicular to the axis).

Conical Tapers

Tapered shanks are used on many small tools, such as drills, reamers, counterbores, and spotfaces to hold them accurately in the machine spindle. Conical taper means the difference in diameter or width in a given length. There are many standard tapers: The Morse taper and the Brown and Sharpe taper are the most common.

The following dimensions may be used in suitable combinations to define the size and form of tapered features:

- the diameter (or width) at one end of the tapered feature
- the length of the tapered feature
- the rate of taper
- the included angle
- the taper ratio

In dimensioning a taper by means of taper ratio, the conical taper symbol should precede the ratio figures, and the vertical leg of the symbol is always shown to the left, Figure 10–2.

Flat Tapers

Flat tapers (slopes) are used as locking devices, such as taper keys and adjusting shims. The methods recommended for dimensioning flat tapers are shown in Figure 10–3. The flat taper symbol should precede the ratio figures, and the vertical leg of the symbol is always shown to the left.

KNURLS

Knurling is the machining of a surface to create uniform depressions. Knurling permits a better

 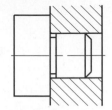 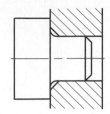

PART CANNOT FIT FLUSH
IN HOLE BECAUSE OF
SHOULDER

SAME PART WITH UNDERCUT
ADDED PERMITS PART
TO FIT FLUSH

CHAMFER ADDED TO HOLE
TO ACCEPT SHOULDER OF
PART

(A) UNDERCUT AND CHAMFER APPLICATION

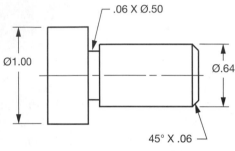

THIS METHOD OF DIMENSIONING
FOR 45° CHAMFERS ONLY

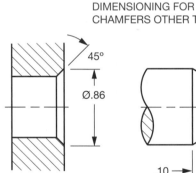 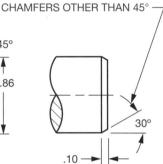

DIMENSIONING FOR
CHAMFERS OTHER THAN 45°

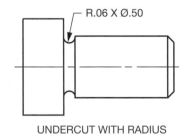

UNDERCUT WITH RADIUS

(B) DIMENSIONING CHAMFERS AND UNDERCUTS

FIGURE 10–1 ■ Chamfers and undercuts

grip. Knurling is shown on drawings as either a straight or diamond pattern. The pitch of the knurl may be specified. It is unnecessary to hatch in the whole area to be knurled if enough is shown to clearly indicate the pattern. Knurls are specified on the drawing by a note calling for the type and pitch. The length and diameter of the knurl are shown as dimensions, Figure 10–4.

REFERENCE

CAN/CSA-B78.2-M91 Dimensioning and Tolerancing of Technical Drawings

INTERNET RESOURCES

Drafting Zone For information on chamfers, tapers, and knurling, see: http://www.draftingzone .com

Integrated Publishing for information on chamfers, undercuts, and knurling, see: http://www.tpub .com/content/draftsman

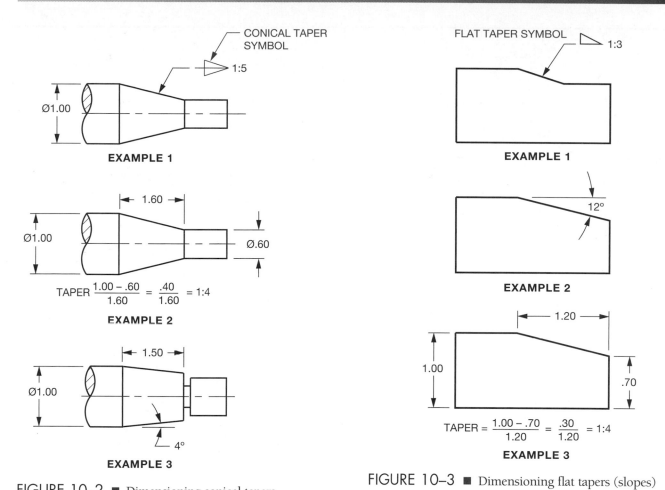

CONICAL TAPER
SYMBOL

1:5

Ø1.00

EXAMPLE 1

1.60

Ø1.00

Ø.60

TAPER $\dfrac{1.00 - .60}{1.60} = \dfrac{.40}{1.60} = 1:4$

EXAMPLE 2

1.50

Ø1.00

4°

EXAMPLE 3

FIGURE 10–2 ■ Dimensioning conical tapers

FLAT TAPER SYMBOL

1:3

EXAMPLE 1

12°

EXAMPLE 2

1.20

1.00

.70

TAPER $= \dfrac{1.00 - .70}{1.20} = \dfrac{.30}{1.20} = 1:4$

EXAMPLE 3

FIGURE 10–3 ■ Dimensioning flat tapers (slopes)

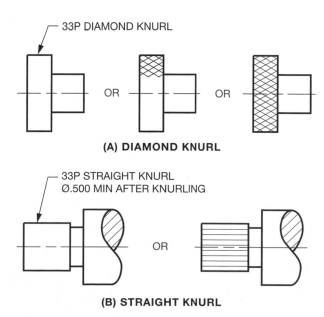

33P DIAMOND KNURL

OR   OR

(A) DIAMOND KNURL

33P STRAIGHT KNURL
Ø.500 MIN AFTER KNURLING

OR

(B) STRAIGHT KNURL

FIGURE 10–4 ■ Knurling

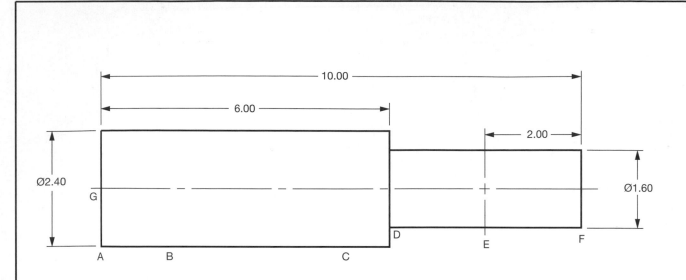

ASSIGNMENT:

ON A ONE-INCH GRID SHEET (.10 IN. SQUARES), SKETCH THE HANDLE SHOWN ABOVE AND ADD THE FOLLOWING FEATURES. ADD DIMENSIONS USING SYMBOLS WHEREVER POSSIBLE. USE A CONVENTIONAL BREAK TO SHORTEN THE LENGTH. SCALE 1:1.

A. 45° X .20 CHAMFER

B. 33P DIAMOND KNURL FOR 1.20 IN. STARTING .80 IN. FROM LEFT END

C. 0.1:1 CIRCULAR TAPER FOR 1.20 IN. LENGTH ON RIGHT END OF Ø2.40

D. .20 X Ø1.40 UNDERCUT ON Ø1.60

E. Ø.20 X .50 DEEP, 4 HOLES EQUALLY SPACED

F. 30° X .30 CHAMFER. THE .30 DIMENSION TAKEN HORIZONTALLY ALONG THE SHAFT

G. Ø.60 HOLE, 1.50 DEEP

| HANDLE | **A-33** |

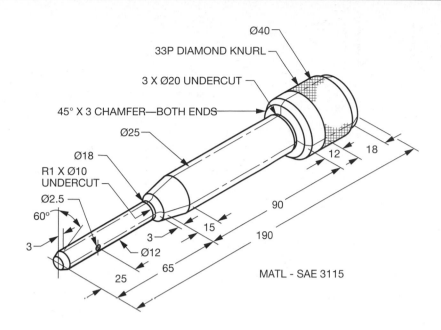

Ø40
33P DIAMOND KNURL
3 X Ø20 UNDERCUT
45° X 3 CHAMFER—BOTH ENDS
Ø25
Ø18
R1 X Ø10 UNDERCUT
Ø2.5
60°
3
12
18
90
15
3
190
Ø12
25
65
MATL - SAE 3115

ASSIGNMENT:

ON A CENTIMETRE GRID SHEET (1 mm SQUARES), SKETCH A ONE-VIEW
DRAWING, COMPLETE WITH DIMENSIONS, OF THE INDICATOR ROD. USE A
CONVENTIONAL BREAK TO SHORTEN THE LENGTH OF THE ROD. SCALE 1:1.

METRIC
DIMENSIONS IN MILLIMETRES

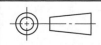

INDICATOR ROD   **A-34M**

11 UNIT

SELECTION OF VIEWS

ONE- AND TWO-VIEW DRAWINGS

Except for complex objects of irregular shapes, it is seldom necessary to draw more than three views; and for simple parts, one- or two-view drawings will often suffice.

In one-view drawings, the third dimension is expressed by a note or by symbols or abbreviations, such as Ø, □, HEX ACR FLT, R, Figure 11–1 and Table 2 of the Appendix. The symmetry symbol shown at the ends of the centre line indicates that the part is symmetrical. Frequently, the drafter will decide that only two views are necessary to explain the shape of an object fully, Figure 11–2. One or two views usually show the shape of cylindrical objects adequately.

MULTIPLE-DETAIL DRAWINGS

Details of parts may be shown on separate sheets, or they may be grouped together on one or more large sheets. Often the details of parts are grouped according to the department in which they are made. Metal parts to be fabricated in the machine shop may appear on one detail sheet; parts to be made in the wood shop are grouped on another. Figure 11–3 shows several details used in assembling a drafting compass.

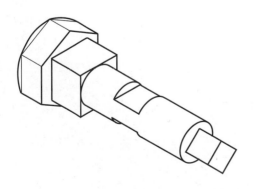

(A) THE PART

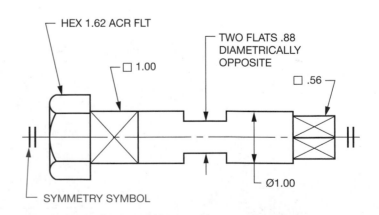

HEX 1.62 ACR FLT

□ 1.00

TWO FLATS .88 DIAMETRICALLY OPPOSITE

□ .56

Ø1.00

SYMMETRY SYMBOL

(B) ONE-VIEW DRAWING OF THE PART

FIGURE 11–1 ■ Words and symbols used to identify shapes and sizes

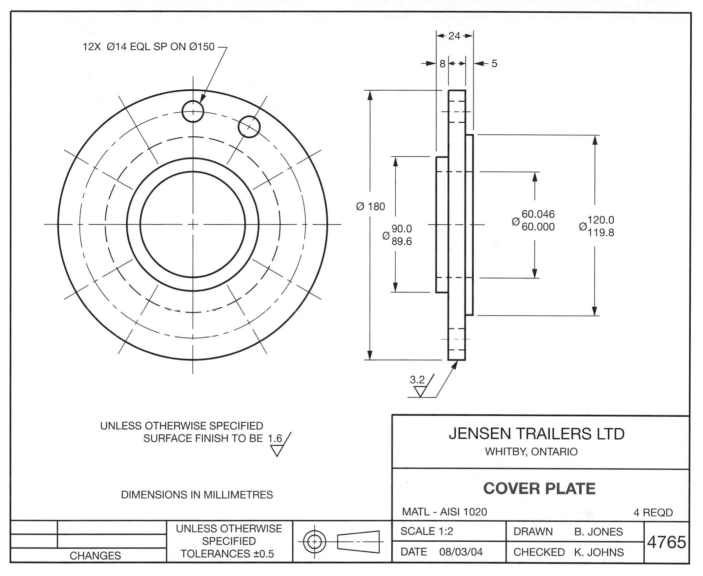

FIGURE 11–2 ■ Two-view drawing

FUNCTIONAL DRAFTING

Since the basic function of the drafting department is to give sufficient information to produce or assemble parts, drafting must use every means to communicate this information in the least expensive manner. There are many ways to reduce drafting time in preparing a drawing:

■ Avoid unnecessary views.

■ Use simplified drawing practices.

■ Use explanatory notes to complement the drawing in order to eliminate views that are time consuming to draw.

■ Eliminate unnecessary lines.

Figure 11–4 shows a simple part in three different ways: The first example uses conventional drawing practices; the second and third examples use simplified drawing techniques.

REFERENCES

CAN/CSA-B78.2-M91 Dimensioning and Tolerancing of Technical Drawings

CAN3-B78.1-M83 Technical Drawings—General Principles

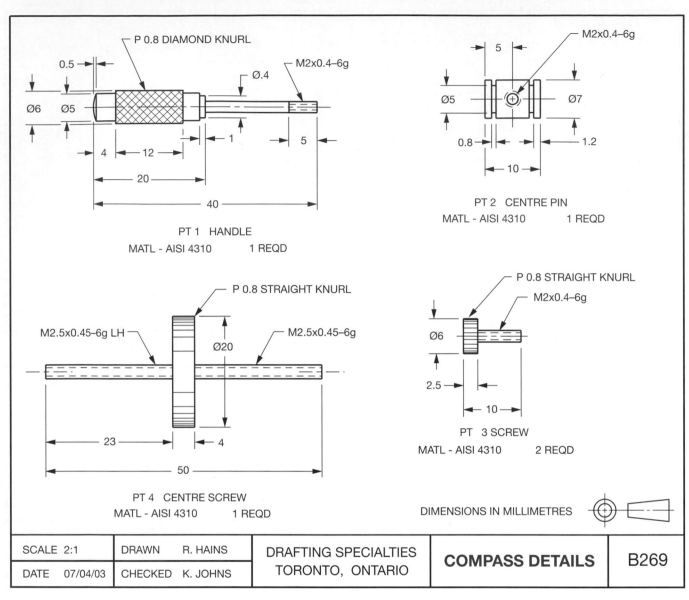

PT 1   HANDLE
MATL - AISI 4310   1 REQD

PT 2   CENTRE PIN
MATL - AISI 4310   1 REQD

PT 4   CENTRE SCREW
MATL - AISI 4310   1 REQD

PT 3 SCREW
MATL - AISI 4310   2 REQD

DIMENSIONS IN MILLIMETRES

SCALE 2:1	DRAWN R. HAINS	DRAFTING SPECIALTIES TORONTO, ONTARIO	COMPASS DETAILS	B269
DATE 07/04/03	CHECKED K. JOHNS			

FIGURE 11–3 ■ Detail drawing containing several parts

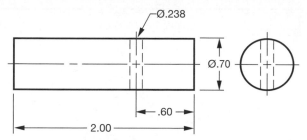

EXAMPLE 1: CONVENTIONAL DRAWING

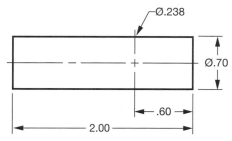

EXAMPLE 2: SIMPLIFIED DRAWING

PT 2   Ø.70 X 2.00 LG
Ø.238 HOLE – .60 FROM END

EXAMPLE 3: PART DESCRIBED BY A NOTE

FIGURE 11–4 ■ Simplified representation for a simple part

INTERNET RESOURCES

American Society of Mechanical Engineers For information on multiview drawings, refer to ASME Y14.3M-1994 (R2003) (*Multi- and Sectional-View Drawings*) at: http://www .asme.org

Drafting Zone For information on functional drafting, see: http://www.draftingzone.com

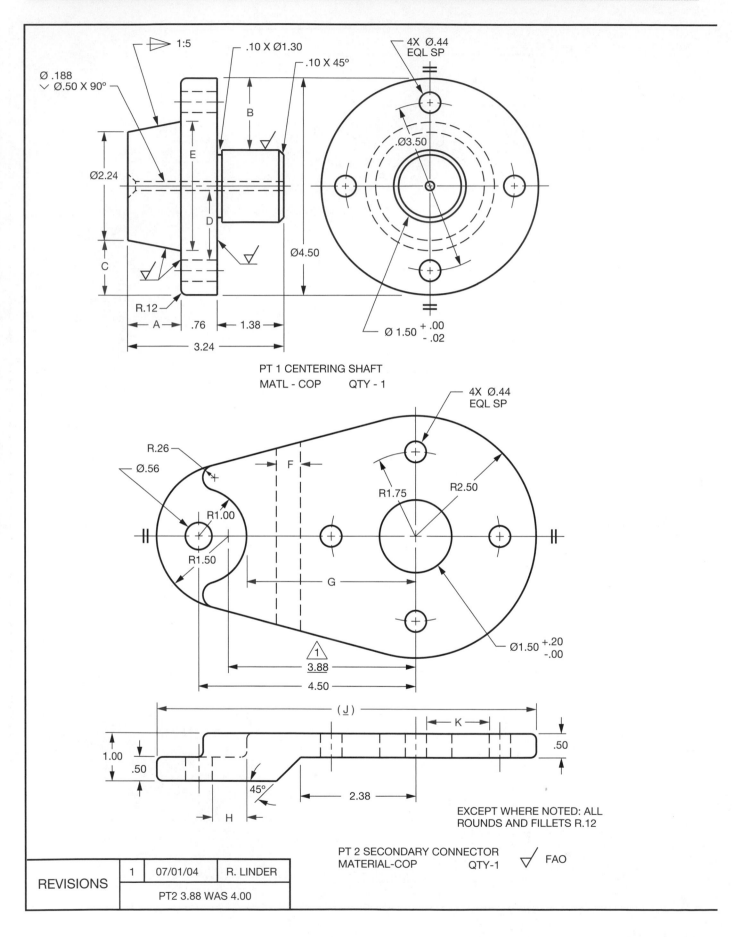

1:5

Ø .188
⌄ Ø.50 X 90°

.10 X 1.30

.10 X 45°

4X Ø.44
EQL SP

B

E

Ø2.24

.Ø3.50

D

Ø4.50

C

Ø 1.50 +.00 -.02

R.12

A   .76   1.38

3.24

PT 1 CENTERING SHAFT
MATL - COP   QTY - 1

4X Ø.44
EQL SP

R.26

Ø.56

F

R1.00

R1.75   R2.50

R1.50

G

Ø1.50 +.20 -.00

1
3.88

4.50

(J)

K

1.00

.50   .50

45°

2.38

H

EXCEPT WHERE NOTED: ALL
ROUNDS AND FILLETS R.12

PT 2 SECONDARY CONNECTOR
MATERIAL-COP   QTY-1   ▽ FAO

REVISIONS	1	07/01/04	R. LINDER
	PT2 3.88 WAS 4.00		

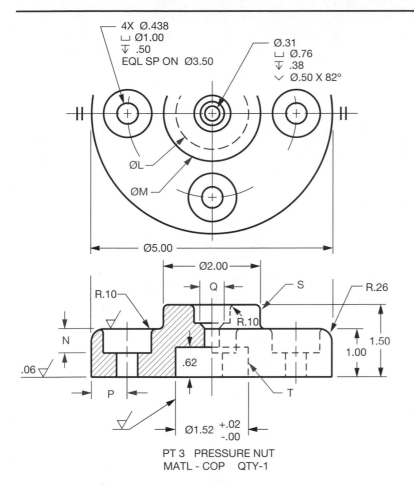

4X Ø.438
⊔ Ø1.00
▽ .50
EQL SP ON Ø3.50

Ø.31
⊔ Ø.76
▽ .38
∨ Ø.50 X 82°

ØL

ØM

Ø5.00

Ø2.00

R.10

Q

S

R.26

R.10

N

1.50
1.00

.06

.62

P

T

Ø1.52 +.02 -.00

PT 3   PRESSURE NUT
MATL - COP   QTY-1

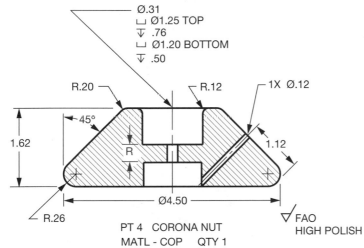

Ø.31
⊔ Ø1.25 TOP
▽ .76
⊔ Ø1.20 BOTTOM
▽ .50

R.20

R.12

1X Ø.12

45°

1.62

R

1.12

Ø4.50

R.26

FAO
HIGH POLISH

PT 4   CORONA NUT
MATL - COP   QTY 1

QUESTIONS:

1. Calculate dimensions A to R. Use nominal sizes. There is no I or O.

Refer to Part 1

2. What is the length of the Ø1.50 shaft? Do not include the undercut or chamfer.

3. What is the length of the Ø.188 hole excluding the countersink?

4. Give two reasons why only part of the end view is drawn.

5. What is the diameter of the undercut?

Refer to Part 2

6. What does FAO mean?

7. What does the line under the 3.88 dimension indicate?

8. What does the abbreviation COP mean?

9. What do the parentheses around dimension J indicate?

10. What is the size of the fillets?

Refer to Part 3

11. How much machining allowance is provided for the bottom of the part?

12. What line in the front view represents line L in the top view?

13. What line in the top view represents line S in the front view?

14. What are the limits of the diameter of the counterbore in the bottom of the part?

15. What type of section view is shown?

Refer to Part 4

16. What is the diameter of the counterbore where the Ø.12 hole terminates?

17. What is the distance between the centre points of the .26 radii?

18. What type of section view is shown?

ASSIGNMENT:

ON A ONE-INCH GRID SHEET (.10 IN. SQUARES), COMPLETE THE TOP VIEW OF PART 4.

SCALE	NTS	
DRAWN	J. LOGAN	DATE 10/11/03

CENTERING CONNECTOR DETAILS

A-35

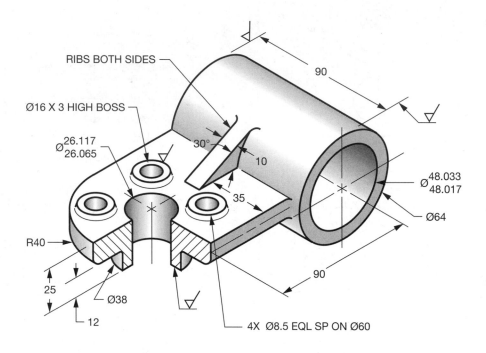

RIBS BOTH SIDES

Ø16 X 3 HIGH BOSS

$Ø^{26.117}_{26.065}$

30°

90

10

$Ø^{48.033}_{48.017}$

35

Ø64

R40

25

Ø38

12

90

4X Ø8.5 EQL SP ON Ø60

MATL - CAST STEEL   ROUNDS AND FILLETS R3

ASSIGNMENT:
ON A CENTIMETRE GRID SHEET (1 mm SQUARES),
SKETCH THE TOP, FRONT, AND LEFT-SIDE VIEWS
OF THE LINK. SCALE 1:2.

METRIC
DIMENSIONS IN MILLIMETRES

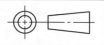

LINK

A-36M

SURFACE TEXTURE

The development of modern, high-speed machines has resulted in higher loadings and faster moving parts. To withstand these severe operating conditions with minimum friction and wear, a particular surface texture is often essential. This requires the designer to accurately describe the needed texture (sometimes called *finish*) to the people who are making the parts.

Rarely are entire machines designed and manufactured in one plant. They are usually designed in one location, manufactured in another, and perhaps assembled in a third.

All surface finish control begins in the drafting room. The designer is responsible for specifying the correct surface finish for maximum performance and service life at the lowest cost. The choice of surface finish for any particular part is based on the designer's experience, field service data, and engineering tests. Factors influencing the designer's choice include function of the parts, type of loading, speed and direction of movement, and operating conditions. Also considered are such factors as physical characteristics of both materials on contact, whether the part is subjected to stress reversals, the type and amount of lubricant, contaminants, and temperature.

The two principal reasons for surface finish control are friction reduction and the control of wear.

Whenever a lubricating film must be maintained between two moving parts, the surface irregularities must be small enough to prevent penetrating the oil film under even the most severe operating conditions. Such parts as bearings, journals, cylinder bores, piston pins, bushings, pad bearings, helical and worm gears, seal surfaces, and machine ways are objects that require this condition.

Surface finish is also important to the wear service of certain pieces subject to dry friction, such as machine tool bits, threading dies, stamping dies, rolls, clutch plates, and brake drums.

Smooth finishes are essential on certain high-precision pieces. In such mechanisms as injectors and high-pressure cylinders, smoothness and lack of waviness are essential for accuracy and pressure-retaining ability. Smooth finishes are also used on micrometer anvils, gauges, gauge blocks, and other items.

Smoothness is often important for the visual appeal of the finished product. For this reason, surface finish is controlled on such articles as rolls, extrusion dies, and precision casting dies.

For gears and other parts, surface finish control may be necessary to ensure quiet operation.

Where boundary lubrication exists or where surfaces are not compatible (for example, two hard surfaces running together), some roughness or character of surface will assist in lubrication.

To meet the requirements for effective control of surface quality under different conditions, there is a system for accurately describing the surface.

Surfaces are usually very complex. Only the height, width, and direction of surface irregularities are covered in this section, because these are of practical importance in specific applications.

Surface Texture Definitions

The following terms relating to surface texture are illustrated in Figure 12–1.

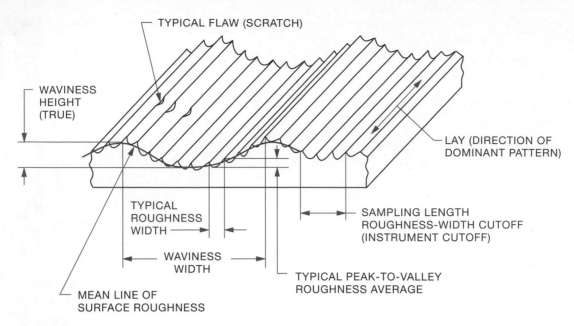

FIGURE 12–1 ■ Surface texture characteristics

Microinch (μin)

A microinch is one millionth of an inch (.000001 inch). For written specifications or reference to surface roughness requirements, microinches may be abbreviated as μin.

Micrometre (μm)

A micrometre is one millionth of a metre (0.000 001 metre.) For written specifications or reference to surface roughness requirements, micrometres may be abbreviated as μm.

Roughness

Roughness is the finer irregularities in the surface texture, usually including those that result from the inherent action of the production process. These include traverse feed marks and other irregularities within the limits of the roughness-width cutoff.

Roughness Average (R_a)

Roughness average is expressed in microinches, micrometres, or roughness grade numbers N1 to N12. The "N" series of roughness grade numbers is often used in lieu of the roughness average values to prevent misinterpretation when drawings are exchanged internationally.

Roughness Width

Roughness width is the distance parallel to the nominal surface between successive peaks or ridges that constitute the predominant pattern of the roughness. Roughness width is rated in inches or millimetres.

Roughness-Width Cutoff

The greatest spacing of repetitive surface irregularities to be included in the measurement of average roughness height is the roughness-width cutoff. This is rated in inches or millimetres and must always be greater than the roughness width to obtain the total roughness height rating.

Waviness

Waviness is the usually widely spaced component of surface texture and is generally spaced farther apart than the roughness-width cutoff. Waviness may result from machine or work deflections, vibration, chatter, heat treatment, or warping strains. Roughness may be considered superimposed on a wavy surface. Although waviness is not currently in ISO standards, it is included as part of the surface texture symbol to follow industrial practices in North America.

Lay

The direction of the predominant surface pattern, which is ordinarily determined by the production method used, is the lay. Symbols for the lay are shown in Figure 12–2.

Flaws

Flaws are surface irregularities at one place or at relatively infrequent or widely varying intervals. Flaws include cracks, blow holes, checks, ridges, and scratches. Unless otherwise specified, the effect

SYMBOL	DESCRIPTION	EXAMPLE
=	LAY PARALLEL TO THE LINE REPRESENTING THE SURFACE TO WHICH THE SYMBOL IS APPLIED	 DIRECTION OF TOOL MARKS
⊥	LAY PERPENDICULAR TO THE LINE REPRESENTING THE SURFACE TO WHICH THE SYMBOL IS APPLIED	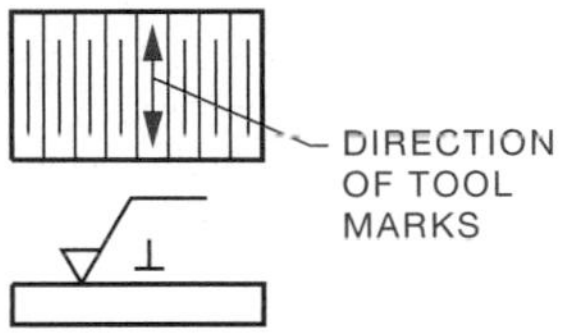 DIRECTION OF TOOL MARKS
X	LAY ANGULAR IN BOTH DIRECTIONS TO THE LINE REPRESENTING THE SURFACE TO WHICH THE SYMBOL IS APPLIED	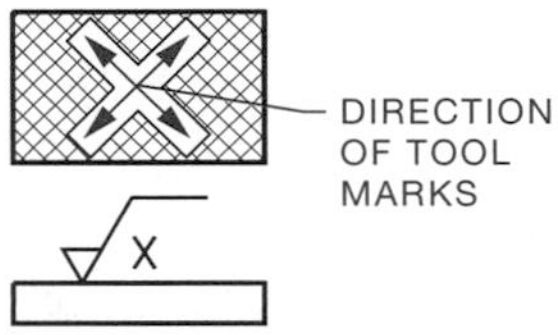 DIRECTION OF TOOL MARKS
M	LAY MULTIDIRECTIONAL	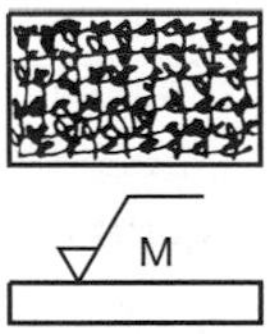
C	LAY APPROXIMATELY CIRCULAR RELATIVE TO THE CENTRE OF THE SURFACE TO WHICH THE SYMBOL IS APPLIED	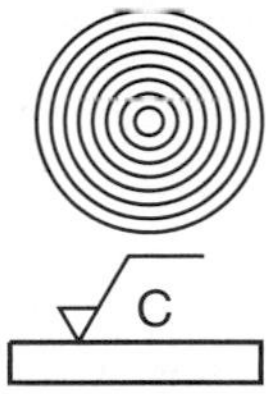
R	LAY APPROXIMATELY RADIAL RELATIVE TO THE CENTRE OF THE SURFACE TO WHICH THE SYMBOL IS APPLIED	
P	LAY NONDIRECTIONAL, PITTED, OR PROTUBERANT	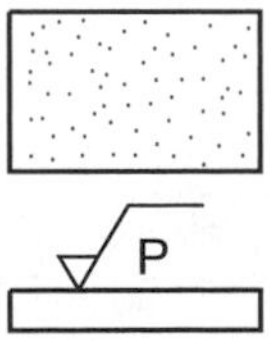

FIGURE 12–2 ■ Lay symbols

of flaws is not included in the roughness height measurements.

SURFACE TEXTURE SYMBOL

The surface texture symbol, Figure 12–3, denotes surface characteristics on the drawing. Roughness, waviness, and lay are controlled by waviness ratings applying the desired values to the surface texture symbol, Figure 12–4, or in a general note. The two methods may be used together. The point of the symbol should be on the line indicating the surface, on an extension line from the surface, or on a leader pointing either to the surface or extension line, Figure 12–5. To be readable from the bottom, the symbol is placed in an upright position when notes or numbers are used. This means the long leg and extension line will be on the right. The symbol applies to the entire surface, unless otherwise specified.

This symbol is the same machining symbol described in Unit 8. As well as identifying which surfaces require machining, other surface characteristics are defined by this symbol.

Like dimensions, the symbol for the same surface should not be duplicated on other views. It should be placed on the view with the dimensions showing size or location of the surfaces. Surface texture symbols designate surface texture characteristics, which include machining of surfaces. The method of indicating machine finishes on surfaces is covered in Unit 8.

Where all the surfaces are to be machined, a general note such as FAO (finish all over) or ✓ ALL OVER may be used and the symbols on the part omitted.

SURFACE TEXTURE RATINGS

Roughness average, which is measured in microinches, micrometres, or roughness grade numbers, is shown to the left of the long leg of the symbol, Figure 12–4. The specification of only one rating defines the maximum value; any lesser value is acceptable. Specifying two ratings defines the minimum and maximum values. Anything within that range is acceptable. The maximum value is placed over the minimum.

Waviness height ratings are indicated in inches or millimetres and positioned as shown in Figure 12–4. Any lesser value is acceptable.

Waviness spacing ratings are indicated in inches or millimetres and positioned as shown in Figure 12–4. Any lesser value is acceptable.

Lay symbols, indicating the directional pattern of the surface texture, are shown in Figure 12–2. The symbol is on the right of the long leg of the symbol.

Roughness sampling length ratings are given in inches or millimetres and are located below the horizontal extension. Unless otherwise specified, roughness-width cutoff is .03 in. (0.8 mm).

See Figure 12–5 for an application of roughness values and waviness ratings.

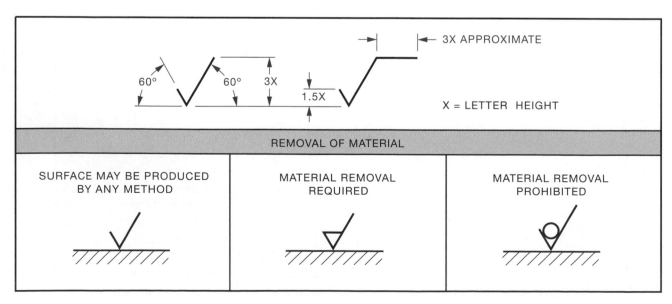

FIGURE 12–3 ■ Basic surface texture symbol

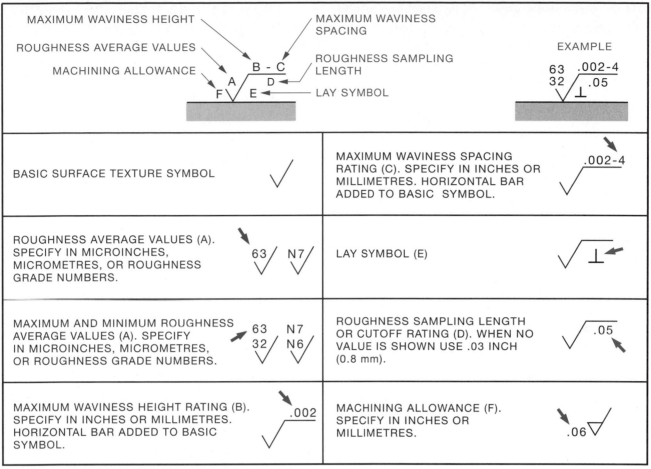

NOTE: WAVINESS IS NOT USED IN ISO STANDARDS.

FIGURE 12–4 ■ Location of ratings and symbols on surface texture symbol

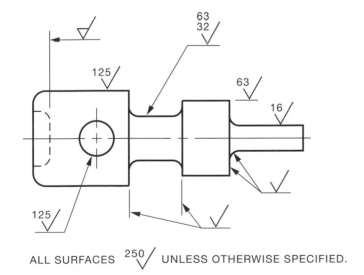

ALL SURFACES 250/ UNLESS OTHERWISE SPECIFIED.

NOTE: VALUES SHOWN ARE IN MICROINCHES.

FIGURE 12–5 ■ Application of roughness values and waviness ratings

Notes

A note is often used where a given roughness requirement applies to either the whole part or the major portion, or before or after plating, Figure 12–6.

CONTROL REQUIREMENTS

Surface texture control should be specified for surfaces where texture is a functional requirement. For example, most surfaces that have contact with a mating part have a certain texture requirement, especially for roughness. The drawing should reflect the texture necessary for optimum part function without depending on the variables of machining practices.

Many surfaces do not need a specification of surface texture because the function is unaffected by the surface quality. Such surfaces should not receive surface quality designations because these could unnecessarily increase the product cost.

Figures 12–7, 12–8, and 12–9 show recommended roughness average ratings, the machining methods used to produce them, and the application of the ratings to the surface texture symbol.

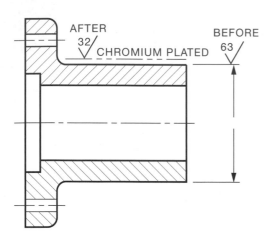

(A) INDICATING SURFACE TEXTURE BEFORE AND AFTER PLATING

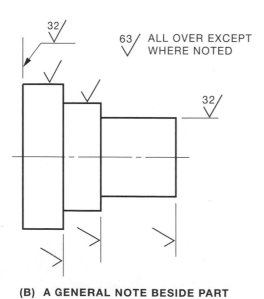

(B) A GENERAL NOTE BESIDE PART

FIGURE 12–6 ■ The use of notes with surface texture symbol

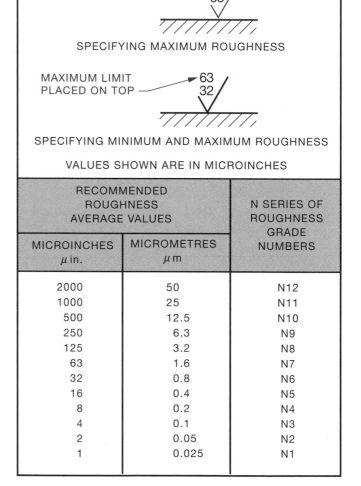

RECOMMENDED ROUGHNESS AVERAGE VALUES		N SERIES OF ROUGHNESS GRADE NUMBERS
MICROINCHES μ in.	MICROMETRES μ m	
2000	50	N12
1000	25	N11
500	12.5	N10
250	6.3	N9
125	3.2	N8
63	1.6	N7
32	0.8	N6
16	0.4	N5
8	0.2	N4
4	0.1	N3
2	0.05	N2
1	0.025	N1

FIGURE 12–7 ■ Recommended roughness average ratings

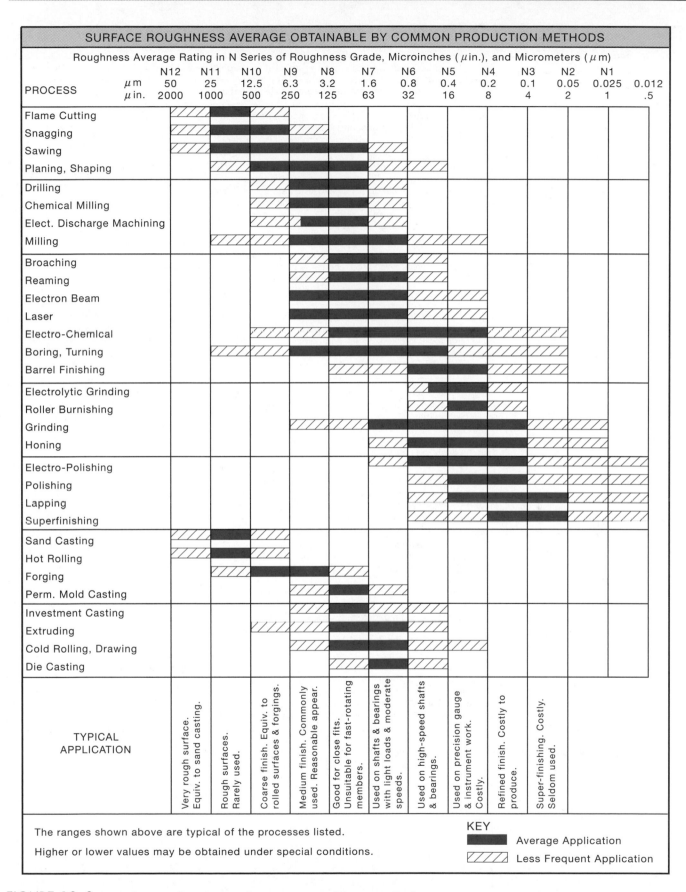

FIGURE 12–8 ■ Surface roughness range for common production methods

MICROMETRES RATING	MICROINCHES RATING	APPLICATION
25	1000	Rough, low-grade surface resulting from sand casting, torch or saw cutting, chipping, or rough forging. Machine operations are not required because appearance is not objectionable. This surface, rarely specified, is suitable for unmachined clearance areas on rough construction items.
12.5	500	Rough, low-grade surface resulting from heavy cuts and coarse feeds in milling, turning, shaping, boring, and rough filing, disc grinding, and snagging. It is suitable for clearance areas on machinery, jigs, and fixtures. Sand casting or rough forging produces this surface.
6.3	250	Coarse production surface, for unimportant clearance and cleanup operation resulting from coarse surface grind, rough file, disc grind, rapid feeds in turning, milling, shaping, drilling, boring, grinding, etc., where tool marks are not objectionable. The natural surfaces of forgings, permanent mould castings, extrusions, and rolled surfaces also produce this roughness. It can be produced economically and is used on parts where stress requirements, appearance, and conditions of operations and design permit.
3.2	125	The roughest surface recommended for parts subject to loads, vibration, and high stress. It is also permitted for bearing surfaces when motion is slow and loads light or infrequent. It is a medium commercial machine finish produced by relatively high speeds and fine feeds taking light cuts with sharp tools. It may be economically produced on lathes, milling machines, shapers, grinders, etc., or on permanent mould castings, die castings, extrusions, and rolled surfaces.
1.6	63	A good machine finish produced under controlled conditions using relatively high speeds and fine feeds to take light cuts with sharp cuttings. It may be specified for close fits and used for all stressed parts, except fast rotating shafts, axles, and parts subject to severe vibration or extreme tension. It is satisfactory for bearing surfaces when motion is slow and loads light or infrequent. It may also be obtained on extrusions, rolled surfaces, die castings, and permanent mould castings when rigidly controlled.
0.8	32	A high-grade machine finish requiring close control when produced by lathes, shapers, milling machines, etc., but relatively easy to produce by centreless, cylindrical, or surface grinders. Also, extruding, rolling, or die casting may produce a comparable surface when rigidly controlled. This surface may be specified in parts where stress concentration is present. It is used for bearings when motion is not continuous and loads are light. When finer finishes are specified, production costs rise rapidly; therefore, such finishes must be analyzed carefully.
0.4	16	A high-quality surface produced by fine cylindrical grinding, emery buffing, coarse honing, or lapping, it is specified where smoothness is of primary importance, such as rapidly rotating shaft bearings, heavily loaded bearings, and extreme tension members.
0.2	8	A fine surface produced by honing, lapping, or buffing. It is specified where packings and rings must slide across the direction of the surface grain, maintaining or withstanding pressures, or for interior honed surfaces of hydraulic cylinders. It may also be required in precision gauges and instrument work, or sensitive value surfaces, or on rapidly rotating shafts and on bearings where lubrication is not dependable.
0.1	4	A costly refined surface produced by honing, lapping, and buffing. It is specified only when the design requirements make it mandatory. It is required in instrument work, gauge work, and where packing and rings must slide across the direction of surface grain such as on chrome-plated piston rods, etc. where lubrication is not dependable.
0.05 / 0.025	2 / 1	Costly refined surfaces produced by only the finest of modern honing, lapping, buffing, and superfinishing equipment. These surfaces may have a satin or highly polished appearance depending on the finishing operation and material. These surfaces are specified only when design requirements make them mandatory. They are specified on fine or sensitive instrument parts or other laboratory items, and certain gauge surfaces, such as precision gauge blocks.

FIGURE 12–9 ■ Surface roughness description and application

REFERENCES

CAN3-B95-1962 (R1996) Surface Texture
ASME Y14.36M-1996 (R2002) Surface Texture
Symbols

INTERNET RESOURCES

Drafting Zone For information on surface texture and surface texture symbols, see: http://www.draftingzone.com

Precision Devices, Inc For information on surface texture and surface texture symbols, see: http://www.predev.com/smg/intro.htm

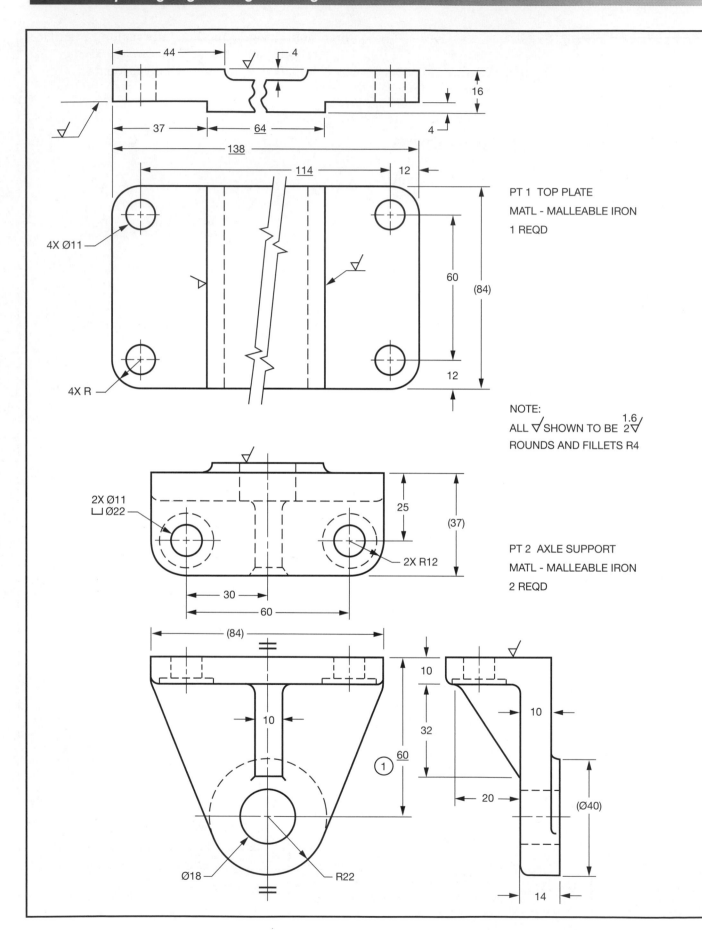

PT 1 TOP PLATE
MATL - MALLEABLE IRON
1 REQD

4X Ø11

4X R

NOTE:
ALL ▽ SHOWN TO BE 2▽ 1.6
ROUNDS AND FILLETS R4

2X Ø11
⌐ Ø22

2X R12

PT 2 AXLE SUPPORT
MATL - MALLEABLE IRON
2 REQD

Ø18

R22

QUESTIONS:

1. How many reference dimensions are shown?

2. How many not-to-scale dimensions are shown?

3. What type of section view is used on part 3?

4. How many machined surfaces are shown? Two surfaces can be on one plane.

5. How many fillets are shown on the drawing?

6. What machining allowance is called for on the machined surfaces?

7. What was the original height of part 2 before machining?

8. What is the height of the caster assembly?

9. What were the overall dimensions of the cast parts before machining?

PT 4 AXLE

MATL - CARBON STEEL

1 REQD

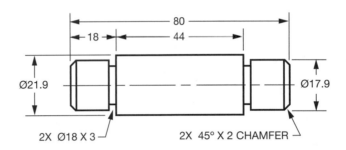

PT 3 WHEEL

MATL - MALLEABLE IRON

1 REQD

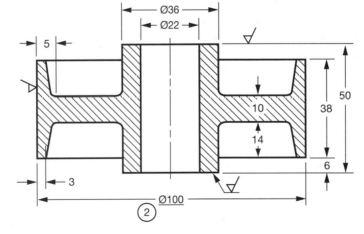

METRIC
DIMENSIONS IN MILLIMETRES

SCALE	1:2	
DRAWN	N. SIKORA	DATE 22/03/04

REV TABLE	1	16/08/04	J. HELSEL	2	16/08/04	J. HELSEL	CASTER DETAILS	A-37M
	60 DIM WAS 65			100 DIM WAS 110				

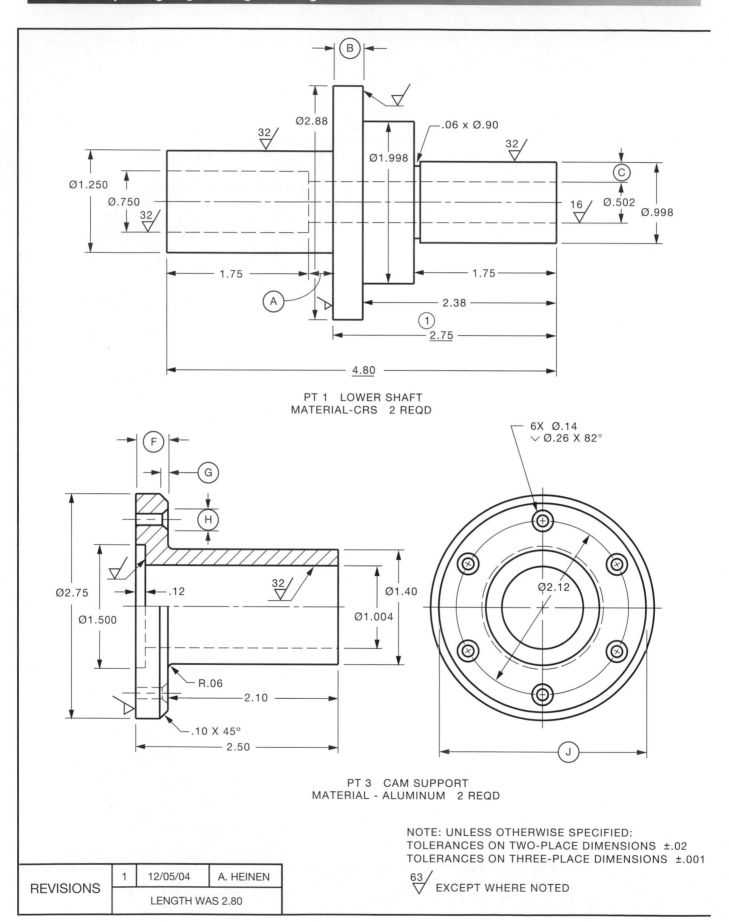

PT 1   LOWER SHAFT
MATERIAL-CRS   2 REQD

PT 3   CAM SUPPORT
MATERIAL - ALUMINUM   2 REQD

NOTE: UNLESS OTHERWISE SPECIFIED:
TOLERANCES ON TWO-PLACE DIMENSIONS ±.02
TOLERANCES ON THREE-PLACE DIMENSIONS ±.001

63 EXCEPT WHERE NOTED

REVISIONS	1	12/05/04	A. HEINEN
	LENGTH WAS 2.80		

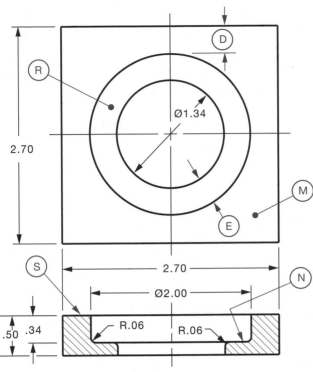

PT 2   WASHER
MATERIAL-MS   4 REQD FAO

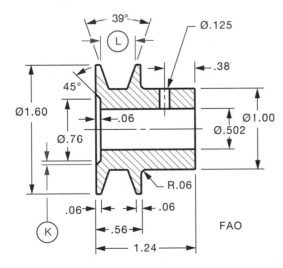

PT 4   V-BELT PULLEY
MATERIAL-CRS   4 REQD

QUESTIONS:

1. Calculate dimensions Ⓐ to Ⓛ.

Referring to Figure 11-8, what production methods would be suitable to produce the surface texture of

2. Ø.502 hole in Part 1?

3. Ø1.250 on Part 1?

4. Ø1.004 hole in Part 3?

5. How many dimensions indicate that the dimension is not drawn to scale?

Refer to Part 1

6. What is the length of the Ø.998 portion?

7. What was the original length of the 2.75 dimension?

8. How many hidden circles would be seen if the right-end view were drawn?

Refer to Part 2

9. Which surface does Ⓡ represent in the front view?

10. Which surface does Ⓢ represent in the top view?

11. If the part that passes through the washer is Ø1.30, what is the clearance per side between the two parts?

12. How many fillets are required?

Refer to Part 3

13. How many degrees apart on the Ø2.12 are the Ø.14 holes?

14. Is the centre line of the countersunk holes in the centre of the flange?

15. What operation is performed to allow the head of the mounting screws to rest flush with the flange?

16. What type of section view is shown?

17. What is the amount and degree of chamfer?

18. What is the (A) size and (B) type of cap screw required to mount the cam support to its mating parts?

Refer to Part 4

19. How deep is the Ø.125 hole?

20. How deep is the belt groove?

21. What does FAO mean?

22. What type of section view is shown?

SCALE	NOT TO SCALE	
DRAWN	J. HELSEL	DATE 14/03/04

HANGER DETAILS

A-38

13 UNIT

TOLERANCES AND ALLOWANCES

The history of engineering drawing as a means of communicating engineering information spans a period of 6000 years. It seems inconceivable that such an elementary practice as the tolerancing of dimensions, which is taken for granted today, was introduced for the first time less than 100 years ago.

Apparently, engineers and workers realized only gradually that exact dimensions and shapes could not be attained in the shaping of physical objects. The skilled handicrafters of the past took pride in the ability to work to exact dimensions. This meant that objects were dimensioned more accurately than they could be measured. The use of modern measuring instruments would have shown the deviations from the sizes that were called exact.

It was soon realized that variations in the sizes of parts had always been present, that such variations could be restricted but not avoided, and that slight variation in the size that a part was originally intended to have could be tolerated without impairment of its correct functioning. It became evident that interchangeable parts need not be identical parts, but rather it would be sufficient if the significant sizes that controlled their fits lay between definite limits. Therefore, the problem of interchangeable manufacture developed from making parts to a supposedly exact size to holding parts between two limiting sizes lying so close together that any intermediate size would be acceptable.

The concept of limits means that a precisely defined basic condition (expressed by one numerical value or specification) is replaced by two limiting conditions. Any result lying between these two limits is acceptable. A workable scheme of interchangeable manufacture, which is indispensable to mass production methods, had been established.

DEFINITIONS

To calculate limits and tolerances, you must understand the following definitions.

Basic Size

The basic size of a dimension is the theoretical size from which the limits for that dimension are derived, by the application of the allowance and tolerance.

Limits of Size

The limits of size are the maximum and minimum sizes permitted for a specific dimension.

Tolerances

The tolerance of a dimension is the total permissible variation in its size. The tolerance is the difference between the limits of size.

Allowance

An allowance is the intentional difference in size of mating parts. It is the minimum clearance (positive allowance) or maximum interference (negative allowance) between such parts. Fits between parts are covered in Units 14 and 15.

All dimensions on a drawing have tolerances. Some dimensions must be more exact than others and consequently have smaller tolerances.

When dimensions require a greater accuracy than the general note provides, individual tolerances or limits must be shown for those dimensions.

Where limit dimensions are used and where either the maximum or the minimum dimension has digits to the right of the decimal point, the other value should have the zeros added so that both limits of size are expressed to the same number of decimal places.

When limit dimensions are used for diameter or radial features and the dimensions are placed one above the other, only one diameter or radius symbol is used and located at midheight.

Where one limit alone is important and where any variations away from that limit in the other direction may be permitted, the MAX (maximum) or MIN (minimum) can be specified. Examples are depth of holes, corner radii, and chamfers. Figures 13–1 and 13–2 show applications of limit dimensioning.

TOLERANCING METHODS

Dimensional tolerances are expressed in one of two ways: limit dimensioning, or plus and minus tolerancing.

Limit Dimensioning

In the limit dimensioning method, only the maximum and minimum dimensions are specified, Figure 13–1. When placed one over the other, the larger dimension is placed on top. When shown with a leader and placed in one line, the smaller size is shown first. A small dash separates the two dimensions. When limit dimensions are used for diameter or radial features, the ∅ or R symbol is centred midway between the two limits, Figure 13–1(A). These rules apply to both inch and metric drawings.

Plus and Minus Tolerancing

In this method the dimension of the specified size is given first and is followed by a plus and minus tolerance expression. The tolerance can be bilateral or unilateral, Figure 13–3.

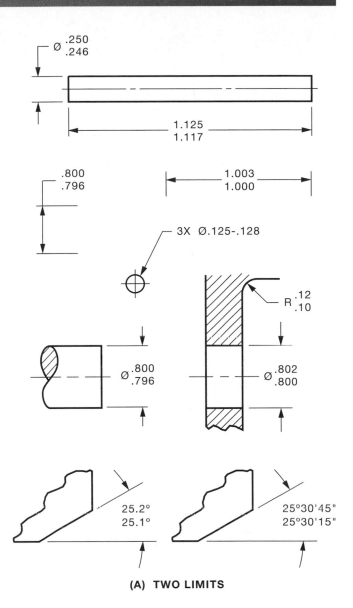

(A) TWO LIMITS

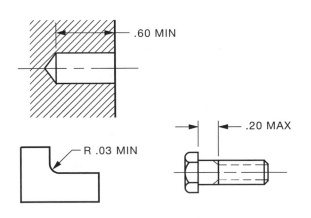

(B) SINGLE LIMITS

FIGURE 13–1 ■ Limit dimensioning

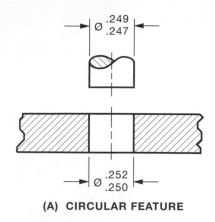

(A) CIRCULAR FEATURE

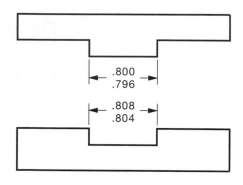

(B) FLAT FEATURE

FIGURE 13–2 ■ Limit dimensioning application

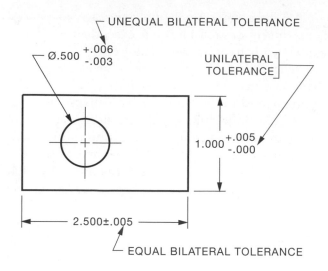

FIGURE 13–3 ■ Plus and minus tolerancing

A *bilateral tolerance* is a tolerance that is expressed as plus and minus values, which need not be the same size.

A *unilateral tolerance* is one that applies in one direction from the specified size, so the permissible variation in the other direction is zero.

Inch Tolerances

Where inch dimensions are used on the drawing, both limit dimensions or the plus and minus tolerance and its dimensions are expressed with the same number of decimal places.

Examples

$$.500 \pm .005 \quad \text{not} \quad .50 \pm .005$$

$$.500 \pm{}^{.005}_{.000} \quad \text{not} \quad .500 \pm{}^{.005}_{.0}$$

$$25.0 \pm .2 \quad \text{not} \quad 25 \pm .2$$

General tolerance notes greatly simplify the drawing. The following examples illustrate the variety of application in this system. The values given in the following examples are typical only:

Example 1
EXCEPT WHERE STATED OTHERWISE:
TOLERANCE ON DIMENSIONS ±.005

Example 2
EXCEPT WHERE STATED OTHERWISE:
TOLERANCES ON FINISHED DIMENSIONS
TO BE AS FOLLOWS:

Example 3

DIMENSION	TOLERANCE
UP TO 3.00	.01
OVER 3.00 TO 12.00	.02
OVER 12.00 TO 24.00	.04
OVER 24.00	.06

Example 4
UNLESS OTHERWISE SPECIFIED:
±.005 TOLERANCE ON MACHINED DIMENSIONS
±.04 TOLERANCE ON CAST DIMENSIONS
ANGULAR TOLERANCE ±30′

Millimetre Tolerances

Where millimetre dimensions are used on the drawings, the following apply:

A. The dimension and its tolerance need not be expressed to the same number of decimal places.

Example

$$15 \pm 0.5 \quad \text{not} \quad 15.0 \pm 0.5$$

B. Where unilateral tolerancing is used and either the plus or minus value is nil, a single zero is shown without a plus or minus sign.

Example

$$32 \begin{smallmatrix} 0 \\ -0.02 \end{smallmatrix} \quad \text{or} \quad 32 \begin{smallmatrix} +0.02 \\ 0 \end{smallmatrix}$$

C. Where bilateral tolerancing is used, both the plus and the minus values have the same number of decimal places, using zeros where necessary.

Example

$$32 \begin{smallmatrix} +0.25 \\ -0.10 \end{smallmatrix} \quad \text{not} \quad 32 \begin{smallmatrix} +0.25 \\ -0.1 \end{smallmatrix}$$

REFERENCES

CAN/CSA-B78.2-M91 Dimensioning and Tolerancing of Technical Drawings

ASME Y14.5M-1994 (R2004) Dimensioning and Tolerancing

INTERNET RESOURCE

Drafting Zone For information on tolerances and allowances, see: http://www.draftingzone.com

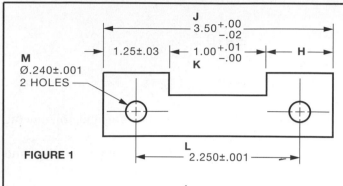

FIGURE 1

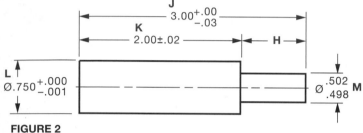

FIGURE 2

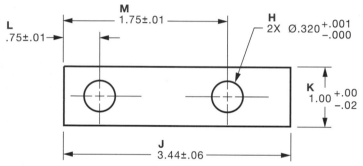

FIGURE 3

QUESTIONS:

1. Refer to Figures 1, 2, and 3, and calculate (A) basic size, (B) tolerance, (C) max. limit, and (D) min. limit for dimensions J, K, L, and M.

2. Refer to Figures 1, 2, and 3, and calculate (A) max. size, (B) min. size for dimension H.

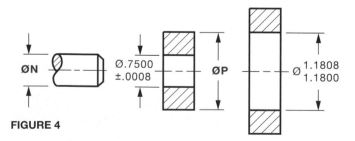

FIGURE 4

3. Refer to Figure 4. What are the limit dimensions for shaft N if it has a tolerance of .0014 and a min. clearance of .0006?

4. Refer to Figure 4. What are the limit dimensions for bushing P if it has a tolerance of .0006 and a min. interference of zero?

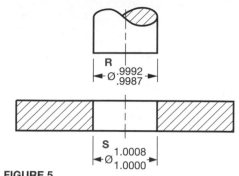

FIGURE 5

Refer to Figure 5.

5. What is the tolerance on shaft R?
6. What is the tolerance on hole S?
7. What is the min. clearance between the parts?
8. What is the max. clearance between the parts?

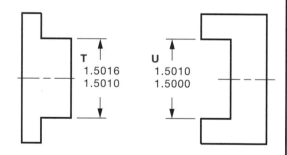

FIGURE 6

Refer to Figure 6.

9. What is the tolerance on part T?
10. What is the tolerance on slot U?
11. What is the min. interference between the parts?
12. What is the max. interference between the parts?

INCH TOLERANCES AND ALLOWANCES **A-39**

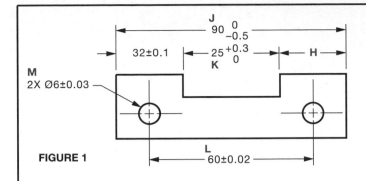

FIGURE 1

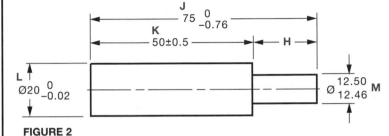

FIGURE 2

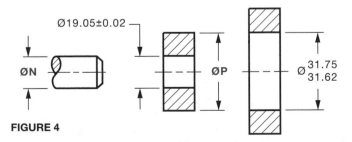

FIGURE 3

QUESTIONS:

1. Refer to Figures 1, 2, and 3, and calculate (A) basic size, (B) tolerance, (C) max. limit, and (D) min. limit for dimensions J, K, L, and M.
2. Refer to Figures 1, 2, and 3, and calculate (A) max. size, (B) min. size for dimension H.

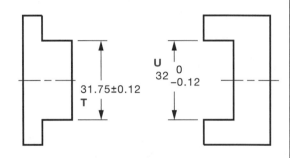

FIGURE 4

3. Refer to Figure 4. What are the limit dimensions for shaft N if it has a tolerance of 0.036 and a min. clearance of 0.015?
4. Refer to Figure 4. What are the limit dimensions for bushing P if it has a tolerance of 0.016 and a min. interference of zero?

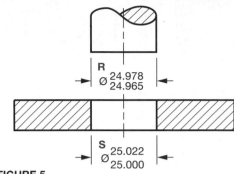

FIGURE 5

Refer to Figure 5.

5. What is the tolerance on shaft R?
6. What is the tolerance on hole S?
7. What is the min. clearance between the parts?
8. What is the max. clearance between the parts?

FIGURE 6

Refer to Figure 6.

9. What is the tolerance on part T?
10. What is the tolerance on slot U?
11. What is the min. clearance between the parts?
12. What is the max. clearance between the parts?

METRIC
DIMENSIONS IN MILLIMETRES

MILLIMETRE TOLERANCES AND ALLOWANCES

A-40M

14 UNIT

INCH FITS

Fit is the general term used to signify the range of tightness or looseness resulting from the application of a specific combination of allowances and tolerances in the design of mating parts. Fits are of three general types: clearance, interference, and transition, Figures 14–1 and 14–2.

Clearance Fits

Refer to Figure 14-2 (A). A Clearance fits have limits of size prescribed so there is always a clearance when mating parts are assembled. Clearance fits are intended for accurate assembly of parts and bearings. The parts can be assembled by hand because the hole is always larger than the shaft.

Interference Fits

Refer to Figure 14-2 (B). Interference fits have limits of size prescribed so there is always that an interference when mating parts are assembled. The hole is always smaller than the shaft. Interference fits are for permanent assemblies of parts that require rigidity and alignment, such as dowel pins and bearings in castings. Parts are usually pressed together with an arbor press.

Transition Fits

Refer to Figure 14-2 (C). Transition fits have limits of size prescribed so that either a clearance or an interference results when mating parts are assembled. A compromise between clearance and interference fits, these are used for applications where

accurate location is important, but either a small amount of clearance or interference is permissible.

DESCRIPTION OF FITS

Running and Sliding Fits

These fits, for which tolerances and clearances are given in Table 17 of the Appendix, represent a special type of clearance fit. These are intended to provide a similar running performance, with suitable lubrication allowance, throughout the range of sizes.

Locational Fits

Locational fits are intended to determine only the location of the mating parts; they may provide

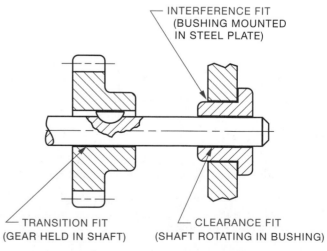

FIGURE 14–1 ■ Application of types of fits

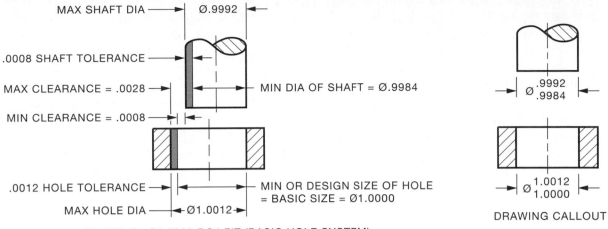

MAX SHAFT DIA → Ø.9992

.0008 SHAFT TOLERANCE →

MAX CLEARANCE = .0028 →

MIN CLEARANCE = .0008 →

MIN DIA OF SHAFT = Ø.9984

.0012 HOLE TOLERANCE →

MAX HOLE DIA → Ø1.0012

MIN OR DESIGN SIZE OF HOLE = BASIC SIZE = Ø1.0000

Ø .9992 / .9984

Ø 1.0012 / 1.0000

DRAWING CALLOUT

EXAMPLE - Ø1.0000 RC4 FIT (BASIC HOLE SYSTEM)

(A) CLEARANCE FIT

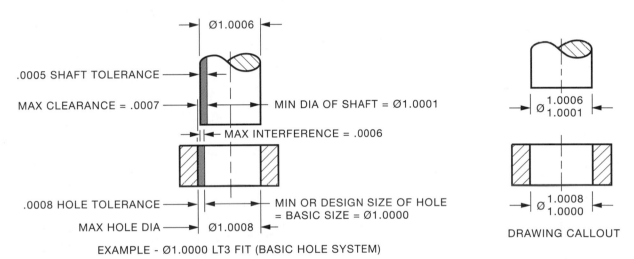

Ø1.0006

.0005 SHAFT TOLERANCE →

MAX CLEARANCE = .0007 →

MIN DIA OF SHAFT = Ø1.0001

MAX INTERFERENCE = .0006

.0008 HOLE TOLERANCE →

MAX HOLE DIA → Ø1.0008

MIN OR DESIGN SIZE OF HOLE = BASIC SIZE = Ø1.0000

Ø 1.0006 / 1.0001

Ø 1.0008 / 1.0000

DRAWING CALLOUT

EXAMPLE - Ø1.0000 LT3 FIT (BASIC HOLE SYSTEM)

(B) TRANSITION FIT

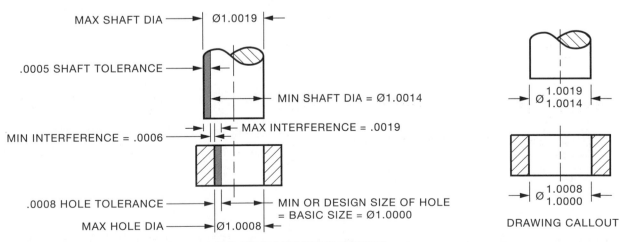

MAX SHAFT DIA → Ø1.0019

.0005 SHAFT TOLERANCE →

MIN SHAFT DIA = Ø1.0014

MAX INTERFERENCE = .0019

MIN INTERFERENCE = .0006 →

.0008 HOLE TOLERANCE →

MAX HOLE DIA → Ø1.0008

MIN OR DESIGN SIZE OF HOLE = BASIC SIZE = Ø1.0000

Ø 1.0019 / 1.0014

Ø 1.0008 / 1.0000

DRAWING CALLOUT

EXAMPLE - Ø1.0000 FN2 FIT (BASIC HOLE SYSTEM)

(C) INTERFERENCE FIT

FIGURE 14–2 ■ Types and examples of inch fits

rigid or accurate location, as with interference fits, or some freedom of location, as with clearance fits. Accordingly, they are divided into three groups: clearance fits, transition fits, and interference fits.

Locational clearance fits are intended for parts that are normally stationary but can be freely assembled or disassembled.

Locational transition fits are a compromise between clearance and interference fits for application where accuracy of location is important, but a small amount of either clearance or interference is permissible.

Locational interference fits are used where accuracy of location is of prime importance and for parts requiring rigidity and alignment.

Drive and Force Fits

Drive and force fits are a special type of interference fit, normally characterized by maintenance of constant bore pressures throughout the range of sizes. The interference varies almost directly with diameter, and the difference between its minimum and maximum values is small to maintain the resulting pressures within reasonable limits.

STANDARD INCH FITS

Standard fits are designated in specifications and on design sketches with the symbols shown in Figure 14–3. These symbols are not intended to be shown directly on shop drawings; instead, the actual limits of size are determined and specified on the drawings. The letter symbols used are as follows:

RC   Running and sliding fit
LC   Locational clearance fit
LT   Locational transition fit
LN   Locational interference fit
FN   Force or shrink fit

Next to these letter symbols are numbers representing the class of fit; for example, FN4 represents a class 4, force fit.

Each of these symbols (two letters and a number) represents a complete fit, for which the minimum and maximum clearance or interference and the limits of size for the mating parts are given directly in Appendix Tables 17 through 21.

Running and Sliding Fits

RC1 Precision Sliding Fit

This fit is intended for the accurate location of parts that must assemble without perceptible play for high-precision work such as gauges.

RC2 Sliding Fit

This fit is intended for accurate location, but with greater maximum clearance than class RC1. Parts made to this fit move and turn easily but are not intended to run freely.

RC3 Precision Running Fit

This fit is about the closest fit that can be expected to run freely, and is intended for precision work for oil-lubricated bearings at slow speeds and light journal pressures.

RC4 Close Running Fit

This fit is intended chiefly as a running fit for grease- or oil-lubricated bearings on accurate machinery with moderate surface speeds and journal pressures, where accurate location and minimum play are desired.

RC5 and RC6 Medium Running Fits

These fits are intended for higher running speeds and/or where temperature variations are likely to be encountered.

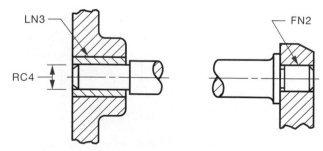

(A) SHAFT IN BUSHED HOLE   **(B) CRANK PIN IN CAST IRON**

FIGURE 14–3 ■ Design sketches showing standard fits

RC7 Free Running Fit

This fit is intended for use where accuracy is not essential, and/or where large temperature variations are likely to be encountered.

RC8 and RC9 Loose Running Fits

These fits are intended for use where materials made to commercial tolerances are involved, such as cold-rolled shafting and tubing.

Locational Clearance Fits

Locational clearance fits are intended for parts that are normally stationary but can be freely assembled or disassembled.

LC1 to LC4

These fits have a minimum zero clearance, but in practice, the fit will always have a clearance.

LC5 and LC6

These fits have a small minimum clearance intended for close location fits for non-running parts.

LC7 to LC11

These fits have progressively larger clearances and tolerances, and are useful for various loose clearances for assembly of bolts and similar parts.

Locational Transition Fits

Locational transition fits are a compromise between clearance and interference fits or for application where accuracy of location is important, but either a small amount of clearance or interference is permissible.

LT1 and LT2

These fits average a slight clearance, giving a light push fit.

LT3 and LT4

These fits average virtually no clearance, and are for use where some interference can be tolerated.

Sometimes referred to as an easy keying fit, they are used for shaft keys and ball race fits. Assembly is generally by pressure or hammer blows.

LT5 and LT6

These fits average a slight interference, although appreciable assembly force will be required.

Locational Interference Fits

Locational interference fits are used where accuracy of location is of prime importance, and for parts requiring rigidity and alignment with no special requirements for bore pressure.

LN1 and LN2

These are light press fits, with very small minimum interference, suitable for such parts as dowel pins, which are assembled with an arbor press in steel, cast iron, or brass. Parts can normally be dismantled and reassembled.

LN3

This is suitable as a heavy press fit in steel and brass, or a light press fit in more elastic materials and light alloys.

LN4 to LN6

While LN4 can be used for permanent assembly of steel parts, these fits are primarily intended as press fits for soft materials.

Force or Shrink Fits

Force or shrink fits constitute a special type of interference fit. The interference varies almost directly with diameter, and the difference between its minimum and maximum values is small to maintain the resulting pressures within reasonable limits.

FN1 Light Drive Fit

This requires light assembly pressure and produces more or less permanent assemblies. It is suitable for thin sections or long fits or in cast-iron external members.

FN2 Medium Drive Fit

This is suitable for heavier steel parts or as a shrink fit on light sections.

FN3 Heavy Drive Fit

This is suitable for heavier steel parts or as a shrink fit in medium sections.

FN4 and FN5 Force Fits

These are suitable for parts that can be highly stressed.

Basic Hole System

In the basic hole system, which is recommended for general use, the basic size will be the design size for the hole, and the tolerance will be plus. The design size for the shaft will be the basic size minus the minimum clearance, or plus the maximum interference, and the tolerance will be minus, as given in the tables in the Appendix. For example (see Table 17), for a 1-inch RC7 fit, values of +.0020, .0025, and −.0012 are given; hence, tolerances will be

$$\text{hole } \varnothing 1.0000 \begin{array}{c} +.0020 \\ -.0000 \end{array}$$

$$\text{shaft } \varnothing.9975 \begin{array}{c} +.0000 \\ -.0012 \end{array}$$

Basic Shaft System

Fits are sometimes required on a basic shaft system, especially where two or more fits are required on the same shaft. This is designated for design purposes by a letter S following the fit symbol; for example, RC7S.

Tolerances for holes and shafts are identical with those for a basic hole system, but the basic size becomes the design size for the shaft, and the design size for the hole is found by adding the minimum clearance or subtracting the maximum interference from the basic size.

For example, for a 1-inch RC7S fit, values of +.0020, .0025, and −.0012 are given; therefore, tolerances will be

$$\text{hole } \varnothing 1.0025 \begin{array}{c} +.0020 \\ -.0000 \end{array}$$

$$\text{shaft } \varnothing 1.000 \begin{array}{c} +.0000 \\ -.0012 \end{array}$$

Additional examples of how to use the inch fits are shown in Tables 17 to 21 in the Appendix.

REFERENCES

CSA-B97.3-1970 (R1992) Standard Fits for Mating Parts, Inch Sizes

CAN/CSA-B78.2-M91 Dimensioning and Tolerancing of Technical Drawings

INTERNET RESOURCE

Drafting Zone For additional information on tolerancing, see: http://www.draftingzone.com

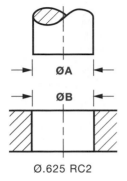

Ø.625 RC2

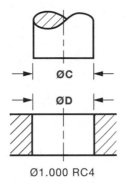

Ø1.000 RC4

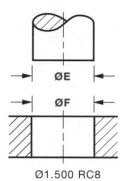

Ø1.500 RC8

RUNNING AND SLIDING FITS

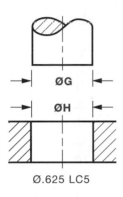

Ø.625 LC5

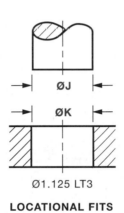

Ø1.125 LT3

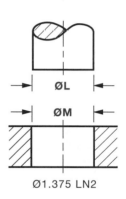

Ø1.375 LN2

LOCATIONAL FITS

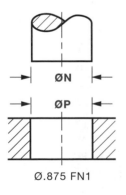

Ø.875 FN1

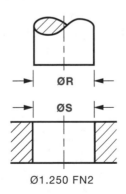

Ø1.250 FN2

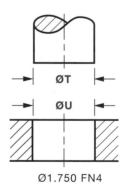

Ø1.750 FN4

FORCE OR SHRINK FITS

ASSIGNMENT:
ON A GRID SHEET, SKETCH A TABLE SHOWING THE
LIMITS OF SIZE FOR EACH OF THE PARTS, AND
THE MINIMUM AND MAXIMUM CLEARANCE OR
INTERFERENCE FOR EACH OF THE FITS SHOWN.

**INCH FITS–
BASIC HOLE SYSTEM**

A-41

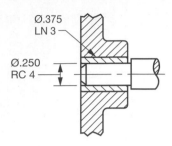

(A) SHAFT IN BUSHED HOLE

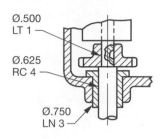

(B) GEAR AND SHAFT IN BUSHED BEARING

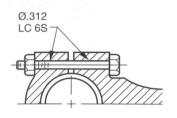

(C) CONNECTING-ROD BOLT

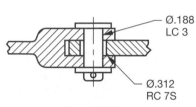

(D) LINK PIN

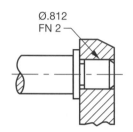

(E) CRANK PIN IN CAST IRON

DESIGN SKETCH	BASIC DIAMETER SIZE (IN)	SYMBOL	BASIS	FEATURE	LIMITS OF SIZE		CLEARANCE OR INTERFERENCE	
					MAX	MIN	MAX	MIN
A	.375		HOLE	HOLE				
				SHAFT				
A	.250		HOLE	HOLE				
				SHAFT				
B	.500		HOLE	HOLE				
				SHAFT				
B	.625		HOLE	HOLE				
				SHAFT				
B	.750		HOLE	HOLE				
				SHAFT				
C	.312		SHAFT	HOLE				
				SHAFT				
D	.188		HOLE	HOLE				
				SHAFT				
D	.312		SHAFT	HOLE				
				SHAFT				
E	.812		HOLE	HOLE				
				SHAFT				

ASSIGNMENT:
PREPARE A CHART SIMILAR TO THE ONE SHOWN ABOVE AND, USING THE INCH FIT TABLES SHOWN IN THE APPENDIX, ADD THE MISSING INFORMATION.

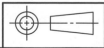

INCH FITS

A-42

UNIT | 15

METRIC FITS

The ISO (metric) system of limits and fits for mating parts is approved and adopted for general use in Canada and the United States. It establishes the designation symbols used to define specific dimensional limits on drawings.

The general terms "hole" and "shaft" can also be taken as referring to the space containing or contained by two parallel faces of any part, such as the width of a slot or the thickness of a key.

An "International Tolerance (IT) Grade" establishes the magnitude of the tolerance zone or the amount of part size variation allowed for internal and external dimensions alike. The smaller the grade number, the smaller the tolerance zone. For general applications of IT grades, see Figure 15–1.

Grades 1 through 4 are very precise grades intended primarily for gauge making and similar precision work, although grade 4 can also be used for very precise production work.

Grades 5 through 16 represent a progressive series suitable for cutting operations, such as turning, boring, grinding, milling, and sawing. Grade 5 is the most precise grade, obtainable by fine grinding and lapping, while grade 16 is the coarsest grade for rough sawing and machining.

Grades 12 through 16 are intended for manufacturing operations such as cold heading, pressing, rolling, and other forming operations.

As a guide to the selection of tolerances, Figure 15–2 has been prepared to show grades that may be expected to be held by various manufacturing processes for work in metals. For work in other materials, such as plastics, it may be necessary to use coarser tolerance grades for the same process.

A fundamental deviation establishes the position of the tolerance zone with respect to the basic size. Fundamental deviations are expressed by "tolerance position letters." Capital letters are used for internal dimensions, and lowercase letters for external dimensions.

Metric Tolerance Symbol

By combining the IT grade number and the tolerance position letter, the tolerance symbol identifies the actual maximum and minimum limits of the part. The toleranced sizes are thus defined by the basic size of the part, followed by the symbol composed of a letter and number, Figure 15–3.

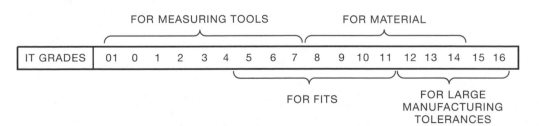

FIGURE 15–1 ■ Application of international tolerance (IT) grades

Hole basis fits have a fundamental deviation of "H" on the hole, and shaft basis fits have a fundamental deviation of "h" on the shaft. Normally, the hole basis system is preferred.

Fit Symbol

A fit is specified by the basic size common to both components, followed by a symbol corresponding to each component, with the internal part symbol preceding the external part symbol, Figure 15–4.

Figure 15–5 shows examples of three common fits.

Hole Basis Fits System

In the hole basis fits system (see Tables 22 and 24 of the Appendix), the basic size will be the minimum size of the hole. For example, for a $\varnothing$25 H8/f7 fit, which is a Preferred Hole Basis Clearance Fit, the limits for the hole and shaft will be:

hole limits = $\varnothing$25.000 and $\varnothing$25.033

shaft limits = $\varnothing$24.959 and $\varnothing$24.980

minimum clearance = 0.020

maximum clearance = 0.074

If a $\varnothing$25 H7/s6 Preferred Hole Basis Interference Fit is required, the limits for the hole and shaft will be:

Hole limits = $\varnothing$25.000 and $\varnothing$25.021

Shaft limits = $\varnothing$25.035 and $\varnothing$25.048

Minimum interference = −0.014

Maximum interference = −0.048

Additional examples of how to use the hole basis fits system are shown in Table 24 of the Appendix.

Shaft Basis Fits System

Where more than two fits are required on the same shaft, the shaft basis fits system is recommended. Tolerances for holes and shafts are identical with those for a basic hole system; however, the basic size becomes the maximum shaft size. For example, for a

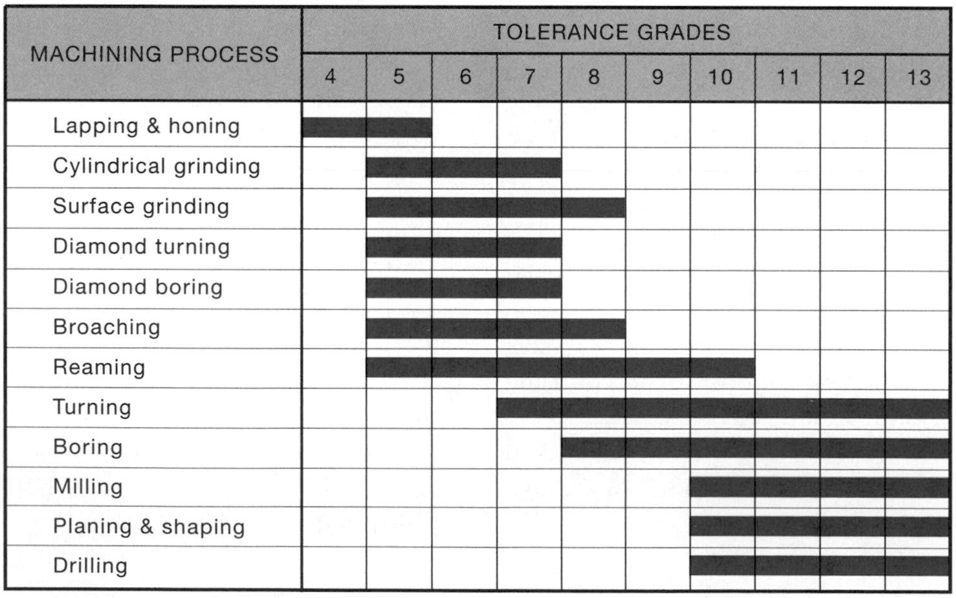

MACHINING PROCESS	TOLERANCE GRADES									
	4	5	6	7	8	9	10	11	12	13
Lapping & honing	███	███								
Cylindrical grinding		███	███	███						
Surface grinding		███	███	███	███					
Diamond turning		███	███	███						
Diamond boring		███	███	███						
Broaching		███	███	███	███					
Reaming		███	███	███	███	███	███			
Turning				███	███	███	███	███	███	███
Boring					███	███	███	███	███	███
Milling							███	███	███	███
Planing & shaping							███	███	███	███
Drilling							███	███	███	███

FIGURE 15–2 ■ Tolerancing grades for machining processes

∅16 C11/h11 fit, which is a Preferred Shaft Basis Clearance Fit, the limits for the hole and shaft will be

hole limits = ∅16.095 and ∅16.205

shaft limits = ∅15.890 and ∅16.000

minimum clearance = 0.095

maximum clearance = 0.315

Additional examples of how to use the shaft basis fits system are shown in Table 25 of the Appendix.

Descriptions of preferred metric fits are shown in Figure 15–6.

Drawing Callout

The method used in Figure 15–7 (A) to specify tolerances is recommended when the system is first introduced. In this case, limit dimensions are specified and the basic size and tolerance symbol are identified as reference.

As experience is gained, the method shown in Figure 15–7 (B) may be used. When the system is established and standard tools, gauges, and stock materials are available with size and symbol identification, the method shown in Figure 15–7 (C) may be used.

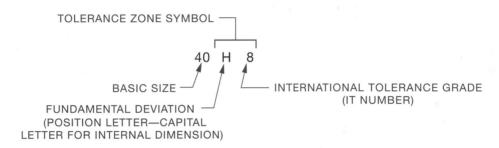

(A) INTERNAL DIMENSION (HOLES)

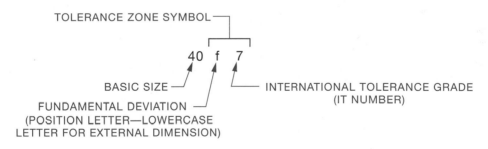

(B) EXTERNAL DIMENSION (SHAFTS)

FIGURE 15–3 ■ Tolerance symbol (hole basis fit)

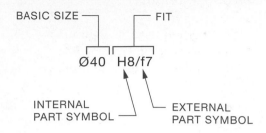

(A) HOLE BASIS

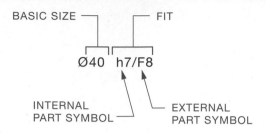

(B) SHAFT BASIS

FIGURE 15–4 ■ Fit symbol

This would result in a clearance between 0.020 and 0.074 mm. A description of the preferred metric fits is shown in Tables 22 and 23 of the Appendix.

REFERENCE

CAN-B97.3-M1982 (R1992) Tolerances and Standard Fits for Mating Parts, Metric Sizes

INTERNET RESOURCE

Maryland Metrics For information and examples of standard millimetre fits, see: http://www .mdmetric.com

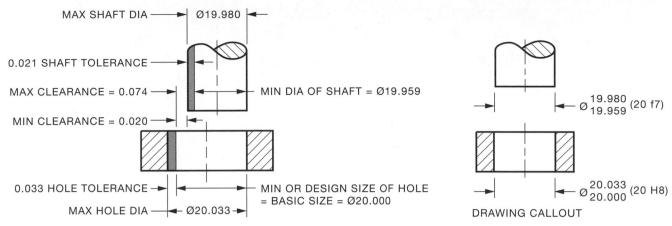

MAX SHAFT DIA — Ø19.980

0.021 SHAFT TOLERANCE

MAX CLEARANCE = 0.074 — MIN DIA OF SHAFT = Ø19.959

MIN CLEARANCE = 0.020

0.033 HOLE TOLERANCE — MIN OR DESIGN SIZE OF HOLE = BASIC SIZE = Ø20.000

MAX HOLE DIA — Ø20.033

Ø $^{19.980}_{19.959}$ (20 f7)

Ø $^{20.033}_{20.000}$ (20 H8)

DRAWING CALLOUT

EXAMPLE—H8/f7 PREFERRED HOLE BASIS FIT FOR A Ø20 HOLE (SEE APPENDIX, TABLE 24)

(A) CLEARANCE FIT

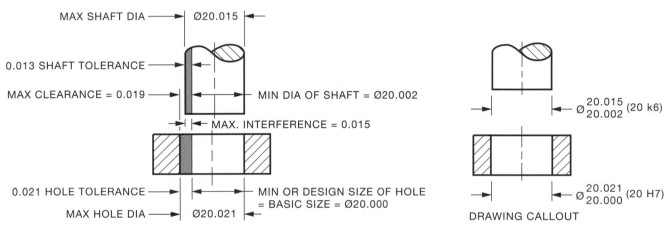

MAX SHAFT DIA — Ø20.015

0.013 SHAFT TOLERANCE

MAX CLEARANCE = 0.019 — MIN DIA OF SHAFT = Ø20.002

MAX. INTERFERENCE = 0.015

0.021 HOLE TOLERANCE — MIN OR DESIGN SIZE OF HOLE = BASIC SIZE = Ø20.000

MAX HOLE DIA — Ø20.021

Ø $^{20.015}_{20.002}$ (20 k6)

Ø $^{20.021}_{20.000}$ (20 H7)

DRAWING CALLOUT

EXAMPLE—H7/k6 PREFERRED HOLE BASIS FIT FOR A Ø20 HOLE (SEE APPENDIX, TABLE 24)

(B) TRANSITION FIT

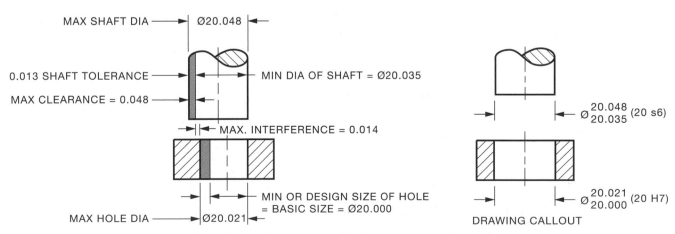

MAX SHAFT DIA — Ø20.048

0.013 SHAFT TOLERANCE — MIN DIA OF SHAFT = Ø20.035

MAX CLEARANCE = 0.048

MAX. INTERFERENCE = 0.014

MIN OR DESIGN SIZE OF HOLE = BASIC SIZE = Ø20.000

MAX HOLE DIA — Ø20.021

Ø $^{20.048}_{20.035}$ (20 s6)

Ø $^{20.021}_{20.000}$ (20 H7)

DRAWING CALLOUT

EXAMPLE—H7/s6 PREFERRED HOLE BASIS FIT FOR A Ø20 HOLE (SEE APPENDIX, TABLE 24)

(C) INTERFERENCE FIT

FIGURE 15–5 ■ Types and examples of millimetre fits

ISO SYMBOL		DESCRIPTION
HOLE BASIS	SHAFT BASIS	
H11/c11	C11/h11	LOOSE RUNNING FIT FOR WIDE COMMERCIAL TOLERANCES OR ALLOWANCES ON EXTERNAL MEMBERS.
H9/d9	D9/h9	FREE RUNNING FIT NOT FOR USE WHERE ACCURACY IS ESSENTIAL, BUT GOOD FOR LARGE TEMPERATURE VARIATIONS, HIGH RUNNING SPEEDS, OR HEAVY JOURNAL PRESSURES.
H8/f7	F8/h7	CLOSE RUNNING FIT FOR RUNNING ON ACCURATE MACHINES AND FOR ACCURATE LOCATION AT MODERATE SPEEDS AND JOURNAL PRESSURES.
H7/g6	G7/h6	SLIDING FIT NOT INTENDED TO RUN FREELY, BUT TO MOVE AND TURN FREELY AND LOCATE ACCURATELY.
H7/h6	H7/h6	LOCATIONAL CLEARANCE FIT PROVIDES SNUG FIT FOR LOCATING STATIONARY PARTS, BUT CAN BE FREELY ASSEMBLED AND DISASSEMBLED.
H7/k6	K7/h6	LOCATIONAL TRANSITION FIT FOR ACCURATE LOCATION, A COMPROMISE BETWEEN CLEARANCE AND INTERFERENCE.
H7/n6	N7/h6	LOCATIONAL TRANSITION FIT FOR MORE ACCURATE LOCATION WHERE GREATER INTERFERENCE IS PERMISSIBLE.
H7/p6	P7/h6	LOCATIONAL INTERFERENCE FIT FOR PARTS REQUIRING RIGIDITY AND ALIGNMENT WITH PRIME ACCURACY OF LOCATION BUT WITHOUT SPECIAL BORE PRESSURE REQUIREMENTS.
H7/s6	S7/h6	MEDIUM DRIVE FIT FOR ORDINARY STEEL PARTS OR SHRINK FITS ON LIGHT SECTIONS, THE TIGHTEST FIT USABLE WITH CAST IRON.
H7/u6	U7/h6	FORCE FIT SUITABLE FOR PARTS THAT CAN BE HIGHLY STRESSED OR FOR SHRINK FITS WHERE THE HEAVY PRESSING FORCES REQUIRED ARE IMPRACTICAL.

FIGURE 15–6 ■ Description of preferred metric fits

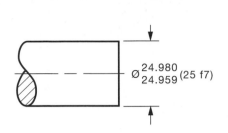
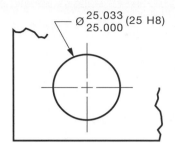

(A) WHEN SYSTEM IS FIRST INTRODUCED

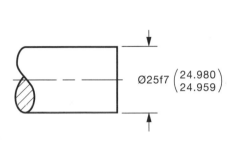
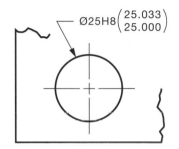

(B) AS EXPERIENCE IS GAINED

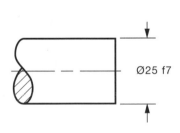
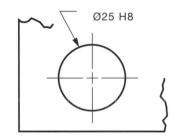

(C) WHEN SYSTEM IS ESTABLISHED

FIGURE 15–7 ■ Metric tolerance symbol shown on drawings

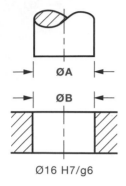

ØA
ØB

Ø16 H7/g6

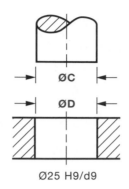

ØC
ØD

Ø25 H9/d9

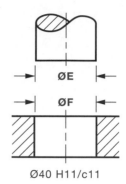

ØE
ØF

Ø40 H11/c11

RUNNING AND SLIDING FITS

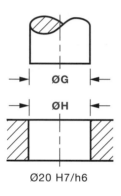

ØG
ØH

Ø20 H7/h6

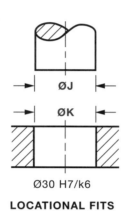

ØJ
ØK

Ø30 H7/k6

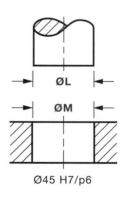

ØL
ØM

Ø45 H7/p6

LOCATIONAL FITS

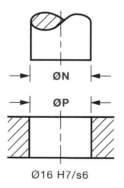

ØN
ØP

Ø16 H7/s6

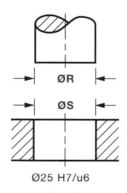

ØR
ØS

Ø25 H7/u6

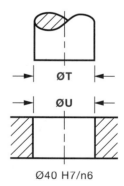

ØT
ØU

Ø40 H7/n6

FORCE OR SHRINK FITS

ASSIGNMENT:
ON A GRID SHEET, SKETCH A TABLE SHOWING THE
LIMITS OF SIZE FOR EACH OF THE PARTS, AND
THE MINIMUM AND MAXIMUM CLEARANCE OR
INTERFERENCE FOR EACH OF THE FITS SHOWN.

| METRIC FITS– BASIC HOLE SYSTEM | A-43M |

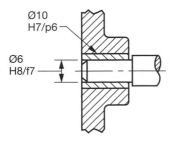

(A) SHAFT IN BUSHED HOLE

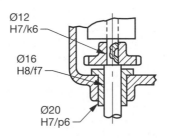

(B) GEAR AND SHAFT IN BUSHED BEARING

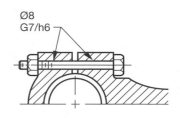

(C) CONNECTING-ROD BOLT

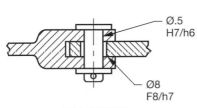

(D) LINK PIN

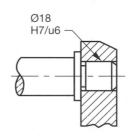

(E) CRANK PIN IN CAST IRON

DESIGN SKETCH	BASIC DIAMETER SIZE (mm)	SYMBOL	BASIS	FEATURE	LIMITS OF SIZE		CLEARANCE OR INTERFERENCE	
					MAX	MIN	MAX	MIN
A	10		HOLE	HOLE				
				SHAFT				
A	6		HOLE	HOLE				
				SHAFT				
B	12		HOLE	HOLE				
				SHAFT				
B	16		HOLE	HOLE				
				SHAFT				
B	20		HOLE	HOLE				
				SHAFT				
C	8		SHAFT	HOLE				
				SHAFT				
D	5		HOLE	HOLE				
				SHAFT				
D	8		SHAFT	HOLE				
				SHAFT				
E	18		HOLE	HOLE				
				SHAFT				

ASSIGNMENT: PREPARE A CHART SIMILAR TO THE ONE SHOWN ABOVE AND, USING THE METRIC FIT TABLES SHOWN IN THE APPENDIX, ADD THE MISSING INFORMATION.

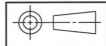 **METRIC FITS** **A-44M**

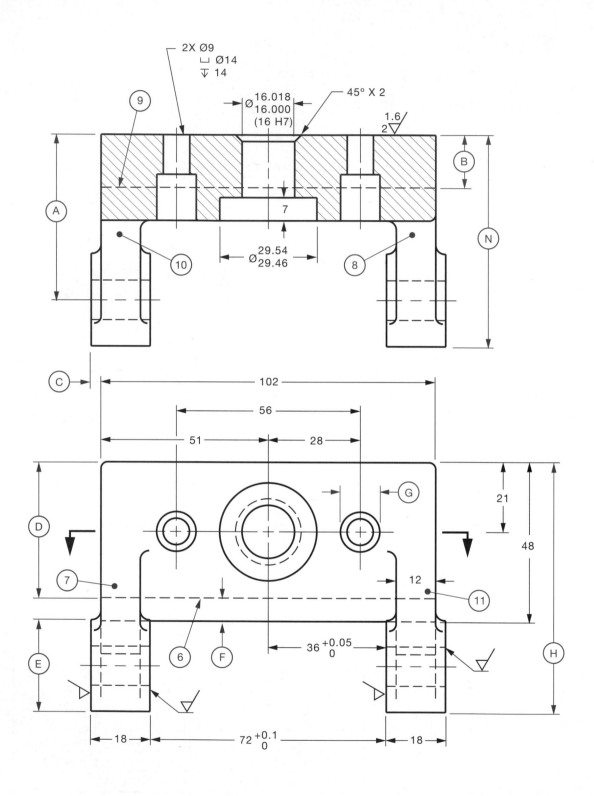

QUESTIONS:

1. How many surfaces are to be finished?

2. Except where noted otherwise, what is the tolerance on all dimensions?

3. What is the tolerance on the Ø12.000-12.018 holes?

4. What are the limit dimensions for the 40.64 dimension shown on the side view?

5. What are the limit dimensions for the 26 dimension shown on the side view?

6. What is the maximum distance permissible between the centres of the Ø9 hole?

7. Express the Ø12.000-12.018 holes as a plus-and-minus tolerance dimension.

8. How many surfaces require a $\frac{6.3}{2\forall}$ finish?

9. What are the limit dimensions for the 7 dimension shown on the top view?

10. Locate surface ④ on the top view.

11. How many bosses are there?

12. Locate line ③ in the top view.

13. Locate line ⑥ in the side view.

14. Which surface in the front view indicates line ④ in the side view?

15. Calculate distances Ⓐ to Ⓝ.

16. What is the (A) size and (B) type of cap screw used to fasten the bracket?

17. Can standard lockwashers be used with the cap screws?

18. What type of tolerance is shown on the (A) 40.64 vertical dimension and (B) 72 horizontal dimension?

19. What type of sectional view is used?

20. What was the size of the casting before finishing?

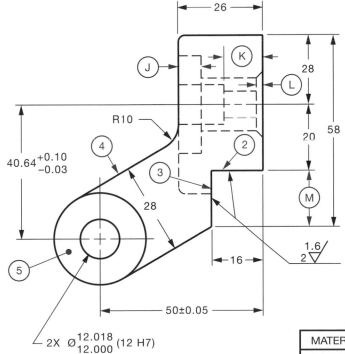

NOTE:
TOLERANCES ON DIMENSIONS ±0.5
$\frac{6.3}{2\forall}$ EXCEPT WHERE NOTED

ROUNDS AND FILLETS R5

METRIC
DIMENSIONS IN MILLIMETRES

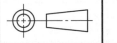

MATERIAL	GI	
SCALE	NOT TO SCALE	
DRAWN	C. JENSEN	DATE 22/09/04

BRACKET	A-45M

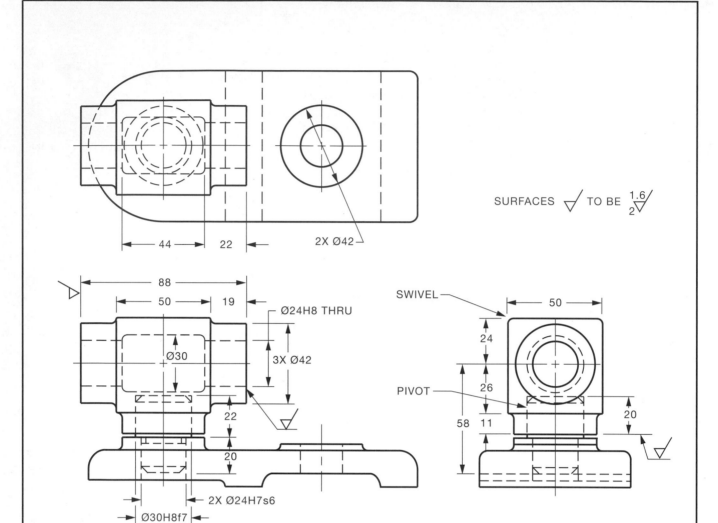

SURFACES ∇ TO BE 1.6/2∇

ASSIGNMENTS:

1. ON A CENTIMETRE GRID SHEET (1 mm SQUARES), SKETCH A FULL SECTION VIEW OF THE SWIVEL SHOWN IN THE ABOVE ASSEMBLY. DIMENSION AND SHOW THE HOLE SIZES AS LIMITS. SCALE 1:1.

2. ON A CENTIMETRE GRID SHEET (1 mm SQUARES), MAKE A ONE-VIEW DRAWING WITH DIMENSIONS OF THE PIVOT. ADD CHAMFERS AT THE END OF THE PIVOT FOR EASE OF ASSEMBLY, AND AN UNDERCUT AT THE SHOULDER. REFER TO UNIT 10 FOR INFORMATION ON CHAMFERS AND UNDERCUTS. SCALE 1:1.

METRIC
DIMENSIONS ARE IN MILLIMETRES

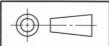 **SWIVEL** | **A-46M**

THREADED FASTENERS

Fastening devices are vital to most aspects of industry. They are used in assembling manufactured products, in the machines and devices used in the manufacturing processes, and in constructing all types of buildings. Fastening devices are standardized.

The two basic types of fasteners are semi-permanent and removable: Rivets are semi-permanent fasteners; bolts, screws, studs, nuts, pins, and keys are removable fasteners. A thorough knowledge of the design and graphic representation of the common fasteners is essential for interpreting engineering drawings, Figure 16–1.

Thread Representation

True representation of a screw thread is seldom provided on working drawings because of the time involved and the drawing cost. Three types of conventions are generally used for screw thread representation: simplified, detailed, and schematic. Simplified representation is used wherever it will clearly describe the requirements. Schematic and detailed representations require more drafting time, but they are sometimes used to avoid confusion with other parallel lines or to more clearly portray particular aspects of threads. Only one method is generally used on any one drawing. When required, however, all three methods may be used. Canadian (ISO) and American (ASME) thread representation vary slightly, Figures 16–2 and 16–3.

When using either simplified or schematic thread representation, on the end views of external threads starting with a chamfer, if the chamfer and minor thread diameter are close to the same size, the minor diameter is not shown.

On the end views showing a countersunk threaded hole, if the countersunk diameter and

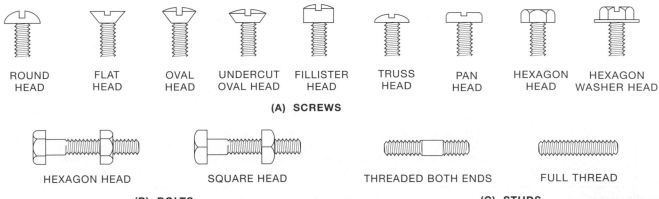

| ROUND HEAD | FLAT HEAD | OVAL HEAD | UNDERCUT OVAL HEAD | FILLISTER HEAD | TRUSS HEAD | PAN HEAD | HEXAGON HEAD | HEXAGON WASHER HEAD |

(A) SCREWS

| HEXAGON HEAD | SQUARE HEAD | THREADED BOTH ENDS | FULL THREAD |

(B) BOLTS **(C) STUDS**

FIGURE 16–1 ■ Threaded fasteners

major thread diameter are close to the same size, the major thread diameter is not shown.

THREADED ASSEMBLIES

Any of the thread representations shown here may be used for assemblies of threaded parts, and two or more methods may be used on the same drawing, Figures 16–2 and 16–3. In sectional views, the externally threaded part is always shown covering the internally threaded part, Figure 16–4.

Thread Standards

With the progress and growth of industry, there is a growing need for uniform, inter-changeable, threaded fasteners. Aside from the threaded forms previously mentioned, the pitch of

the thread and the major diameters are factors affecting standards.

THREADED HOLES

When a small threaded hole is required on a part, a common method of producing it is to drill a hole (tap drill size), then add threads to the hole by means of a threading tool called a tap. The tap drill size is not specified on the drawing. For a blind hole, the drilled hole is made a little deeper than the depth required for the threads. The tap drill sizes for inch and metric threads are shown in Tables 5 and 6 of the Appendix.

INCH THREADS

Until 1976, nearly all threaded assemblies in North America were designed using inch-sized threads. In this system, the pitch is equal to the

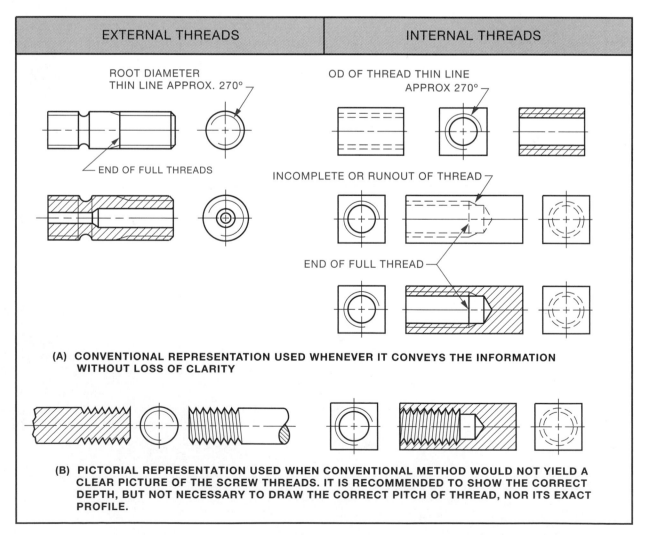

FIGURE 16–2 ■ Standard thread conventions (Canadian and ISO)

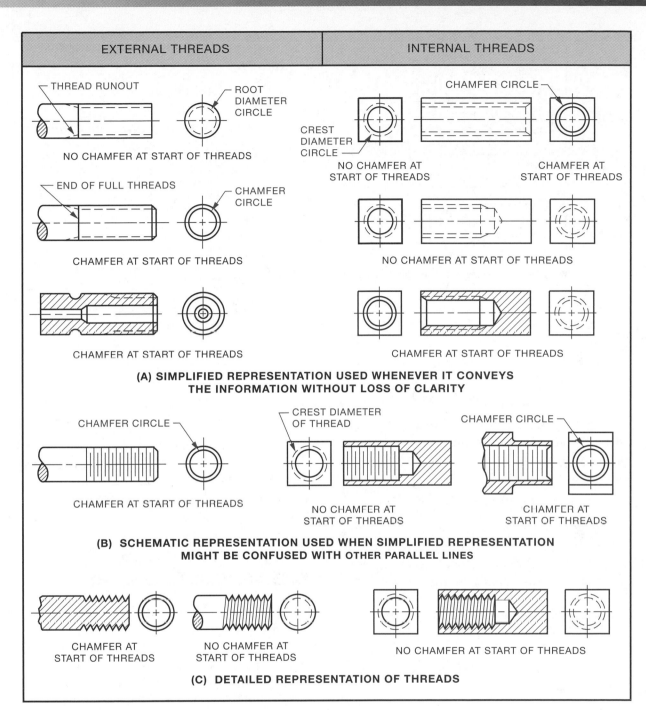

FIGURE 16–3 ■ Standard thread conventions (ASME)

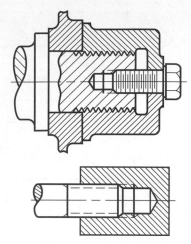

FIGURE 16–4 ■ Showing threads on assembly drawings

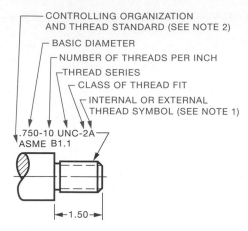

CONTROLLING ORGANIZATION
AND THREAD STANDARD (SEE NOTE 2)
BASIC DIAMETER
NUMBER OF THREADS PER INCH
THREAD SERIES
CLASS OF THREAD FIT
INTERNAL OR EXTERNAL
THREAD SYMBOL (SEE NOTE 1)

.750-10 UNC-2A
ASME B1.1

←1.50→

NOTE 1:
LETTER "A" DESIGNATES EXTERNAL THREAD
LETTER "B" DESIGNATES INTERNAL THREAD

NOTE 2:
MAY BE SHOWN AS A GENERAL
NOTE ON THE DRAWING

FIGURE 16–5 ■ Specifications for inch threads

distance between corresponding points on adjacent threads and is expressed as

$$\text{pitch} = \frac{1}{\text{number of threads per inch}}$$

The number of threads per inch is set for different diameters in a thread "series." For the Unified National system, there are the coarse thread series and the fine thread series.

There is also an extra-fine thread series, UNEF, for use where a small pitch is wanted, such as on thin-walled tubing. For special work and diameters larger than those specified in the coarse and fine series, the Unified Thread system has three series that allow the same number of threads per inch regardless of the diameter. These are the 8-thread series, the 12-thread series, and the 16-thread series.

Three classes of external thread (Classes 1A, 2A, and 3A) and three classes of internal thread (Classes 1B, 2B, and 3B) are provided.

Classes 1A and 1B

These classes produce the loosest fit, that is, the most play (free motion) in assembly. They are useful for work where ease of assembly and disassembly is essential, such as for some automotive work and for stove bolts and other rough bolts and nuts.

Classes 2A and 2B

These classes are designed for the ordinary good grade of commercial products, including machine screws and fasteners and most interchangeable parts.

Classes 3A and 3B

These classes are intended for exceptionally high-grade commercial products needing a particularly close or snug fit. Classes 3A and 3B are used only when the high cost of precision tools and machines is warranted.

INCH THREAD DESIGNATION

Thread designation for both external and internal 60° inch threads is expressed in this order: nominal thread diameter in inches, a dash, the number of threads per inch, a space, the letter symbol of the thread series, a dash, the number and letter of the thread class symbol, a space follows by any qualifying information, such as the letters LH for left-hand threads, or the gauging system number, Figure 16–5. For example:

- Standard Unified External Screw Thread

  .250-20 UNC-2A

- Standard Unified Internal Screw Thread, Gauging System 21

  .500-20 UNF-2B (21)

For multiple start threads, the number of threads per inch is replaced by the following: pitch in inches (P), a dash, lead in inches (L), and the number of starts in inches. For example:

- Standard Unified External Multiple Start Screw Thread

  .750-.0625P-.1875L(3 STARTS)UNF-2A

- Standard Unified Internal Multiple Start Screw Thread, Gauging System 21

  .500-.050P-.100L(2 STARTS)UNF-2B (21)

Where decimal-inch sizes are shown for fractional-inch threads, they are shown to four decimal places (not showing zero in the fourth place). For example, a 1/2" thread would be shown as .500; and a 9/16" thread as .5625.

Due to established drawing practices, numbered sizes may be shown as the nominal thread diameter. Where a number is used, a three-place decimal inch equivalent, enclosed in parentheses, is placed after the number. For example:

- Standard Unified External Thread

  10 (.190)-32 UNF-2A

- Standard Unified Internal Thread

  5 (.125)-20 UNC-2B

Typical clearance holes and thread callouts are shown in Figure 16–6.

RIGHT- AND LEFT-HANDED THREADS

Unless designated otherwise, threads are right-handed. A bolt being threaded into a tapped hole would be turned in a right-hand (clockwise) direction. For some special uses, such as turnbuckles, left-hand threads are required. When such a thread is necessary, the letters "LH" are added to the thread designation.

METRIC THREADS

Metric threads are grouped into diameter-pitch combinations differentiated by the pitch applied to specified diameters. The pitch for metric threads is the distance between corresponding points on adjacent teeth. In addition to a coarse and fine pitch series, a series of constant pitches is available.

For each of the two main thread elements, pitch diameter and crest diameter, there are numerous tolerance grades. The number of the tolerance grade reflects the tolerance size. For example, grade 4 tolerances are smaller than grade 6 tolerances; grade 8 tolerances are larger than grade 6 tolerances.

In each case, grade 6 tolerances should be used for medium-quality length of engagement applications. The tolerance grades below grade 6 are intended for applications involving fine quality and/or short lengths of engagement. Tolerance grades above grade 6 are intended for coarse quality and/or long lengths of engagement.

In addition to the tolerance grade, positional tolerance is required. The positional tolerance defines the maximum-material limits of the pitch and crest diameters of the external and internal threads and indicates their relationship to the basic profile.

In conformance with current coating (or plating) thickness requirements and the demand for ease of assembly, a series of tolerance positions reflecting the application of varying amounts of allowance has been established.

For External Threads:
   Tolerance position "e" (large allowance)
   Tolerance position "g" (small allowance)
   Tolerance position "h" (no allowance)
For Internal Threads:
   Tolerance position "G" (small allowance)
   Tolerance position "H" (no allowance)

METRIC THREAD DESIGNATION

The two types of metric 60° screw threads in common use are the **M** and the **MJ** forms. The MJ form is similar to the standard metric (M) threads, except the sharp V at the root diameter of the external thread has been replaced with a large radius, which strengthens this stress point. Since the radius increases the minor diameter of the external thread, the minor diameter of the internal thread is enlarged to clear the radius. All other dimensions are the same as in the M threads. The MJ thread form is predominantly used in applications requiring high fatigue strength, as in the aerospace and automotive industries.

Metric 60° screw thead designation is expressed in this order: the metric thread symbol "M," the thread form symbol "J," where applicable, the nominal diameter in millimetres, a lowercase "x," the pitch in millimetres, a dash, the pitch diameter tolerance symbol, the crest diameter tolerance symbol (if different from that of the pitch diameter), and a space followed by any qualifying information, Figure 16–7. For example:

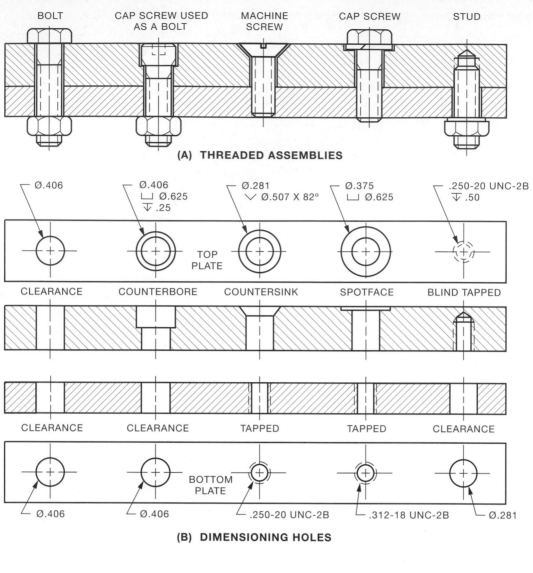

(A) THREADED ASSEMBLIES

BOLT CAP SCREW USED AS A BOLT MACHINE SCREW CAP SCREW STUD

Ø.406
Ø.406
⊔ Ø.625
▽ .25
Ø.281
⌄ Ø.507 X 82°
Ø.375
⊔ Ø.625
.250-20 UNC-2B
▽ .50

TOP PLATE

CLEARANCE COUNTERBORE COUNTERSINK SPOTFACE BLIND TAPPED

CLEARANCE CLEARANCE TAPPED TAPPED CLEARANCE

BOTTOM PLATE

Ø.406 Ø.406 .250-20 UNC-2B .312-18 UNC-2B Ø.281

(B) DIMENSIONING HOLES

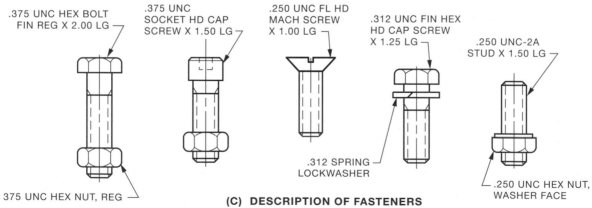

.375 UNC HEX BOLT FIN REG X 2.00 LG

.375 UNC SOCKET HD CAP SCREW X 1.50 LG

.250 UNC FL HD MACH SCREW X 1.00 LG

.312 UNC FIN HEX HD CAP SCREW X 1.25 LG

.250 UNC-2A STUD X 1.50 LG

375 UNC HEX NUT, REG

.312 SPRING LOCKWASHER

.250 UNC HEX NUT, WASHER FACE

(C) DESCRIPTION OF FASTENERS

FIGURE 16–6 ■ Common threaded fasteners

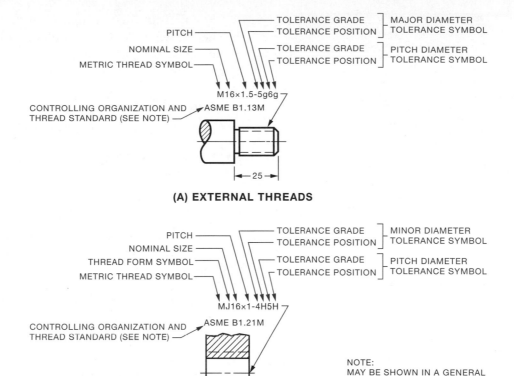

(A) EXTERNAL THREADS

(B) INTERNAL THREADS

NOTE:
MAY BE SHOWN IN A GENERAL
NOTE ON THE DRAWING.

FIGURE 16–7 ■ Metric Screw Thread Designation, Right-Hand Threads

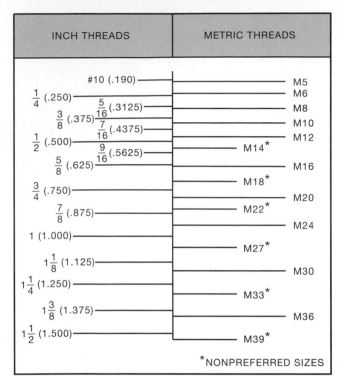

FIGURE 16–8 ■ Size comparison of inch and metric threads

- Standard Metric M Screw Thread

  M6x1-4h6h

- Standard Metric MJ Thread

  MJ6x1-4H

The metric thread size and the pitch should include a zero before the decimal if the value is less than one, and should not show a zero as the last number of the value. (e.g., M10x1.5 and MJ 2.5x0.45).

For multiple start threads the pitch is replaced by the following: L (lead in millimetres), P (pitch in millimetres), and the number of starts in parentheses. For example:

- Standard Metric MJ Thread

  MJ20xL7.5P2.5(3 STARTS)-4h6h

- Standard Metric M Multiple Start Thread, Gauging System 21

  M16xL4P2(2 STARTS)-6g (21)

An earlier metric thread designation, taken from European standards, is found on many older drawings still in use. The thread designation is identical for both external and internal threads, and expressed in this order: M denoting the ISO metric thread symbol, the nominal diameter in millimetres, a capital "X" followed by the pitch in millimetres. For the coarse thread series, the pitch was not shown. For example, a 10 mm diameter, 1.25 pitch, fine thread series was expressed as M10 X 1.25. A 10 mm diameter, 1.5 pitch, coarse thread series was expressed as M10. If the length of thread was added to the callout, the pitch was added to avoid confusion. When specifying the length of the thread in the callout, an "X" separates the length of the thread from the rest of the designation. In the latter example, if a thread length was 25 mm, the thread callout would be M10 X 1.25 X 25.

In addition to this basic designation, a tolerance class identification was often added. A dash separated the tolerance identification from the basic designation.

Figure 16–8 shows a comparison of metric threads and inch threads.

Pipe thread designation is explained in Unit 24.

REFERENCES

CAN3-B78.1-M83 (R1990) Technical Drawings— General Principles

ASME Y14.6-2001 SCREW THREAD REPRESENTATION

ASME B1.1-2003 Unified Inch Screw Threads (UN and UNR Thread Form)

ASME B1.13M-2001 Metric Screw Threads—M Profile

ASME B1.21M-1997 Metric Screw Threads—MJ Profile

INTERNET RESOURCES

Drafting Zone For information on drafting symbols and reference dimensioning, see: http://www.draftingzone.com

Fastener Hut For information on all types of fasteners, see: http://www.fastenerhut.com

Industrial Fasteners Institute For both general and specific information on fasteners, see: http://www.ifi-fasteners.org

Machine Design For information on threaded fasteners and joining processes, see *Machine Design*, *Fastening/Joining Reference* at: http://www.machinedesign.com

Metrication.com For information on metric threads, see: http://www.metrication/engineering

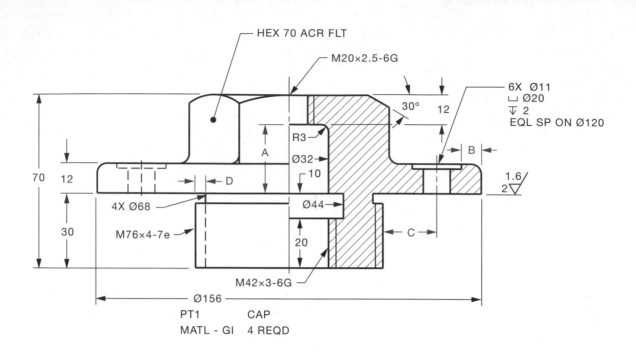

HEX 70 ACR FLT

M20×2.5-6G

6X Ø11
⊔ Ø20
↧ 2
EQL SP ON Ø120

30°
12

R3
Ø32
10
A

B

70
12
D

4X Ø68
Ø44

30
M76×4-7e

1.6
2 ▽

20
C

M42×3-6G

Ø156

PT1   CAP
MATL - GI   4 REQD

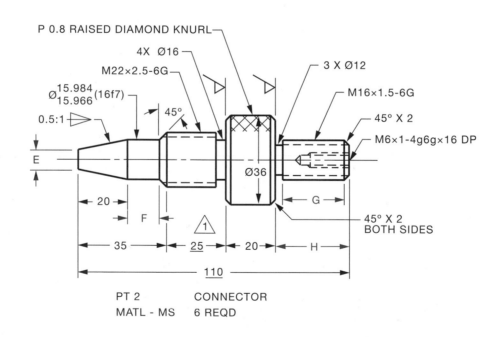

P 0.8 RAISED DIAMOND KNURL

4X Ø16

M22×2.5-6G

3 X Ø12

M16×1.5-6G

Ø $^{15.984}_{15.966}$ (16f7)

45°

45° X 2

0.5:1 ▷

M6×1-4g6g×16 DP

E

Ø36

20

F

G

35

25

20

H

45° X 2
BOTH SIDES

△ 1

110

PT 2   CONNECTOR
MATL - MS   6 REQD

ROUNDS AND FILLETS R4
EXCEPT WHERE OTHERWISE SHOWN

REVISIONS	1	27/05/03	R. HINES
		25 IN PT 2 WAS 30	

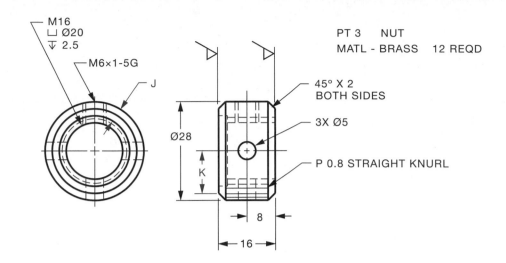

M16
⊔ Ø20
▽ 2.5

M6×1-5G

J

Ø28

K

8

16

PT 3   NUT
MATL - BRASS   12 REQD

45° X 2
BOTH SIDES

3X Ø5

P 0.8 STRAIGHT KNURL

QUESTIONS:

1. Calculate dimensions A to K.

Refer to Part 1

2. How many holes are in the part?

3. How many external threads are on the part?

4. How many internal threads are on the part?

5. What is the pitch for the M20 thread?

6. How many complete threads are on the M20 hole?

7. How many complete threads are on the M42 section?

8. What is the diameter of the spotface?

9. What is the angle between the Ø11 holes?

10. How much extra metal was provided on the surface that required machining?

Refer to Part 2

11. What is the pitch of the internal thread?

12. What is the pitch of the M22 thread?

13. What provides for better gripping when rotating the part?

14. How many undercuts are shown?

15. How many chamfers are required?

16. If the 16f7 fit was replaced with a 16g6 fit, what would be the limit dimensions?

Refer to Part 3

17. How many holes are in the part?

18. What is the depth of the counterbore?

19. How many full threads has the M16 hole?

20. What surface finish is required for the sides of the knurled portion?

21. What are the tap drill sizes required for the (A) M6 and (B) M16 threaded holes?

NOTE: UNLESS OTHERWISE SHOWN:
—TOLERANCES ON DIMENSIONS ±0.5
—TOLERANCES ON ANGLES ±30'

SURFACES  TO BE $\frac{3.2}{}$ ▽

THREAD CONTROLLING ORGANIZATION
AND STANDARD – ASME B1.13M-2001

METRIC
DIMENSIONS ARE IN MILLIMETRES

DRIVE SUPPORT
DETAILS

A-47M

QUESTIONS:

1. Calculate dimensions (A) to (J).
2. How many threaded holes or shafts are shown?
3. How many chamfers are shown?
4. How many necks are shown?
5. How many surfaces require finishing?

Refer to Part 1

6. What are the limit sizes for the Ø.64 dimension?
7. What operation provides better gripping when turning the clamping nut?
8. What is the tap drill size?
9. Is the thread right-hand or left-hand?
10. What is the depth of the threads?

Refer to Part 2

11. What is the length of the thread?
12. Is the thread pitch fine or coarse?
13. What heat treatment does the part undergo?
14. How many threads are there in a one-inch length?

NOTE: EXCEPT WHERE NOTED:

TOLERANCE ON TWO-PLACE
DIMENSIONS ±.02

TOLERANCE ON THREE-PLACE
DIMENSIONS ±.005

TOLERANCE ON ANGLES ±0.5°

$\frac{32}{\bigtriangledown}$ EXCEPT WHERE NOTED

THREAD CONTROLLING ORGANIZATION
AND STANDARD – ASME B1.1-2003

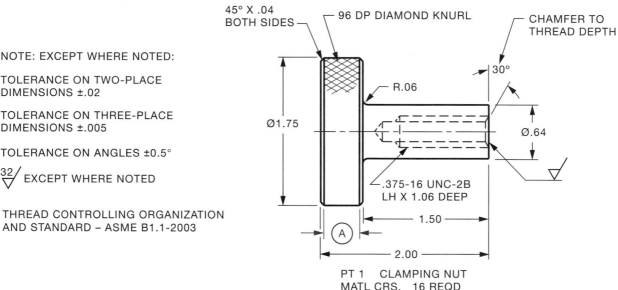

PT 1   CLAMPING NUT
MATL CRS,   16 REQD

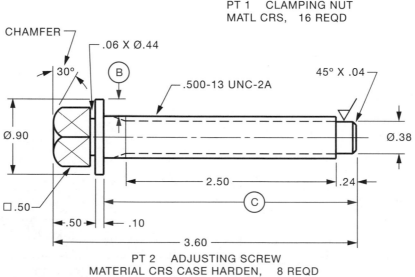

PT 2   ADJUSTING SCREW
MATERIAL CRS CASE HARDEN,   8 REQD

Refer to Part 3

15. What is the maximum permissible width of the part?

16. What size thread is cut on the underside of the piece?

17. What is the distance between the last thread and the flange?

18. What is the tolerance on the centre-to-centre distance between the tapped holes?

19. What is the smallest diameter to which the hole through the stuffing box can be made?

20. What are the limits of the Ø2.000 dimension?

21. What is the tap drill size for the threaded holes?

22. What would be the new limit dimensions for the Ø1.250 hole in Part 3 if an RC3 fit is required?

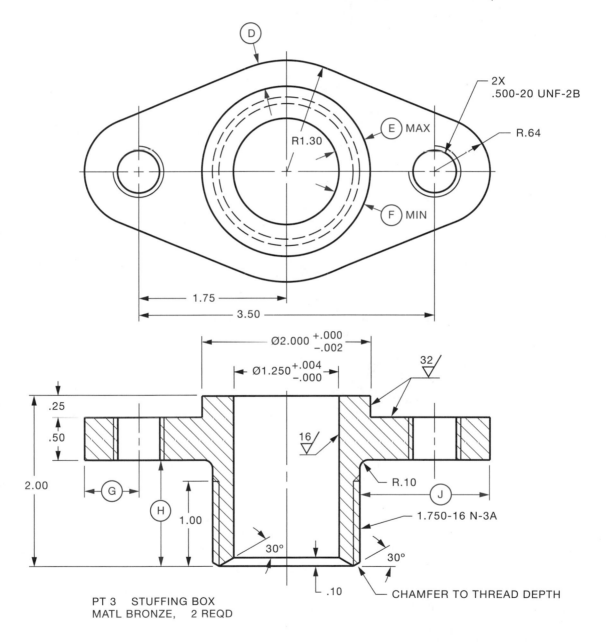

PT 3   STUFFING BOX
MATL BRONZE,   2 REQD

SCALE	NOT TO SCALE	
DRAWN	D. PERONI	DATE 14/02/04

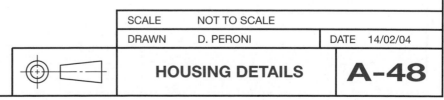

HOUSING DETAILS   **A-48**

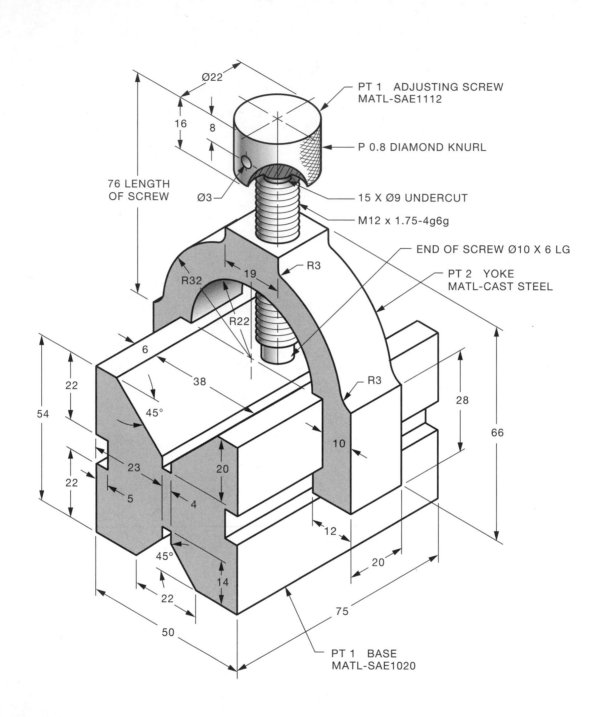

Ø22

16

8

76 LENGTH
OF SCREW

Ø3

PT 1   ADJUSTING SCREW
MATL-SAE1112

P 0.8 DIAMOND KNURL

15 X Ø9 UNDERCUT

M12 x 1.75-4g6g

END OF SCREW Ø10 X 6 LG

PT 2   YOKE
MATL-CAST STEEL

R32

R22

19

R3

6

38

45°

23

5

4

22

54

22

20

45°

22

14

50

R3

10

28

66

12

20

75

PT 1   BASE
MATL-SAE1020

ASSIGNMENT:
ON A CENTIMETRE GRID SHEET (1 mm SQUARES),
PREPARE DETAIL DRAWINGS OF THE PARTS
SHOWN IN THE V-BLOCK ASSEMBLY. A CLEARANCE
OF 1 mm FOR A SLIDING FIT IS TO BE ADDED TO THE
APPROPRIATE YOKE DIMENSIONS.

THREAD CONTROLLING ORGANIZATION
AND STANDARD – ASME B1.13M-2001

METRIC
DIMENSIONS ARE IN MILLIMETRES

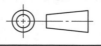

**V-BLOCK
ASSEMBLY**

A-49M

REVOLVED AND REMOVED SECTIONS

Unit 9 introduced full, half, and offset sectional views and how they are shown on drawings. For many drawings, only a portion of a complete view needs to be shown in section to improve their clarity. Two additional types of sectional views are introduced in this unit.

Revolved and removed sections are used to show the cross-sectional shape of ribs, spokes, or arms when the shape is not obvious in the regular views. End views are often not needed when a revolved section is used.

Revolved Sections

For a *revolved section*, a centre line is drawn through the shape on the plane to be described, the part is imagined to be rotated 90°, and the view that would be seen when rotated is superimposed on the view. If the revolved section does not interfere with the view on which it is revolved, then the view is not

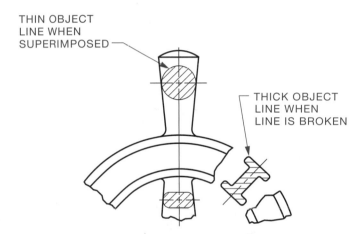

THIN OBJECT LINE WHEN SUPERIMPOSED

THICK OBJECT LINE WHEN LINE IS BROKEN

FIGURE 17–1 ■ Revolved section

broken unless it would facilitate clearer dimensioning. When the revolved section interferes with or passes through lines on the view on which it is revolved, the view is usually broken. Often the break is used to shorten the length of the object. When superimposed on the view, the outline of the revolved section is a thin continuous line, Figure 17–1.

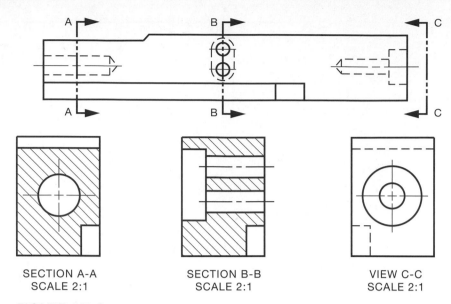

SECTION A-A
SCALE 2:1

SECTION B-B
SCALE 2:1

VIEW C-C
SCALE 2:1

FIGURE 17–2 ■ Removed sections and removed view

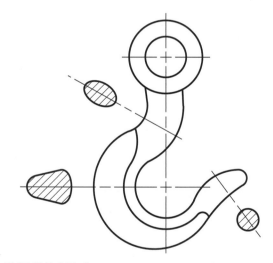

FIGURE 17–3 ■ Removed sectional views of crane hook

Removed Sections

The *removed section* differs from the revolved section in that the section is removed to an open area on the drawing instead of being drawn directly on the view. Whenever practical, sectional views should be projected perpendicular to the cutting plane and be placed in the normal position for third-angle projection, Figures 17–2, 17–3, 17–4, and 17–5. Frequently, the removed section is drawn to an enlarged scale for clarification and easier dimensioning.

Removed sections of symmetrical parts are placed on the extension of the centre line where possible.

Broken-out and partial sections are explained and illustrated in Unit 30.

REFERENCES

CAN3-B78.1-M83 (R1990) Technical Drawings—General Principles
ASME Y14.3M-1994 (R2003) Multi and Sectional View Drawings

INTERNET RESOURCES

American Society of Mechanical Engineers For information on revolved and removed sections, refer to ASME Y14.3M-1994 (R1999) (*Multi- and Sectional-View Drawings*) at: http://www.asme.org

Drafting Zone For information on revolved and removed sections, see: http://www.draftingzone.com

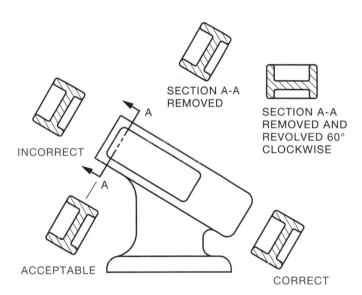

SECTION A-A REMOVED

SECTION A-A REMOVED AND REVOLVED 60° CLOCKWISE

INCORRECT

ACCEPTABLE

CORRECT

FIGURE 17–4 ■ Placement of removed sectional views

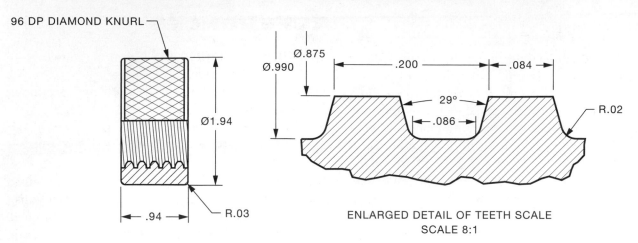

96 DP DIAMOND KNURL

Ø1.94

.94

R.03

Ø.875

Ø.990

.200

.084

29°

.086

R.02

ENLARGED DETAIL OF TEETH SCALE
SCALE 8:1

FIGURE 17–5 ■ Removed section of thread detail

ASSIGNMENT:
ON A ONE-INCH GRID SHEET (.10 IN. SQUARES),
SKETCH THE BOTTOM VIEW OF THE SHAFT
INTERMEDIATE SUPPORT. IF CERTAIN DIMENSIONS
CAN BE SHOWN BETTER ON THE BOTTOM VIEW,
DUPLICATE THEM. SCALE 1:1.

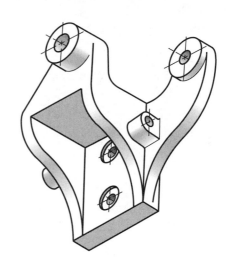

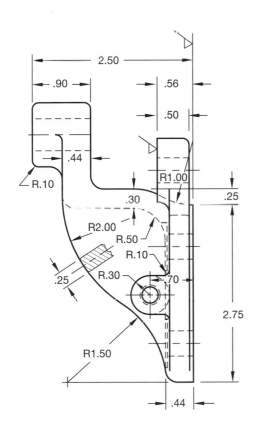

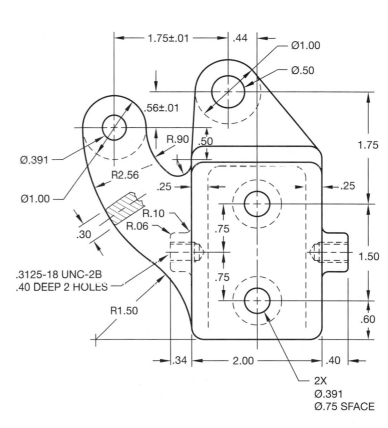

NOTES:
ROUNDS AND FILLETS R.25 EXCEPT
WHERE OTHERWISE NOTED.

THREAD CONTROLLING ORGANIZATION
AND STANDARD – ASME B1.1-2003

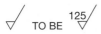 TO BE

MATERIAL	GI		
SCALE	1:1		
DRAWN	D. SMITH	DATE	22/10/04

SHAFT INTERMEDIATE SUPPORT

A-50

QUESTIONS:

1. Calculate distances (A) to (K).

2. How many surfaces require finishing?

3. What tolerance is permitted on dimensions that do not specify a certain tolerance?

4. What type of section view is used on Part 1?

5. What type of section view is used on Part 2?

6. How many holes are there?

Refer to Part 1

7. What is the maximum centre-to-centre distance between the holes?

8. Which surface finish is required?

9. Using the smallest permissible hole size as the basic size, replace the limit dimensions for the larger hole with plus-and-minus tolerances.

10. What is the maximum permissible wall thickness at the larger hole?

11. What is the minimum permissible wall thickness at the smaller hole?

12. What is the height of the shaft support casting before machining?

13. The size of the two holes is to be changed to accommodate an LN3 fit with mating bushings. What are the new limit dimensions for these holes?

Refer to Part 2

14. Name two machines that could produce the type of finish required for the slot.

15. What is the nominal size machine screw used in the counterbored holes?

16. What type of cap screw should be used if the heads of these screws must not protrude above the surface of the support?

17. What is the distance between the centre of the counterbored holes and the centre of the slot?

18. What type of tolerance is used on the (A) Ø1.250 hole and (B) 3.250 horizontal dimension?

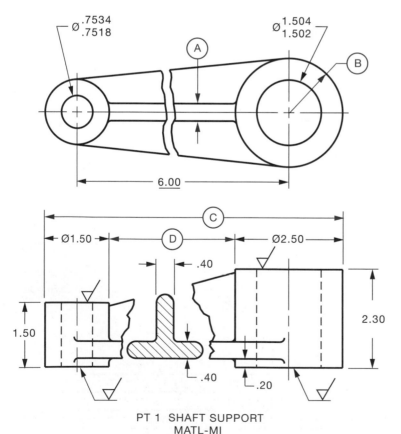

NOTE:
EXCEPT WHERE STATED OTHERWISE:
TOLERANCES ON
    TWO-DECIMAL DIMENSIONS ±.02
    THREE-DECIMAL DIMENSIONS ±.005
ROUNDS AND FILLETS R.10

MACHINE FINISH

PT 1 SHAFT SUPPORT
MATL-MI

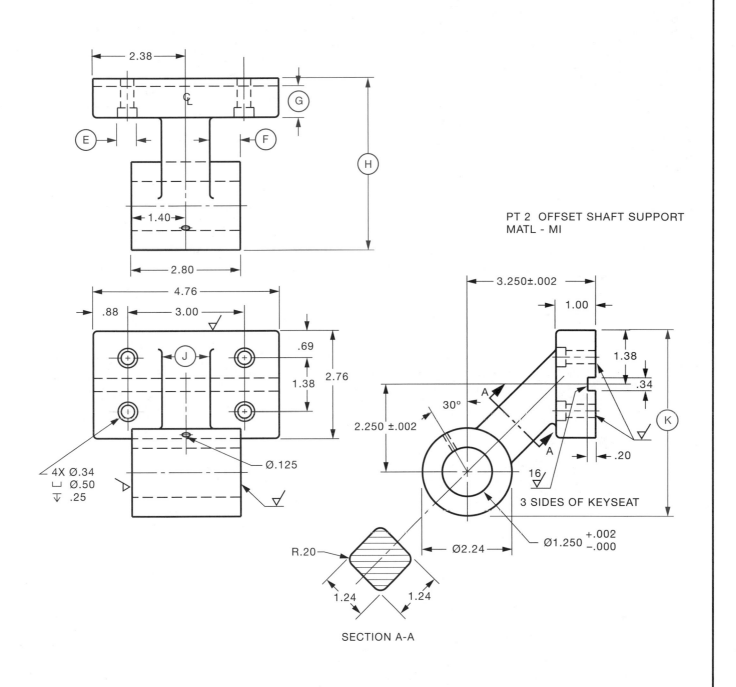

2.38

C L

E

F

G

H

1.40

2.80

4.76

.88

3.00

J

.69

1.38

2.76

4X Ø.34
⊔ Ø.50
▽ .25

Ø.125

PT 2 OFFSET SHAFT SUPPORT
MATL - MI

3.250±.002

1.00

1.38

.34

30°

A

A

2.250 ±.002

16

3 SIDES OF KEYSEAT

K

.20

Ø1.250 $^{+.002}_{-.000}$

R.20

1.24

1.24

Ø2.24

SECTION A-A

SCALE	NOT TO SCALE	
DRAWN	G. HIDER	DATE 16/11/04

SHAFT SUPPORTS

A-51

18 UNIT

KEYS

A *key* is a piece of metal lying partly in a groove in the shaft and extending into another groove in the hub, Figure 18–1. The groove in the shaft is referred to as a keyseat, and the groove in the hub or surrounding part as a keyway. Keys are used to secure gears, pulleys, cranks, handles, and similar machine parts to shafts, so that the motion of the part is transmitted to the shaft or the motion of the shaft is transmitted to the part without slippage. The key is also a safety feature. Because of its size, when overloading occurs, the key shears or breaks before the part or shaft breaks.

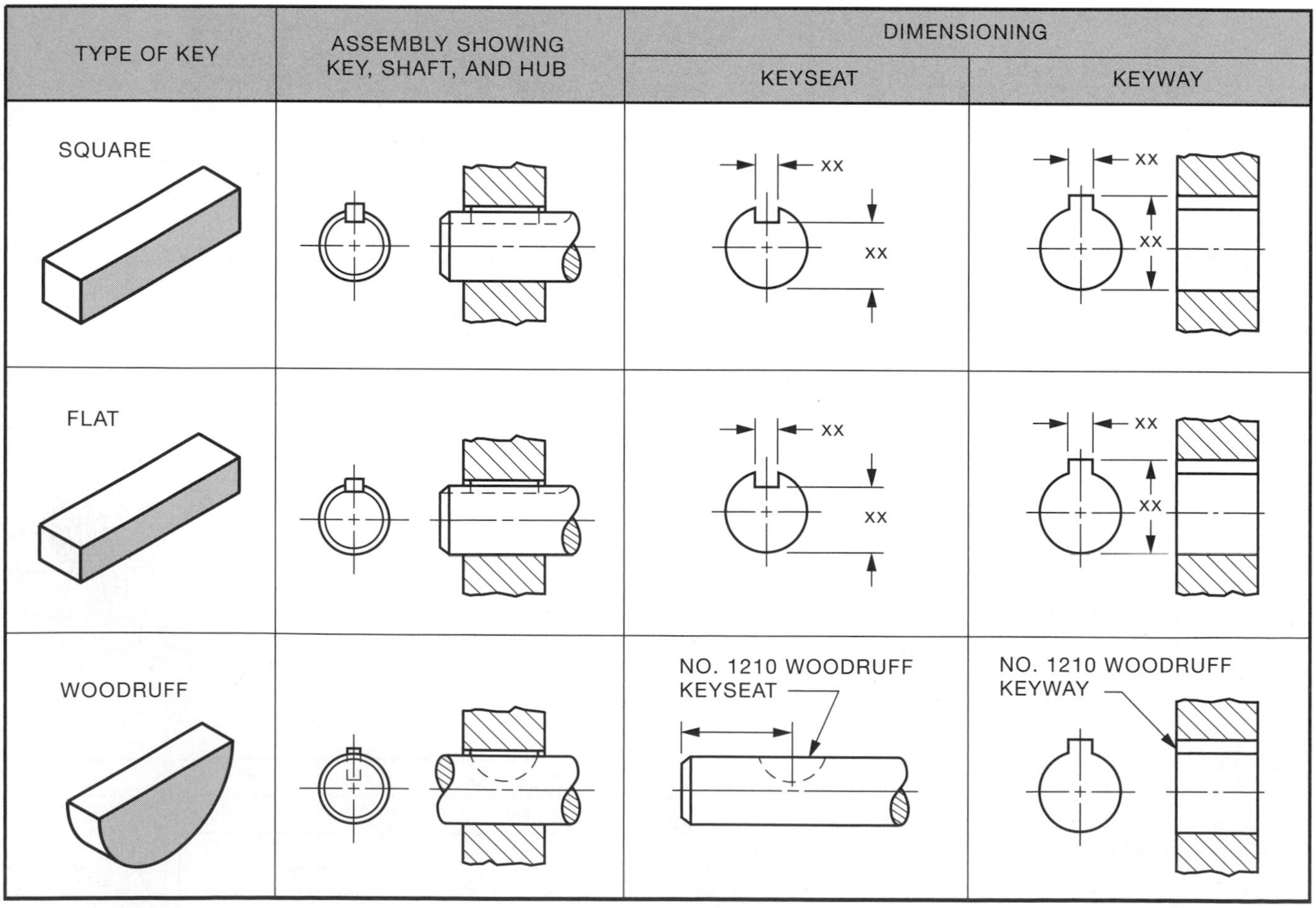

TYPE OF KEY
ASSEMBLY SHOWING KEY, SHAFT, AND HUB
DIMENSIONING
KEYSEAT
KEYWAY
SQUARE
XX
XX
XX
XX
FLAT
XX
XX
XX
XX
WOODRUFF
NO. 1210 WOODRUFF KEYSEAT
NO. 1210 WOODRUFF KEYWAY

FIGURE 18–1 ■ Common keys

Common key types are square, flat, and Woodruff. Tables 13 and 14 in the Appendix give standard square and flat key sizes recommended for various shaft diameters and the necessary dimensions for Woodruff keys.

The *Woodruff key* is semicircular and fits into a semicircular keyseat in the shaft and a rectangular keyway in the hub. Woodruff keys are available only in inch sizes, and are identified by a three- or four-digit key number. The last two numbers give the nominal diameter in eighths of an inch. The digit or digits preceding the last two digits gives the key width in thirty-second of an inch. For example, a No. 1210 Woodruff key indicates a key 12/32 inch wide by 10/8 (1.25) inches in diameter.

Dimensioning of Keyways and Keyseats

Keyway and keyseat dimensions, usually given in limit dimensions to ensure proper fits, are located from the opposite side of the hole or shaft, Figure 18–2. Alternatively, for unit production where the machinist is expected to fit the key into the keyseat, a leader pointing to the keyseat, specifying first the width and then the depth of the keyseat, may be used. See Part 2, Rack Details, Assignment A–54.

SET SCREWS

Set screws are used as semipermanent fasteners to hold a collar, sheave, or gear on a shaft against rotational or translational forces. In contrast to most fastening devices, the set screw is essentially a compression device. Forces developed by the screw point during tightening produce a strong clamping action, which resists relative motion between assembled parts. The basic problem in set screw selection is finding the best combination of set screw form, size, and point style providing the required holding power.

Set screws are categorized by their forms and the desired point style, Figure 18–3. Each of the standardized set screw forms is available in a variety of point styles. Selection of a specific form or point is influenced by functional and other considerations.

The selection of the type of driver, and thus the set screw form, is usually determined by factors other than tightening. Despite higher tightening ability, the protrusion of the square head is a major disadvantage. Compactness, weight saving, safety,

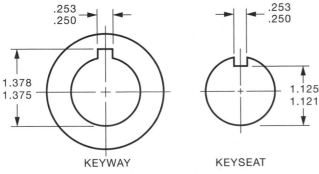

FIGURE 18–2 ■ Dimensioning of keyways and keyseats

STANDARD POINTS	
	CUP Most generally used. Suitable for quick and semipermanent location of parts on soft shafts, where cutting in of edges of cup shape on shaft is not objectionable.
	FLAT Used where frequent resetting is required, on hard steel shafts, and where minimum damage to shafts is necessary. Flat is usually ground on shaft for better contact.
	CONE For setting machine parts permanently on shaft, which should be spotted to receive cone point. Also used as a pivot or hanger.
	OVAL Should be used against shafts spotted, splined, or grooved to receive it. Sometimes substituted for cup point.
	HALF DOG For permanent location of machine parts, although cone point is usually preferred for this purpose. Point should fit closely to diameter of drilled hole in shaft. Sometimes used in place of a dowel pin.

STANDARD HEADS			
HEXAGON SOCKET	SLOTTED SOCKET	FLUTED SOCKET	SQUARE HEAD

FIGURE 18–3 ■ Set screws

and appearance may dictate the use of flush-seating socket or slotted headless forms.

The conventional approach to set screw selection is usually based on a rule-of-thumb procedure: the set screw diameter should be roughly equal to one-half the shaft diameter. This rule of thumb often gives satisfactory results, but it has a limited range of usefulness. When a set screw and key are used together, the screw diameter should be equal to the width of the key.

Standard set screws are shown in Table 9 of the Appendix.

FLATS

A *flat* is a slight depression usually cut on a shaft to serve as a surface on which the point of a set screw can rest when holding an object in place, Figure 18–4.

BOSSES AND PADS

A *boss* is a relatively small cylindrical projection above the surface of an object. A *pad* is a slight, noncircular projection above the surface of an object, Figure 18–5. Bosses and pads are generally found on castings and provide additional clearance or strength in the area where they are used. They also minimize the amount of machining required.

DIMENSION ORIGIN SYMBOL

This symbol is used to indicate that a toleranced dimension between two features originates from one of these features, Figures 18–6 and 18–7.

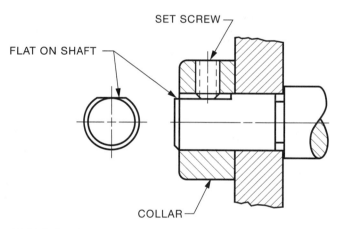

FIGURE 18–4 ■ Flat and set screw application

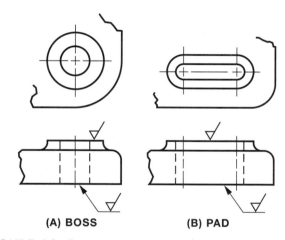

(A) BOSS   **(B) PAD**

FIGURE 18–5 ■ Bosses and pads

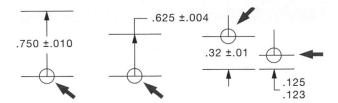

FIGURE 18–6 ■ Dimension origin symbol

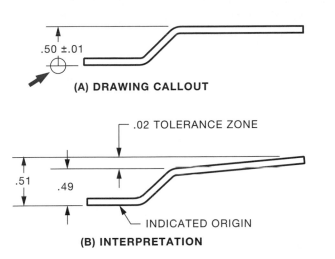

(A) DRAWING CALLOUT

.50 ±.01

.02 TOLERANCE ZONE

.51   .49

INDICATED ORIGIN

(B) INTERPRETATION

.02 TOLERANCE ZONE

.51   .49

LONGEST ARM USED AS ORIGIN

(C) INCORRECT INTERPRETATION

FIGURE 18–7 ■ Relating dimensional limits to an origin

RECTANGULAR COORDINATE DIMENSIONING WITHOUT DIMENSION LINES

To avoid having many dimensions extending away from the object, *rectangular coordinate dimensioning* without dimension lines may be used, Figure 18–8. In this system, the "zero" lines are used as reference lines and each of the dimensions shown without arrowheads indicates the distance from the zero line. There is never more than one zero line in each direction.

This type of dimensioning is particularly useful when such features are produced on a general-purpose machine, such as a jig borer, a numerically controlled drill, or a turret-type press. Drawings for numerical control are covered in Unit 32.

RECTANGULAR COORDINATE DIMENSIONING IN TABULAR FORM

When there are many holes or repetitive features, such as in a chassis or a printed-circuit board, and where the multitude of centre lines would make a drawing difficult to read, rectangular coordinate dimensioning in tabular form is recommended. In

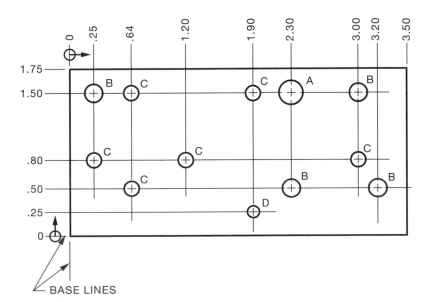

HOLE SYMBOL	HOLE Ø
A	.250
B	.188
C	.156
D	.125

FIGURE 18–8 ■ Rectangular coordinate dimensioning without dimension lines

this system, each hole or feature is assigned a letter or a letter with a numeral subscript. The feature dimensions and the feature location along the X, Y, and Z axes are given in a table, Figure 18–9.

REFERENCES

CAN3-B232-75 (R1996) Keys, Keyseats and Keyways
CAN/CSA-B78.2-M91 Dimensioning and Tolerancing
    of Technical Drawings

INTERNET RESOURCES

Drafting Zone For information on rectangular coordinate dimensioning, see: http://www.draftingzone.com

Machine Design For information on keys and set screws, see *Machine Design, Fastening/Joining Reference* at: http://www.machinedesign.com

Maryland Metric For information on dimensioning metric and inch keyways and keyseats see: http://www.mdmetric.com/tech/keywayspecify.pdf

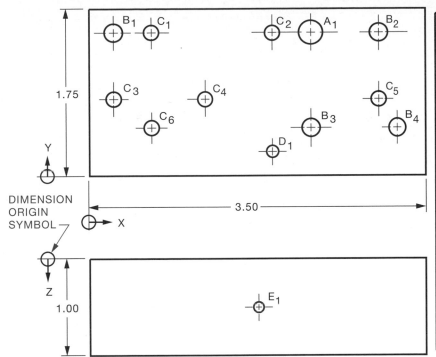

HOLE SYMBOL	Ø HOLE	LOCATION		
		X	Y	Z
A_1	.250	2.30	1.50	.62
B_1	.188	.25	1.50	THRU
B_2	.188	3.00	1.50	THRU
B_3	.188	2.30	.50	THRU
B_4	.188	3.20	.50	THRU
C_1	.156	.64	1.50	THRU
C_2	.156	1.90	1.50	THRU
C_3	.156	.25	.80	THRU
C_4	.156	.120	.80	THRU
C_5	.156	3.00	.80	THRU
C_6	.156	.64	.50	THRU
D_1	.125	1.90	.25	.50
E_1	.109	1.75	1.00	.50

FIGURE 18–9 ■ Rectangular coordinate dimensioning in tabular form

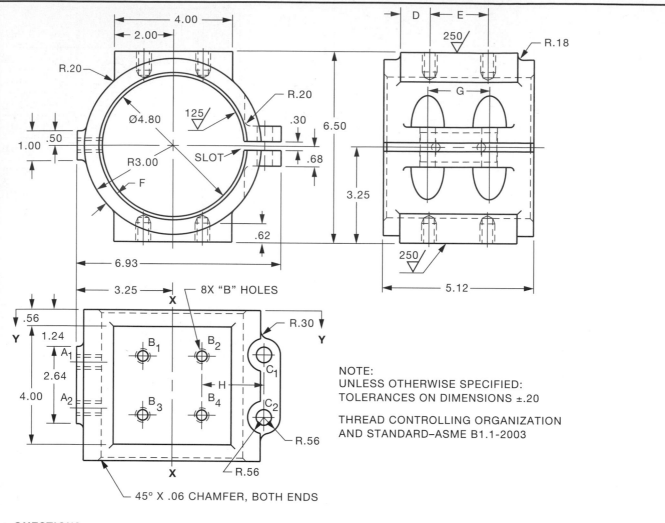

NOTE:
UNLESS OTHERWISE SPECIFIED:
TOLERANCES ON DIMENSIONS ±.20

THREAD CONTROLLING ORGANIZATION
AND STANDARD–ASME B1.1-2003

QUESTIONS:

1. What is the centre-to-centre distance between the following?
   (A) the A_1 and A_2 holes
   (B) the B_1 and B_2 holes
   (C) the C_1 and C_2 holes

2. What finish is required on the 4.00 x 4.00 face?

3. How far apart are the two surfaces of the slot?

4. How many surfaces require finishing?

5. What is the depth of the A holes?

6. What is the length of the threading in the B holes?

7. What size bolts would be used in the C holes if .031 clearance is used?

8. How many chamfers are called for?

9. How many tapped holes are there?

10. Calculate distances D, E, F, G, and H.

11. What is the tap drill size required for the (A) #10 and (B) .500 threaded holes?

HOLE	HOLE SIZE	LOCATION	
		X-X	Y-Y
A_1	10(.190)-24 UNC-2B		1.78
A_2	10(.190)-24 UNC-2B		3.32
B_1	.500-13 UNC-2B	1.00	1.56
B_2	.500-13 UNC-2B	1.00	1.56
B_3	.500-13 UNC-2B	1.00	3.56
B_4	.500-13 UNC-2B	1.00	3.56
C_1	Ø.531	3.12	1.50
C_2	Ø.531	3.12	3.62

MATERIAL	COPPER	
SCALE	NOT TO SCALE	
DRAWN	J. MEANS	

DATE 06/10/04

TERMINAL BLOCK

A-52

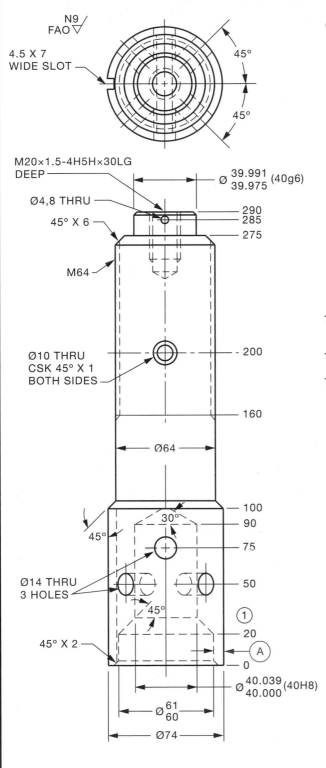

N9
FAO ▽

4.5 X 7
WIDE SLOT

45°

45°

M20×1.5-4H5H×30LG
DEEP

Ø4,8 THRU

45° X 6

M64

Ø $\frac{39.991}{39.975}$ (40g6)

290
285
275

Ø10 THRU
CSK 45° X 1
BOTH SIDES

200

160

Ø64

100
90
75

30°

45°

Ø14 THRU
3 HOLES

50

45°

①

20

45° X 2

Ⓐ

0

Ø $\frac{40.039}{40.000}$ (40H8)

Ø $\frac{61}{60}$

Ø74

NOTE: HOLES AND THREADS TO BE CLEAN AND BRIGHT

QUESTIONS:

1. How long is the Ø64 threaded section?

2. (A) How deep is the M20 thread?
   (B) How many full threads are in the tapped hole?

3. How many Ø14 holes are there?

4. What is the maximum thickness of the wall at the Ø14 holes?

5. What series of thread is required for the tapped hole?

6. What is the vertical centre distance between the top Ø14 hole and the Ø4.8 hole?

7. What is the length of the Ø64 unthreaded section?

8. What are the minimum and maximum thicknesses at Ⓐ?

9. What is the maximum overall diameter of the finished stud?

10. What is the tolerance on the largest hole at the bottom of the stud?

11. Give the quality of surface texture in micrometres.

12. What is the angle between the slot and the Ø14 hole located 75 from the base of the stud?

NOTE:
UNLESS OTHERWISE SPECIFIED:
—TOLERANCES ON DIMENSIONS ±0.5
—TOLERANCES ON ANGLES ±0.5°

THREAD CONTROLLING ORGANIZATION
AND STANDARD – ASME B1.13M-2001

METRIC
DIMENSIONS ARE IN MILLIMETRES

MATERIAL	COPPER	
SCALE	NOT TO SCALE	
DRAWN	T. FURMAN	DATE 22/05/04

REVISIONS	1	DN. T. FURMAN	CHK. C. JENSEN		TERMINAL STUD	A-53M
		DIMENSION WAS 24	11/10/04			

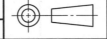

QUESTIONS:

Refer to Part 1

1. What was the original length of the shaft?

2. What symbol is used to indicate that the 1.90 dimension is not to scale?

3. At how many places are threads being cut?

4. Specify for any left-hand threads the thread diameter and the number of threads per inch.

5. What is the pitch for the 1.000 thread?

6. What is the length of that portion of the shaft that includes the (A) .875 thread (include chamfer), (B) 1.250 thread (do not include undercut), and (C) 1.000 thread (include chamfer)?

7. What distance is there between the last thread and the shoulder of the Ø.875 portion of the shaft?

8. How many dimensions have been changed?

9. What type of sectional view is used?

10. What is the largest size to which the Ø1.250 shaft can be turned?

11. What is the minimum permissible size for dimension A?

Refer to Part 2

12. How many holes are there?

13. What are the overall width and depth dimensions of the base?

14. How many surfaces are to be finished?

15. What is the diameter of the bosses on the base?

16. How wide is the pad on the upright column?

17. What is the depth of the keyseat?

18. How far does the horizontal hole overlap the vertical hole? (Use maximum sizes of holes and minimum centre-to-centre distances.)

19. How much material was added when the change to the bosses was made?

20. Which scale was used on this part?

21. What is the maximum permissible centre-to-centre distance of the two large holes?

22. If limit dimensions (refer to Tables 13 M and W of the Appendix) were to replace the keyseat dimensions shown, what would they be?

23. What dimension is indicated as not drawn to scale?

24. What was the overall height of the casting before finishing?

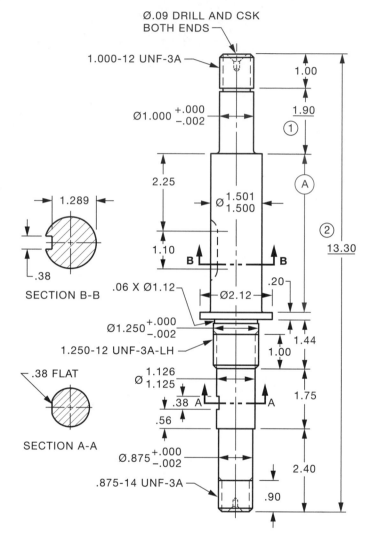

PT 1   SPINDLE SHAFT,   SCALE 1:2,   MATL - CRS
2 REQD

NOTES:

CHAMFER STARTING END OF ALL THREADS 45° TO THREAD DEPTH

ALL FILLETS R.10

THREAD CONTROLLING ORGANIZATION AND STANDARD – ASME B1.1-2003

REVISIONS	1	15/08/03	R.H.	2	12/10/03	C.J.	3	07/01/04	G.H.
		1.90 WAS 2.00			13.30 WAS 13.40			.80 WAS .75	

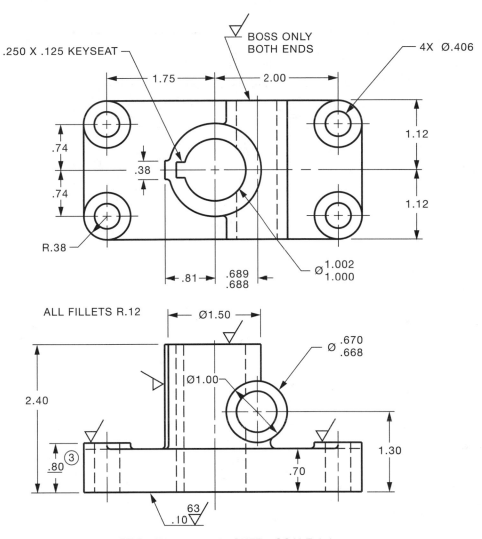

.250 X .125 KEYSEAT

BOSS ONLY
BOTH ENDS

4X Ø.406

1.75

2.00

1.12

.74

.38

.74

1.12

R.38

.81

.689
.688

Ø 1.002
1.000

ALL FILLETS R.12

Ø1.50

Ø .670
.668

2.40

Ø1.00

1.30

.80 ③

.70

63
.10

PT 2   COLUMN BRACKET,   SCALE 1:1
MATL - WROUGHT IRON   4 REQD

NOTES:
UNLESS OTHERWISE SPECIFIED:
TOLERANCE ON DIMENSIONS ±.02

TO BE .06 63

SCALE	AS SHOWN	
DRAWN	D. SMITH	DATE 24/06/03
	RACK DETAILS	A-54

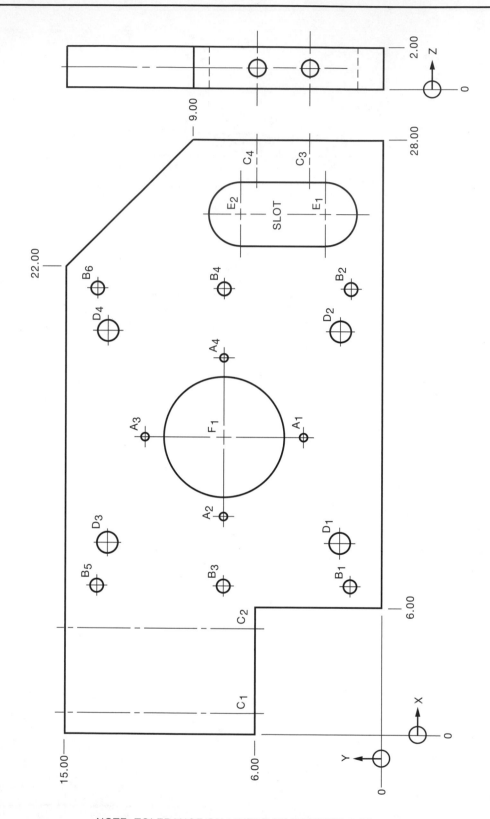

NOTE: TOLERANCE ON LINEAR DIMENSIONS ±.02

QUESTIONS:

1. What is the width of the part?

2. What is the height of the part?

3. What are the width and height of the chamfer on the corner?

4. What is the width of slot E?

5. What is the height of slot E?

6. How much wood is left between slot E and the right-hand edge of the part?

7. What are the centre distances between holes?

   (A) B_5 and B_6 (D) D_3 and D_4 (F) C_1 and C_2

   (B) D_2 and D_4 (E) A_1 and A_3 (G) B_3 and A_4

   (C) B_1 and B_5

8. How much wood is left between holes?

   (A) A_1 and E_1 (C) C_1 and C_2

   (B) A_2 and B_3 (D) B_3 and B_4

9. How many places is the origin symbol used?

HOLE SYMBOL	HOLE DIA	LOCATION		
		X	Y	Z
A_1	.375	14.00	3.75	
A_2	.375	10.25	7.50	
A_3	.375	14.00	11.25	
A_4	.375	17.75	7.50	
B_1	.625	7.00	1.50	
B_2	.625	21.00	1.50	
B_3	.625	7.00	7.50	
B_4	.625	21.00	7.50	
B_5	.625	7.00	13.50	
B_6	.625	21.00	13.50	
C_1	.812	1.00		1.00
C_2	.812	5.00		1.00
C_3	.812		3.50	1.00
C_4	.812		6.00	1.00
D_1	1.000	9.00	2.00	
D_2	1.000	19.00	2.00	
D_3	1.000	9.00	13.00	
D_4	1.000	19.00	13.00	
E_1	3.000	24.50	2.75	
E_2	3.000	24.50	6.75	
F_1	5.688	14.00	7.50	

NOTE: ALL HOLES THRU UNLESS OTHERWISE SPECIFIED

MATERIAL	DRY MAPLE	
SCALE	NOT TO SCALE	
DRAWN	D. GRAY	
		DATE 15/09/04

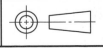

SUPPORT BRACKET

A-55

19 UNIT

OBLIQUE SURFACES

When a surface is sloped so that it is not parallel to any of the six principal orthographic views or planes, it will appear as a surface in all normal views but never in its true shape. This is referred to as an *oblique* surface, Figure 19–1.

In many cases, the exact shape of an oblique surface is not necessary. This is especially true if there is no detail (e.g., hole, slot) on the oblique surface that requires size or location dimensions. However, if a true representation of an oblique surface is required, two successive auxiliary views, primary and secondary, need to be projected and developed.

Auxiliary views are a type of orthographic projection used to develop the true size and shape of inclined and oblique surfaces of objects by projecting them onto imaginary auxiliary planes. These auxiliary planes are not parallel to any of the normal planes, but are parallel to either the primary or secondary auxiliary surface. Auxiliary views are explained in greater detail in Units 20 and 21. Figure 19–2 shows additional examples of parts having oblique surfaces.

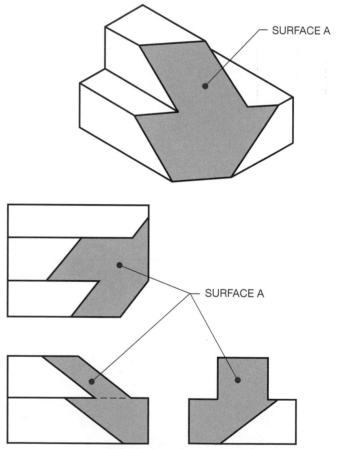

FIGURE 19–1 ■ Surface A is not parallel to any of the three normal planes and is therefore considered oblique

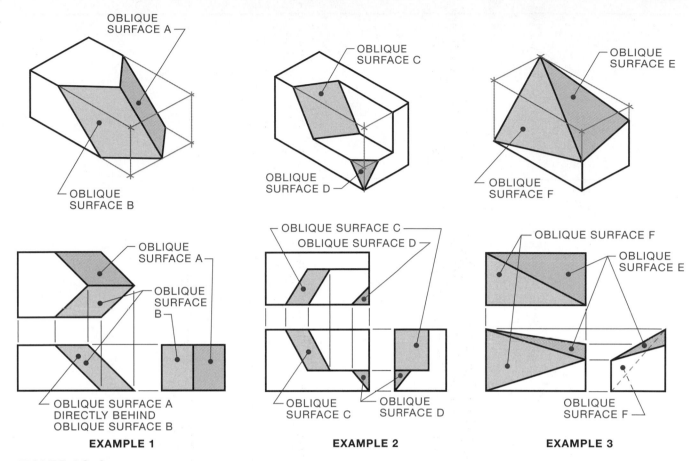

FIGURE 19–2 ■ Parts with oblique surfaces

REFERENCES

CAN3-B78.1-M83-Technical Drawings—General Principles

ASME Y14. 3M-1994 (R2003) Multi- and Sectional-View Drawings

INTERNET RESOURCE

Michigan Tech For additional information and practice on oblique surfaces, see http://www.geneng.mtu.edu/courses/1102/current/g04_teaming_oblique.pdf

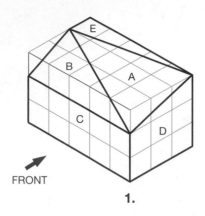

1.

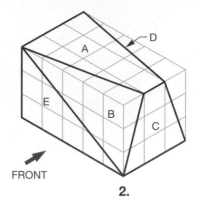

2.

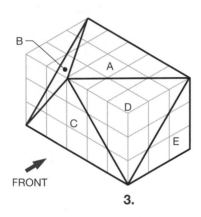

3.

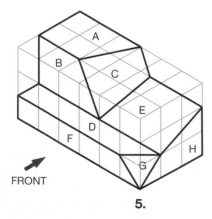

4.

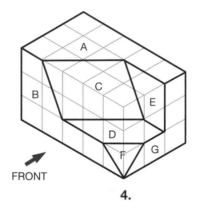

5.

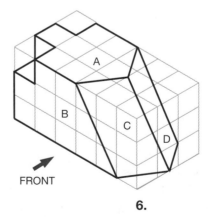

6.

ASSIGNMENT:

ON A ONE-INCH GRID SHEET (.25 IN. SQUARES) AND USING
THE SAME NUMBER OF SQUARES SHOWN ON THE PARTS,
SKETCH THE FRONT, TOP, AND SIDE VIEWS. IDENTIFY THE
OBLIQUE SURFACES ON THE THREE VIEWS WITH THE
APPROPRIATE LETTERS.

| IDENTIFYING OBLIQUE SURFACES | A-56 |

FIGURE 19–3 ■ Examples of auxiliary-view drawings

ARROW INDICATES DIRECTION OF FRONT VIEW.

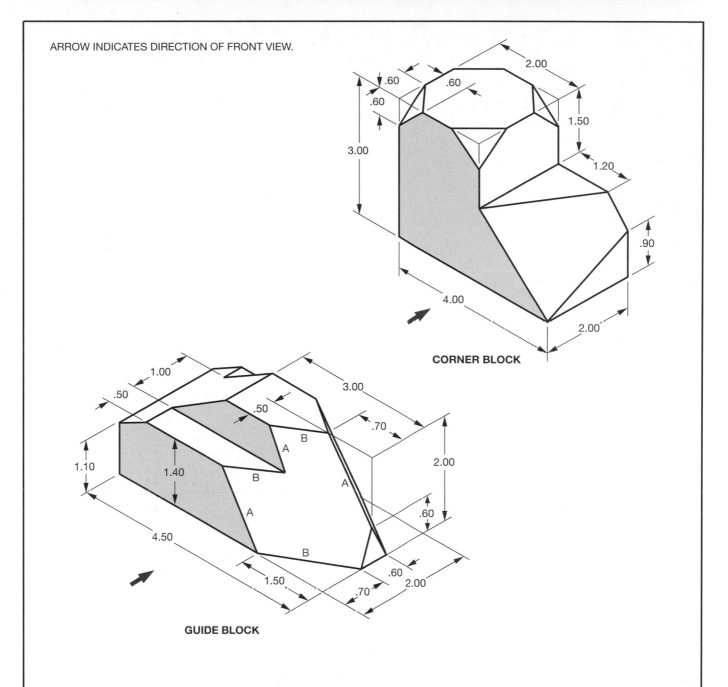

CORNER BLOCK

GUIDE BLOCK

ASSIGNMENT:
ON A 1.00-IN. GRID SHEET (.10 IN. SQUARES), MAKE A
THREE-VIEW DRAWING OF ONE OF THE PARTS SHOWN
ABOVE. ALLOW 1.00 IN. BETWEEN VIEWS. USING
LETTERS, IDENTIFY THE OBLIQUE SURFACES ON ALL
THREE VIEWS.

NOTE: LINES MARKED "A" ARE PARALLEL
LINES MARKED "B" ARE PARALLEL

| COMPLETING OBLIQUE SURFACES | A-57 |

20 UNIT

PRIMARY AUXILIARY VIEWS

Many objects have surfaces that are perpendicular to only one plane of projection. These surfaces are called *inclined sloped surfaces*. In the remaining two orthographic views, such surfaces appear to be foreshortened, and their true shape is not shown, Figure 20–1. When an inclined surface has important characteristics that must be shown clearly and without distortion, an auxiliary view is used to completely explain the shape of the object.

One of the regular views will have a true-length line representing the edge of the inclined surface. The auxiliary view is projected from this edge line, at right angles, and is drawn parallel to this edge line.

For example, Figure 20–2 clearly shows why an auxiliary view is required. The circular features on the sloped surface on the front view cannot be seen in their true shape on either the top or side view. The auxiliary view is the only view that shows the actual shape of these features, Figure 20–3. Note that only the sloped surface details are shown. Background detail is often omitted on auxiliary views and regular views to simplify the drawing and avoid confusion. A break line is used to signify the break in an incomplete view. The break line is not required if only the exact surface is drawn for either an auxiliary view or a partial regular view.

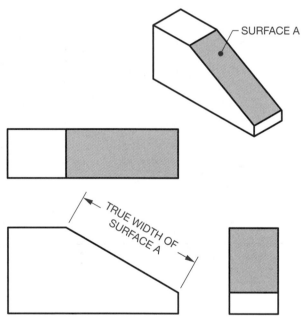

FIGURE 20–1 ■ Object with an inclined surface

This procedure is recommended for functional and production drafting, when drafting costs are an important consideration. However, complete views of the part are often used on catalogue and standard parts drawings.

One of the basic rules for dimensioning is to dimension the feature where it can be seen in its true shape and size. Thus the auxiliary view should show only dimensions pertaining to features for which it was drawn.

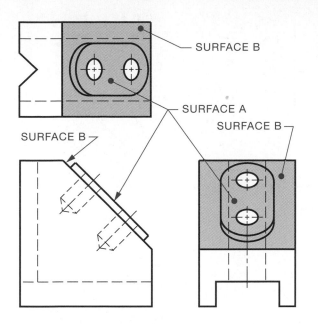

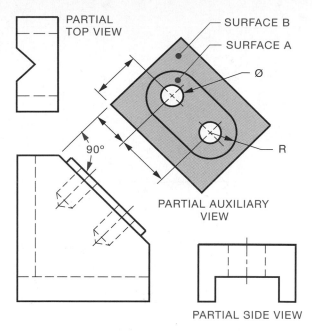

(A) REGULAR VIEWS DO NOT SHOW TRUE FEATURES OF SURFACES A AND B.

(B) AUXILIARY VIEW ADDED TO SHOW TRUE FEATURES OF SURFACES A AND B.

FIGURE 20–2 ■ The need for auxiliary views

REFERENCES

CAN3-B78.1-M83 (R1990) Technical Drawings—General Principles

ASME Y14.3M-1994 (R2003) Multi and Sectional View Drawings

INTERNET RESOURCE

Wikipedia, the Free Encyclopedia For information on auxiliary views, see: http://en.wikipedia.org/wiki/Orthographic_projection

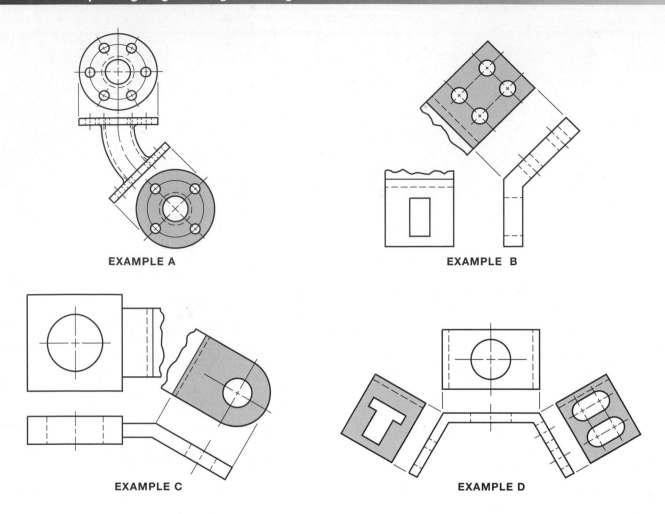

EXAMPLE A

EXAMPLE B

EXAMPLE C

EXAMPLE D

NOTE: ONLY CONVENTIONAL BREAK OR PROJECTED SURFACE
NEED BE SHOWN ON PARTIAL VIEWS.

FIGURE 20–3 ■ Auxiliary-view drawings

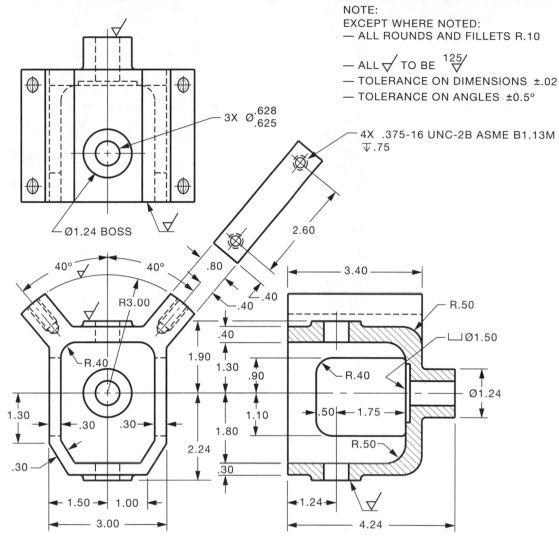

NOTE:
EXCEPT WHERE NOTED:
— ALL ROUNDS AND FILLETS R.10
— ALL ∇ TO BE 125/ ∇
— TOLERANCE ON DIMENSIONS ±.02
— TOLERANCE ON ANGLES ±0.5°

3X Ø .628 / .625

4X .375-16 UNC-2B ASME B1.13M ▽.75

Ø1.24 BOSS

40°   40°

R3.00

.80   .40   .40

R.40

1.30   .30   .30

.30

1.50   1.00

3.00

2.24

1.90   1.30   .90   1.10   1.80   .30

3.40

R.50

⌴Ø1.50

R.40

Ø1.24

.50   1.75

R.50

1.24

4.24

QUESTIONS:

1. What is the diameter of the bosses?

2. What is the tolerance on the holes in the bosses?

3. What are the width and height of the cutouts in the sides of the box?

4. What are the width and depth of the legs of the box?

5. How many degrees are there between the legs of the box?

6. What is the maximum surface roughness in microinches permitted on the machined surfaces?

7. How many surfaces are machined?

8. What would be the inside diameter of the mating part that this box fits into?

9. If the mating part is .44 thick and a lockwasher is used, what diameter and longest standard length of socket head cap screws could be used to fasten the parts together? (See Tables 7 and 12 in the Appendix.) Lengths available in .25 in. increments.

10. What is the thickness of (A) the side walls of the box, (B) the top of the box, and (C) the bottom of the box? Disregard the bosses.

11. Give overall inside dimensions of the box.

12. What are the overall outside dimensions of the box? (Do not include legs or bosses.)

13. Of what material is the box made?

14. How many screws are required to fasten the box to the mating part?

15. What is the tap drill size required for the four threaded holes?

THREAD CONTROLLING ORGANIZATION
AND STANDARD – ASME B1.1-2003

MATERIAL	GREY IRON	
SCALE	NOT TO SCALE	
DRAWN	B. LONG	DATE 07/09/04

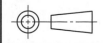

GEAR BOX   **A-58**

ASSIGNMENT:
ON A ONE-INCH GRID SHEET (.10 IN. SQUARES), SKETCH A PARTIAL
RIGHT-SIDE VIEW AND AN AUXILIARY VIEW OF THE INCLINED STOP.
THE DRAWING BELOW SHOWS THE VIEWING DIRECTION FOR
THESE VIEWS. ADD APPROPRIATE DIMENSIONS. SCALE 1:1.

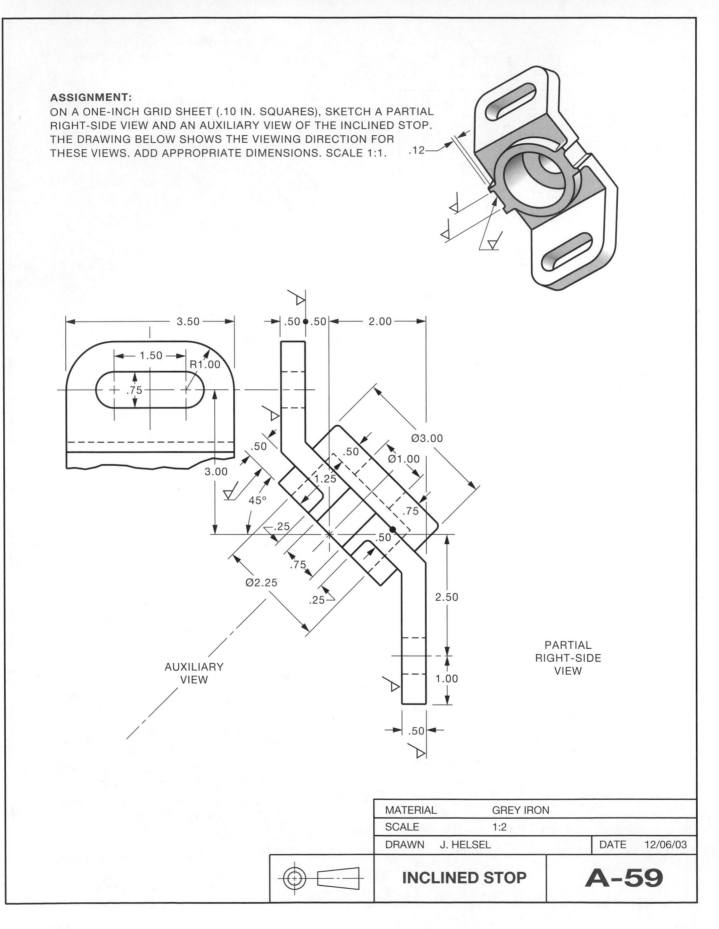

AUXILIARY
VIEW

PARTIAL
RIGHT-SIDE
VIEW

MATERIAL	GREY IRON	
SCALE	1:2	
DRAWN J. HELSEL		DATE 12/06/03

INCLINED STOP   **A-59**

NEL

SECONDARY AUXILIARY VIEWS

As mentioned in Unit 20, auxiliary views show the true lengths of lines and the true shapes of surfaces that cannot be described in the ordinary views.

A primary auxiliary view is drawn by projecting lines from a regular view where the inclined surface appears as an edge.

The auxiliary view in Figure 21–1(A), which shows the true projection and true shape of surface X, is called a *primary auxiliary view* because it is projected directly from the regular front view.

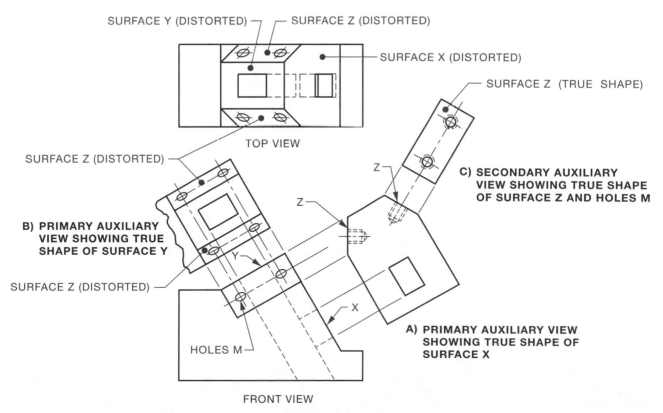

SURFACE Y (DISTORTED) — SURFACE Z (DISTORTED)

SURFACE X (DISTORTED)

SURFACE Z (TRUE SHAPE)

TOP VIEW

SURFACE Z (DISTORTED)

Z

C) SECONDARY AUXILIARY VIEW SHOWING TRUE SHAPE OF SURFACE Z AND HOLES M

B) PRIMARY AUXILIARY VIEW SHOWING TRUE SHAPE OF SURFACE Y

Z

SURFACE Z (DISTORTED)

Y

X

A) PRIMARY AUXILIARY VIEW SHOWING TRUE SHAPE OF SURFACE X

HOLES M

FRONT VIEW

NOTE: MANY HIDDEN LINES ARE OMITTED FOR CLARITY

FIGURE 21–1 ■ Primary and secondary auxiliary views

Some surfaces are inclined so that they are not perpendicular to any of the three viewing planes. In this case, they appear as a surface in all three views but never in their true shape. These are called *oblique surfaces*. Surface Z shown in Figure 21–1 is an oblique surface. To show the true shape of surface Z, and the true shape and location of holes M located on surface Z, a secondary auxiliary view must be shown, as at (C). This auxiliary view is projected from the first or primary auxiliary view, and is known as a *secondary auxiliary view*. The view at (B) is a primary auxiliary view because it is projected from one of the regular views. The side view is not drawn because the auxiliary views provide the information usually shown on this view.

REFERENCES

CAN3-B78.1-M83 (R1990) Technical Drawings—General Principles
ASME Y14.3M-1994 (R2003) Multi and Sectional View Drawings

INTERNET RESOURCE

Drafting Zone For more information on secondary auxiliary views, see: http://www.draftingzone .com

ASSIGNMENT

A-60 Hexagon Bar Support
A-61 Control Block

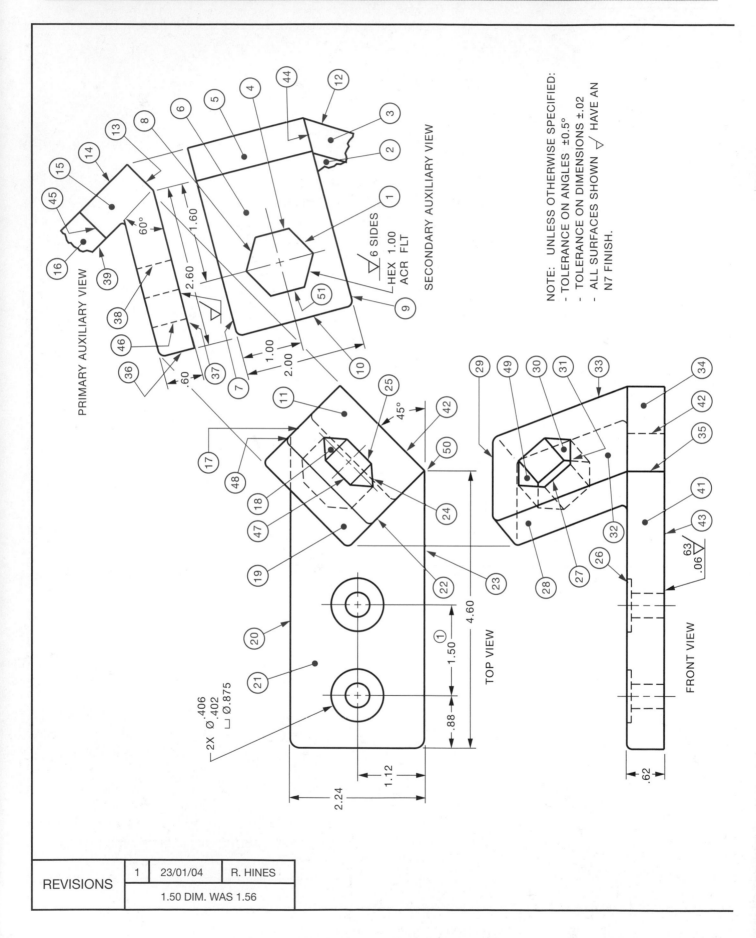

PRIMARY AUXILIARY VIEW

SECONDARY AUXILIARY VIEW

60°

6 SIDES
HEX 1.00
ACR FLT

45°

TOP VIEW

FRONT VIEW

2X Ø .406
Ø .402
⌴ Ø .875

2.24

1.12

.88

1.50 ①

4.60

.62

1.00
2.00
2.60
1.60
.60

.06
63

NOTE: UNLESS OTHERWISE SPECIFIED:
- TOLERANCE ON ANGLES ±0.5°
- TOLERANCE ON DIMENSIONS ±.02
- ALL SURFACES SHOWN ▽ HAVE AN
  N7 FINISH.

REVISIONS	1	23/01/04	R. HINES
		1.50 DIM. WAS 1.56	

ASSIGNMENT:

A FEATURE IS IDENTIFIED IN ONE OF THE VIEWS BY A NUMBER. PREPARE A CHART SIMILAR TO THE ONE SHOWN AND IDENTIFY THE FEATURE IN THE OTHER VIEWS BY ADDING THE APPROPRIATE NUMBERS TO THE CHART.

QUESTIONS:

1. Which other view(s) show the true height of the .62 dimension shown in the front view?

2. Which other view(s) show the true width of the .60 dimension shown in the primary auxiliary view?

3. What view(s) show the (A) height, (B) depth, and (C) width of the top portion of the support?

4. How many surfaces require finishing?

5. List the surface(s) or feature(s) that are shown in their true shape or size in the primary auxiliary view but are distorted in all other views.

6. List the surface(s) or feature(s) that are shown in their true shape or size in the secondary auxiliary view but are distorted in all other views.

7. What would be the thickness of the base ㉖ before machining?

8. If the hexagon bar support was fastened to another member (refer to the Appendix when necessary):

   (A) How many cap screws would be used?

   (B) What would be the cap screw size?

   (C) What would be the cap screw length if the supporting member had tapped holes that were .80 deep and flat washers were used under the cap screw heads?

   (D) If the cap screws were of the fine-thread series, how would you call out these cap screws?

   (E) What would be the I.D. and O.D. of the flat washers used under these cap screws?

Front View	Top View	Primary Auxiliary View	Secondary Auxiliary View
26	23	—	1
		—	
—	20	38	9
33		—	—
	—	14	5
35	19	37	8
42	47	—	—

MATERIAL	GREY IRON	
SCALE	NOT TO SCALE	
DRAWN	S. KLINGER	DATE 15/10/03

HEXAGON BAR SUPPORT

A-60

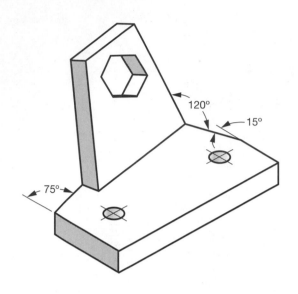

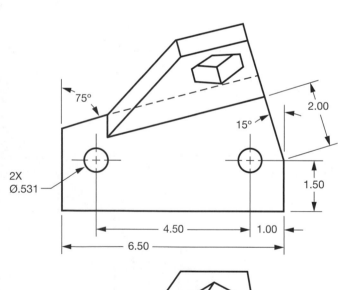

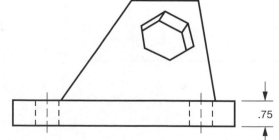

PRIMARY AUXILIARY VIEW

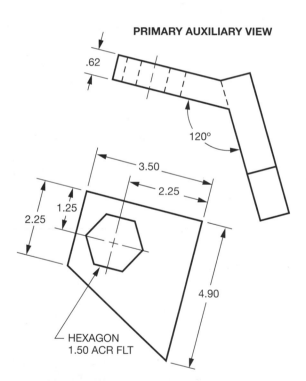

120°

3.50

2.25

1.25

2.25

4.90

HEXAGON
1.50 ACR FLT

SECONDARY AUXILIARY VIEW

NOTE: MANY UNNECESSARY HIDDEN
LINES ARE OMITTED FOR CLARITY.

ASSIGNMENT:
THE HEXAGONAL HOLE HAS BEEN REPLACED BY THE
TRIANGULAR HOLE AS SHOWN IN THE CONTROL BLOCK
BELOW. PHOTOCOPY THE DRAWING AND ACCURATELY
SKETCH ON THIS DRAWING THE SIZE AND POSITION
OF THE HOLE IN THE OTHER VIEWS. LEAVE THE
CONSTRUCTION LINES ON YOUR DRAWING.

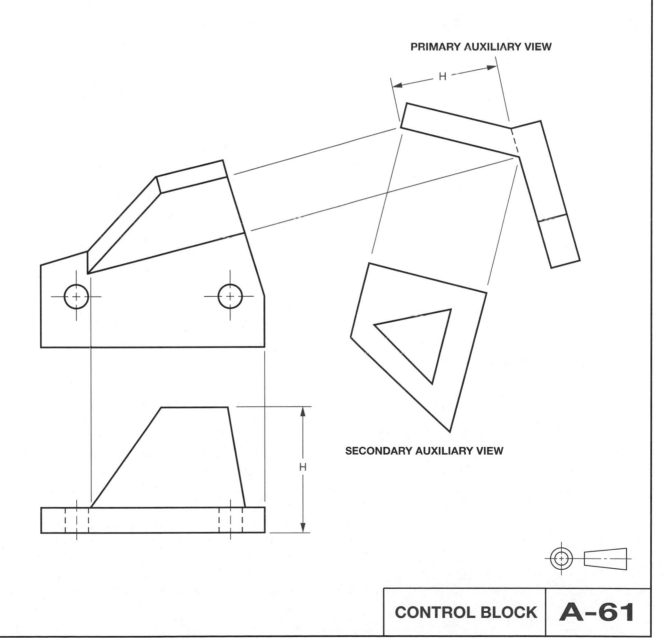

PRIMARY AUXILIARY VIEW

H

SECONDARY AUXILIARY VIEW

H

CONTROL BLOCK | A-61

22 UNIT

DEVELOPMENT DRAWINGS

Many objects, such as metal and cardboard boxes, ductwork for heating, funnels, gutters, and downspouts, are made from sheet material that is cut so that, when folded, formed, or rolled, it will take the shape of the object. Because a definite shape and size are desired, a regular orthographic drawing of the object is first made; then a development drawing is made to show the complete surface or surfaces laid out in a flat plane, Figure 22–1.

A development drawing is sometimes referred to as a pattern drawing, because the layout, when made of heavy cardboard, metal, or wood, is used as a pattern for tracing out the developed shape on flat material. Such patterns are used extensively in sheet metal shops.

JOINTS, SEAMS, AND EDGES

Additional material is required for assembly and design. When two or more pieces of material or surfaces are joined, extra material must be provided for the joint or seam. The type of joint or seam for joining metal, Figure 22–2, depends on such design criteria as strength, waterproofing, and appearance. Rivets or solder can be added to these joints if required. Exposed edges of metal parts may also be reinforced by the addition of extra material for hemming or for containing a wire. A round metal wastebasket, Figure 22–3, for example, requires extra material for the joint, seam, and edge.

SHEET METAL SIZES

Metal thicknesses up to .25 in. (6 mm) are usually designated by a series of gauge numbers. The more common gauges are shown in Table 16 of the Appendix. Metal .25 in. and over is given in inch or millimetre sizes. In calling for the material size of sheet metal developments, customary practice is to give the gauge number, its type of gauge, and the inch or millimetre equivalent in parentheses, followed by the developed width and length, Figure 22–4.

STRAIGHT LINE DEVELOPMENT

This is the term given to the development of an object that has surfaces on a flat plane of projection. The true size of each side of the object is known, and these sides can be laid out in successive order. Figure 22–1 shows the development of a simple rectangular box having a bottom and four sides. In the development of the box, an allowance is made for lap seams at the corners and for folded edges around the top. All lines for each surface are parallel or perpendicular to the other surfaces. The bottom corners of the lap joints are chamfered to facilitate assembly. The fold lines on the development are shown as thin unbroken lines.

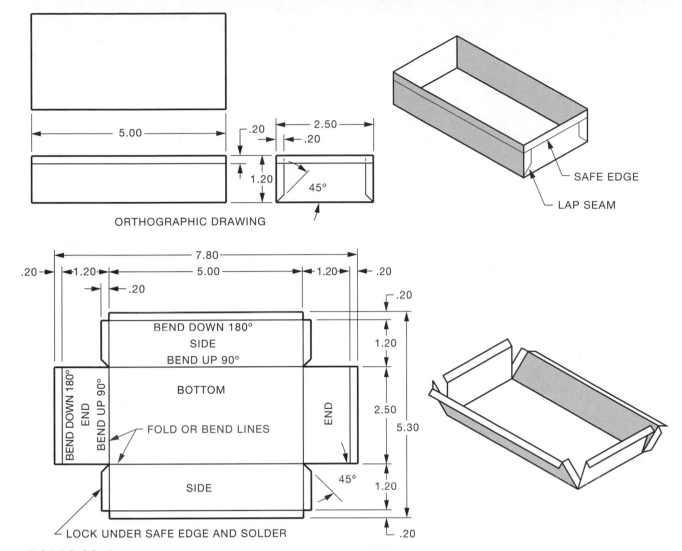

ORTHOGRAPHIC DRAWING

SAFE EDGE

LAP SEAM

7.80

.20 1.20 5.00 1.20 .20

.20

.20

1.20

BEND DOWN 180°
SIDE
BEND UP 90°

BOTTOM

1.20

BEND DOWN 180°
END
BEND UP 90°

END

2.50

5.30

FOLD OR BEND LINES

SIDE

45°

1.20

LOCK UNDER SAFE EDGE AND SOLDER

.20

FIGURE 22–1 ■ Development drawing with a complete set of folding instructions

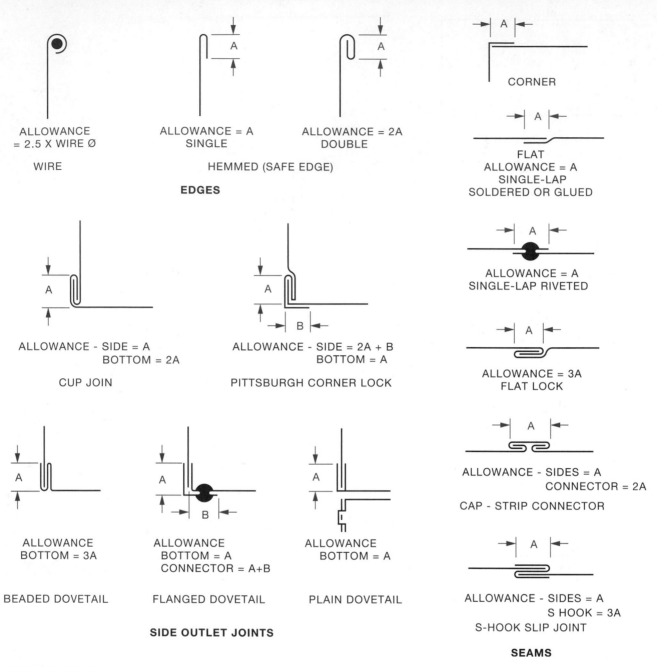

FIGURE 22–2 ■ Joints, seams, and edges

STAMPINGS

Stamping is the art of press-working sheet metal to change its shape by the use of punches and dies. It may involve punching out a hole or the product itself from a sheet of metal. It may also involve bending or forming, Figures 22–5 and 22–6.

Stamping may be divided into two general classifications: *forming* and *shearing*.

Forming

Forming includes stampings made by forming sheet metal to the shape desired without cutting or shearing the metal. For thicker sheet metal plates, bending allowances must be taken into consideration.

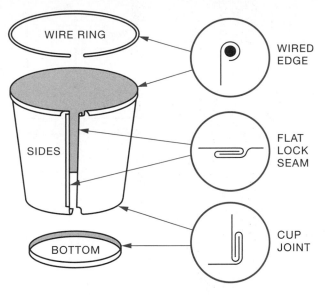

(A) CONSTRUCTION DETAILS

WIRE RING

WIRED EDGE

SIDES

FLAT LOCK SEAM

BOTTOM

CUP JOINT

(B) METAL WASTEBASKET ASSEMBLY

WIRED EDGE

FLAT LOCK SEAM

CUP JOINT

FIGURE 22–3 ■ Wastebasket construction

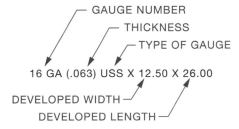

GAUGE NUMBER
THICKNESS
TYPE OF GAUGE

16 GA (.063) USS X 12.50 X 26.00

DEVELOPED WIDTH
DEVELOPED LENGTH

FIGURE 22–4 ■ Callout of sheet metal material

Shearing

Shearing includes stampings made by shearing the sheet metal either to change the outline or to cut holes in the interior of the part. Punching forms a hole or opening in the part.

Height and width dimensions should normally be given to the same side of the metal, either the punch side or the die side. Because larger radii facilitate production, inside radii on stampings should not be less than 1.5 times stock thickness.

The following formula may be used for blank development, Figure 22–7:

$$\text{total length} = A + B + \text{bend allowance}$$

where length for 90° bend =

$$\frac{\pi}{2}\,(R + .33T\ \text{min}) = 1.57\,(R + .33T\ \text{min})$$

PUNCH
DIE SHOE
PRESS BED

FIGURE 22–5 ■ Punch press used to punch and form metal. (*Courtesy of Whitney Metal Tool Co.*)

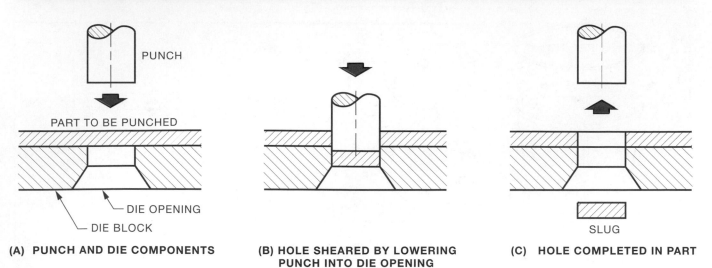

(A) **PUNCH AND DIE COMPONENTS**

(B) **HOLE SHEARED BY LOWERING PUNCH INTO DIE OPENING**

(C) **HOLE COMPLETED IN PART**

FIGURE 22–6 ■ Punching a hole in sheet metal

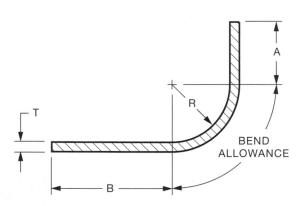

FIGURE 22–7 ■ Dimensions for calculating blank development

It is not general practice to show the blank development on production drawings.

INTERNET RESOURCES

Design & Technology Online For information on packaging, see: http://www.dtonline.org/ (packaging)

Drafting Zone For information on sheet metal practices, see: http://www.draftingzone.com

efunda For information on sheet metal and sheet metal processes, see: http://www.efunda.com/home.cfm

Sheetmetal Shop For information on sheet metal layout, see: http://www.thesheetmetalshop.com

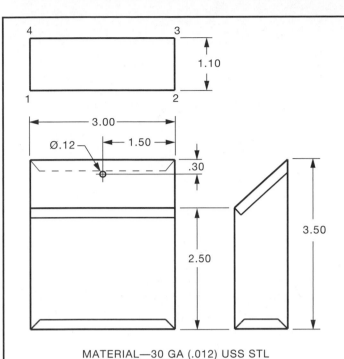

3.00

Ø.12

1.50

1.10

.30

2.50

3.50

MATERIAL—30 GA (.012) USS STL

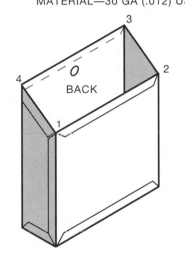

BACK

QUESTIONS:

1. What are the overall dimensions of the back?

2. What are the overall dimensions of the side?

3. What are the overall dimensions of the bottom?

4. What are the overall dimensions of the front?

5. How much has the width of the development been increased due to the edge of the seam allowance?

6. How much has the height of the development been increased due to the edge of the seam allowance?

7. What are the overall sizes of the development?

ASSIGNMENT:
ON A 1.00-IN. GRID SHEET (.10 IN. SQUARES), SKETCH THE DEVELOPMENT OF THE LETTER BOX AND SHOW THE OVERALL DIMENSIONS. SCALE 1:2

.20 SINGLE HEMMED EDGES AND 90° LAP SEAMS

SIDE SEAM AT CORNER 1

FOLD LINE BETWEEN BOTTOM AND BACK

| | **LETTER BOX** | **A-62** |

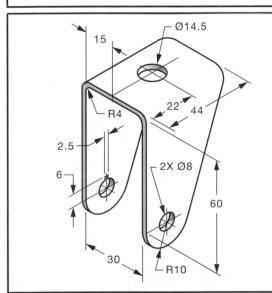

Ø14.5

15

R4

22

44

2.5

6

2X Ø8

60

30

R10

ASSIGNMENT:
ON A CENTIMETRE GRID SHEET (1 mm SQUARES), SKETCH THE FRONT AND SIDE VIEWS PLUS THE DEVELOPMENT DRAWING OF THE BRACKET BEFORE THE THREE HOLES ARE PRODUCED. THE BRACKET IS PART OF THE CASTER ASSEMBLY SHOWN IN ASSIGNMENT A-73M. SCALE 1:1. ADD DIMENSIONS TO ALL VIEWS.

NOTE:
MATL - 13 GA (2.38) USS STL

METRIC
DIMENSIONS ARE IN MILLIMETRES

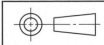

| | **BRACKET** | **A-63M** |

23 UNIT

ARRANGEMENT OF VIEWS

The shape of an object and its complexity influence the possible choices and arrangement of views for that particular object. Because one of the main purposes of making drawings is to give the worker enough information to make the object, only the views that will aid in the interpretation of the drawing should be drawn.

The drafter chooses the view of the object that gives the viewer the clearest idea of the purpose and general contour of the object, and then calls this the front view. The choice of the front view may have no relationship to the actual front of the piece when it is used; it does not have to be the actual front of the object.

The selection of views to best describe the mounting plate shown in Assignment A-64 is shown in Figure 23–1. These views could be called front,

right-side, and bottom views as illustrated in Arrangement A. These views could also be designated top, front, and right-side views, as shown in Arrangement B. In Arrangement B, the right-side view is projected from the top view and not the front view as shown in Arrangement A. The designation of names is not important; what is important is that these views, in the opinion of the drafter or designer, give the necessary information in the most understandable way.

Once the basic views that best describe the mounting plate have been established, auxiliary (helper) views, when required, are added to completely describe the part (Assignment A-64). The No. 1 auxiliary view is added by projecting it from the front view: This is the only one of the five views that shows this surface and the slotted hole in their true shape. The No. 2 auxiliary view is added by projecting it from the top view. This is the only view that shows this surface and the hexagon hole in

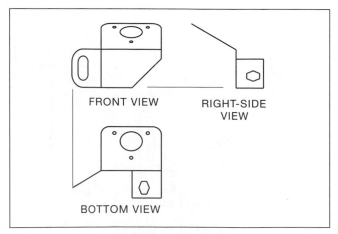

ARRANGEMENT A

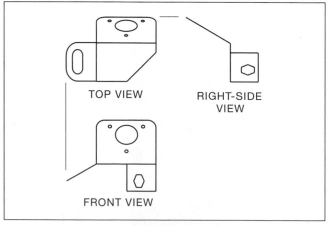

ARRANGEMENT B

FIGURE 23–1 ■ Naming of views for mounting plate, Assignment A-64

their true shape. Dimensions are then added to the views or surfaces that are not shown distorted.

A variety of arrangements and naming of views to describe the index pedestal in Assignment A-65 is shown in Figure 23–2. Arrangements A and B are identical except for the naming of the views. Although Arrangements C and D are acceptable, they are not as easily read.

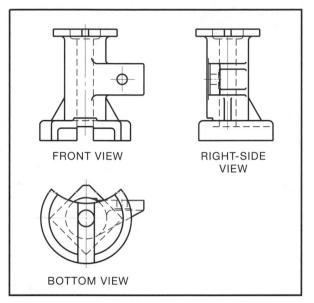

ARRANGEMENT A

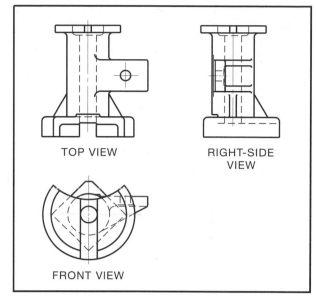

ARRANGEMENT B

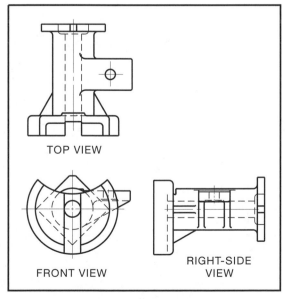

ARRANGEMENT C

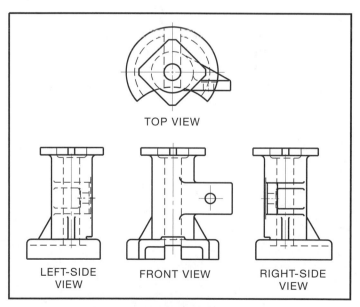

ARRANGEMENT D

FIGURE 23–2 ■ Arrangements and naming of views for index pedestal, Assignment A-65

REFERENCE

CAN3-B78.1-M83 Technical Drawings—General
    Principles

INTERNET RESOURCE

American Society of Mechanical Engineers For
    information on the arrangement of views, refer
    to ASME Y14.3M-1994 (R2003) (*Multi- and
    Sectional-View Drawings*) at: http://www
    .asme.org

QUESTIONS:

1. Calculate the dimensions and hole sizes of Ⓐ through Ⓣ on the pictorial sketch.
2. What would be the overall size of the sheet used to make the part? Sizes to be in .10-inch increments.
3. On which view(s) are the Ø.68 slots shown?
4. Which view(s) show the true shape of the (A) Ø.68 slot, (B) Ø1.30 hole, (C) .75 hex hole?
5. How would the holes in the part be produced?

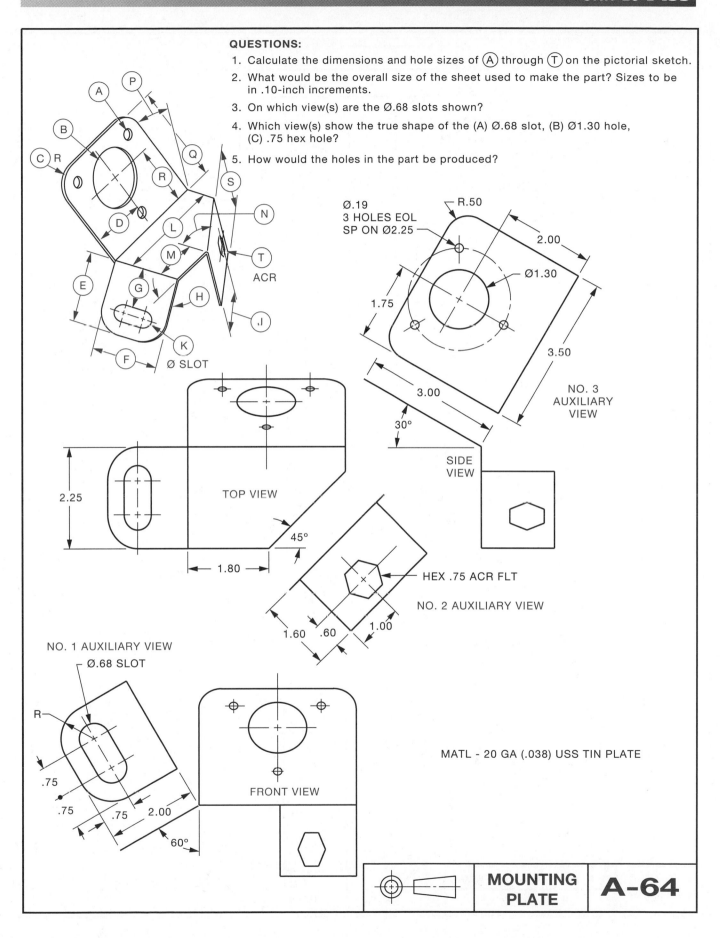

Ø.19
3 HOLES EOL
SP ON Ø2.25

R.50

2.00

Ø1.30

1.75

3.50

NO. 3
AUXILIARY
VIEW

3.00

30°

SIDE
VIEW

TOP VIEW

45°

1.80

HEX .75 ACR FLT

NO. 2 AUXILIARY VIEW

1.60 .60 1.00

2.25

NO. 1 AUXILIARY VIEW
Ø.68 SLOT

R

.75
.75 .75 2.00

60°

FRONT VIEW

MATL - 20 GA (.038) USS TIN PLATE

MOUNTING PLATE   A-64

ACR
Ø SLOT

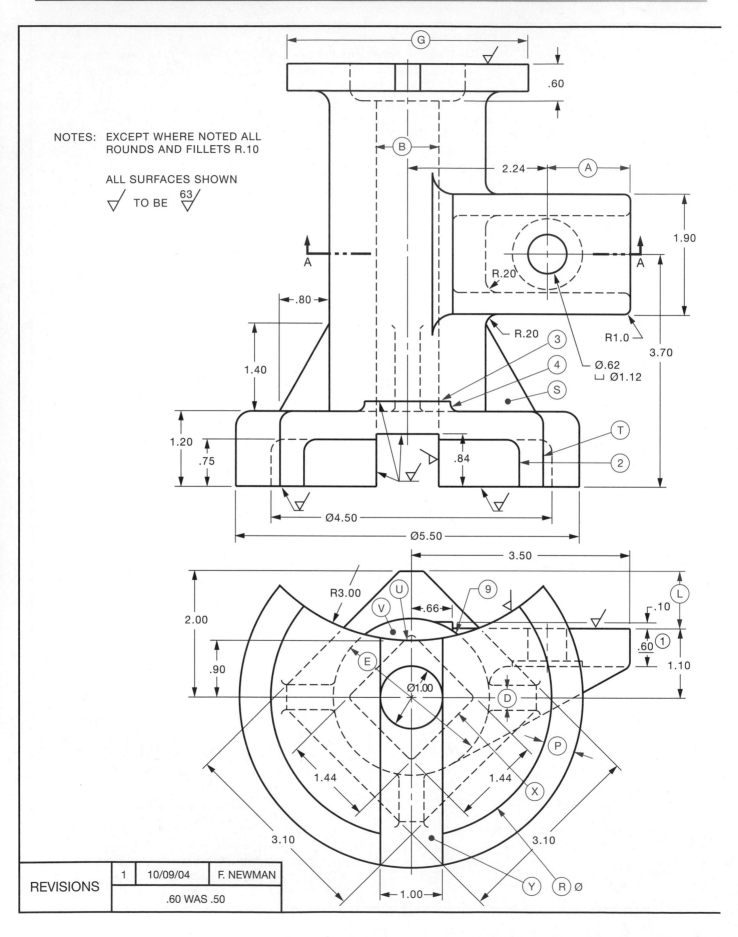

NOTES: EXCEPT WHERE NOTED ALL
ROUNDS AND FILLETS R.10

ALL SURFACES SHOWN
TO BE 63

REVISIONS	1	10/09/04	F. NEWMAN
		.60 WAS .50	

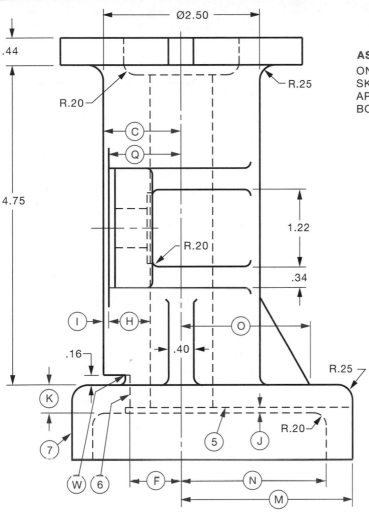

ASSIGNMENT:
ON A 1.00-INCH GRID SHEET (.10 IN. SQUARES),
SKETCH SECTION A-A. SCALE 1:1. PLACE THE
APPROPRIATE DIMENSIONS SHOWN ON THE
BOTTOM VIEW OF THIS SECTION.

QUESTIONS:

1. Determine distances Ⓐ through Ⓡ.

2. Which line in the front view does line ⑦ represent?

3. Which line or surface in the front view represents the surface at Ⓥ?

4. Locate line ④ in the right side view.

5. Locate line ④ in the bottom view.

6. How deep is the square Ⓧ hole?

7. How many different finished surfaces are indicated?

8. From which point on the bottom view is line ⑥ projected?

9. Determine overall height of the pedestal.

10. Which line or surface in the right view represents surface Ⓨ?

FRONT
VIEW

RIGHT-
SIDE
VIEW

BOTTOM
VIEW

ARRANGEMENT OF VIEWS

MATERIAL	WROUGHT IRON	
SCALE	NOT TO SCALE	
DRAWN J. HELSEL		DATE 12/06/04

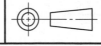

**INDEX
PEDESTAL**

A-65

24 UNIT

....

PIPING

Until about 100 years ago, water was the only important fluid conveyed from place to place through pipe. Today numerous fluids are handled in pipe during their production, processing, transportation, or utilization. See Figure 24–1. Liquid metals, sodium, and nitrogen have recently been added to the list of more common fluids such as oil, water, and acids being transported through pipe. Many gases are also being stored and delivered through piping systems.

Pipe is also used for hydraulic and pneumatic mechanisms; for the controls of machinery and other equipment; and as a structural element for columns and handrails.

The nominal size of pipe and the inside diameter, outside diameter, and wall thickness are given in fractional inches.

Kinds of Pipe

Steel and Wrought-Iron Pipe

These pipes carry water, steam, oil, and gas and are commonly used under high temperatures and pressures. Standard steel or cast-iron pipe is specified by the nominal diameter, which is always less than the actual inner diameter (ID) of the pipe. Until recently, this pipe was available in only three weights: standard, extra strong, and double extra strong, Figure 24–2. To use common fittings with these different pipe weights, the outside diameter (OD) of each of the different pipes remained the same. The extra metal was added to the ID to increase the wall thickness of the extra strong and double extra strong pipe.

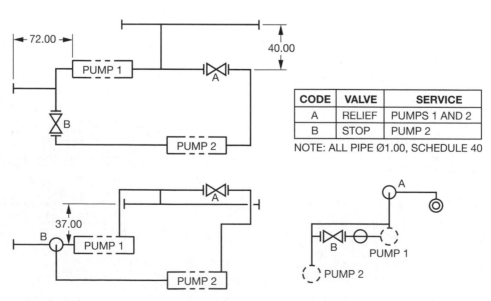

CODE	VALVE	SERVICE
A	RELIEF	PUMPS 1 AND 2
B	STOP	PUMP 2

NOTE: ALL PIPE Ø1.00, SCHEDULE 40

FIGURE 24–1 ■ Piping for a heating system for a building

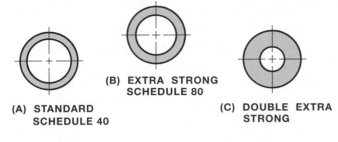

FIGURE 24–2 ■ Comparison of wall thicknesses

The demand for a greater variety of pipe for use under increased pressure and temperature led to the introduction of 10 pipe weights, each designated by a schedule number. Standard pipe is now called schedule 40 pipe. Extra strong pipe is schedule 80.

Cast-Iron Pipe

This is often installed underground to carry water, gas, and sewage.

Seamless Brass and Copper Pipe

These pipes are used extensively in plumbing because of their ability to withstand corrosion.

Copper Tubing

This is used in plumbing and heating and where vibration and misalignment are factors, such as in automotive, hydraulic, and pneumatic design.

Plastic Pipe

This pipe or tubing, because of its resistance to corrosion and chemicals, is often used in the chemical industry. It is easily installed. However, it is not recommended where heat or pressure is a factor.

Pipe Joints and Fittings

Parts joined to pipe are called *fittings*. They may be used to change size or direction and to join or provide branch connections. There are three general classes of fitting: screwed, welded, and flanged. Other methods such as soldering, brazing, and gluing are used for cast-iron pipe, and copper and plastic tubing.

Pipe fittings are specified by the nominal pipe size, the name of the fitting, and the material.

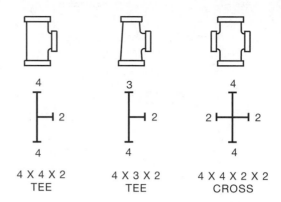

NOTE: NOMINAL PIPE SIZES IN INCHES

FIGURE 24–3 ■ Order of specifying the openings of reducing fittings

Some fittings, such as tees, crosses, and elbows, connect different sizes of pipe. These are called *reducing fittings*. Their nominal pipe sizes must be specified. The largest opening of the through run is given first, followed by the opposite end and the outlet. Figure 24–3 illustrates the method of designating sizes of reducing fittings.

Screwed Fittings

Screwed fittings are generally used on small pipe design of 2.50-inch nominal pipe size or less.

There are two types of American Standard Pipe Thread: tapered and straight. The tapered thread is more common. Straight threads are used for special applications.

Tapered threads are designated on drawings as NPT (National Pipe Thread) or whichever standard is used and may be drawn with or without the taper, Figure 24–4. When drawn in tapered form, the taper is exaggerated. Straight pipe threads are designated on drawings as NPTS and standard thread symbols are used. Pipe threads are assumed to be tapered unless specified otherwise.

Pipe thread designation for drawings is covered in the following sequence: the nominal size in fractional inches, a dash, the number of threads per inch, a space, the thread series symbol, and the thread class if applicable, Figure 24–4(c).

Welded Fittings

Welded fittings are used where connections will be permanent and on high pressure and temperature lines. The ends of the pipe and pipe fittings are usually bevelled to accommodate the weld.

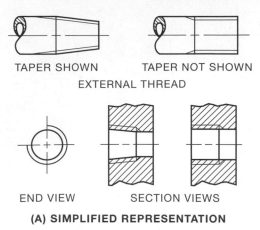

(A) SIMPLIFIED REPRESENTATION

TAPER SHOWN — TAPER NOT SHOWN
EXTERNAL THREAD

END VIEW — SECTION VIEWS

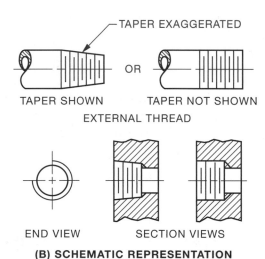

TAPER EXAGGERATED

TAPER SHOWN — OR — TAPER NOT SHOWN
EXTERNAL THREAD

END VIEW — SECTION VIEWS

(B) SCHEMATIC REPRESENTATION

1/8-27 NPT

(C) PIPE THREAD DESIGNATION

FIGURE 24–4 ■ Pipe thread conventions

Flanged Fittings

Flanged joint fittings provide a quick way to disassemble pipe. Flanges are attached to the pipe ends by welding, screwing, or lapping.

Valves

Valves are used in piping systems to stop or regulate the flow of fluids and gases. These are a few of the more common types.

Gate Valves

These are used to control the flow of liquids. The wedge, or gate, lifts to allow full, unobstructed flow and lowers to stop the flow. They are generally used where operation is infrequent and are not intended for throttling or close control.

Globe Valves

These are used to control the flow of liquids or gases. The design of the globe valve produces two changes in the direction of flow, slightly reducing the pressure in the system. The globe valve is recommended for the control of air, steam, gas, and other compressibles where instantaneous on-and-off operation is essential.

Check Valves

Check valves permit flow in one direction, but check all reverse flow. They are operated by pressure and velocity of line flow alone and have no external means of operation.

PIPING DRAWINGS

Piping drawings show the size and location of pipes, fittings, and valves. Because of the detail required to accurately describe these items, there is a set of symbols to indicate them on drawings.

There are two types of piping drawings in use, single line and double line, Figure 24–5. Double-line drawings take longer to draw and are therefore not recommended for production drawings. They are, however, suitable for catalogues and other applications where the appearance is important.

Single-Line Drawings

Single-line drawings, also known as simplified representation, of pipe lines provide substantial time savings without loss of clarity or reduction of comprehensiveness of information. Therefore, the simplified method is used whenever possible.

These drawings use a single line to show the arrangement of the pipe and fittings. The centre line of the pipe, regardless of pipe size, is drawn as a thick line to which symbols are added. The size of the symbol is left to the discretion of the drafter. When pipelines carry different liquids, such as cold or hot water, a coded line symbol is often used.

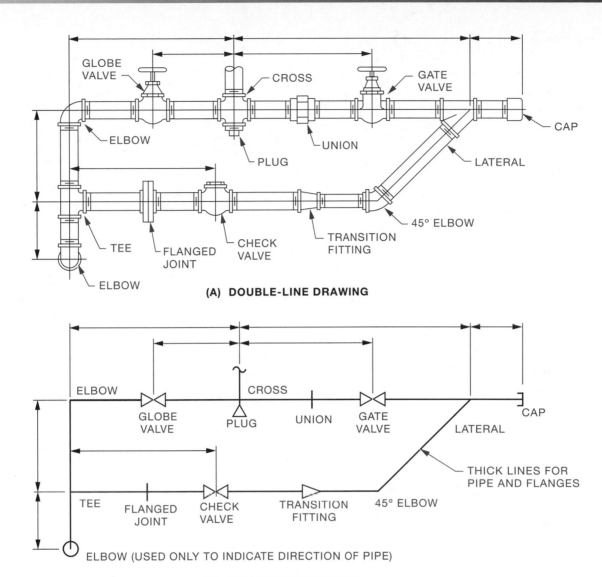

(A) DOUBLE-LINE DRAWING

(B) SINGLE-LINE DRAWING

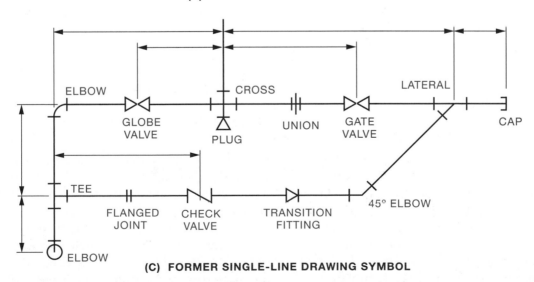

(C) FORMER SINGLE-LINE DRAWING SYMBOL

FIGURE 24–5 ■ Piping drawing symbols

Drawing Projection

Two methods of projection are used, orthographic and isometric, Figure 24–6. Orthographic projection is recommended for the representation of single pipes that are either straight or bent in one plane only. However, this method is also used for more complicated pipings.

Isometric projection is recommended for all pipes bent in more than one plane and for assembly and layout work because the finished drawing is easier to understand.

Crossings

The crossing of pipes without connections is usually drawn without interrupting the line representing the hidden line, Figure 24–7(A). But when it is desirable to indicate that one pipe must pass behind the other, the line representing the pipe farthest away from the viewer will be shown with a break or interruption where the other pipe passes in front of it.

Connections

Permanent connections or junctions, whether made by welding or other processes, are indicated on the drawing by a heavy dot, Figure 24–7(F). A general note or specification may indicate the process used.

Detachable connections or junctions are indicated by a single thick line. Specifications, a general note, or the bill of material will indicate the type of fitting, for example, flanges, union, or coupling. The specifications will also indicate whether the fittings are flanged, threaded, or welded.

PIPE DRAWING SYMBOLS

If specific symbols are not standardized, fittings such as tees, elbows, and crosses are not specially drawn but are represented, like pipe, by a continuous line. The circular symbol for a tee or elbow may be used when necessary to indicate whether the piping is viewed from the front or back, Figure 24–7(H). Elbows on isometric drawings may be shown without the radius. However, if this method is used, the direction change of the piping must be shown clearly.

Adjoining Apparatus

If needed, adjoining apparatus, such as tanks, and machinery not belonging to the piping itself, are shown by outlining them with a thin phantom line.

Dimensioning

■ Dimensions of pipe and pipe fittings are always given from centre to centre of pipe and to the outer face of the pipe end or flange, Figure 24–7(C).

■ Individual pipe lengths are usually cut to suit by the pipe fitter. However, the total length of pipe required is usually called for in the bill of material.

■ Pipe and fitting sizes and general notes are placed on the drawing beside the part concerned or, where space is restricted, with a leader.

■ A bill of material is usually provided with the drawing.

■ Pipes with bends are dimensioned from vertex to vertex.

■ Radii and angles of bends are dimensioned, Figure 24–7(C) and (D). Whenever possible, the smaller of the supplementary angles is specified.

■ The outer diameter and wall thickness of the pipe are indicated on the line representing the pipe, or in the bill of material, general note, or specifications, Figure 24–7(E).

Orthographic Piping Symbols

Pipe Symbols

If flanges are not attached to the ends of the pipelines when drawn in orthographic projection, pipeline symbols indicating the direction of the pipe are required. If the pipeline direction is toward the front (or viewer), it is shown by two concentric circles, the smaller one of which is a large solid dot, Figure 24–7(G). If the pipeline direction is toward the back (or away from the viewer), it will be shown by one solid circle. No extra lines are required on the other views.

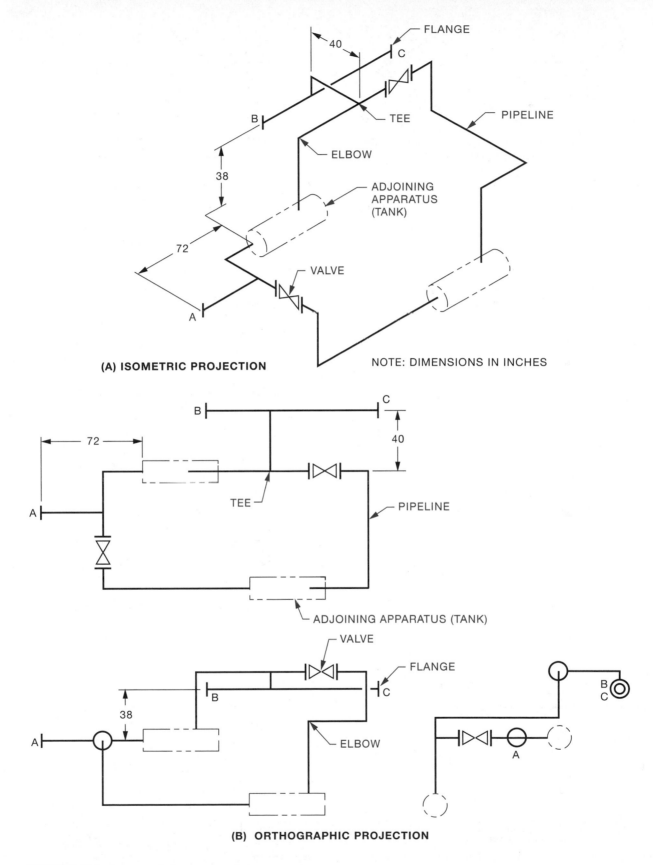

FLANGE

40

C

B

TEE

PIPELINE

ELBOW

38

ADJOINING
APPARATUS
(TANK)

72

VALVE

A

(A) ISOMETRIC PROJECTION

NOTE: DIMENSIONS IN INCHES

B C

72

40

TEE

PIPELINE

A

ADJOINING APPARATUS (TANK)

VALVE

FLANGE

C

B

38

ELBOW

A

B
C

A

(B) ORTHOGRAPHIC PROJECTION

FIGURE 24–6 ■ Single-line piping drawing

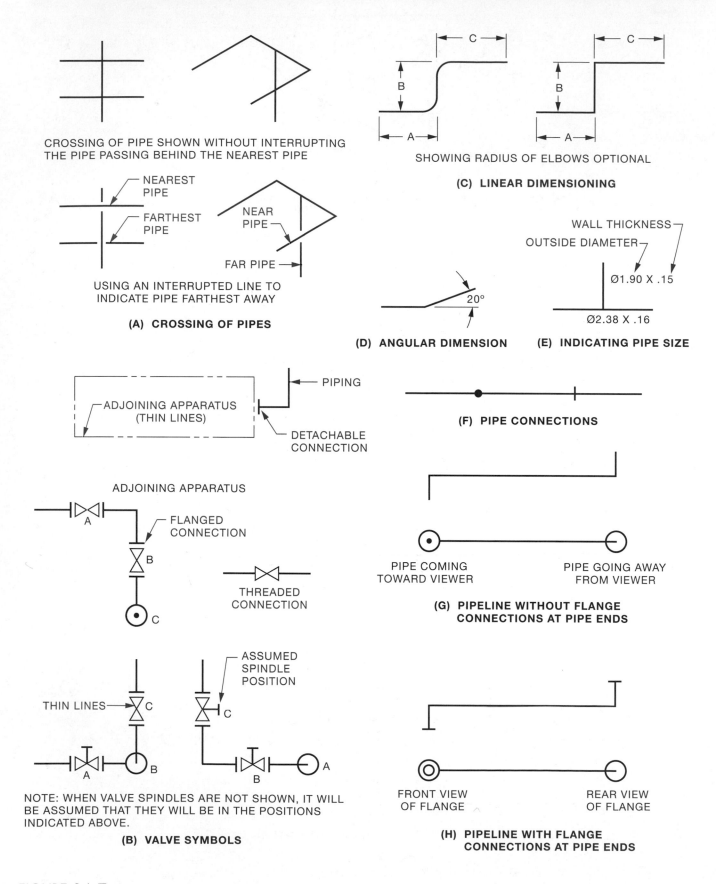

CROSSING OF PIPE SHOWN WITHOUT INTERRUPTING THE PIPE PASSING BEHIND THE NEAREST PIPE

NEAREST PIPE

FARTHEST PIPE

NEAR PIPE

FAR PIPE

USING AN INTERRUPTED LINE TO INDICATE PIPE FARTHEST AWAY

(A) CROSSING OF PIPES

C

B

A

SHOWING RADIUS OF ELBOWS OPTIONAL

(C) LINEAR DIMENSIONING

20°

(D) ANGULAR DIMENSION

WALL THICKNESS
OUTSIDE DIAMETER
Ø1.90 X .15
Ø2.38 X .16

(E) INDICATING PIPE SIZE

ADJOINING APPARATUS (THIN LINES)

PIPING

DETACHABLE CONNECTION

ADJOINING APPARATUS

A

FLANGED CONNECTION

B

C

THREADED CONNECTION

(F) PIPE CONNECTIONS

PIPE COMING TOWARD VIEWER

PIPE GOING AWAY FROM VIEWER

(G) PIPELINE WITHOUT FLANGE CONNECTIONS AT PIPE ENDS

THIN LINES

C

A

B

ASSUMED SPINDLE POSITION

C

B

A

NOTE: WHEN VALVE SPINDLES ARE NOT SHOWN, IT WILL BE ASSUMED THAT THEY WILL BE IN THE POSITIONS INDICATED ABOVE.

(B) VALVE SYMBOLS

FRONT VIEW OF FLANGE

REAR VIEW OF FLANGE

(H) PIPELINE WITH FLANGE CONNECTIONS AT PIPE ENDS

FIGURE 24–7 ■ Single-line piping drawing symbols and dimension

Flange Symbols

Irrespective of their type and sizes, flanges are to be represented by:

■ two concentric circles for the front view

■ one circle for the rear view

■ a short stroke for the side view

while using lines of equal thickness as chosen for the representation of pipes, Figure 24–7(H).

Valve Symbols

Symbols representing valves are drawn with continuous thin lines (not thick lines as for piping and flanges). The valve spindles should be shown only if it is necessary to define their positions. It will be assumed that unless otherwise indicated, the valve spindle is in the position shown in Figure 24–7(B).

REFERENCES

ASME Y32.2.3-1994 (R1999) Graphic Symbols for Pipe Fittings, Valves and Piping
ASME Y14.6-2001 Screw Thread Representation

INTERNET RESOURCES

American Design and Drafting Association For information on piping drafting practices in their *Drafting Reference Guide*, see: http://www.adda.org/

Crane Valve Group For information on valves and links to related sites, see: http://www.cranevalve.com/

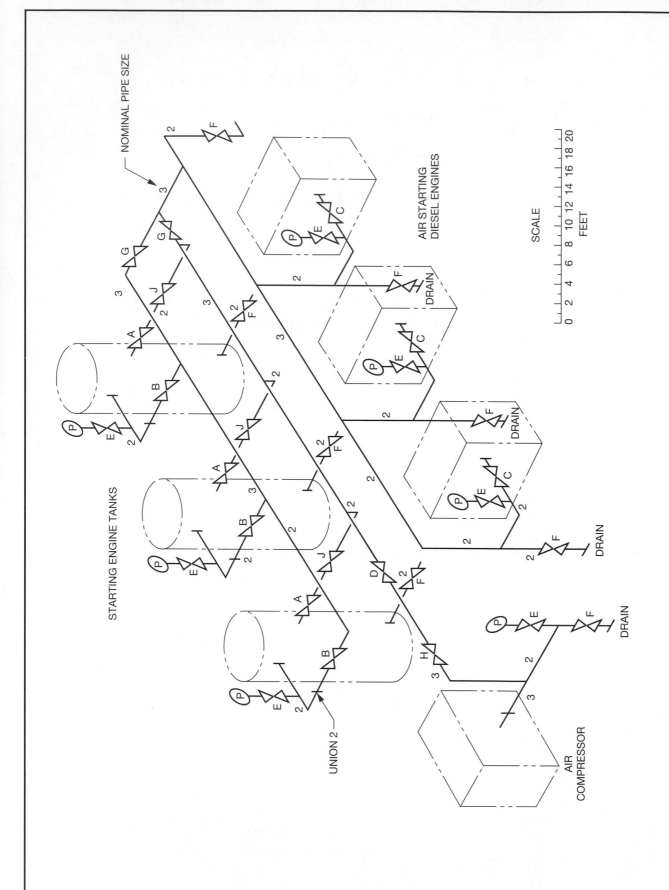

NOMINAL PIPE SIZE

STARTING ENGINE TANKS

UNION 2

AIR STARTING
DIESEL ENGINES

SCALE

FEET

0 2 4 6 8 10 12 14 16 18 20

DRAIN

DRAIN

DRAIN

DRAIN

AIR
COMPRESSOR

Safety valves are provided for the compressor and the air storage tanks. Check valves are installed on the air storage tank feed lines and the compressor discharge lines to prevent accidental discharge of the tanks.

Piping is arranged so that the compressor will fill the storage tanks and/or pump directly to the engines. Any of the three storage tanks may be used for starting, and pressure gauges indicate their readiness. The engines are fitted with quick-opening valves to admit air quickly at full pressure and shut it off at the instant rotation is obtained. A bronze globe valve is installed to permit complete shutdown of the engine for repairs, and regulation of the flow of air. Drains are provided at low points to remove condensate from the air storage tanks, lines, and engine feed.

Globe valves are recommended throughout this hookup except on the main shutoff lines where gate valves are used because of infrequent operation.

QUESTIONS (Use scale provided where necessary):

1. What size pipe is used for the main feed line?
2. Disregarding the length of the valves and fittings, what is the total approximate length of
   (A) the 3-inch pipe?   (B) the 2-inch pipe?
3. What is the approximate centre line to centre line spacing of the diesel engines?
4. Calculate the approximate number of cubic feet of each of the air tanks.
5. Why are diesel engines used instead of public power supply?
6. What is the purpose of the air compressor?
7. Why is a gate valve used instead of a globe valve in the main line shutoff?
8. State the uses of the following parts.
   (A) Valve C   (C) Gauge P   (E) Valve D
   (B) Valve G   (D) Valve F   (F) Valve H

CODE	VALVE	SERVICE
A	BRONZE GLOBE	AIR STORAGE TANK FEED LINES
B	BRONZE GLOBE	AIR STORAGE TANK DISCHARGE LINES
C	BRONZE GLOBE	DIESEL ENGINE SHUTOFF CONTROL
D	BRONZE GLOBE	AIR COMPRESSOR DISCHARGE
E	BRONZE GLOBE	PRESSURE GAUGE SHUTOFF
F	BRONZE GLOBE	DRAIN VALVES
G	SPINDLE GATE	MAIN LINE SHUTOFF
H	BRONZE CHECK	AIR COMPRESSOR CHECK
J	BRONZE CHECK	AIR STORAGE TANK FEED LINES
P	PRESSURE GAUGE	DISCHARGE OR FEED LINES

SCALE	AS SHOWN	
DRAWN A. JAMES		DATE 22/10/04

ASSIGNMENT:
PREPARE A BILL OF MATERIAL SHOWING ALL THE VALVES AND FITTINGS.

ISOMETRIC PROJECTION

ENGINE STARTING AIR SYSTEM

A-66

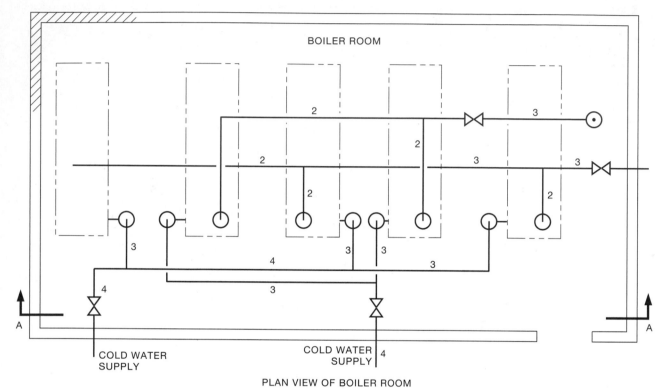

PLAN VIEW OF BOILER ROOM

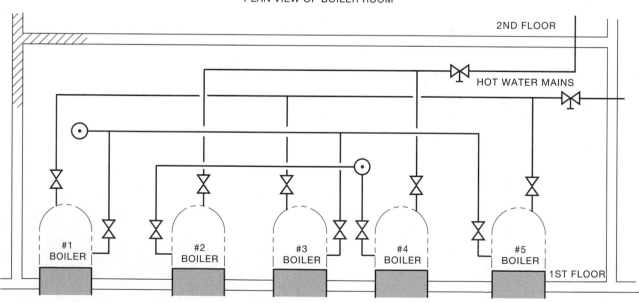

SECTIONAL ELEVATION A-A

ASSIGNMENTS:

1. ON A .25 INCH ISOMETRIC GRID SHEET, MAKE A SKETCH OF THE BOILER ROOM
   PIPING SHOWN ON THE OPPOSITE PAGE. THE PARTIAL ISOMETRIC VIEW BELOW
   SHOWS THE VIEWING DIRECTION AND THE POSITIONING OF THE BOILERS ON THE
   GRID SHEET. A SCALE IS PROVIDED FOR MEASURING THE DISTANCES ON THE
   DRAWING. THE .25 IN. SQUARES ON THE GRID SHEET REPRESENT ONE FOOT ON
   THE DRAWING (SCALE 1:48).

2. ON A SEPARATE SHEET, PREPARE A BILL OF MATERIAL CALLING FOR ALL VALVES,
   FITTINGS, AND PIPE. GIVE THE APPROXIMATE TOTAL LENGTH OF EACH SIZE OF PIPE.

3. FROM THE BILL OF MATERIAL, ADD PART NUMBERS TO THE ISOMETRIC SKETCH
   FOR THE VALVES AND FITTINGS.

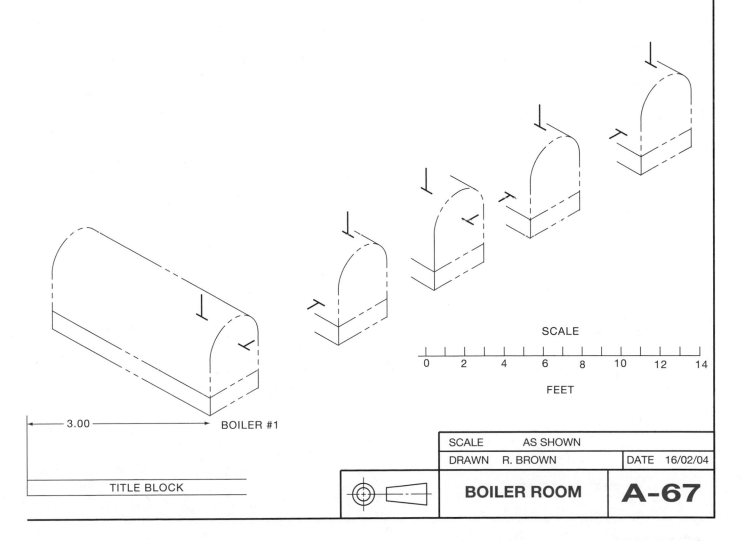

SCALE

0   2   4   6   8   10   12   14

FEET

|← 3.00 →| BOILER #1

SCALE	AS SHOWN	
DRAWN R. BROWN		DATE 16/02/04
TITLE BLOCK	**BOILER ROOM**	**A-67**

BEARINGS

All rotating machinery parts are supported by *bearings*. Each bearing type and style has advantages and disadvantages. Bearings are classified into two groups: plain bearings and anti-friction bearings.

PLAIN BEARINGS

Plain bearings have many uses. They are available in a variety of shapes and sizes, Figure 25–1. Because of their simplicity, plain bearings are versatile. There are several plain bearing categories. The most common are journal (sleeve) bearings and thrust bearings, available in a number of standard sizes and shapes.

Journal, or Sleeve, Bearings

Journal bearings are the simplest and most economical means of supporting moving parts. Journal bearings are usually made of one or two pieces of metal enclosing a shaft. They have no moving parts. The journal is the supporting portion of the shaft.

Speed, mating materials, clearances, temperature, lubrication, and type of loading affect the performance of bearings. The maintenance of an oil film between the bearing surfaces is important. The oil film reduces friction, dissipates heat, and retards wear by minimizing metal-to-metal contact, Figure 25–2. Starting and stopping are the most critical periods of

SLEEVE — SPLIT

FLANGED — SPLIT

(A) JOURNAL TYPE

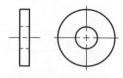

(B) THRUST TYPE

FIGURE 25–1 ■ Plain bearings

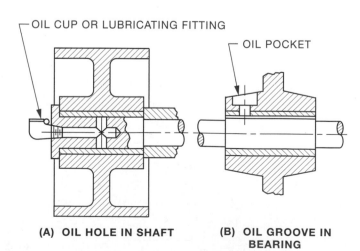

OIL CUP OR LUBRICATING FITTING

OIL POCKET

(A) OIL HOLE IN SHAFT   **(B) OIL GROOVE IN BEARING**

FIGURE 25–2 ■ Common methods of lubricating journal bearings

operation, because the load may cause the bearing surfaces to touch each other.

The shaft should have a smooth finish and be harder than the bearing material. The bearing will perform best with a hard, smooth shaft. The length of the bearing should be between one and two times the shaft diameter. The outside diameter should be about 25 percent larger than the shaft diameter.

Cast bronze and porous bronze are usually used for journal bearings.

Bearings are sometimes split to facilitate assembly and permit adjustment and replacement of worn parts. Split bearings allow the shaft to be set in one half of the bearing, while the other half, or cover, is later secured in position, Figure 25–3.

If the bearings shown in Figure 25–3 are to be made from two parts, they must be fastened together before the hole is bored or reamed. This will facilitate the machining operation and make a perfectly round bearing. An incorrectly assembled bearing is shown in Figure 25–4.

To give longer life to the bearing, very thin strips of metal are inserted between the base and cover halves before boring. These thin strips of varying thickness are called *shims*, Figure 25–5.

When a bearing is shimmed, the same number of pieces of corresponding thickness are used on both sides of it. As the hole wears, one or more pairs of these shims may be removed for wear compensation.

Thrust Bearings

Plain thrust bearings, or thrust washers, are available in various materials, including sintered metal, plastic, woven TFE fabric on steel backing, sintered Teflon-bronze-lead on metal backing, aluminum alloy on steel, aluminum alloy, and carbon-graphite.

Antifriction Bearings

Ball, roller, and needle bearings are classified as antifriction bearings, because friction has been reduced to a minimum. These are covered in detail in Unit 42.

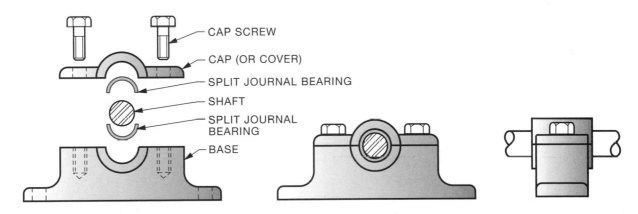

FIGURE 25–3 ■ Pillow block with split journal bearing

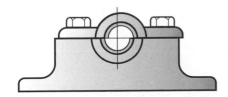

FIGURE 25–4 ■ Bearing halves incorrectly matched

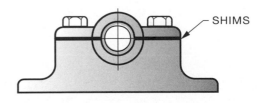

FIGURE 25–5 ■ Shimmed bearing

Premounted Bearings

Premounted bearing assemblies consist of a bearing element and a housing, usually assembled to permit convenient adaptation to a machine frame. All components are incorporated within a single unit to ensure proper protection, lubrication, and operation of the bearing. Both plain and roller element bearing units are available in a variety of housing design and shaft sizes. An example of an adjustable premounted bearing is shown in Assignment A-68M.

REFERENCES

A. O. De Hart, "Basic Bearing Types," *Machine Design*, 40, No. 14

W. A. Glaeser, "Plain Bearings," *Machine Design*, 40, No. 14

INTERNET RESOURCES

eFunda For information on all types of bearings, see: http://www.efunda.com/home.cfm

Emerson Power Transmission For information on all types of bearings and related products, see: http://www.emerson-ept.com

HowStuffWorks For information on the theory and practical applications of various types of bearings, see: http://science.howstuffworks.com/bearing.htm

Machine Design For information on bearings and related products, see: http://www.machinedesign.com

TechSourcer.com For information on various types of bearings, see: http://www.techsourcer.com/industrial_supply/bearings.html

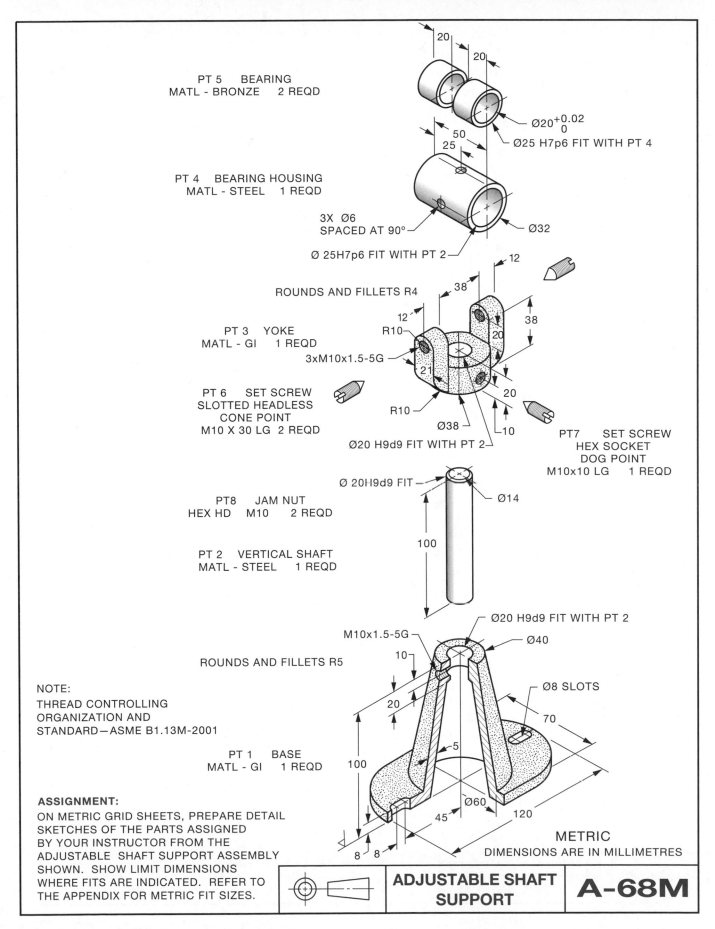

PT 5   BEARING
MATL - BRONZE   2 REQD

20
20

Ø20$^{+0.02}_{0}$

50
25

Ø25 H7p6 FIT WITH PT 4

PT 4   BEARING HOUSING
MATL - STEEL   1 REQD

3X Ø6
SPACED AT 90°

Ø32

Ø 25H7p6 FIT WITH PT 2

ROUNDS AND FILLETS R4

12
38
12
38

PT 3   YOKE
MATL - GI   1 REQD

R10
3xM10x1.5-5G

20
21

PT 6   SET SCREW
SLOTTED HEADLESS
CONE POINT
M10 X 30 LG 2 REQD

20

R10

Ø38
10

Ø20 H9d9 FIT WITH PT 2

PT7   SET SCREW
HEX SOCKET
DOG POINT
M10x10 LG   1 REQD

Ø 20H9d9 FIT

Ø14

PT8   JAM NUT
HEX HD   M10   2 REQD

PT 2   VERTICAL SHAFT
MATL - STEEL   1 REQD

100

Ø20 H9d9 FIT WITH PT 2

M10x1.5-5G

Ø40

ROUNDS AND FILLETS R5

10

Ø8 SLOTS

20

70

NOTE:
THREAD CONTROLLING
ORGANIZATION AND
STANDARD—ASME B1.13M-2001

100

5

PT 1   BASE
MATL - GI   1 REQD

Ø60

45
120

Ø8

METRIC
DIMENSIONS ARE IN MILLIMETRES

8   8

ASSIGNMENT:
ON METRIC GRID SHEETS, PREPARE DETAIL
SKETCHES OF THE PARTS ASSIGNED
BY YOUR INSTRUCTOR FROM THE
ADJUSTABLE SHAFT SUPPORT ASSEMBLY
SHOWN. SHOW LIMIT DIMENSIONS
WHERE FITS ARE INDICATED. REFER TO
THE APPENDIX FOR METRIC FIT SIZES.

**ADJUSTABLE SHAFT
SUPPORT**

A-68M

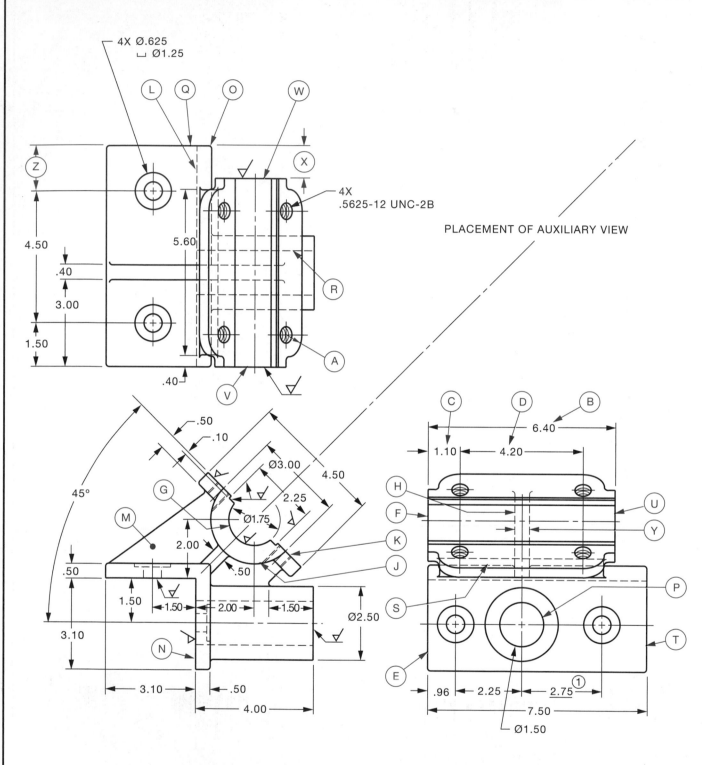

4X Ø.625
⊔ Ø1.25

L   Q   O   W

X

4X
.5625-12 UNC-2B

PLACEMENT OF AUXILIARY VIEW

Z

4.50

.40

3.00

5.60

1.50

R

A

.40

V

.50
.10

45°

Ø3.00

4.50

2.25

G

Ø1.75

M

2.00

K

J

.50

2.00

.50

3.10

1.50

1.50

Ø2.50

N

3.10   .50

4.00

ROUNDS AND FILLETS R.10

C   D   B

6.40

1.10   4.20

H

F

U

Y

S

P

E

T

.96   2.25   2.75   ①

7.50

Ø1.50

REVISIONS	1	12/10/04	B. JENSEN
		2.75 WAS 2.60	

THREAD CONTROLLING ORGANIZATION
AND STANDARD—ASME B1.1-2003

QUESTIONS:

1. How many definite finished surfaces are on the casting?
   (Note: Two or more surfaces could lie on one plane.)
2. What is the size of (P) hole?
3. What size spotface is used on the mounting holes?
4. What are the number and size of the mounting holes?
5. Note that surface (E) is not finished, but surface (F) is to be finished. Allowing .06 in. for finishing, what would be the depth of the rough casting?
6. Would (G) hole be bored before or after bearing cap is assembled?
7. Which line in the side view shows surface (M)?
8. In which view, and by what line, is the surface represented by (N) shown?
9. Which line or surface in the side view shows the projection of point (J)?
10. Which point or surface in the side view does line (R) represent?
11. Locate surface (T) in the top view.
12. Locate surface (U) in the top view.
13. Locate surface (V) in the side view.
14. What are dimensions (X), (Y), and (Z)?
15. What size is the round (O)?
16. Determine dimension (A) and place it correctly on the sketch of the auxiliary view.
17. Place dimensions (B), (C), and (D) correctly on the sketch of the auxiliary view.
18. What is the tap drill size required for the four threaded holes?
19. How many dimensions indicated are not drawn to scale?
20. The Ø1.50 hole is to be revised to accommodate a plain bearing with an LN3 fit. What would be the limits of size for the hole?

ASSIGNMENT:

ON A ONE-INCH GRID SHEET (.10 IN. SQUARES), SKETCH THE AUXILIARY VIEW OF THE CORNER BRACKET. ADD THE APPROPRIATE DIMENSIONS TO THE DRAWING. SCALE 1:1.

NOTE: UNLESS OTHERWISE SPECIFIED:

- TOLERANCE ON DIMENSIONS ±.02
- TOLERANCE ON ANGLES ±0.5°

- ⩔ TO BE ⩔63

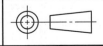

MATERIAL	GREY IRON	
SCALE	NOT TO SCALE	
DRAWN	L. DAVIS	DATE 15/07/04

CORNER BRACKET   **A-69**

26 UNIT

MANUFACTURING MATERIALS

One of the first decisions a designer must make is the choice of manufacturing material. This is influenced by many factors, such as the end use of the product and the properties of the selected material. Would plastic be a better choice of material than rubber or metal? Would one metal be better than another? Will the part come in contact with water or chemicals? Is strength a factor? If so, what material will meet the stresses required? What material is in stock or readily obtainable? Is the material the correct choice if a plating or coating is required?

Just as steel composition may vary—tool steel and stainless steel, for example—so do other manufacturing materials. This unit covers some of the commonly used manufacturing materials.

CAST IRONS

Iron and the large family of iron alloys called steel are the most frequently specified metals. All commercial forms of iron and steel contain carbon, which is an integral part of the manufacture of iron and steel.

Types of Cast Iron

Grey Iron

Grey iron is a supersaturated solution of carbon in an iron matrix. Generally, grey iron serves well in any machinery applications because of its fatigue resistance. Typical applications of grey iron include automotive engine blocks, flywheels, brake disks and drums, machine bases, and gears.

Ductile or Nodular Iron

Ductile iron is less available than grey iron and is more difficult to control in production. However, ductile iron can be used where higher ductility or strength is required than is available in grey iron. Typical applications of ductile iron include crank shafts, heavy-duty gears, and automotive door hinges.

White Iron

White iron is produced by a process called chilling, which prevents carbon from precipitating out. Because of its extreme hardness, white iron is used primarily for applications requiring wear and abrasion resistance, such as mill liners and shot-blasting nozzles. Other uses include brick-making equipment, crushers, and pulverizers. The disadvantage of white iron is that it is very brittle.

Malleable Iron

Malleable iron is a white iron that has been converted to a malleable condition. It is a commercially cast material that is similar to steel in many respects. It is strong and ductile, has good impact and fatigue properties, and has excellent machining characteristics.

The two basic types of malleable iron are ferritic and pearlite. Ferritic grades are more machinable and ductile, whereas the pearlite grades are stronger and harder. For design information on the casting process for cast iron see Unit 27.

STEEL

Carbon steels are the workhorses of product design. They account for over 90 percent of total steel production. More carbon steels are used in product manufacturing than all other metals combined. Far more research is going into carbon steel metallurgy and manufacturing technology than into all other steel mill products. Various technical societies and trade associations have issued specifications covering the composition of metals. They serve as a selection guide and are a way for the buyer to conveniently specify certain known and recognized requirements. The main technical societies and trade associations concerned with metal identification in Canada and the United States are the Canadian Institute of Steel Construction (CISC), the American Iron and Steel Institute (AISI), and the Society of Automotive Engineers (SAE).

SAE and AISI Systems of Steel Identification

The specifications for steel bar are based on a number code listing the composition of each type of steel. They include both plain carbon and alloy steels. The code is a four-number system: The first two figures indicate the alloy series, and the last two figures the carbon content in hundredths of a percent, Figure 26–1. The numbering code or symbol XX15 indicates 0.15 of 1 percent carbon.

For example, AISI 4830 is a molybdenum-nickel steel containing 0.2–0.3 percent molybdenum, 3.25–3.75 percent nickel, and 0.3 percent carbon. In addition to the four-number designation, the suffix "H" is used to specify hardenability limits, and the prefix "E" indicates a steel made by the basic electric-furnace method.

Originally, the second figure indicated the percentage of the major alloying element. However, this had to be varied to classify all the steels that became available.

Alloying materials are added to steel to improve such properties as strength, hardness, machinability, corrosion resistance, electrical conductivity, and ease of forming. Figure 26–2 lists many types of steel, their properties, and their uses.

Effect of Alloys on Steel

Carbon

Increasing the carbon content increases the tensile strength and hardness.

Sulphur

When sulphur content is over 0.06 percent, the metal tends toward *red shortness* (brittleness in steel when it is red hot). Free-cutting steel, for threading and screw machine work, is obtained by increasing sulphur content from about 0.075 to 0.1 percent.

Phosphorus

Phosphorus produces brittleness and general cold shortness. It also strengthens low carbon steel, increases resistance to corrosion, and improves machinability.

Manganese

Manganese is added during the making of steel to prevent red shortness and increase hardenability.

Chromium

Chromium increases hardenability, corrosion resistance, oxidation, and abrasion.

Nickel

Nickel strengthens and toughens ferrite and pearlite steels.

Silicon

Silicon is a general-purpose deoxidizer. It strengthens low-alloy steels and increases hardenability.

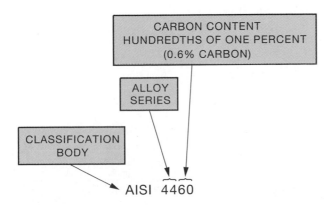

FIGURE 26–1 ■ Steel designation system

TYPE OF STEEL	AISI SYMBOL	PRINCIPAL PROPERTIES	COMMON USES
CARBON STEELS			
- Nonresulphurized (Basic Carbon Steel)	10XX		
- Plain Carbon	10XX		
- Low Carbon Steel (0.06% to 0.2% Carbon)	1006 to 1020	Toughness and less strength	Chains, rivets, shafts, pressed steel products
- Medium Carbon Steel (0.2% to 0.5% Carbon)	1020 to 1050	Toughness and strength	Gears, axles, machine parts, forgings, bolts, and nuts
- High Carbon Steel (Over 0.5% Carbon)	1050 and over	Less toughness and greater hardness	Saws, drills, knives, razors, finishing tools, music wire
- Resulphurized (Free Cutting)	11XX	Improves machinability	Threads, splines, machined parts
- Phosphorized	12XX	Increases strength and hardness but reduces ductility	
- Manganese Steels	13XX	Improves surface finish	
MOLYBDENUM STEELS			
0.15%–0.30% Mo	40XX		
0.08%–0.35% Mo 0.4%–1.1% Cr	41XX	High strength	Axles, forgings, gears, cams, mechanism parts
1.65%–2.00% NI 0.2%–0.3% Mo			
0.4%–0.9% Cr	43XX		
0.45%–0.60% Mo	44XX		
0.7%–2.0% Ni 0.15%–0.30% Mo	46XX		
0.90%–1.2% Ni 0.15%–0.40% Mo			
0.35%–0.55% Cr	47XX		
3.25%–3.75% Ni 0.2%–0.3% Mo	48XX		
CHROMIUM STEELS			
0.3%–0.5% Cr	50XX	Hardness, great strength, and toughness	Gears, shafts, bearings springs, connecting rods
0.70%–1.15% Cr	51XX		
1.00% C 0.90%–1.15% Cr	E51100		
1.00% C 0.90%–1.15% Cr	E52100		
CHROMIUM VANADIUM STEELS			
0.5%–1.1% Cr 0.10%–0.15% V	61XX	Hardness and strength	Punches, piston rods, gears, axles
NICKEL - CHROMIUM - MOLYBDENUM STEELS			
0.4%–0.7% Ni 0.4%–0.6% Cr			
0.15%–0.25% Mo	86XX	Rust resistance, hardness, and strength	Food containers, surgical equipment
0.4%–0.7% Ni 0.4%–0.6% Cr			
0.2%–0.3% Mo	87XX		
0.4%–0.7% Ni 0.4%–0.6% Cr			
0.3%–0.4% Mo	88XX		
SILICON STEELS			
1.8%–2.2% Si	92XX	Springiness and elasticity	Springs

FIGURE 26–2 ■ Designations, uses, and properties of steel

Copper

Copper is used to increase atmospheric corrosion resistance.

Molybdenum

Molybdenum increases hardenability and coarsening temperature.

Boron

Boron increases hardenability of lower carbon steels and has better machinability than standard alloy steels.

Vanadium

Vanadium elevates coarsening temperatures, increases hardenability, and is a strong deoxidizer.

Structural Steel

Structural steel shapes and their drawing callouts are covered in Unit 34.

PLASTICS

Plastics are nonmetallic materials capable of being formed or moulded with the aid of heat, pressure, chemical reactions, or a combination of these.

Plastics are strong, tough, durable materials that solve many problems in machine and equipment design. Metals are hard and rigid. This means they can be machined to a very close tolerance into cams, bearings, bushings, and gears that will work smoothly under heavy loads for long periods. Although some come close, no plastic has the hardness or creep resistance of steel, for example. However, metals have many weaknesses that engineering plastics do not. Metals corrode or rust, they must be lubricated, their working surfaces wear readily, they cannot be used as electrical or thermal insulators, they are opaque and noisy, and where they must flex, they fatigue rapidly.

Engineering plastics can be run at low speeds and loads and, without lubrication, are among the world's slipperiest solids, being comparable with ice.

Plastics are a family of materials, not a single material. Each material has its advantage. Being manufactured, plastic raw materials are capable of being combined to give almost any property desired in an end product.

The advantages of plastics are light weight, range of colour, good physical properties, adaptability to mass production methods, and often lower cost. They are usually classified as either *thermoplastic* or *thermosetting*. See Figure 26–3.

Thermoplastics

Thermoplastics soften or liquefy and flow when heat is applied. Removal of the heat causes these materials to set or solidify. They can be

reheated or reformed or reused. This group includes the ABS, acetals, acrylics, celluloses, fluorocarbons, nylons, polyethylene, polystyrene, vinyls, and polycarbonates.

Thermosetting Plastics

Thermosetting plastics undergo an irreversible chemical change when heat is applied or when a catalyst or reactant is added. They become hard, insoluble, and infusible, and they do not soften upon reapplication of heat. Thermosetting plastics include phenolics, amino plastics (melamine and urea), polyesters, epoxies, silicones, alkydes, allylics, and casein.

RUBBER

Rubber is advantageous when design considerations involve one or more of these factors: electrical insulation, vibration isolation, sealing surfaces, chemical resistance, or flexibility.

Elastomers, rubber-like substances, are derived from natural or synthetic sources. Rubber can be formed into useful rigid or flexible shapes, usually with the aid of heat, pressure, or both. The most outstanding characteristics of rubber are its low modulus of elasticity and its ability to withstand large deformation and quickly recover its shape when released.

Rubber parts are produced in mechanical (solid) or cellular form, depending on the desired performance of the part. The two kinds of rubber are natural and synthetic. Synthetic rubbers are further classified into several kinds.

Mechanical Rubber

Mechanical rubber is used in pressure-moulded, cast, or extruded form. Parts typically produced by these methods are tires, belts, and bumpers. Mechanical rubber is preferable to sponge rubber because of its superior physical properties.

Cellular Rubber

Cellular rubber can be produced with open or closed cells. Open-cell sponge rubber is made by including a gas-forming chemical compound in the mixture before vulcanization. The heat of the vulcanizing process causes a gas to form in the rubber, making a cellular structure. Typical applications are pads and weather stripping. Foam rubber is a specialized type of open cell.

BASE RESIN	OUTSTANDING CHARACTERISTICS	TYPICAL APPLICATIONS
THERMOPLASTICS		
ABS	Colourability, toughness	Instrument panels, telephone housings
Acetal	Impact strength, chemical resistance	Replaces zinc or aluminum for castings
Acrylic	Clarity, weather resistance	Windows, lenses, dials, shoe heels, medallions
Cellulosic	Clarity, toughness	Safety glass, film, knobs, bowling balls, electrical parts
Fluorocarbon	Chemical and heat resistance, self-lubrication	Low-pressure bearings, chemical linings, gaskets, bushings
Polyamide (Nylon)	Impact strength, cold flow	Gears, cams, bushings, gaskets, cable clamps, housings, electrical insulation
Polycarbonate	Dimensional stability, clarity, good electrical insulation	Switch housings, terminal blocks
Polyethylene	Ease of forming, chemical resistance	Bottles, toys, underground pipe, electrical insulation, ducts
Polypropylene	Ease of forming, chemical resistance	Bottles, toys, replacement for zinc or aluminum for castings
Polystyrene	Ease of forming, transparent	Toys, display and jewellery cases, foamed insulation, refrigerator parts, containers
Polyvinyl Chloride	Flexibility, toughness, chemical resistance	Toys, upholstery, lawn hose, floor tile, electrical insulation, gaskets, ducts
THERMOSETTING PLASTICS		
Epoxy	Room temperature cure, no pressure	Adhesives, coatings, terminal boards, electrical potting, tooling fixtures
Phenol	Dark colours only, good electrical insulation	Knobs, terminal boards, adhesives, distributor caps and housings
Polyester	Room temperature cure on pressure	Car bodies, boat hulls, heater ducts
Silicone	Relative constancy of properties over wide temperature range	Terminal boards, electrical mouldings
Urea & Melamine	Indoor use, pastel colours	Knobs, jars, buttons, dishes, electrical mouldings

FIGURE 26–3 ■ Common plastics

Both open- and closed-cell sponge rubber is available in block or sheet form that can be cut to size and shape.

INTERNET RESOURCES

Animated Worksheets For information on materials and their properties, see: http://www.sheets.co.uk (materials)

Bayer MaterialScience AG For information on industrial plastics, see: http://www.bayermaterialscience.com

eFunda For information on manufacturing materials, see: http://www.efunda.com/home.cfm

Gates Rubber Company For information on automotive and industrial drive belts, see: http://www.gates.com

Howstuffworks For information on the drop forging of steel products, see: http://science.howstuffworks.com/question376.htm

Machine Design For information on manufacturing materials and related processes, see: http://www.machinedesign.com

TechStudent.Com For information on manufacturing materials, see: http://www.technologystudent.com (equipment and accessories)

Rubber-Cal For information on various types of rubber and their applications, see: http://www.rubbercal.com

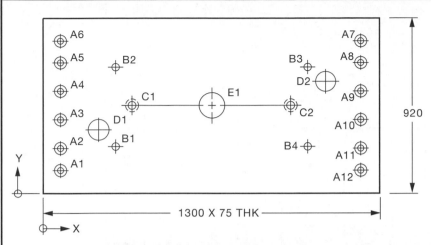

HOLE	LOCATION		HOLE SIZE	HOLE	LOCATION		HOLE SIZE
	X	Y			X	Y	
A1	50	120	12X Ø32	B1	280	260	
A2	50	235	⊔ Ø64	B2	280	660	4X Ø16
A3	50	385	⩒ 30	B3	1020	660	
A4	50	535	NEAR SIDE ONLY	B4	1020	260	
A5	50	685		C1	342	460	2X M20 X 2.5-6G
A6	50	800		C2	958	460	
A7	1250	800		D1	210	334	2X Ø$^{80.30}_{80.00}$ (80 H7)
A8	1250	685		D2	1090	586	
A9	1250	535		E1	650	460	1X Ø$^{100.054}_{100.000}$ (100 H8)
A10	1250	385					
A11	1250	235					
A12	1250	120					

THREAD CONTROLLING ORGANIZATION
AND STANDARD—ASME B1.13M-2001

NOTE: UNLESS OTHERWISE SPECIFIED,
TOLERANCE ON DIMENSIONS ±0.5

QUESTIONS:

1. What is the thickness of the material?

2. In what classification of plastic does the material for the crossbar belong?

3. What indicates that the drawing is not to scale?

4. How many counterbored holes are there?

5. What is the diameter of the CBORE?

6. How many D holes are there?

7. Which letter specifies the Ø100 hole?

8. How many different-sized holes are specified?

9. What is the total number of tapped holes?

10. What is the thread pitch of the C holes?

11. What is the tolerance given for the E hole?

12. What is the high limit of the E hole?

13. What is the maximum size for the D holes?

14. Is the E hole on the centre of the part?

15. What is the centre distance between the C holes?

16. What type of dimensioning is used to establish the location of the holes?

METRIC
DIMENSIONS ARE IN MILLIMETRES

MATERIAL	ABS PLASTIC	
SCALE	NOT TO SCALE	
DRAWN	D. GRAY	DATE 29/09/04

CROSSBAR   **A-70M**

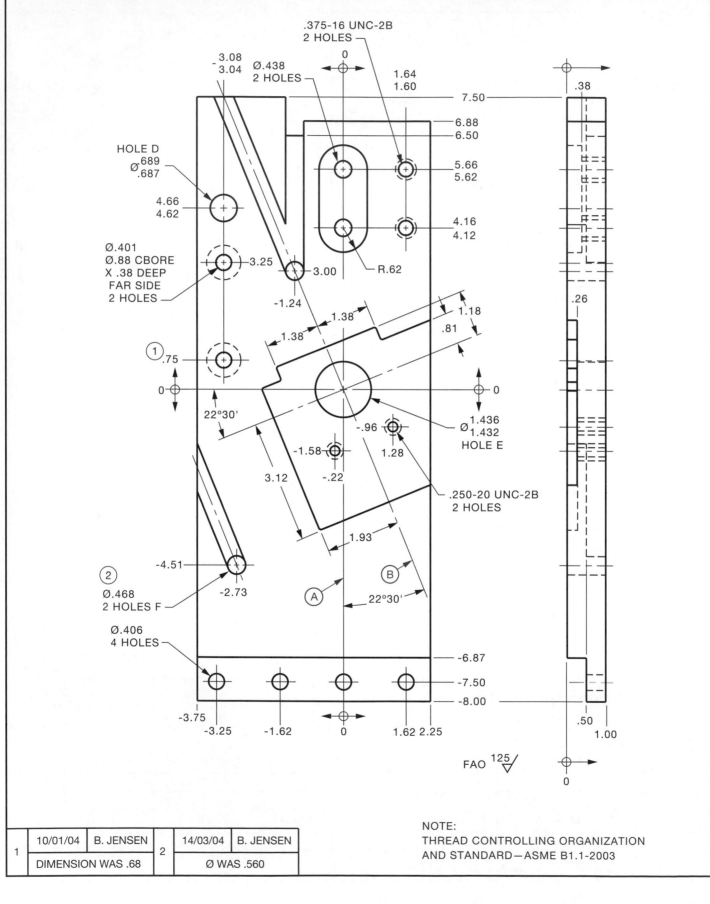

.375-16 UNC-2B
2 HOLES

Ø.438
2 HOLES

-3.08
-3.04

0

1.64
1.60

7.50

6.88
6.50

HOLE D
Ø.689
.687

5.66
5.62

4.66
4.62

4.16
4.12

Ø.401
Ø.88 CBORE
X .38 DEEP
FAR SIDE
2 HOLES

3.25

3.00

R.62

-1.24

1.38

1.18

1.38

.81

① .75

22°30'

Ø 1.436
1.432
HOLE E

0

0

-.96

-1.58

1.28

3.12

-.22

.250-20 UNC-2B
2 HOLES

1.93

②

-4.51

Ⓑ

-2.73

Ⓐ

22°30'

Ø.468
2 HOLES F

Ø.406
4 HOLES

-6.87

-7.50

-8.00

-3.75

-3.25

-1.62

0

1.62 2.25

.38

.26

.50

1.00

FAO ¹²⁵⟍

0

1	10/01/04	B. JENSEN	2	14/03/04	B. JENSEN
	DIMENSION WAS .68			Ø WAS .560	

NOTE:
THREAD CONTROLLING ORGANIZATION
AND STANDARD—ASME B1.1-2003

NEL

QUESTIONS:

1. What are the overall (A) width, (B) height, and (C) depth of the part?

2. What is the width of the slots that are cut out from the F holes?

3. What is their depth?

4. What class of surface texture is required?

5. What is the depth of the recess adjacent to the (A) E hole, (B) Ø.438 holes?

6. How many degrees are there between line B and the horizontal?

7. What was the size of the F holes before they were changed?

8. What tolerance is required on the E hole?

9. What is the low limit on the E hole?

10. What tolerance is required on the D hole?

11. What is the high limit on the D hole?

12. Determine the maximum vertical centre distance from the D hole to the four Ø.406 holes located at the bottom of the part.

13. Determine the (A) maximum and (B) minimum centre distance between the two .375 tapped holes.

14. Determine the maximum horizontal centre distance between the D hole and the .375 tapped holes.

15. What are the width, height, and depth of the cutout at the bottom of the part?

16. How deep are (A) the Ø.438 holes and (B) the Ø.401 holes?

17. How deep are the .250-20 UNC holes?

18. How many full threads do the .250-20 UNC tapped holes have?

19. What are the main alloys in the material?

20. What type of steel is specified for the oil chute?

ASSIGNMENT:

ON A ONE-INCH GRID SHEET (.10 IN. SQUARES), SKETCH TWO SECTION VIEWS, ONE ALONG LINE "A," THE OTHER ALONG LINE "B." SCALE 1:2. ADD THE RECTANGULAR COORDINATE DIMENSIONS WITHOUT DIMENSION LINES TO THESE SECTION VIEWS.

NOTES:

- ALL RADII R.06 UNLESS OTHERWISE SPECIFIED.

- DIMENSIONS ARE TAKEN FROM PLANES DESIGNATED 0-0, AND ARE PARALLEL TO THESE PLANES.

- UNLESS OTHERWISE STATED:

    ±.02 ON DIMENSIONS
    ±0.5 ° ON ANGLES

MATERIAL	SAE 4020 STEEL	
SCALE	NOT TO SCALE	
DRAWN	B. JENSEN	DATE 20/03/03

OIL CHUTE **A-71**

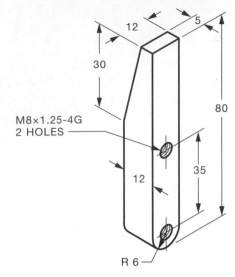

12 5

30

80

M8×1.25-4G
2 HOLES

12

35

R 6

PT 1   MOVABLE JAW
1 REQD   MATL - SAE 1020

PT 6 CAP SCREW
M3×8 LG
RD HD   1 REQD

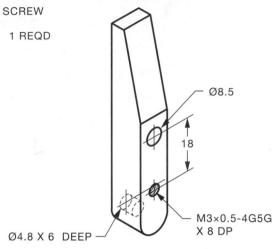

Ø8.5

18

Ø4.8 X 6 DEEP

M3×0.5-4G5G
X 8 DP

AS SHOWN, OTHERWISE SAME
AS PART 1.

PT 2   STATIONARY JAW
1 REQD   MATL - SAE 1020

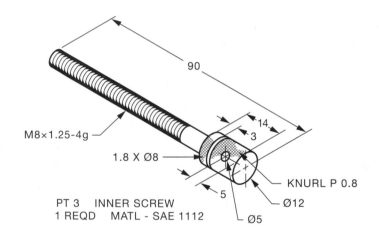

90

M8×1.25-4g

14
3

1.8 X Ø8

KNURL P 0.8

5

Ø12

Ø5

PT 3   INNER SCREW
1 REQD   MATL - SAE 1112

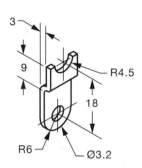

3

9

R4.5

18

R6

Ø3.2

PT 5   CLIP   1 REQD
MATL - 16 GA (1.6) USS STL

ASSIGNMENT:

ON METRIC GRID PAPER, PREPARE WORKING
(DETAIL) DRAWINGS OF THE PARALLEL CLAMP
DETAILS SHOWN. USE SIMPLIFIED THREAD
CONVENTIONS (UNIT 16) AND REFER TO UNIT 11
FOR THE SELECTION OF VIEWS. SCALE 1:1.

THREAD CONTROLLING ORGANIZATION
AND STANDARD – ASME B1.13M-2001

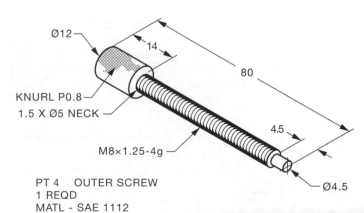

Ø12

14

80

KNURL P0.8

1.5 X Ø5 NECK

M8×1.25-4g

4.5

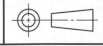

Ø4.5

PT 4   OUTER SCREW
1 REQD
MATL - SAE 1112

METRIC
DIMENSIONS ARE IN MILLIMETRES

**PARALLEL CLAMP
DETAILS**

A-72M

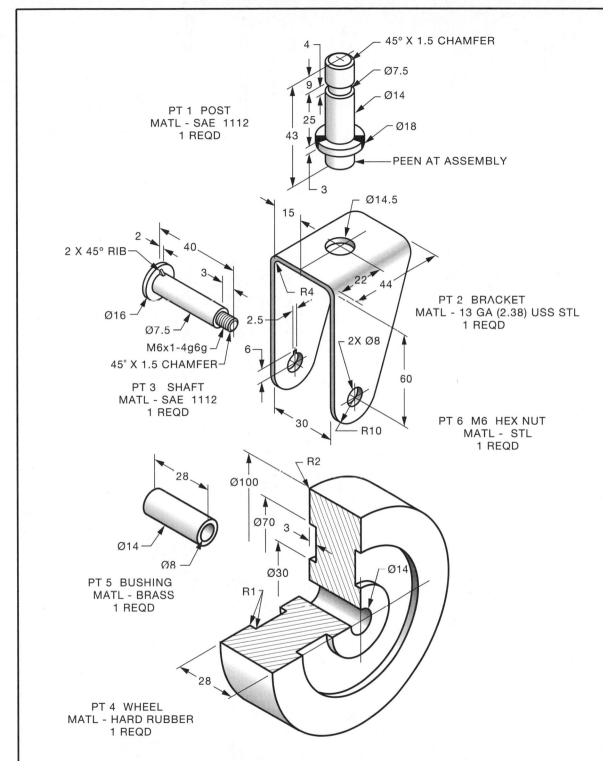

PT 1 POST
MATL - SAE 1112
1 REQD

45° X 1.5 CHAMFER
Ø7.5
Ø14
Ø18
PEEN AT ASSEMBLY

4
9
25
43
3

Ø14.5
15

2 X 45° RIB
2
40
3
Ø16
Ø7.5
M6x1-4g6g
45° X 1.5 CHAMFER

PT 3 SHAFT
MATL - SAE 1112
1 REQD

R4
2.5
6
22
44
2X Ø8
30
R10
60

PT 2 BRACKET
MATL - 13 GA (2.38) USS STL
1 REQD

PT 6 M6 HEX NUT
MATL - STL
1 REQD

28
Ø14
Ø8

PT 5 BUSHING
MATL - BRASS
1 REQD

Ø100
Ø70
Ø30
R1
R2
3
Ø14

PT 4 WHEEL
MATL - HARD RUBBER
1 REQD

ASSIGNMENT:
ON CENTIMETRE GRID SHEETS (1 mm SQUARES), PREPARE DETAILED SKETCHES,
COMPLETE WITH DIMENSIONS, OF THE CASTER PARTS SHOWN. SCALE 1:1.

METRIC
DIMENSIONS ARE IN MILLIMETRES

THREAD CONTROLLING ORGANIZATION
AND STANDARD—ASME B1.13M-2001

CASTER ASSEMBLY | A-73M

27 UNIT

CASTING PROCESSES

Irregular or odd-shaped parts that are difficult to make from metal plate or bar stock may be cast to the desired shape. Casting processes for metals can be classified by either the type of mould or pattern or the pressure of force used to fill the mould. Conventional sand, shell, and plaster moulds utilize a durable pattern, but the mould is used only once. Permanent moulds and die-casting dies are machined in metal or graphite sections and are used for a large number of castings. Investment casting and the full mould process involve both an expendable mould and a pattern.

Sand Mould Casting

The most widely used casting process for metals uses a permanent pattern of metal or wood that shapes the mould cavity when loose moulding material is compacted around the pattern. This material consists of a fine sand plus a binder that serves as the adhesive.

Figures 27–1 and 27–2 show a typical sand mould, with the various provisions for pouring the molten metal and compensating for contraction of the solidifying metal, and a sand core for forming a cavity in the casting. Sand moulds are prepared in flasks, which consist of two or more sections: bottom (drag), top (cope), and intermediate sections (cheeks) when required.

The cope and drag are equipped with pins and lugs to ensure the alignment of the flask. Molten metal is poured into the sprue, and connecting runners provide flow channels for the metal to enter the

mould cavity through gates. Riser cavities are located over the highest section of the casting.

The gating system, besides providing a way for the molten metal to enter the mould, functions as a venting system for the removal of gases from the mould and acts as a riser to furnish liquid metal to the casting during solidification.

In producing sand moulds, a metal or wooden pattern must first be made. The pattern is slightly

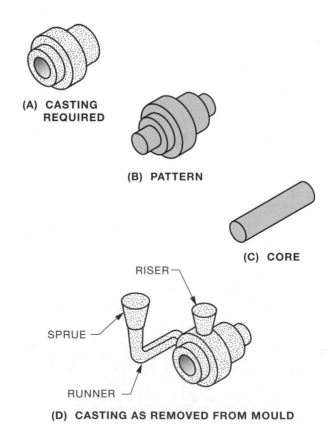

(A) CASTING REQUIRED

(B) PATTERN

(C) CORE

RISER

SPRUE

RUNNER

(D) CASTING AS REMOVED FROM MOULD

FIGURE 27–1 ■ Sand casting parts

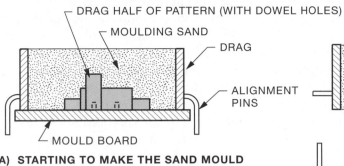

(A) STARTING TO MAKE THE SAND MOULD

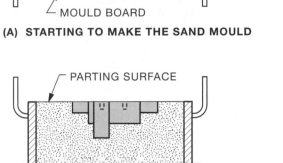

(B) AFTER ROLLING OVER THE DRAG

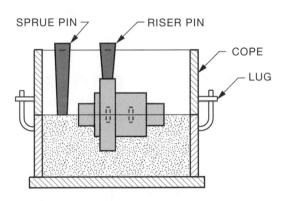

(C) PREPARING TO RAM MOULDING SAND IN COPE

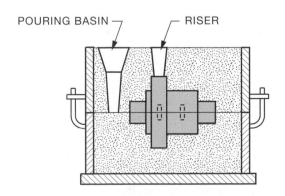

(D) REMOVING RISER AND GATE SPRUE PINS AND ADDING POURING BASIN

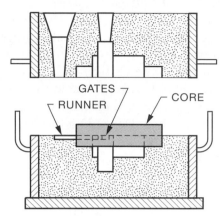

(E) PARTING COPE AND DRAG TO REMOVE PATTERN AND TO ADD CORE AND RUNNER

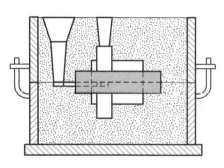

(F) SAND MOULD READY FOR POURING

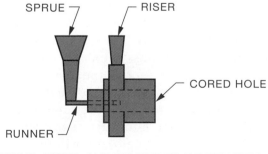

SPRUE, RISER, AND RUNNER TO BE REMOVED FROM CASTING

(G) CASTING AS REMOVED FROM THE MOULD

FIGURE 27–2 ■ Sequence in preparing a sand casting

larger in every dimension than the part to be cast to allow for shrinkage when the casting cools. This is known as *shrinkage allowance,* which the patternmaker allows for by using a shrink rule for each of the cast metals. Because shrinkage and draft are taken care of by the patternmaker, they are of no concern to the drafter.

Additional metal, known as machining or finish allowance, must be provided on the casting where a surface is to be finished. Depending on the material being cast, from .06 to .12 in. is usually allowed on small castings for each surface requiring finishing.

When casting a hole or recess, a core is often used. A core is a mixture of sand and a bonding agent that is baked and hardened to the desired shape of the cavity in the casting, plus an allowance to support the core in the sand mould. In addition to the shape of the casting desired, the pattern must be designed to produce areas in the mould cavity to locate and hold the core. The core must be solidly supported in the mould, permitting only that part of the core that corresponds to the shape of the cavity in the casting to project into the mould.

Preparation of Sand Moulds

The drag portion of the flask is first prepared in an upside-down position with the pins pointing down, Figure 27–2(A). The drag half of the pattern is placed in position on the mould board, and a light coating of parting compound is used as a release agent. The moulding sand is then rammed or pressed into the drag flask. A bottom board is placed on the drag, the whole unit is rolled over, and the mould board is removed.

The cope half of the pattern is placed over the drag half, and the cope portion of the flask is placed in position over the pins, Figure 27–2(C). A light coating of parting compound is sprinkled throughout. Next, the sprue pin and riser pin, tapered for easy removal, are located, and the moulding sand is rammed into the cope flask. The sprue pin and riser pin are then removed, Figure 27–2(D). A pouring basin may or may not be formed at the top of the gate sprue.

Now the cope is lifted carefully from the drag, and the pattern is exposed. The runner and gate, passageways for the molten metal into the mould cavity, are formed in the drag sand. The pattern is removed, and the core is placed in position, Figure 27–2(E). The cope is then put back on the drag, Figure 27–2(F).

The molten metal is poured into the pouring basin and runs down the sprue to a runner and through the gate and into the mould cavity. When the mould cavity is filled, the metal will begin to fill the sprue and riser. Once the sprue and riser have been filled, the pouring should stop.

When the metal has hardened, the sand is broken and the casting removed, Figure 27–2(G). The excess metal, gates, and risers are removed and later remelted.

Full Mould Casting

The characteristic feature of the *full mould process* is the use of gasifiable patterns made of foamed plastic. These are not extracted from the mould but are vaporized by the molten metal.

The full mould process is suitable for individual castings and for a small series of up to five castings. Very economical, it reduces the delivery time required for prototypes, articles urgently needed for repair jobs, and individual large machine parts.

CASTING DESIGN

Simplicity of Moulding from Flat Back Patterns

Simple shapes such as the one shown in Figure 27–3, are very easy to mould. In this case, the flat face of the pattern is at the parting line and lies perfectly flat on the moulding board. In this position, no moulding sand sifts under the flat surface to interfere with the drawing of the pattern. The simplicity with which flat back patterns of this type may be drawn from the mould is illustrated in Figure 27–3.

Irregular or Odd-Shaped Castings

When a casting is to be made for an odd-shaped piece, such as the offset bracket, Figure 27–4, it is necessary to make the pattern for the bracket in one or more parts to facilitate making the mould. The difficulties with this are mostly due to the removal of the pattern from the sand mould.

The pattern for the bracket is made in two parts, Figure 27–5. The two adjacent flat surfaces of the divided pattern come together at the parting lines.

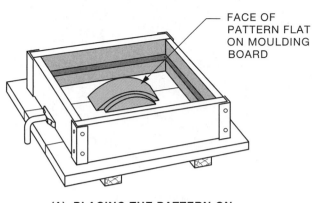

(A) PLACING THE PATTERN ON THE MOULDING BOARD

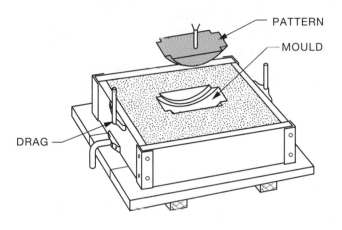

(B) DRAWING THE PATTERN

FIGURE 27-3 ■ Making a mould of a flat back pattern

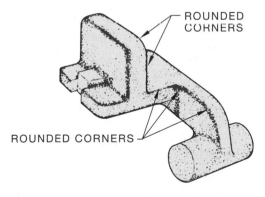

FIGURE 27-4 ■ Casting of offset bracket shown in Assignment A-74

Set Cores

The offset bracket, Assignment A-74, has rounded corners. To mould these rounded corners, a block must be added to the pattern for ease in removing the piece from the mould.

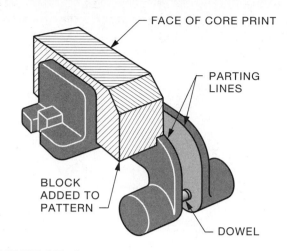

FIGURE 27-5 ■ Two-piece pattern for offset bracket

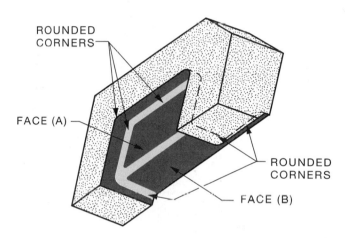

FIGURE 27-6 ■ Set core for offset bracket

This block, which becomes an integral part of the pattern, also acts as a core print for a *set core* (a core that has been baked hard), Figure 27-6. It is made to conform to the shape of the faces of the casting, including the rounded corners.

The face of the core print also forms the parting line for one side of the two-part pattern. In making the mould for the bracket casting, this face, which corresponds to the flat face of a flat back pattern, is laid on the moulding board with the drag of the flask in position. The sand is then rammed around the pattern.

When the drag is reversed, the cope, or upper part of the flask, is placed in position. The other half of the pattern is then joined with the first part, and the sand is rammed into the cope to flow around that part of the pattern that projects into it.

After the pattern is removed, the set core, which is formed in a core box and baked hard, is set in the impression in the mould made by the core print of

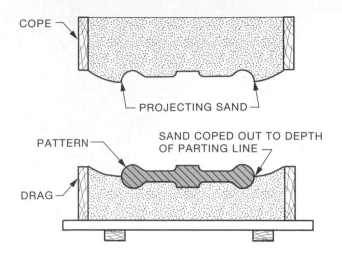

FIGURE 27–7 ■ Coping down

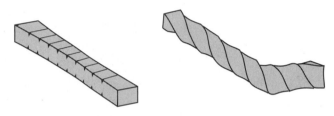

FIGURE 27–8 ■ Soldiers and gaggers

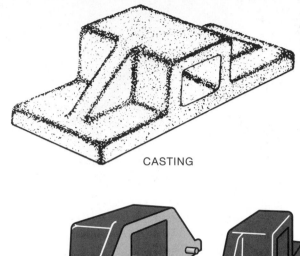

CASTING

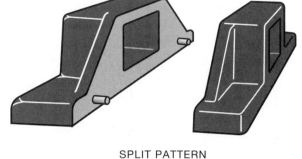

SPLIT PATTERN

FIGURE 27–9 ■ Application of split pattern

the pattern. When poured into the mould, the molten metal fills the cavity made by the pattern and the faces of the set core as shown in Figure 27–6 at A and B to form the casting.

Coping Down

The core print is made as part of the pattern to avoid removing moulding sand in the drag, which would correspond to the shape of the core print.

If this sand were dug out or *coped out,* as shown in Figure 27–7, the remaining cavity would be again filled with moulding sand when the cope was rammed. The sand would then hang below the parting line of the cope down into the drag.

When the mould is made by coping down, the hanging portion of the cope is supported by soldiers or gaggers embedded within the sand to hold the projecting part in position for subsequent operations, Figure 27–8.

Coping down requires skill and takes time. The set core principle is at times preferred to coping down to avoid delay and ensure a more even parting line on the casting.

Split Patterns

Irregularly shaped patterns that cannot be drawn from the sand are sometimes split so that one half of the pattern can be rammed in the drag as a simple flat back pattern while the cope half, when placed in position on the drag half, forms the mould in the cope. Patterns of this type are called *split patterns* and do not require coping down to the parting line, which would be necessary if the pattern were made solid.

The pattern for the casting in Figure 27–9, when made without a print for a core, can be drawn from the sand only by splitting the pattern on the parting line as illustrated.

The drawing must be examined to determine how the pattern should be constructed. This is important, because the parting line must be located in a position that permits the halves of the pattern to be drawn from the sand without interference.

CORED CASTINGS

Cored castings have certain advantages over solid castings. Where practical, castings are designed with cored holes or openings for economy, appearance, and accessibility to interior surfaces.

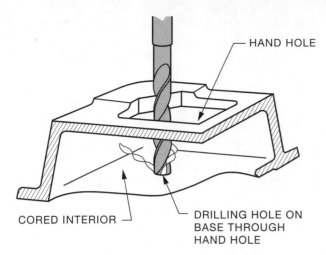

FIGURE 27-10 ■ Section of cored casting

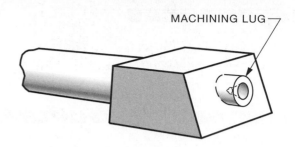

FIGURE 27-11 ■ Application of machining lug

Cored openings often improve the appearance of a casting. In most instances, cored castings are more economical than solid castings because of the savings in metal. Although cored castings are lighter, they are designed without sacrificing strength. The openings cast in the part eliminate unnecessary machining.

Hand holes can also be formed by coring to provide an opening through which the interior of the casting can be reached. These openings also permit machining an otherwise inaccessible surface of the part, Figure 27-10 and Assignment A-76.

MACHINING LUGS

It is difficult to hold and machine certain parts without using lugs, an integral part of the casting. They are sometimes removed to avoid interference with the functioning of the part. Lugs are usually represented in phantom outline.

An example of a machining lug is shown in Figure 27-11. This lug is used to provide a flat surface on the end of the casting for centring and to give a uniform centre bearing for other machining operations. Both the function of the lug and its appearance determine whether it is removed after machining.

SURFACE COATINGS

Machined parts are frequently finished either to protect the surfaces from oxidation or for appearance. The type of finish depends on the use of the part. It may be a protective coat of paint or lacquer, or a metallic plating. In some cases, only a surface finish such as polishing or buffing may be specified.

The finish may be applied before any machining is done, between the various stages of machining, or after the piece has been completed. The type of finish is usually specified on the drawing in a notation similar to the one indicated on the auxiliary pump base, Assignment A-76, which reads CASTING TO BE PAINTED WITH ALUMINUM BEFORE MACHINING.

The decorative finish on the drive housing should be added before machining so that there will be no accidental deposit of paint on the machined surfaces. This will ensure the desired accuracy when assembled with other parts.

REFERENCES

Machine Design, Materials Reference Issue, Mar. 1981
ASME Y14.8M-1996 (R2002) Castings and Forgings

INTERNET RESOURCES

eFunda For information on castings, surface coatings, and related topics, see: http://www.efunda.com/home.cfm

Intermet For links to a variety of casting industries and associations, see: http://www.intermet.com

Industrial Coaters List For a complete list of industrial coating manufacturers with links to specific sites, see: http://industrialquicksearch.com

Machine Design For information on forming processes, see: http://www.machinedesign.com

Meehanite Metal Corp. For information on cast iron and related materials, see: http://www.meehanite.com

TechStudent.com For information on cast iron and casting processes, see: http://www.technologystudent.com (equipment and accessories)

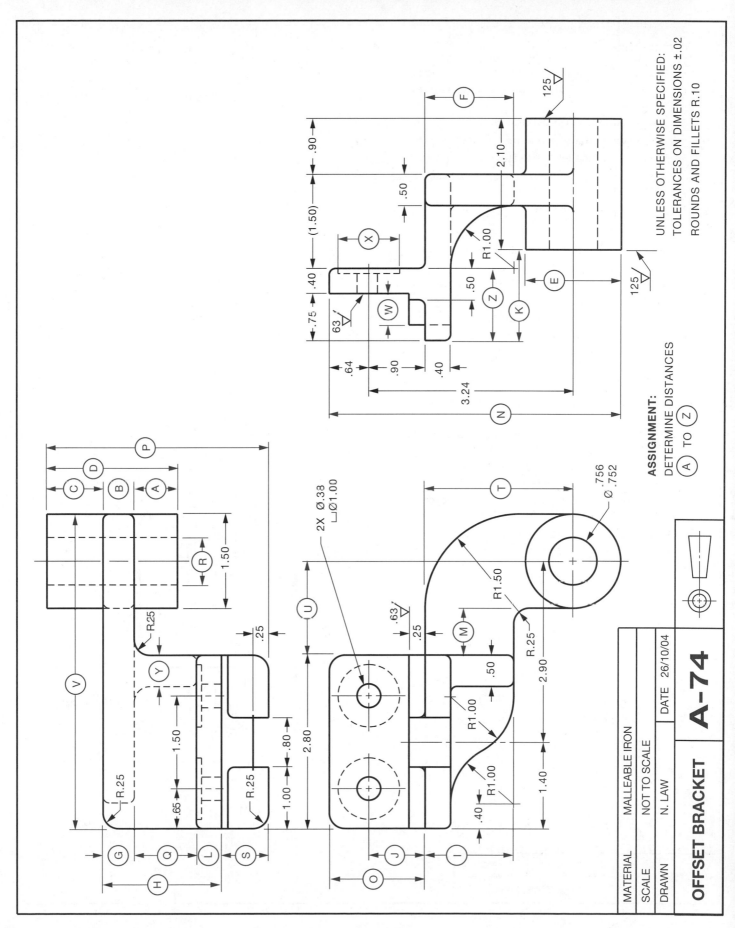

UNLESS OTHERWISE SPECIFIED:
TOLERANCES ON DIMENSIONS ±.02
ROUNDS AND FILLETS R.10

ASSIGNMENT:
DETERMINE DISTANCES
Ⓐ TO Ⓩ

MATERIAL	MALLEABLE IRON	
SCALE	NOT TO SCALE	
DRAWN	N. LAW	DATE 26/10/04
OFFSET BRACKET		**A-74**

QUESTIONS:

1. Which line in the top view represents surface ①?

2. Locate surface Ⓐ in the left-side view and the front view.

3. Locate surface ⑧ in the front view.

4. How many surfaces are to be finished?

5. Which line in the left-side view represents surface ③?

6. What is the vertical centre distance between holes Ⓑ and Ⓞ in the front view?

7. Determine distances at ④, ⑤, ⑥, and ⑪.

8. Locate surface Ⓙ in the top view.

9. Which surface of the left-side view does line ⑭ represent?

10. What point in the front view is represented by line ⑮?

11. What is the thickness of boss Ⓔ ?

12. Locate surface Ⓖ in the left-side view.

13. Locate point Ⓚ in the top view.

14. Locate surface Ⓓ in the top view.

15. Determine distances at Ⓜ, Ⓝ, Ⓢ, and Ⓣ.

16. What point or line in the top view is represented by point ⑯?

17. What is the maximum horizontal centre distance between the holes in the front view?

18. What would be the (A) width, (B) height, and (C) depth of the part before machining?

19. Name the machining processes that could provide the surface quality required for this part.

20. What was the vertical distance from the centre of the Ø.38 hole and the top of the trip box before the drawing revision was made?

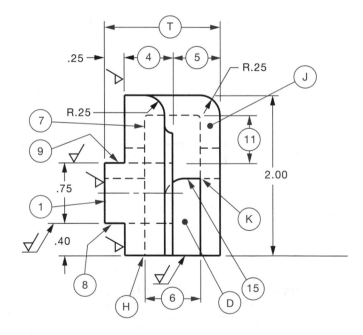

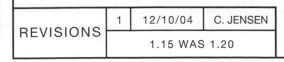

REVISIONS	1	12/10/04	C. JENSEN
		1.15 WAS 1.20	

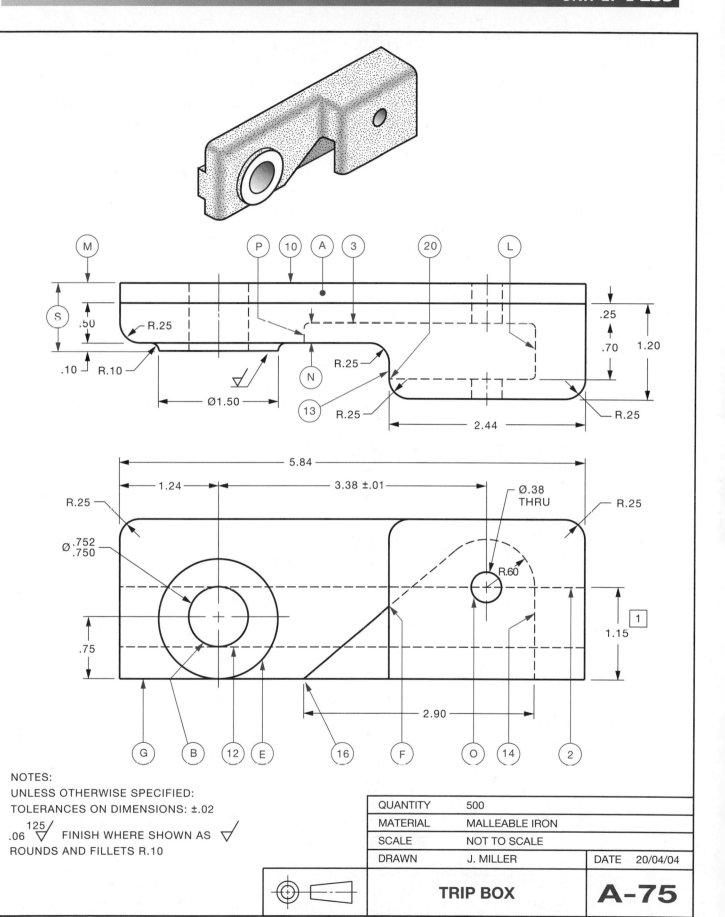

NOTES:
UNLESS OTHERWISE SPECIFIED:
TOLERANCES ON DIMENSIONS: ±.02

.06 $\overset{125}{\underset{\nabla}{/}}$ FINISH WHERE SHOWN AS ∇
ROUNDS AND FILLETS R.10

QUANTITY	500	
MATERIAL	MALLEABLE IRON	
SCALE	NOT TO SCALE	
DRAWN	J. MILLER	DATE 20/04/04
	TRIP BOX	**A-75**

QUESTIONS:

1. Of what material is the base made?
2. How many finished surfaces are indicated?
3. Which circled letters on the drawing indicate the spaces that were cored when the casting was made?
4. Locate the surface in the top view that is represented by line ⑥.
5. What is the height of the cored area Ⓧ?
6. What surface finish is used on the top pad?
7. What is the horizontal length of the pad ⑤?
8. What reason might be given for openings Ⓠ, Ⓢ, Ⓨ, and Ⓩ?
9. Determine distance Ⓦ.
10. What radius fillet would be used at Ⓡ?
11. What is the total allowance added to the height for machining?
12. Determine distances Ⓒ through Ⓝ.
13. What is the size of the tap drill required for the four threaded holes?
14. What size bolts would be used to hold the auxiliary pump base in position?
15. What must be done to the casting before machining?

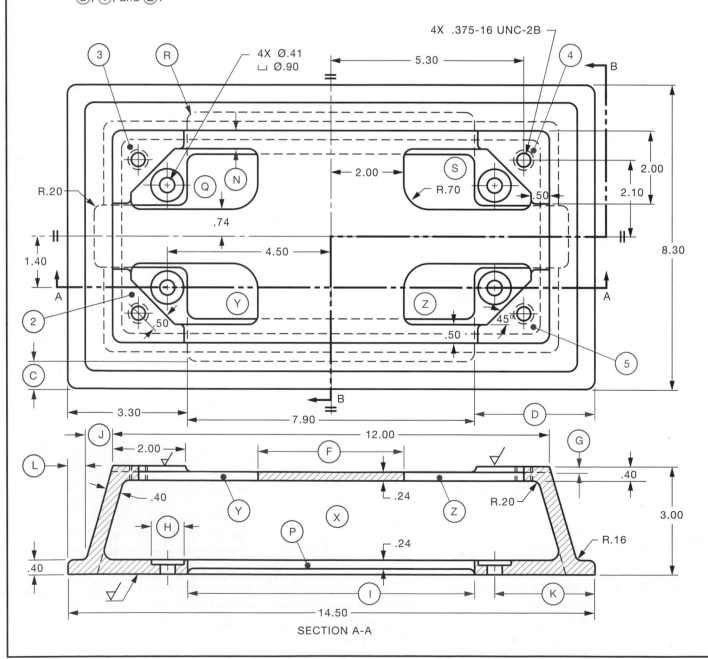

SECTION A-A

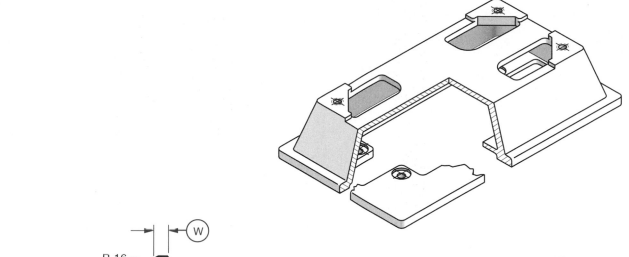

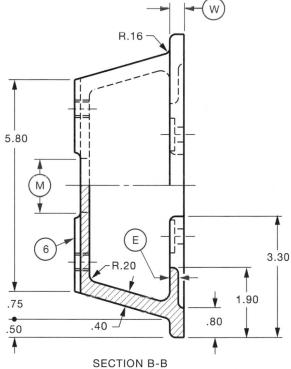

SECTION B-B

R.16

W

M

6

R.20

E

5.80

3.30

1.90

.75

.50

.40

.80

NOTES:

UNLESS OTHERWISE SPECIFIED:

— TOLERANCES ON DIMENSIONS ±.02

— ROUNDS AND FILLETS R.25

— FINISHES $\frac{125}{.06}$✓

— CASTING TO BE PAINTED ALUMINUM BEFORE MACHINING

THREAD CONTROLLING ORGANIZATION
AND STANDARD – ASME B1.1-2003

MATERIAL	GREY IRON	
SCALE	NOT TO SCALE	
DRAWN	T. WRIGHT	DATE 25/10/04

**AUXILIARY PUMP
BASE**

A-76

NOTE:

1. TOLERANCE ON DIMENSIONS ±0.5

2. ▽ TO BE 1.6/▽

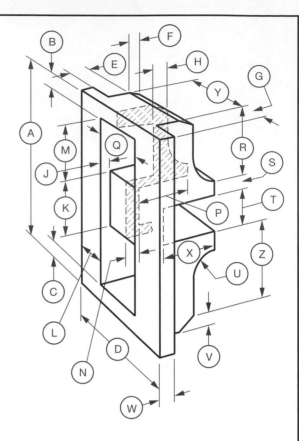

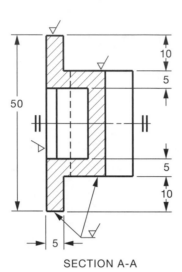

SECTION A-A

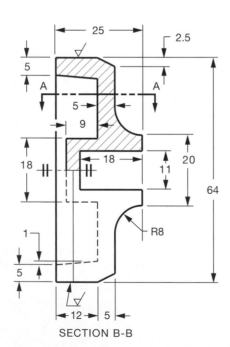

SECTION B-B

ASSIGNMENTS:

1. ON A CENTIMETRE GRID (1 mm SQUARES), SKETCH THE RIGHT-SIDE VIEW OF THE SLIDE VALVE AND SHOW THE POSITION OF THE CUTTING PLANE FOR SECTION B-B.

2. DETERMINE DISTANCES (A) TO (Z).

METRIC
DIMENSIONS ARE IN MILLIMETRES

MATERIAL	GREY IRON	
SCALE	NOT TO SCALE	
DRAWN	J. HELSEL	DATE 10/11/04

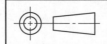

SLIDE VALVE **A-77M**

CHAIN DIMENSIONING

Most linear dimensions are intended to apply on a point-to-point basis. *Chain* dimensioning is applied directly from one feature to another, Figure 28–1(A). Such dimensions locate surfaces and features directly between the points indicated, or between corresponding points on the indicated surfaces.

For example, a diameter applies to all diameters of a cylindrical surface (not merely to the diameter at the end where the dimension is shown), a thickness applies to all opposing points on the surfaces, and a hole-locating dimension applies from the hole axis perpendicular to the edge of the part on the same centre line.

BASE LINE (DATUM) DIMENSIONING

When several dimensions extend from a common data point or points on a line or surface, Figure 28–1(B), this is called *base line,* or *datum,* dimensioning. This form of dimensioning is preferred for parts to be manufactured by numerical control machines.

A *feature* is a physical portion of a part, such as a surface, hole, tab, or slot.

A *datum* is a theoretically exact point, axis, or plane derived from the true geometric counterpart of a specified datum feature. A datum may be the centre of a circle, the axis of a cylinder, or the axis of symmetry.

The location of a series of holes or step features, Figure 28–1(B), is an example of the use of base line dimensioning. Figure 28–2 illustrates more complex uses of base line dimensioning. Without this type of dimensioning, the distances between the first and last steps could vary considerably because of the buildup of tolerances permitted between the adjacent holes. Dimensioning from a datum line controls the tolerances between the holes to the basic general tolerance of ±.02 inch.

In this system, the tolerance from the common point to each of the features must be held to half the tolerance acceptable between individual features. For example, in Figure 28–1(C), if a tolerance between two individual holes of ±.02 inch were desired, each of the dimensions shown would have to be held to ±.01 inch.

REFERENCES

CAN/CSA-B78.2-M91 Dimensioning and Tolerancing of Technical Drawings
ASME Y14.5M-1994 (R2004) Dimensioning and Tolerancing

INTERNET RESOURCE

American Society of Mechanical Engineers
For information on chain and base line dimensioning, refer to ASME Y14.5M-1994 (R2004), at: http://www.asme.org

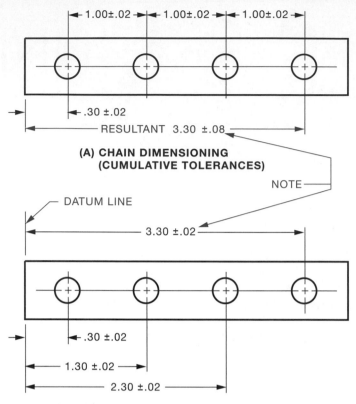

**(A) CHAIN DIMENSIONING
(CUMULATIVE TOLERANCES)**

**(B) BASE LINE DIMENSIONING
(NON-CUMULATIVE TOLERANCES)**

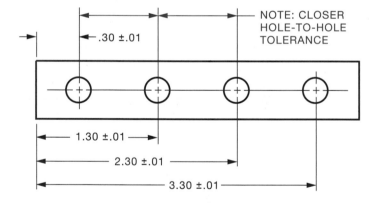

**(C) MAINTAINING SAME DIMENSION BETWEEN
HOLES AS (A) OR (B) BUT WITH CLOSER
TOLERANCE**

FIGURE 28–1 ■ Comparison between chain dimensioning and base line dimensioning

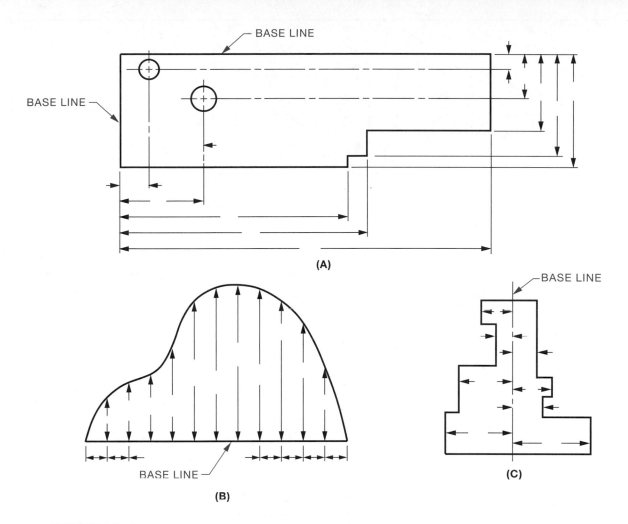

FIGURE 28–2 ■ Applications of base line dimensioning

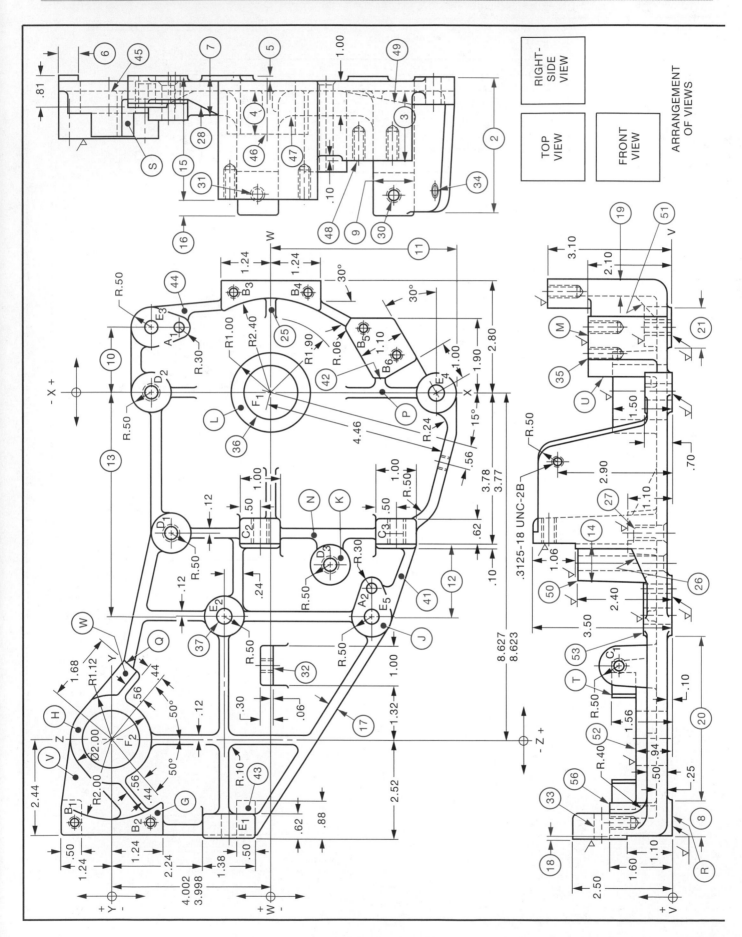

RIGHT-SIDE VIEW

TOP VIEW

FRONT VIEW

ARRANGEMENT OF VIEWS

QUESTIONS:

NOTE: Where limit dimensions are given, use larger limit.

1. What are the overall dimensions of the casting?

2. Show a roughness symbol meaning the same as the one given on the drawing.

3. Identify surfaces (G) to (U) on one of the other views.

4. Locate rib (25) on the front view.

5. Locate rib (26) on the top view.

6. Locate rib (27) on the right-side view.

7. Locate rib (28) on the top view.

8. Give the sizes of the following holes: (30),(31),(32), (33), and (34).

9. How deep is hole (35)?

10. What is the tolerance on hole (36)?

11. How deep is hole (37)?

12. Determine distances (2) to (21). (Where limit dimensions are shown, use maximum size.)

13. What is the (A) largest and (B) smallest permissible hole on the part? Do not include tapped holes.

14. What is the smallest tap drill size used?

HOLE SIZE AND LOCATION

HOLE	DISTANCE FROM					SIZE
	V-V	W-W	X-X	Y-Y	Z-Z	
A₁		2.32	1.62			Ø .252 / .250
A₂		-2.50	-4.88			
B₁				.94	-2.12	.3125-18 UNC-2B X.75 DEEP
B₂				-.94	-2.12	
B₃		.94	2.50			
B₄		-.94	2.50			
B₅		-2.28	1.62			
B₆		-3.08	.96			
C₁	1.38				-1.82	.375-16 UNC-2B
C₂	3.00	.28				
C₃	3.00	-3.12				
D₁		2.50	-3.50			Ø .3752 / .3750 CSK .06X95°
D₂		3.00	.00			
D₃		-1.48	-4.28			
E₁	1.80			-2.94		Ø.406
E₂		1.18	-5.56			
E₃		3.00	1.62			
E₄		-4.12	.00			
E₅		-2.50	-5.56			
F₁		.00	.00			Ø 1.3765 / 1.3745
F₂				.00	.00	

NOTES:
— ALL FILLETS R.06 UNLESS OTHERWISE SHOWN
— ALL RIBS AND WEBS .24 THICK
— TOLERANCE ON DIMENSIONS ±.02
— TOLERANCE ON ANGLES ±0.5°

— SURFACES MARKED ∀ TO BE ²⁵⁰∀

THREAD CONTROLLING ORGANIZATION
AND STANDARD—ASME B1.1-2003

MATERIAL	GREY IRON	
SCALE	NOT TO SCALE	
DRAWN	P. JENSEN	DATE 19/10/04

INTERLOCK BASE

A-78

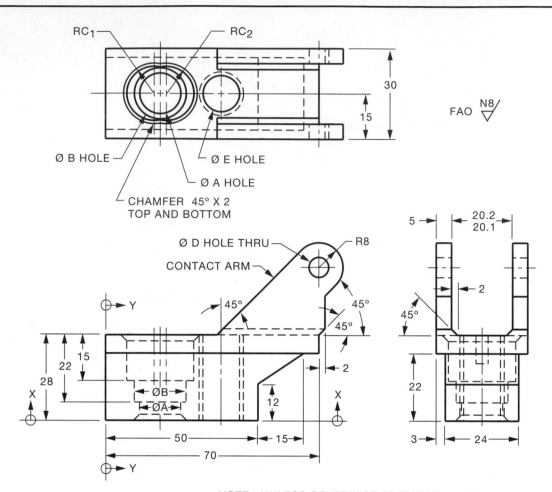

RC₁ · RC₂

Ø B HOLE
Ø E HOLE
Ø A HOLE
CHAMFER 45° X 2
TOP AND BOTTOM

30
15

FAO N8

Ø D HOLE THRU · R8
CONTACT ARM

45° · 45° · 45°

Y
15
22
28
ØB
ØA
X
X
50 · 15
70
12 · 2
Y

5 · 20.2 / 20.1
2
45°
22
3 · 24

NOTE: UNLESS OTHERWISE SPECIFIED
- TOLERANCE ON DIMENSIONS ±0.5
- TOLERANCE ON ANGLES ±0.5°

QUESTIONS:

1. What is the overall width?
2. What is the overall height?
3. Give the chamfer angle for the C hole.
4. What is the distance from Y-Y to E hole?
5. What is the distance from X-X to D hole?
6. How many complete threads does the tapped hole have?
7. Which thread series is the tapped hole?
8. What is the surface finish in micrometres?
9. How deep is the C counterbore from the top of the surface?
10. How long is the B counterbored hole?
11. Give the distance between the C_L of C_1 and C_2 radii.
12. What is the nominal thickness of the contact arm at D hole?
13. What is the tolerance on the distance between the contact arms?
14. What is the tolerance on D hole?
15. What is the centre distance between A and E holes?
16. What type of dimensioning is shown on the front view?
17. What type of dimensioning is used to locate the holes?

HOLE SYMBOL	HOLE SIZE	LOCATION	
		X-X	Y-Y
A	Ø13.5		18
B	Ø17		18
C_1	R9		16
C_2	R9		20
D	Ø6.5-6.6	50	70
E	M12x1.25-4G6G		38

METRIC
DIMENSIONS ARE IN MILLIMETRES

MATERIAL	MANGANESE BRONZE	
SCALE	NOT TO SCALE	
DRAWN	D. SMITH	DATE 09/06/04

THREAD CONTROLLING ORGANIZATION
AND STANDARD—ASME B1.13M-2001

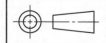

CONTACT ARM | **A-79M**

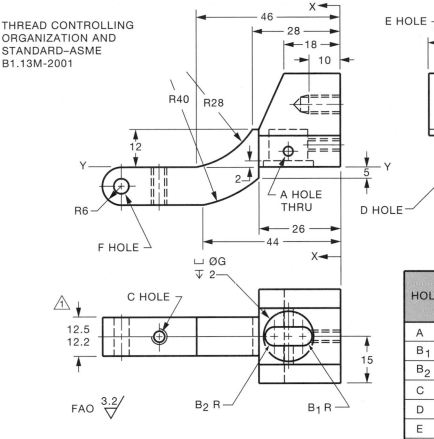

THREAD CONTROLLING
ORGANIZATION AND
STANDARD—ASME
B1.13M-2001

NOTE: UNLESS OTHERWISE SPECIFIED:
TOLERANCE ON DIMENSIONS ±0.5

HOLE	HOLE SIZE	DISTANCE FROM	
		X-X	Y-Y
A	Ø3	16.5	5
B$_1$	R3	12	
B$_2$	R3	21	
C	M5x0.8-5G	58	
D	M4x0.7-5G		7
E	M6x1-5G		20
F	Ø4.78-4.80	70	6
G	Ø16	16.5	

QUESTIONS:

1. What is the overall width?

2. What is the overall height?

3. What is the distance from X-X to the centre of hole F?

4. What is the distance from Y-Y to the centre of hole E?

5. What is the horizontal distance between the centre lines of A and F holes?

6. What is the horizontal distance between the centre lines of C hole and B$_1$ radius?

7. How many full threads does E hole have? (See Table 6 of the Appendix.)

8. How deep is the spotface?

9. What is the tolerance on F hole?

10. What type of projection is used on this drawing?

11. How far apart are the D and E holes?

12. What is the tolerance on the thickness of the material at F hole?

13. What is the length of the C hole?

14. What is the distance from line X-X to the termination of both ends of the 28 radius?

15. What does FAO mean?

16. What is the tap drill size required for the (A) C, (B) D, and (C) E threaded holes?

17. What is the length of the slot?

18. What type of dimensioning is shown on the three views?

19. What type of dimensioning is used to locate the holes?

METRIC
DIMENSIONS ARE IN MILLIMETRES

MATERIAL	ALUMINUM	
SCALE	NOT TO SCALE	
DRAWN	D. SMITH	DATE 09/06/04

REVISIONS	1	F. NEWMAN 15/10/04	CH A. HEINEN			CONTACTOR	A-80M
		DIMENSION WAS 12.4-12.7					

29 UNIT

ALIGNMENT OF PARTS AND HOLES

Two important factors you must consider when drawing an object are the number of views to draw and the time required to draw them. If possible, use time-saving devices, such as templates, for drawing standard features.

To simplify the representation of common features, a number of conventional drawing practices are used, Figure 29–1. Conventions deviate from true projection either for clarity or to save

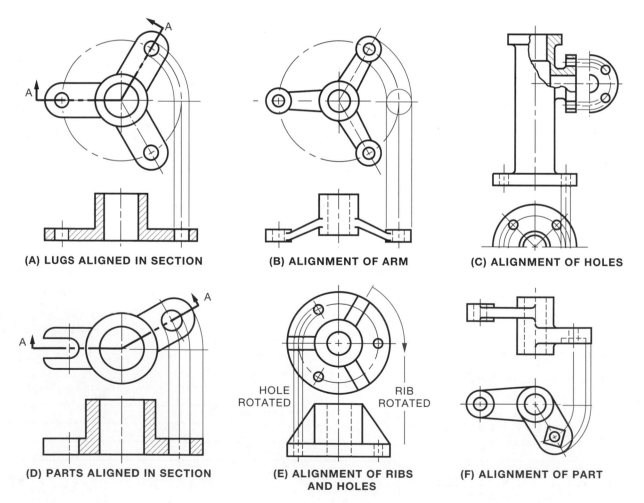

(A) LUGS ALIGNED IN SECTION

(B) ALIGNMENT OF ARM

(C) ALIGNMENT OF HOLES

(D) PARTS ALIGNED IN SECTION

(E) ALIGNMENT OF RIBS AND HOLES

HOLE ROTATED   RIB ROTATED

(F) ALIGNMENT OF PART

FIGURE 29–1 ■ Alignment of holes and parts to show their true relationship

drafting time. These conventions must be executed carefully; clarity is even more important than speed.

Foreshortened Projection

When the true projection of ribs or arms results in confusing foreshortening, these parts should be rotated until parallel to the line of the section or projection, Figure 29–1.

Holes Revolved to Show True Centre Distance

Drilled flanges in elevation or section should show the holes at their true distance from centre, rather than the true projection, Figure 29–1.

PARTIAL VIEWS

Partial views, which show a limited portion of the object with remote details omitted, should be used when necessary to clarify the meaning of the drawing, Figure 29–2. Such views are used to avoid the necessity of drawing many hidden features.

On drawings of objects where two side views can be used to better advantage than one, each need not be complete if together they depict the shape. Show only the hidden lines of features immediately behind the view, Figure 29–2(C).

Another type of partial view is shown in Figure 29–3. However, sufficient information is given in the partial view to complete the description of the object. The partial view is limited by a break line. Partial views are used because:

■ They save time in drawing.

■ They conserve space that might otherwise be needed for drawing the object.

■ They sometimes permit the drawing to be made to scale large enough to bring out all details clearly, whereas if the whole view were drawn, lack of space might make it necessary to draw to a smaller scale, resulting in the loss of detail clarity.

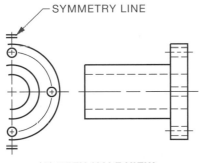

(A) WITH HALF VIEW

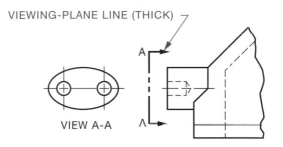

(B) PARTIAL VIEW WITH A VIEWING-PLANE LINE USED TO INDICATE DIRECTION

VIEW A-A

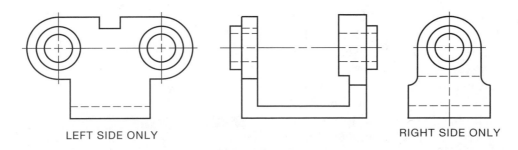

LEFT SIDE ONLY

RIGHT SIDE ONLY

(C) PARTIAL SIDE VIEWS

FIGURE 29–2 ■ Partial views

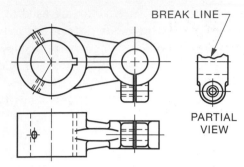

FIGURE 29–3 ■ Partial view

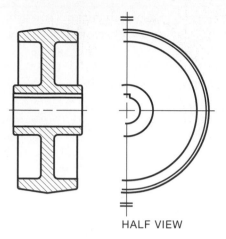

HALF VIEW

FIGURE 29–4 ■ Half views

■ If the part is symmetrical, a partial view (referred to as a half view) may be drawn on one side of the centre line, Figure 29–4. In the case of the coil frame (Assignment A-84), a partial view was used so that the object could be drawn to a larger scale for clarity, thus saving time and space.

NAMING OF VIEWS FOR SPARK ADJUSTER

The drawing of the spark adjuster, Assignment A-81, illustrates several violations of true projection. The names of the views of the spark adjuster could be questionable. The importance lies not in the names, however, but in the relationship of the views to each other. This means that the right view must be on the right side of the front, the left view must be on the left side of the front view, and so on. Any combination in Figure 29–5 may be used for naming the views of the spark adjuster.

DRILL SIZES

Twist drills are the most common tools used in drilling. They are made in many sizes. Inch size twist drills are grouped according to decimal-inch sizes; by number sizes, from 1 to 80, which correspond to the Stubbs steel wire gauge; by letter sizes A to Z; and by fractional sizes from 1/64th up. Inch twist drill sizes are listed in Table 3 of the Appendix.

Metric twist drill sizes are in millimetres and are classified as *preferred* and *available*. These sizes will eventually replace the fractional-inch, letter, and number size drills that are in existence. Metric twist drill sizes are listed in Table 4 of the Appendix.

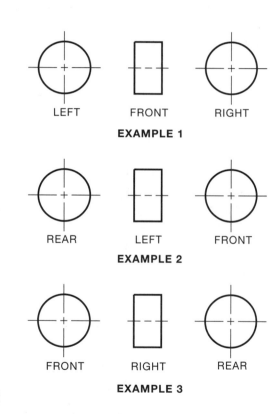

LEFT FRONT RIGHT
EXAMPLE 1

REAR LEFT FRONT
EXAMPLE 2

FRONT RIGHT REAR
EXAMPLE 3

FIGURE 29–5 ■ Naming of views for spark adjuster, Assignment A-81

REFERENCES

CAN3B78.1-M83 (R1990) Technical Drawings— General Principles
ASME Y14.3M-1994 (R2003) Multi and Sectional View Drawings

NEL

INTERNET RESOURCES

Amazon New York, Industrial Press For information on twist drill sizes, see: http://www.sizes.com/tools/twist_drills.htm

American Society of Mechanical Engineers For information on multiview drawings, refer to ASME Y14.3M-1994 (R2003) (*Multi- and Sectional-View Drawings*) at: http://www.asme.org

QUESTIONS:

1. What is radius (A)?
2. What surface is line (2) in the rear view?
3. Locate surface (B) in section A-A.
4. Locate surface (C) in section A-A.
5. Locate line (D) in the rear view.
6. Locate a surface or line in the front view that represents line (F).
7. Which point or line in the front view does line (G) represent?
8. What size cap screw would be used in (N) hole?
9. What is the diameter of (K) hole?
10. Locate the point or line in the rear view from which line (1) is projected.
11. Locate surface (H) in the rear view.
12. Locate (Z) in section A-A.
13. How thick is lug (7)?
14. What are the diameters of holes (L), (M), and (N)?
15. Determine angles (O) and (P).
16. Determine distances (Q) through (U).
17. What standard size drill could be used to produce the Ø.386 hole? Refer to Table 3 in the Appendix.

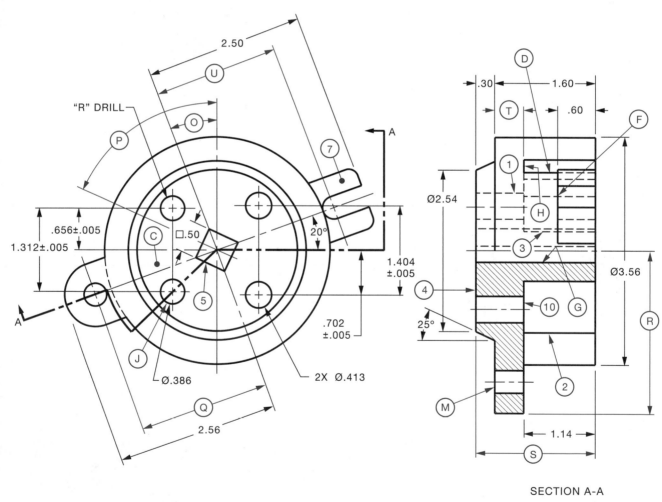

FRONT VIEW

SECTION A-A

RIGHT-SIDE VIEW

NOTE: UNLESS OTHERWISE STATED:
TOLERANCE ON TWO-PLACE DIMENSIONS ±.02
TOLERANCE ON THREE-PLACE DIMENSIONS ±.010
TOLERANCE ON ANGLES ±0.5°

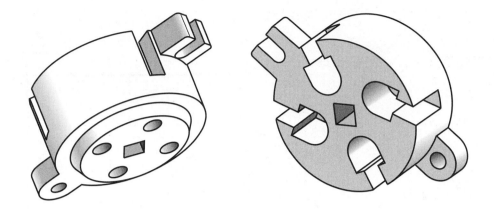

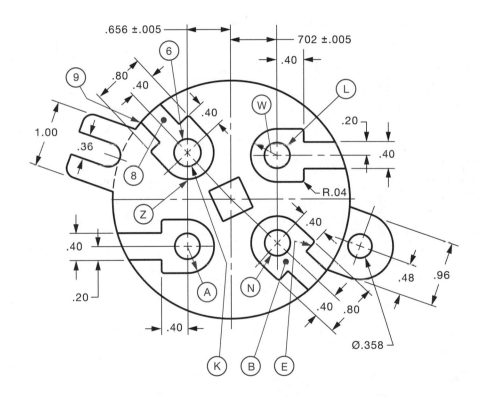

REAR VIEW

MATERIAL	BAKELITE	
SCALE	1:1	
DRAWN	T. JOHNSON	DATE 11/11/04

SPARK ADJUSTER

A-81

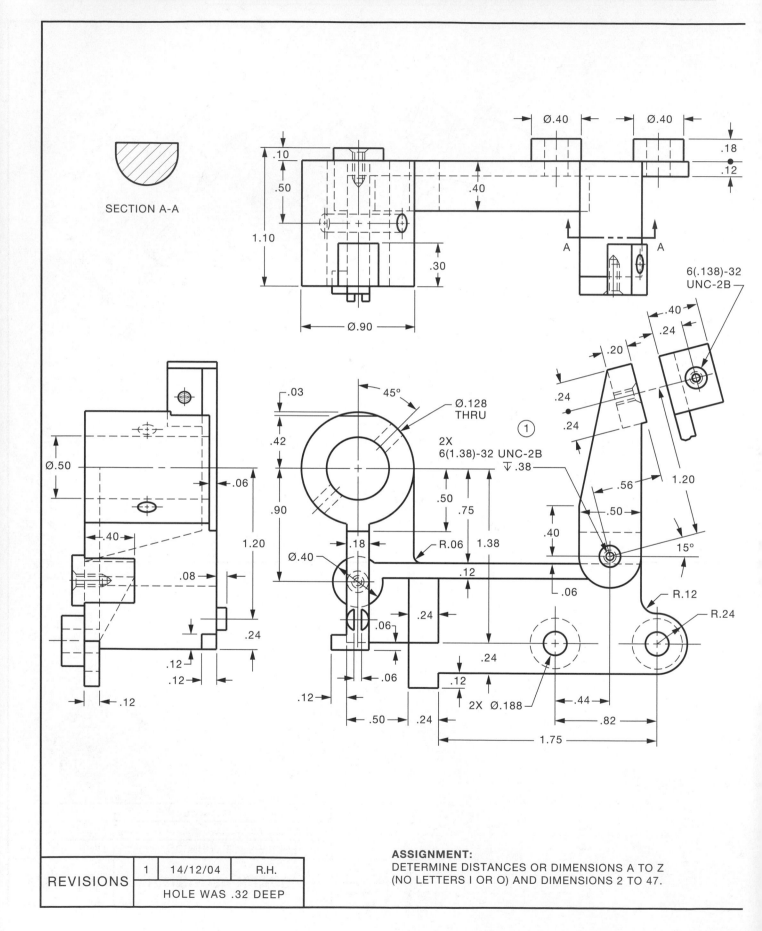

SECTION A-A

6(.138)-32
UNC-2B

Ø.128
THRU

2X
6(1.38)-32 UNC-2B
▽ .38

R.06

Ø.40

Ø.50

Ø.90

R.12

R.24

2X Ø.188

| REVISIONS | 1 | 14/12/04 | R.H. |
| | HOLE WAS .32 DEEP | | |

ASSIGNMENT:
DETERMINE DISTANCES OR DIMENSIONS A TO Z
(NO LETTERS I OR O) AND DIMENSIONS 2 TO 47.

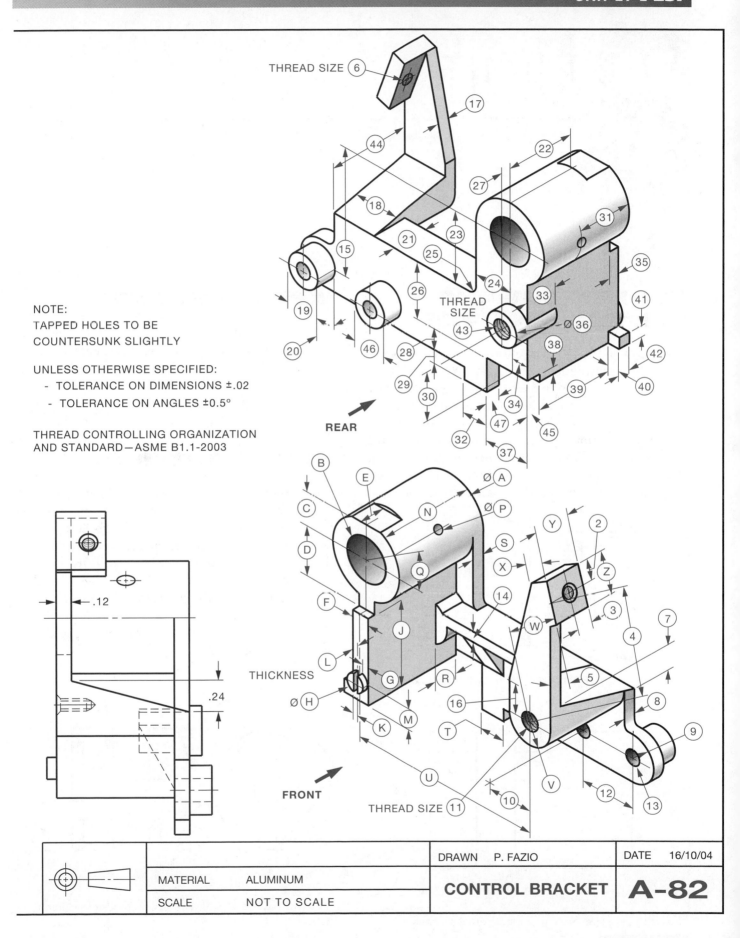

THREAD SIZE ⑥

⑰
㊹
⑰
⑱
⑮
㉑
㉓
㉕
㉖
THREAD SIZE
⑲
⑳
㊼
㉘
㉙
㉚
㉒
㉗
㉛
㉝
Ø ㊱
㉔
㊸
㉟
㊶
㊳
㊲
㊵
㊴
㉜
㊻
㊹
㊺
㊷

NOTE:
TAPPED HOLES TO BE
COUNTERSUNK SLIGHTLY

UNLESS OTHERWISE SPECIFIED:
- TOLERANCE ON DIMENSIONS ±.02
- TOLERANCE ON ANGLES ±0.5°

THREAD CONTROLLING ORGANIZATION
AND STANDARD—ASME B1.1-2003

REAR

B
C
D
F
L
Ø H
E
Ø A
Ø P
N
S
X
Q
J
G
R
M
K
14
16
T
Y
2
Z
3
W
4
5
7
8
9
V
12
13
U
10
THREAD SIZE 11

.12

THICKNESS

.24

FRONT

DRAWN	P. FAZIO	DATE 16/10/04
MATERIAL	ALUMINUM	CONTROL BRACKET	A-82
SCALE	NOT TO SCALE		

30 UNIT

BROKEN-OUT AND PARTIAL SECTIONS

Broken-out and partial sections are used to show certain internal and external features of an object without drawing another view, Figure 30–1. A break or cutting-plane line indicates where the section is taken. On the raise block, Assignment A-83M, two broken-out sections are shown in the front view, and one broken-out section is shown in the left-side view. Although this method of showing a partial section is not commonly used, it is an accepted practice.

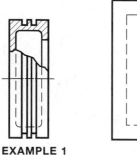 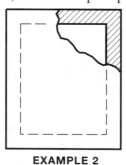

EXAMPLE 1   **EXAMPLE 2**

(A) BROKEN-OUT SECTIONS

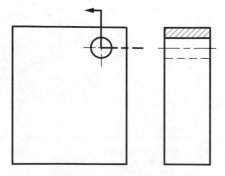

(B) PARTIAL SECTION

FIGURE 30–1 ■ Broken-out and partial sections

WEBS IN SECTION

The conventional methods of representing a section of a part having webs or partitions are shown in Figure 30–2. Although the conventional methods are a violation of true projection, they are preferred over true projection because of clarity and ease in drawing.

RIBS IN SECTION

A true projection section view, Figure 30–3(A), would be misleading when the cutting plane passes longitudinally through the centre of a rib. To avoid this impression of solidity, a preferred section, not showing the ribs section-lined or crosshatched, is used. When there is an odd number of ribs, Figure 30–3(B), the top rib is aligned with the bottom rib to show its true relationship with the hub and flange. If the rib is not aligned or revolved, it appears distorted on the section view and is misleading.

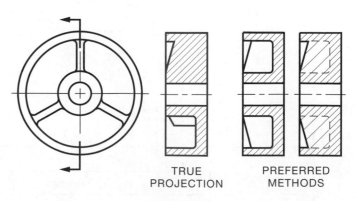

TRUE PROJECTION   PREFERRED METHODS

FIGURE 30–2 ■ Conventional methods of sectioning webs

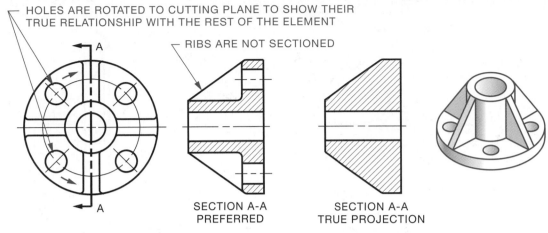

HOLES ARE ROTATED TO CUTTING PLANE TO SHOW THEIR
TRUE RELATIONSHIP WITH THE REST OF THE ELEMENT

RIBS ARE NOT SECTIONED

A

A

SECTION A-A
PREFERRED

SECTION A-A
TRUE PROJECTION

(A) CUTTING PLANE PASSING THROUGH TWO RIBS

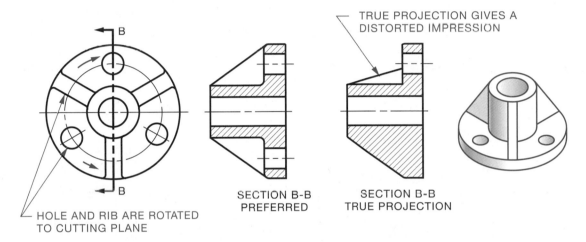

B

B

TRUE PROJECTION GIVES A
DISTORTED IMPRESSION

SECTION B-B
PREFERRED

SECTION B-B
TRUE PROJECTION

HOLE AND RIB ARE ROTATED
TO CUTTING PLANE

(B) CUTTING PLANE PASSING THROUGH ONE RIB AND ONE HOLE

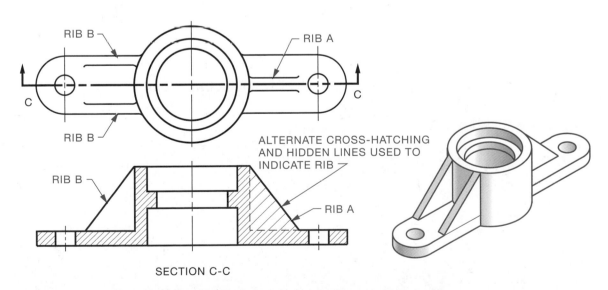

RIB B

RIB A

C

C

RIB B

ALTERNATE CROSS-HATCHING
AND HIDDEN LINES USED TO
INDICATE RIB

RIB B

RIB A

SECTION C-C

(C) ALTERNATE METHOD OF SHOWING RIBS IN SECTION

FIGURE 30–3 ■ Ribs in section

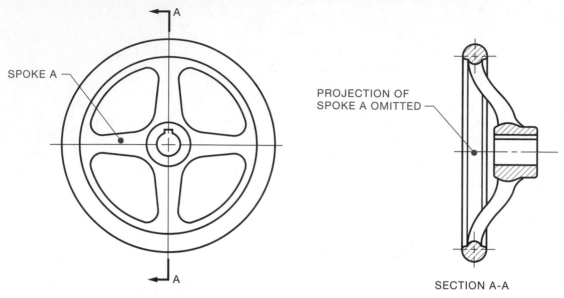

SPOKE A

PROJECTION OF
SPOKE A OMITTED

SECTION A-A

(A) CUTTING PLANE PASSING THROUGH TWO SPOKES

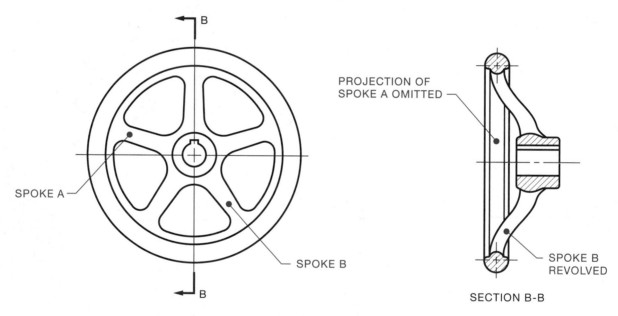

SPOKE A

SPOKE B

PROJECTION OF
SPOKE A OMITTED

SPOKE B
REVOLVED

SECTION B-B

(B) CUTTING PLANE PASSING THROUGH ONE SPOKE

FIGURE 30–4 ■ Spokes in section

An alternate method of identifying ribs in a section view is shown in Figure 30–3(C). If rib A of the base were not sectioned as previously mentioned, it would appear exactly like B in the section view and would be misleading. To distinguish between the ribs on the base, alternate section lining on the ribs is used. The line between the rib and solid portions is shown as a broken line.

SPOKES IN SECTION

Spokes in section are represented in the same manner as ribs. Figure 30–4 shows the preferred method of representing spokes in section for aligned and unaligned designs. The spokes are not sectioned in either case.

REFERENCES

CAN3-B78.1-M83 (R1990) Technical Drawings—General Principles
ASME Y14.3M-1994 (R2003) Multi and Sectional View Drawings

INTERNET RESOURCE

American Society of Mechanical Engineers For information on multiview drawings, refer to ASME Y14.3M-1994 (R2003) (*Multi- and Sectional-View Drawings*) at: http://www.asme.org

QUESTIONS:

1. What is the diameter of the largest unthreaded hole?

2. What size is the smallest threaded hole?

3. What size are the smallest unthreaded holes?

4. Which surface does ⑦ represent in the top view?

5. Which surface does ① represent in the top view?

6. Which line or surface does ⑥ represent in the left view?

7. Which line or surface does Ⓥ represent in the front view?

8. What was the original width of the part?

9. By which line or surface is Ⓗ represented in the left view?

10. Locate in the left view the line or surface that is represented by line Ⓖ.

11. Which line or surface represents Ⓕ in the top view?

12. Which line represents surface Ⓙ in the left view?

13. Determine the overall depth of the raise block.

14. Determine distances Ⓐ through Ⓔ.

15. Determine distances ⑧ through ⑲.

16. What is the tap drill size required for the (A) M10x1.5, (B) M16x2 threaded holes?

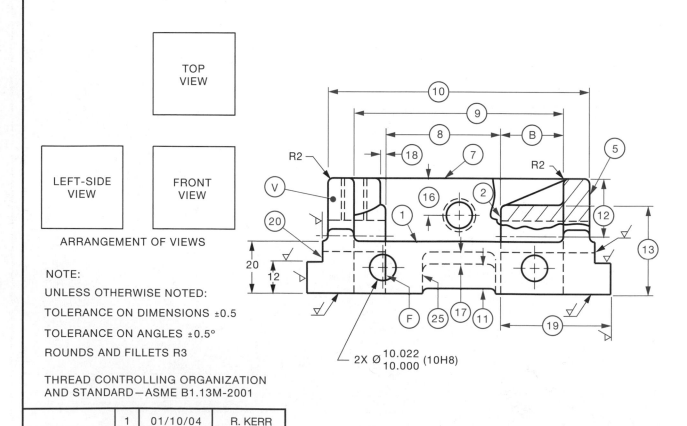

ARRANGEMENT OF VIEWS

NOTE:

UNLESS OTHERWISE NOTED:

TOLERANCE ON DIMENSIONS ±0.5

TOLERANCE ON ANGLES ±0.5°

ROUNDS AND FILLETS R3

THREAD CONTROLLING ORGANIZATION
AND STANDARD—ASME B1.13M-2001

2X Ø 10.022 / 10.000 (10H8)

REVISIONS	1	01/10/04	R. KERR
		166 WAS 170	

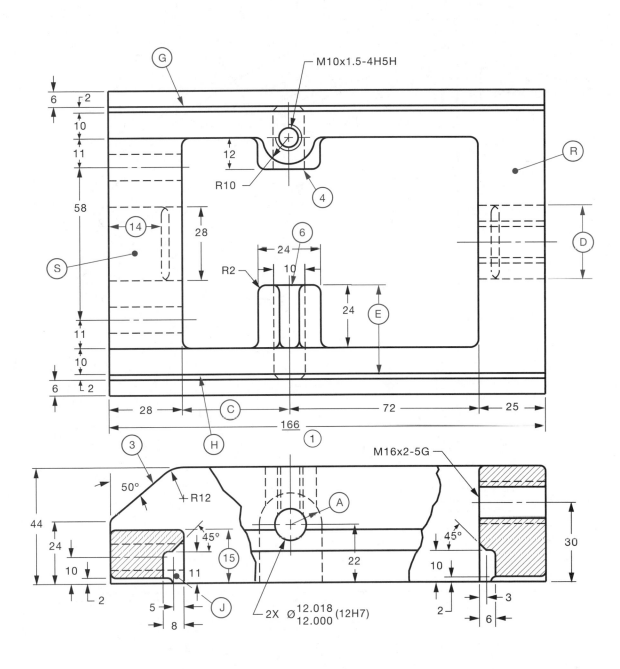

M10x1.5-4H5H

R

R10

4

12

G

6

58

14

28

S

24

R2

10

6

24

E

D

11

11

10

6   2

6   2

28

C

72

25

166

H

1

3

50°

R12

M16x2-5G

A

44

45°

24

15

22

45°

30

10

10

J

2

5

2

3

8

2X Ø 12.018 / 12.000 (12H7)

6

MATERIAL	GREY IRON	
SCALE	NOT TO SCALE	
DRAWN	W. JONES	DATE 18/09/04

METRIC
DIMENSIONS ARE IN MILLIMETRES

RAISE BLOCK

A-83M

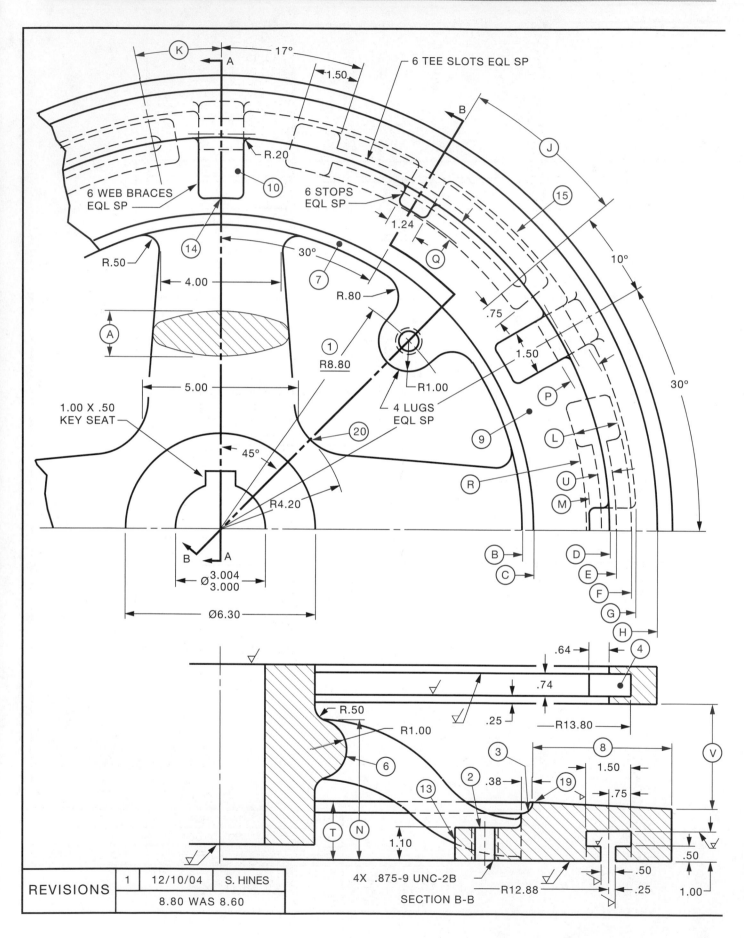

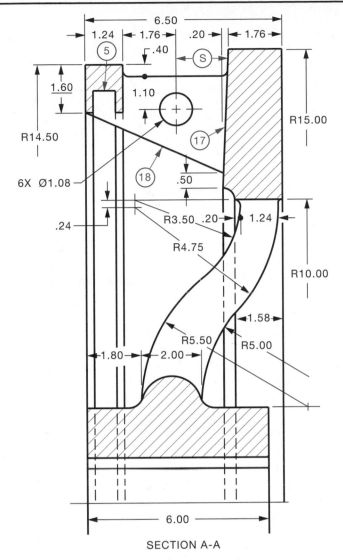

SECTION A-A

EXCEPT WHERE OTHERWISE SPECIFIED:

- FEATURES SYMMETRICAL AROUND
  CENTRE POINT
- TOLERANCE ON DIMENSIONS ±.02
- TOLERANCE ON ANGLES ±0.5°

 TO BE $\sqrt{\dfrac{125}{}}$

THREAD CONTROLLING ORGANIZATION
AND STANDARD—ASME B1.1-2003

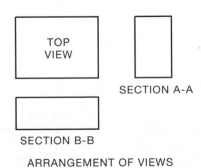

TOP
VIEW

SECTION A-A

SECTION B-B

ARRANGEMENT OF VIEWS

QUESTIONS:

1. What surface texture is required on the machined surfaces?
2. What is thickness (A), assuming the revolved section was taken at the middle of the arm?
3. What is the total quantity of feature (4)?
4. Locate surface (5) in the top view.
5. Locate point (6) in the top view.
6. How far is line (14) from the centre point?
7. How far is surface (13) from the centre point?
8. Which point on section B-B represents radius (C)?
9. Which line on section B-B represents surface (7)?
10. What is the radius of surface (3)?
11. Determine distance (8).
12. Which line on section A-A represents surface (10)?
13. Which surface in the top view represents line (17)?
14. What is the angle at (J)?
15. What is the angle at (K)?
16. What is the thickness of the web brace?
17. What would be the length of key used between the shaft and coil frame?
18. What is the thickness of the lugs?
19. What is the distance from the Ø1.00 hole to the centre of the coil frame?
20. Determine distances (L), (N), (P), (Q), (S), (T), (U), and (V).
21. Determine radii (B), (C), (D), (E), (F), (G), (H), (M), and (R).
22. What type of sectional view is used to show the shape of the spokes?
23. What type of section view is section B-B?

MATERIAL	GREY IRON		
SCALE	NOT TO SCALE		
DRAWN	C. LANG	DATE	19/04/04

COIL FRAME | **A-84**

31 UNIT

PIN FASTENERS

Pin fasteners offer an inexpensive and effective approach to assembly where loading is primarily in shear. They can be divided into two groups: semi-permanent and quick release.

Semipermanent pin fasteners require application of pressure or the aid of tools for installation or removal. Representative types are machine pins (dowel, straight, taper, clevis, and cotter pins) and radial locking pins (grooved surface and spring).

Quick-release fasteners are more elaborate, self-contained pins that are used for rapid manual assembly or disassembly. They use a form of spring-loaded mechanism to provide a locking action in assembly.

Machine Pins

Five types of machine pins are commonly used: ground dowel pins; commercial straight pins; taper pins; clevis pins; and standard cotter pins, Figure 31–1.

Dowel Pins

Dowel pins, or small straight pins, have many uses. They are used to hold parts in alignment and to guide parts into desired positions. Dowel pins are most commonly used for the alignment of parts that are fastened with screws or bolts and must be accurately assembled.

When two pieces are to be fitted together, like the part in Figure 31–2, one method of alignment is to clamp the two pieces in the desired location,

drill and ream the dowel holes, insert the dowel pins, and then drill and tap for the screw holes.

Drill jigs are frequently used when the interchangeability of parts is required or when the nature of the piece does not permit the transfer of the dowelled holes from one piece to the other. A drill jig, Figure 31–3, was used in drilling the dowel holes for the spider in Assignment A-85M.

Taper Pins

Holes for taper pins are usually sized by reaming. A through hole is formed by step drills and straight fluted reamers. The trend is toward the use of helically fluted taper reamers, which provide more accurate sizing and require only a pilot hole the size of the small end of the taper pin. The pin is usually driven into the hole until it is fully seated. The taper of the pin aids hole alignment in assembly.

A tapered hole in a hub and a shaft is shown in Figure 31–4. If the hub and shaft are drilled and reamed separately, a misalignment might occur, Figure 31–5. To prevent misalignment, the hub and shaft should be drilled at the same time as the parts are assembled. Each of the detailed parts should carry a note similar to: DRILL AND REAM FOR NO. 1 TAPER PIN AT ASSEMBLY.

Cotter Pins

The cotter pin is a standard machine pin commonly used as a fastener in the assembly of machine parts where great accuracy is not required, Figure 31–6. There is no standard way to represent

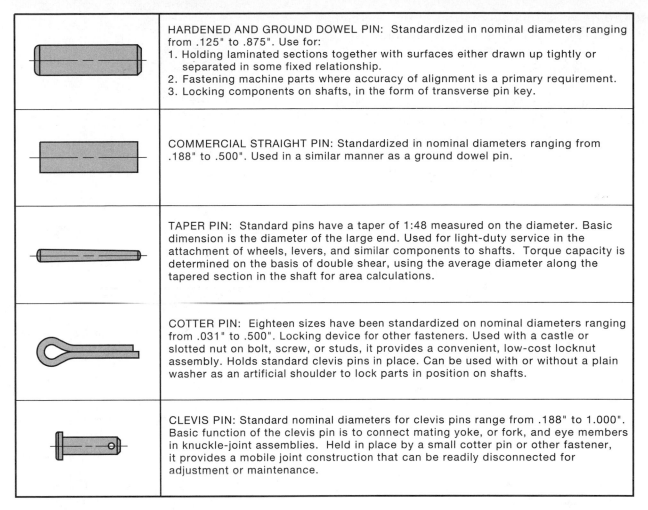

	HARDENED AND GROUND DOWEL PIN: Standardized in nominal diameters ranging from .125" to .875". Use for: 1. Holding laminated sections together with surfaces either drawn up tightly or separated in some fixed relationship. 2. Fastening machine parts where accuracy of alignment is a primary requirement. 3. Locking components on shafts, in the form of transverse pin key.
	COMMERCIAL STRAIGHT PIN: Standardized in nominal diameters ranging from .188" to .500". Used in a similar manner as a ground dowel pin.
	TAPER PIN: Standard pins have a taper of 1:48 measured on the diameter. Basic dimension is the diameter of the large end. Used for light-duty service in the attachment of wheels, levers, and similar components to shafts. Torque capacity is determined on the basis of double shear, using the average diameter along the tapered section in the shaft for area calculations.
	COTTER PIN: Eighteen sizes have been standardized on nominal diameters ranging from .031" to .500". Locking device for other fasteners. Used with a castle or slotted nut on bolt, screw, or studs, it provides a convenient, low-cost locknut assembly. Holds standard clevis pins in place. Can be used with or without a plain washer as an artificial shoulder to lock parts in position on shafts.
	CLEVIS PIN: Standard nominal diameters for clevis pins range from .188" to 1.000". Basic function of the clevis pin is to connect mating yoke, or fork, and eye members in knuckle-joint assemblies. Held in place by a small cotter pin or other fastener, it provides a mobile joint construction that can be readily disconnected for adjustment or maintenance.

FIGURE 31–1 ■ Machine pins

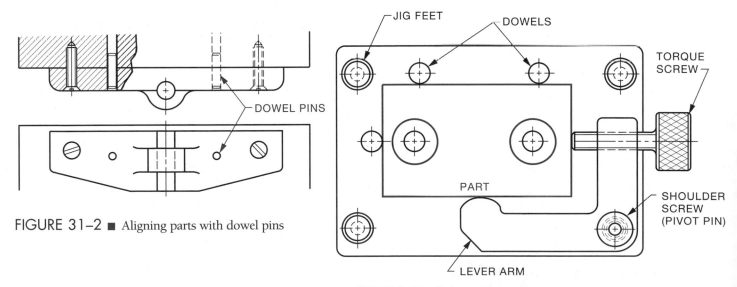

FIGURE 31–2 ■ Aligning parts with dowel pins

FIGURE 31–3 ■ Dowel pins used to align part during drillings

cotter pins in assembly drawings; the method shown in Figure 31–7 is, however, commonly used.

Radial-Locking Pins

Low cost, ease of assembly, and high resistance to vibration and impact loads are common attributes of this group of commercial pin devices designed primarily for semipermanent fastening service. Two basic pin forms are used: solid with grooved surfaces and

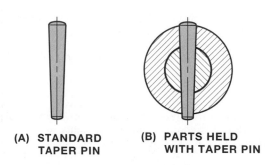

(A) STANDARD TAPER PIN   **(B) PARTS HELD WITH TAPER PIN**

FIGURE 31–4 ■ Taper pin application

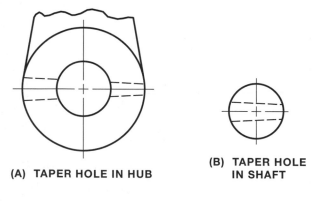

(A) TAPER HOLE IN HUB   **(B) TAPER HOLE IN SHAFT**

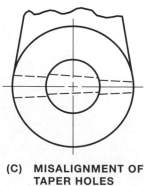

(C) MISALIGNMENT OF TAPER HOLES

FIGURE 31–5 ■ Possibility of hole misalignment if holes are not drilled at assembly

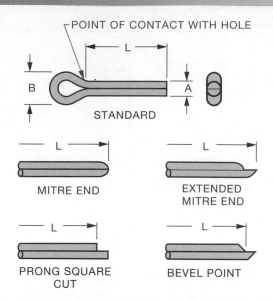

POINT OF CONTACT WITH HOLE

STANDARD

MITRE END   EXTENDED MITRE END

PRONG SQUARE CUT   BEVEL POINT

NOMINAL THREAD SIZE		NOMINAL COTTER PIN SIZE		COTTER PIN HOLE		END CLEARANCE*	
in.	(mm)	in.	(mm)	in.	(mm)	in.	(mm)
.250	(6)	.062	(1.5)	.078	(1.9)	.12	(3)
.312	(8)	.078	(2)	.094	(2.4)	.12	(3)
.375	(10)	.094	(2.5)	.109	(2.8)	.14	(4)
.500	(12)	.125	(3)	.141	(3.4)	.18	(5)
.625	(14)	.156	(3)	.172	(3.4)	.25	(5)
.750	(20)	.156	(4)	.172	(4.5)	.25	(7)
1.000	(24)	.188	(5)	.203	(5.6)	.31	(8)
1.125	(27)	.188	(5)	.203	(5.6)	.39	(8)
1.250	(30)	.219	(6)	.234	(6.3)	.44	(10)
1.375	(36)	.219	(6)	.234	(6.3)	.44	(11)
1.500	(42)	.250	(6)	.266	(6.3)	.50	(12)
1.750	(48)	.312	(8)	.312	(8.5)	.55	(14)

*DISTANCE FROM EXTREME POINT OF BOLT OR SCREW TO CENTRE OF COTTER PIN HOLE.

FIGURE 31–6 ■ Cotter pin data

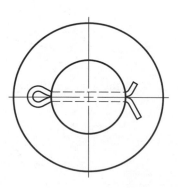

FIGURE 31–7 ■ Cotter pin in an assembly drawing

hollow spring pins, which may be either slotted or spiral wrapped, Figure 31–8. In assembly, radial forces produced by elastic action at the pin surface develop a secure, frictional-locking grip against the hole wall. These pins are reusable and can be removed and reassembled many times without appreciable loss of fastening effectiveness. Live spring action at the pin surface also prevents loosening under shock and vibration loads. The need for accurate sizing of holes is reduced because the pins accommodate variations.

Solid Pins with Grooved Surfaces

The locking action of groove pins is provided by parallel, longitudinal grooves uniformly spaced around the pin surface. Rolled or pressed into solid pin stock, the grooves expand the effective diameter of the pin. When the pin is driven into a drilled hole corresponding in size to the nominal pin diameter, elastic deformation of the raised groove edges produces a secure interference fit with the hole wall. Figure 31–8 shows the six standardized constructions of grooved pins.

Hollow Spring Pins

Spiral-wrapped and slotted-tubular pin forms are made to controlled diameters greater than the holes into which they are pressed. Compressed when driven into the hole, the pins exert spring pressure against the hole wall along their entire engaged length to develop a strong locking action.

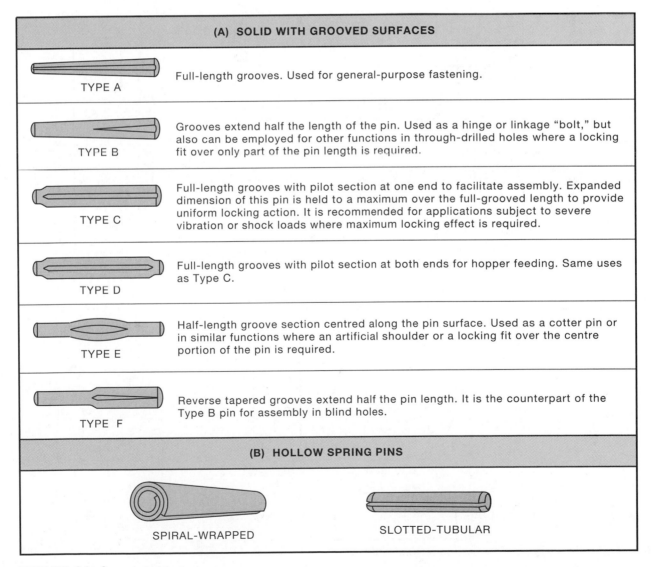

(A) SOLID WITH GROOVED SURFACES	
TYPE A	Full-length grooves. Used for general-purpose fastening.
TYPE B	Grooves extend half the length of the pin. Used as a hinge or linkage "bolt," but also can be employed for other functions in through-drilled holes where a locking fit over only part of the pin length is required.
TYPE C	Full-length grooves with pilot section at one end to facilitate assembly. Expanded dimension of this pin is held to a maximum over the full-grooved length to provide uniform locking action. It is recommended for applications subject to severe vibration or shock loads where maximum locking effect is required.
TYPE D	Full-length grooves with pilot section at both ends for hopper feeding. Same uses as Type C.
TYPE E	Half-length groove section centred along the pin surface. Used as a cotter pin or in similar functions where an artificial shoulder or a locking fit over the centre portion of the pin is required.
TYPE F	Reverse tapered grooves extend half the pin length. It is the counterpart of the Type B pin for assembly in blind holes.
(B) HOLLOW SPRING PINS	
SPIRAL-WRAPPED SLOTTED-TUBULAR	

FIGURE 31–8 ■ Radial locking pins

SECTION THROUGH SHAFTS, PINS, AND KEYS

Shafts, bolts, nuts, rods, rivets, keys, pins, and similar solid parts, the axes of which lie on the cutting plane, are sectioned only when a broken-out section of the shaft is used to clearly indicate the key, keyseat, and pin, Figure 31–9.

ARRANGEMENT OF VIEWS OF ASSIGNMENT A-85M

Parts that are to be fitted over shafts as a single unit are sometimes made in two or more pieces to facilitate assembly and replacement on the main structure of a machine rather than manufacture. Assignment A-85M shows two parts that are bolted and dowelled together to form one unit.

The arrangement of views of the spider is illustrated in Figure 31–10. By comparing this figure with the drawing of the spider, note that the two halves together represent the top view. The right view is a full section of each half (**A** and **B**). The front view is a drawing of the front of part A only.

Although the front and side views are incomplete, the manner in which they are drawn and the arrangement of the views satisfies the demand for clearness and economy in time and space.

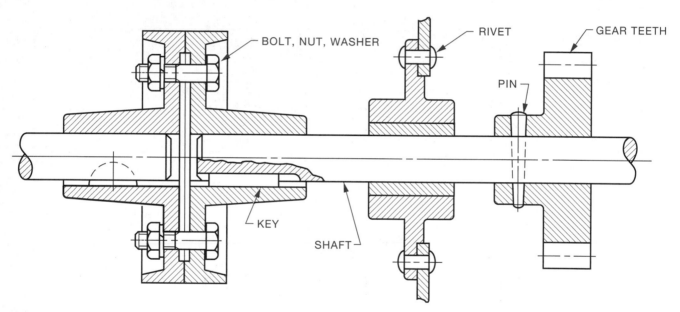

FIGURE 31–9 ■ Parts that are not section lined in section drawings

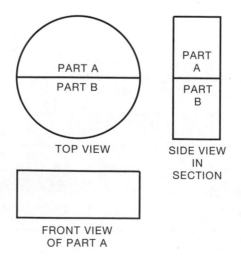

FIGURE 31–10 ■ Arrangement of views for spider, Assignment A-85M

REFERENCES

Machine Design, Fasteners Reference Issue, Nov. 1981

ASME B18.8.2-2000 Taper Pins, Dowel Pins, Straight Pins, Grooved Pins and Spring Pins

INTERNET RESOURCES

American Society of Mechanical Engineers For information on sections through shafts, pins, and keys, refer to ASME Y14.3M-1994 (R2003) (*Multi- and Sectional-View Drawings*) at: http://www.asme.org

Machine Design For information on pin fasteners, see *Machine Design, Fastening/Joining Reference* at: http://www.machinedesign.com

Vogelsang Corporation For information on machine pin fasteners, see http://www.rollpin.com/ProductInformation/Springpins/Coiled/index.html

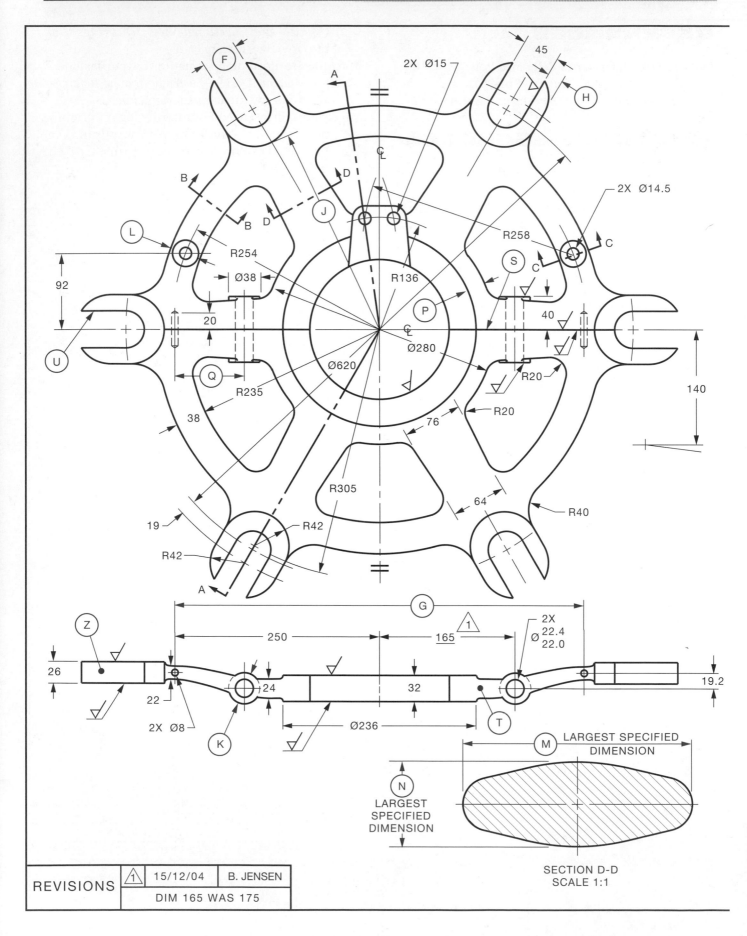

2X Ø15

45

H

2X Ø14.5

R258

C

R254

R136

S

Ø38

R305

P

Ø280

40

20

Ø620

R235

R20

R20

Q

38

76

R40

19

64

R42

R42

92

140

L

U

A

B

B

D

D

J

F

A

G

1

2X
Ø 22.4
22.0

250

165

Z

26

24

32

19.2

22

2X Ø8

K

Ø236

T

2X Ø14.5

M — LARGEST SPECIFIED
DIMENSION

N

LARGEST
SPECIFIED
DIMENSION

SECTION D-D
SCALE 1:1

REVISIONS	△ 1	15/12/04	B. JENSEN
		DIM 165 WAS 175	

NEL

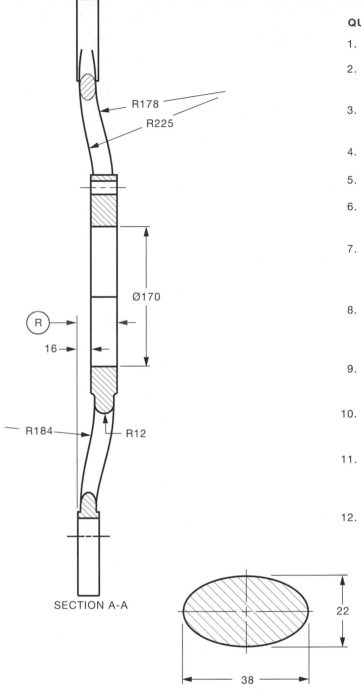

R178

R225

Ø170

R

16

R184

R12

SECTION A-A

QUESTIONS:

1. How far were the Ø22 holes moved?

2. What type of projection does the ISO projection symbol (E) indicate?

3. How many different size scales were used to make the drawing?

4. Locate surface (T) in the top view.

5. Locate surface (Z) in the top view.

6. What is the approximate outside diameter of the spider?

7. What will be the rough dimension of casting at (F), assuming that 1.5 mm have been added for each surface to be finished?

8. What would be used to position both halves of the spider together before bolting? What is their size?

9. What size bolts would be used to fasten the spider together?

10. Calculate distances (G), (H), (J), (K), (L), (M), (N), (P), (Q), and (R).

11. With reference to the scales used on the drawings, what percentage larger are the removed sections over the main drawings?

12. What type of section view is section A-A?

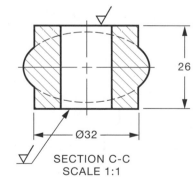

22

38

SECTION B-B
SCALE 1:1

26

Ø32

SECTION C-C
SCALE 1:1

METRIC
DIMENSIONS ARE IN MILLIMETRES

NOTE:
- TOLERANCE ON DIMENSIONS ±0.5
- TOLERANCE ON ANGLES ±0.5°
- ∇ TO BE 1.6∇

E

MATERIAL	GREY IRON	
SCALE	1:5 EXCEPT WHERE NOTED	
DRAWN	T. LOGAN	DATE 29/09/04

SPIDER

A-85M

QUESTIONS:

1. Which views are shown?

2. How many surfaces are to be finished?

3. How many scraped surfaces are indicated?

4. How many holes are to be tapped?

5. What is the purpose of tapped hole (R)?

6. Which surface is (3) in the left-side view?

7. Which surface is (2) in the left-side view?

8. Which surface is (4) in the front view?

9. Which surface is (14) in the left-side view?

10. Which surface in the top view and front view is (9)?

11. Which surface in the top view and front view is (8)?

12. Which surface is (12) in the top view?

13. What is the name of part (V)?

14. What is the purpose of part (V)?

15. What do dotted lines at (W) represent?

16. Which surface is line (6) in the front view?

17. Which top view line indicates point (Z)?

18. What is the depth of the tapped hole at (X)?

19. Which edges or surfaces in the left and front views does line (T) represent?

20. What is the diameter of tap drill (Y)? (See Appendix.)

21. Determine dimensions or operations at (A) to (Q), (20) to (52).

22. What is the largest permissible diameter of the largest hole?

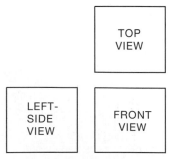

ARRANGEMENT OF VIEWS

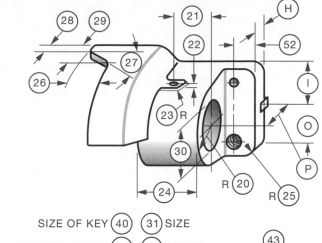

SIZE OF KEY (40) (31) SIZE

LENGTH OF KEY (41) (32) DEPTH

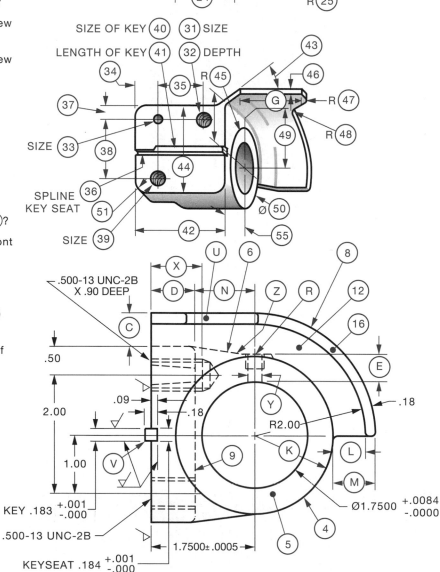

REVISIONS | 1 | |

NOTES: UNLESS OTHERWISE SPECIFIED:
- TOLERANCE ON DIMENSIONS ±.02
- TOLERANCE ON ANGLES ±0.5°
- FINISH AND SCRAPE SURFACE BEFORE
  CUTTING SPLINE KEY SEAT

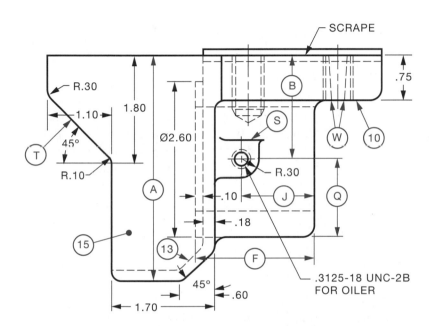

SCRAPE

R.30
1.10
45°
R.10
1.80
Ø2.60
.10
.18
.60
45°
1.70

B
S
T
A
15
13
F
W
10
J
Q

R.30

.3125-18 UNC-2B
FOR OILER

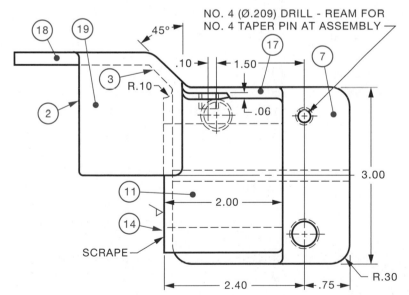

NO. 4 (Ø.209) DRILL - REAM FOR
NO. 4 TAPER PIN AT ASSEMBLY

18 19 45° 17 7
.10 1.50
3   R.10
2
.06
11   2.00
14
SCRAPE
2.40   .75
3.00
R.30

NOTE:
THREAD CONTROLLING ORGANIZATION
AND STANDARD–ASME B1.1-2003

MATERIAL	GREY IRON	
SCALE	NOT TO SCALE	
DRAWN	C. JENSEN	DATE 06/10/04

HOOD **A-86**

NEL

DRAWINGS FOR NUMERICAL CONTROL

Numerical control (NC) is a means of automatically directing the functions of a machine using electronic instructions. Originally, information was fed to NC machines through punched tapes. Improvements in technology have led to the integration of computers with manufacturing machinery, called computer numerical control (**CNC**). The machine interprets digitally coded instructions and directs various operations of the cutting tool.

Because of the consistent high accuracy of numerically controlled machines and elimination of human error, scrap has been considerably reduced.

NC machines also improve the quality or accuracy of the work. In most cases, a machine can produce parts more accurately at no additional cost, resulting in reduced assembly time and better interchangeability of parts, especially important when spare parts are required.

Computer-aided design (CAD) and computer-aided manufacturing (CAM) techniques are now widely used in conjunction with NC processes in industry.

DIMENSIONING FOR NUMERICAL CONTROL

Common guidelines have been established for dimensioning for NC that enable dimensioning and tolerancing practices to be used effectively for both NC and conventional fabrication. The NC concept is based on the system of rectangular, or Cartesian, coordinates, in which any position can be described in terms of distance from an origin point along either two or three perpendicular axes. Each object is prepared using base line (or datum) dimensioning methods described in Unit 28. First, the selection of an absolute (0, 0, 0) or (0, 0) coordinate origin is made, depending on whether the control is three axis or two axis. All part dimensions would be referenced from that origin.

After a working drawing is produced, the information is transferred to manufacturing equipment. This allows the NC computer to compile instructions from programs in its memory. The result is a detailed program plan for tool-path generation.

DIMENSIONING FOR TWO-AXIS COORDINATE SYSTEM

Two dimensional coordinates (**X, Y**) define points in a plane, Figure 32–1. Examples of parts using rectangular coordinates were shown in Unit 18.

The **X** axis is horizontal and considered the first and basic reference axis. The **Y** axis is vertical and perpendicular to the **X** axis in the plane of a drawing showing **XY** relationships. The position where the **X** and **Y** axes cross is called the origin, or zero point. Distances to the right of the origin are considered positive values and those to the left of the origin are negative values. Distances above the origin are considered positive **Y** values, and below the origin negative values.

For example, four points lie in a plane, Figure 32–1. The plane is divided into four quadrants, the origin being in the centre. Point **A** lies in quadrant 1 and is located at position (6, 3) with the X coordinate first, followed by the Y coordinate. Point **B** lies in quadrant 2 and is located at position (−6, 5). Point **C** lies in quadrant 3 and is located at position (−5, −2). Point **D** lies in quadrant 4 and is located at position (2, −3).

Designing for NC would be greatly simplified if all work were done in the first quadrant because all of the values would be positive, and the plus and minus signs would not be required. For that reason many NC systems place the origin (0, 0) to the lower left of the part. This way only the positive values apply. However, any of the four quadrants may be used.

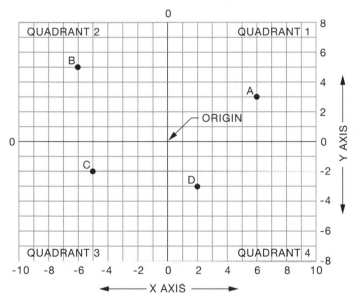

FIGURE 32–1 ■ Two-dimensional coordinates

Some NC machines, called two-axis machines, are designed for locating points in only the X and Y directions. The function of these machines is to move the machine table or tool to a specified position to perform work, Figure 32–2.

With the fixed spindle and movable table, Figure 32–2(B), hole **A** is drilled, then the table moves to the left, positioning point **B** below the drill. This is the most frequently used method. With the fixed table and movable spindle, Figure 32–2(C), hole **A** is drilled, then the spindle moves to the right, positioning the drill above point **B**. This changes the direction of the motion, but the movement of the cutter as delivered to the work remains the same.

Origin (Zero Point)

This is the position from which all coordinate dimensions are measured. A setup point is located on the part or the fixture holding the part. It may be the intersection of two finished surfaces, the centre of a previously machined hole in the part, or a feature of the fixture.

Setup Point

The part must be accurately positioned on the fixture before any work is performed. This establishes a setup point, which is accurately located in relationship to the origin. It may be located on the part or on the fixture holding the part. It may be the intersection of two finished surfaces, the centre of a previously machined hole, or a feature of the fixture.

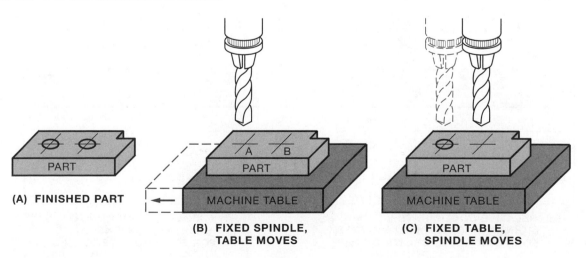

FIGURE 32–2 ■ Positioning the work

Relative Coordinate (Point-to-Point) Programming

With point-to-point programming, each new position is given from the last position. To compute the next position wanted, it is necessary to establish the sequence in which the work is to be done.

Absolute Coordinate Programming

Many systems use absolute coordinate programming instead of the point-to-point method of dimensioning. With this, all dimensions are taken from the origin; as such, base line or datum dimensioning is used.

Examples of both of these dimensioning techniques are shown in Figure 32–3. In these examples, two of the outer surfaces of the part are positioned on the fixture by three locating pins.

This establishes the setup point at 80 mm above and 80 mm to the right of the origin. Regardless of which dimensioning method is to be used, the coordinates for the first hole (hole 1) to be machined are the same and are taken from the origin (zero point). The X coordinate is 100 (80 + 20), and the Y coordinate is also 100 (80 + 20).

INTERNET RESOURCES

Machine Design For information on CAD/CAM, see *Machine Design, CAD/CAM Reference* at: http://www.machinedesign.com

MMS Online For information on CNC and related subjects, see: http://www.mmsonline.com

TechStudent.Com For information, including illustrations, on the basics of NC/CNC, see: http://www.technologystudent.com/cam/camex.htm

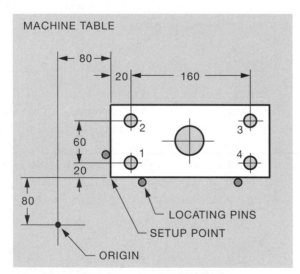

POINT-TO-POINT DIMENSIONING

HOLE	X	Y
1	100	100
2	0	60
3	160	0
4	0	-60

(A) RELATIVE COORDINATE (POINT-TO-POINT) DIMENSIONING FOR FOUR HOLES SHOWN ON PART

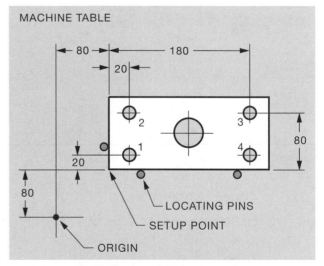

BASE LINE DIMENSIONING

HOLE	X	Y
1	100	100
2	100	160
3	260	160
4	260	100

(B) ABSOLUTE COORDINATE (BASE LINE) DIMENSIONING FOR FOUR HOLES SHOWN ON PART

FIGURE 32–3 ■ Dimensioning for numerical control

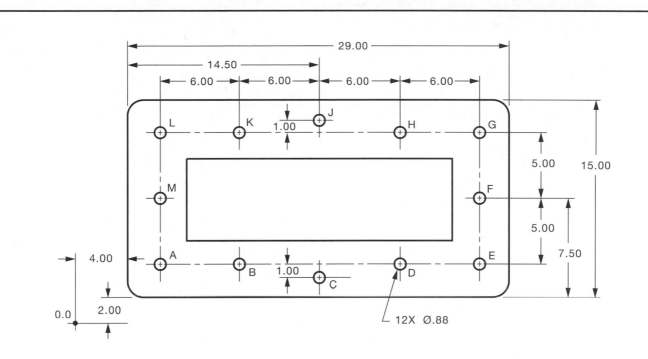

ABSOLUTE COORDINATES		HOLE	RELATIVE COORDINATES	
X	Y		X	Y
		A		
		B		
		C		
		D		
		E		
		F		
		G		
		H		
		J		
		K		
		L		
		M		

ASSIGNMENT:
PREPARE A CHART SIMILAR TO THE ONE SHOWN ABOVE AND PLACE
THE COORDINATES FOR EACH OF THE CIRCULAR HOLES IN THE CHART.
THE LETTERS AT THE HOLES INDICATE THE SEQUENCE IN WHICH THEY
ARE TO BE DRILLED. NOTE THE LOCATION OF THE ORIGIN AND THAT THE
LETTER "I" IS NOT USED TO IDENTIFY A HOLE.

COVER PLATE	A-87

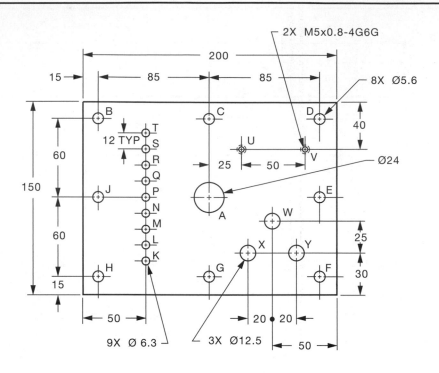

POINT-TO-POINT PROGRAMMING		
HOLE	X AXIS	Y AXIS
A		
B		
C		
D		
E		
F		
G		
H		
J		
K		
L		
M		

POINT-TO-POINT PROGRAMMING		
HOLE	X AXIS	Y AXIS
N		
P		
Q		
R		
S		
T		
U		
V		
W		
X		
Y		

ASSIGNMENT:
PREPARE A CHART SIMILAR TO THE ONE SHOWN ABOVE AND
PLACE THE X AND Y COORDINATES FOR EACH OF THE HOLES
IN THE CHART. POINT-TO-POINT PROGRAMMING IS TO BE USED
TO LOCATE EACH HOLE. THE LETTERS AT THE HOLES INDICATE
THE SEQUENCE IN WHICH THEY ARE TO BE DRILLED. ORIGIN
FOR THE X AND Y COORDINATES IS THE BOTTOM LEFT-HAND
CORNER OF THE PART. NOTE THAT THE LETTERS "I" AND "O"
ARE NOT USED TO IDENTIFY HOLES.

THREAD CONTROLLING ORGANIZATION
AND STANDARD—ASME B1.13M-2001

METRIC
DIMENSIONS ARE IN MILLIMETRES

TERMINAL BOARD	**A-88M**

ASSEMBLY DRAWINGS

In an *assembly drawing,* the various parts of a machine or structure are drawn in their relative positions in the completed unit. In addition to showing how the parts fit together, the assembly drawing has other uses:

- It represents the working relationships of the mating parts of a machine or structure and the function of each.

- It gives a general idea of how the finished product should look.

- It aids in securing overall dimensions and centre distances in assembly.

- It gives the detailer data needed to design the smaller units of a larger assembly.

- It supplies illustrations that may be used for catalogues, maintenance manuals, or other illustrations.

To show the working relationship of interior parts, the principles of projection may be violated and details omitted for clarity. Assembly drawings should not be overly detailed because precise information describing part shapes is provided on detail drawings.

Detail dimensions that would confuse the assembly drawing should be omitted. Only such dimensions as centre distances, overall dimensions, and dimensions showing the relationship of the parts as they apply to the mechanism as a whole should be included. Sometimes a simple assembly drawing may be dimensioned so that no other detail drawings are needed; in this case, it becomes a working assembly drawing.

Sectioning is used more extensively on assembly drawings than on detail drawings. The conventional method of section lining is used on assembly drawings to show the relationships between the various parts, Figure 33–1.

Symbolic section lining, Figure 9-4, may be used for many purposes:

- to represent the material of the part or parts

- to represent conductive or nonconductive materials

- to represent moving and stationary parts

Subassembly Drawings

Subassembly drawings are often made of smaller mechanical units. When combined in final assembly, they make a single machine. For a lathe, subassembly drawings would be furnished for the headstock, the apron, and other units of the carriage. These units might be machined and assembled in different departments by following the subassembly drawings. The individual units would later be combined in final assembly according to the assembly drawing.

Identifying Parts of an Assembly Drawing

When a machine is designed, an assembly drawing or design layout is first drawn to visualize clearly the method of operation, shape, and clearances of the various parts. From this assembly drawing, the detail drawings are made, and each part is given a part number.

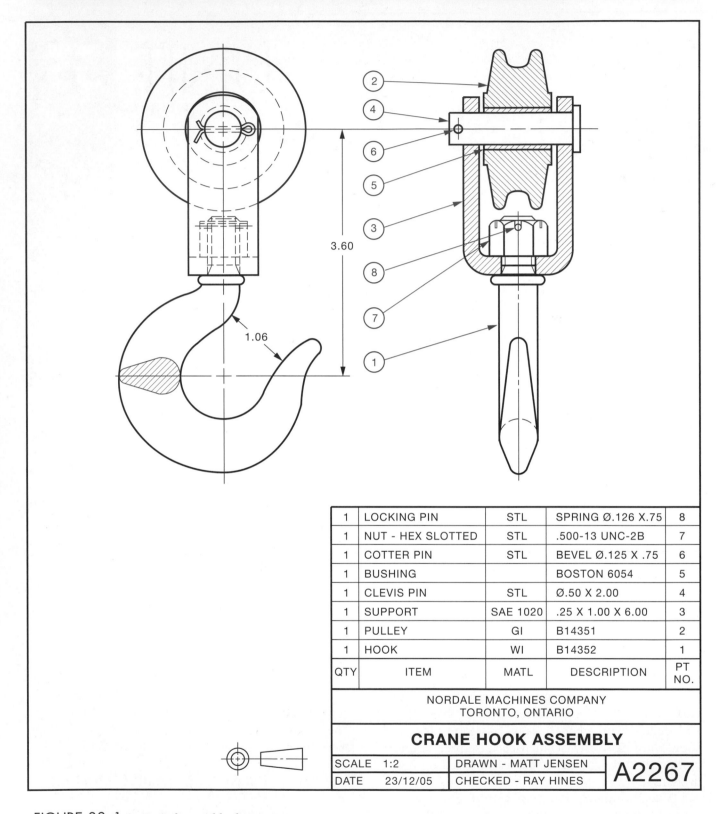

1	LOCKING PIN	STL	SPRING Ø.126 X .75	8
1	NUT - HEX SLOTTED	STL	.500-13 UNC-2B	7
1	COTTER PIN	STL	BEVEL Ø.125 X .75	6
1	BUSHING		BOSTON 6054	5
1	CLEVIS PIN	STL	Ø.50 X 2.00	4
1	SUPPORT	SAE 1020	.25 X 1.00 X 6.00	3
1	PULLEY	GI	B14351	2
1	HOOK	WI	B14352	1
QTY	ITEM	MATL	DESCRIPTION	PT NO.

NORDALE MACHINES COMPANY
TORONTO, ONTARIO

CRANE HOOK ASSEMBLY

SCALE 1:2	DRAWN - MATT JENSEN	A2267
DATE 23/12/05	CHECKED - RAY HINES	

FIGURE 33–1 ■ Typical assembly drawing

To assist in the assembly of the machine, item numbers corresponding with the part numbers of various details are placed on the assembly drawing attached to the corresponding part with a leader. The part number is often enclosed in a small circle, called a balloon, which helps distinguish part numbers from dimensions, Figure 33–1.

BILL OF MATERIAL (ITEMS LIST)

A *bill of material,* or *items list,* is an itemized list of all the components shown on an assembly drawing or a detail drawing, Figure 33–2. Often, a bill of material is placed on a separate sheet of paper for handling and duplicating. For castings, a pattern number appears in the size column instead of the physical size of the part.

Standard components, which are purchased rather than fabricated, including bolts, nuts, and bearings, should have a part number and appear on the bill of material. There should be sufficient information in the descriptive column to enable the purchasing agent to order these parts.

Standard components are incorporated in the design of machine parts for economical production. These parts are specified on the drawing according to the manufacturer's specification. The use of manufacturers' catalogues is essential for determining detailing standards, characteristics of a special part, methods of representation, and so on. However, these catalogues are very unreliable for specifying parts; they should be used as a guide only. To protect the integrity of a design, a purchase part drawing must be made. This avoids the problem when the component supplier makes changes, unknown to the user, which may affect the design.

The four-wheel trolley (Assignment A-91) includes many standard parts—grease cups, lockwashers, Hyatt roller bearings, rivets, and nuts—all of which are standard purchased items. These parts are not detailed but are listed in the bill of material. However, the special countersunk head bolts and the taper washers, commonly called Dutchmen, are not standard parts and must therefore be made especially for this particular assembly.

HELICAL SPRINGS

The *coil,* or *helical spring,* is commonly used in machine design and construction. It may be cylindrical or conical or a combination of the two, Figure 33–3.

Because of the labour and time involved, the true projection of a helical spring is usually not drawn. Instead, a schematic or simplified drawing is used. All the required information can be given on such a drawing.

On assembly drawings, springs are usually shown in section; either crosshatched lines or simplified, symbolic representation is recommended, depending on the size of the wire diameter, Figure 33–4.

4	NUT-HEX REG	STL	.375-16 UNC	7
4	BOLT-HEX REG	STL	.375-16 UNC X 1.50	6
1	KEY	MS	WOODRUFF 608	5
2	BEARINGS	SKF	RADIAL BALL 620	4
1	SHAFT	CRS	Ø1.00 X 6.50 LG	3
1	SUPPORT	MST	.375 X 2.00 X 5.50	2
1	BASE	GI	PATTERN - A3154	1
QTY	ITEM	MATL	DESCRIPTION	PT NO.

FIGURE 33–2 ■ Bill of material for assembly drawing

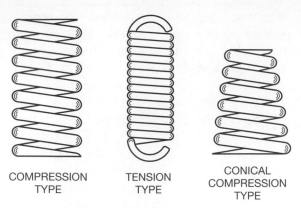

COMPRESSION
TYPE

TENSION
TYPE

CONICAL
COMPRESSION
TYPE

(A) PICTORIAL REPRESENTATION

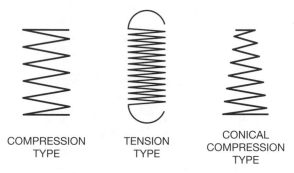

COMPRESSION
TYPE

TENSION
TYPE

CONICAL
COMPRESSION
TYPE

(B) SCHEMATIC OR SIMPLIFIED REPRESENTATION

FIGURE 33–3 ■ Helical springs

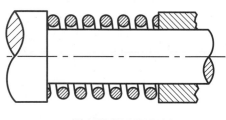

(A) LARGE SPRINGS

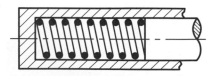

(B) SMALL SPRINGS

FIGURE 33–4 ■ Showing helical springs on assembly drawings

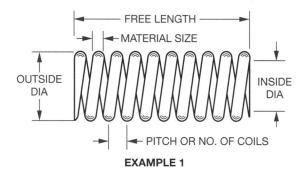

FREE LENGTH

MATERIAL SIZE

OUTSIDE
DIA

INSIDE
DIA

PITCH OR NO. OF COILS

EXAMPLE 1

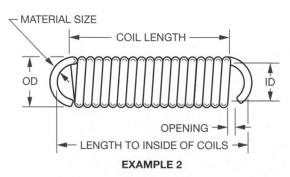

MATERIAL SIZE

COIL LENGTH

OD

ID

OPENING

LENGTH TO INSIDE OF COILS

EXAMPLE 2

FIGURE 33–5 ■ Information given on spring drawings

The following information must be given on a drawing of a spring, Figure 33–5:

- size, shape, and the type of material used in the spring

- diameter (outside or inside)

- pitch or number of coils

- pitch or number of coils

- shape of ends

- length

For example, ONE HELICAL TENSION SPRING 3.00 LG (OR NUMBER OF COILS), .50 ID, PITCH .18, 18 B&S GA. SPRING BRASS WIRE clearly states the required information.

The *pitch* of a coil spring is the distance from the centre of one coil to the centre of the next. The sizes of spring wires are designated by inch sizes or in gauge numbers. The tables for these are found in handbooks.

Springs are made to a dimension of either outside diameter (if the spring works in a hole) or inside diameter (if the spring works on a rod). In some cases, the mean diameter is specified for computation.

REFERENCES

ASME Y14.24-1999 (R2004) Types and Applications of Engineering Drawings
CAN3-B78.1-M83 (R1990) Technical Drawings—General Principles

INTERNET RESOURCES

eFunda For information on spring and spring applications, see: http://www.efunda.com/home.cfm

Integrated Publishing For information on bill of materials, see: www.tpub.com

Machine Design For information on universal joints, see *Machine Design, Mechanical Reference* at: http://www.machinedesign.com

TechStudent.Com For information on springs and spring applications, see: www.technologystudent.com (Mechanisms)

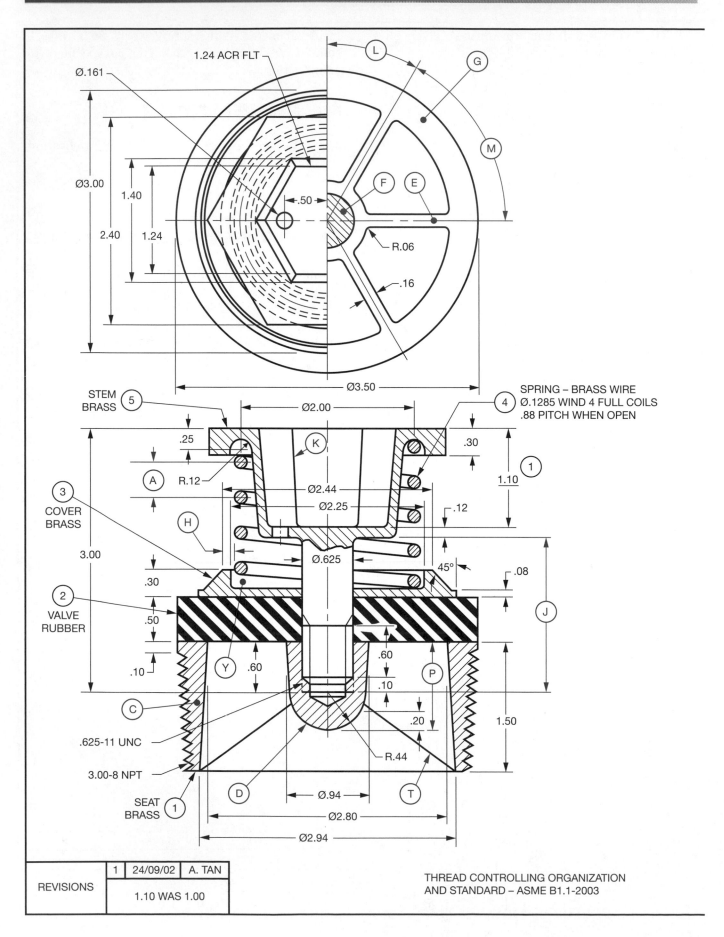

1.24 ACR FLT

Ø.161

Ø3.00

2.40

1.40

1.24

.50

L

G

M

F E

R.06

.16

STEM
BRASS 5

COVER
BRASS 3

3.00

VALVE
RUBBER 2

.625-11 UNC

3.00-8 NPT

SEAT
BRASS 1

Ø3.50

SPRING – BRASS WIRE
Ø.1285 WIND 4 FULL COILS
.88 PITCH WHEN OPEN

Ø2.00

.25

A R.12

H

.30

K

Ø2.44

Ø2.25

Ø.625

45°

.30

.50

.10

Y .60

C

D

Ø.94

Ø2.80

Ø2.94

R.44

T

P

.60

.10

.20

4

1

1.10

.12

.08

J

1.50

REVISIONS | 1 | 24/09/02 | A. TAN
1.10 WAS 1.00

THREAD CONTROLLING ORGANIZATION
AND STANDARD – ASME B1.1-2003

QUESTIONS:

1. How many separate parts are shown on the valve assembly?
2. What is the length of the spring when the valve is closed?
3. Determine distances (A), (J), and (P).
4. What is the overall free length of the spring?
5. Locate (E) in the front view.
6. How many supporting ribs are there connecting (D) to (C)?
7. How thick are these ribs?
8. Identify the part numbers to which features (F) and (G) belong.
9. How many full threads are there on the stem (part 5)?
10. What is the nominal size of the pipe thread?
11. Give the length of the pipe thread.
12. Determine clearance distance (H).
13. Determine angles (L) and (M).

NOTES: UNLESS OTHERWISE SPECIFIED:
– TOLERANCES ON DIMENSIONS ±.02
– TOLERANCES ON ANGLES ±0.5°

ASSIGNMENTS:

1. ON A 1.00-INCH GRID SHEET (.10 IN. SQUARES), SKETCH THE TOP VIEW AND THE FRONT VIEW OF THE STEM IN FULL SECTION, PT 5. USE A CONVENTIONAL BREAK TO SHORTEN THE HEIGHT OF THE FRONT VIEW. ADD DIMENSIONS. SCALE 1:1.

2. ON A 1.00-INCH GRID SHEET (.10 IN. SQUARES), SKETCH A PARTIAL TOP VIEW AND THE FRONT VIEW OF THE SEAT IN FULL SECTION, PT 1. ADD DIMENSIONS. SCALE 1:1.

SCALE	NOT TO SCALE	
DRAWN	J. LUCY	DATE 15/10/01

FLUID PRESSURE VALVE — A-89

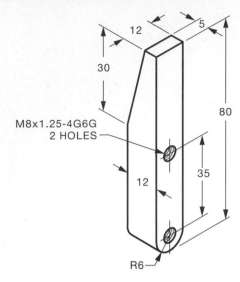

M8x1.25-4G6G
2 HOLES

12

30

12

5

80

35

R6

PT 1   MOVABLE JAW
1 REQD   MATL - SAE 1020

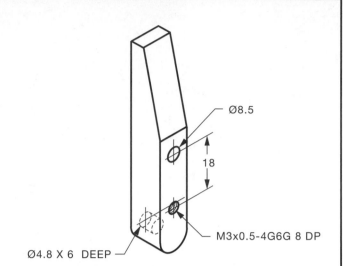

Ø8.5

18

M3x0.5-4G6G 8 DP

Ø4.8 X 6 DEEP

AS SHOWN, OTHERWISE SAME
AS PART 1.
PT 2   STATIONARY JAW
1 REQD   MATL - SAE 1020

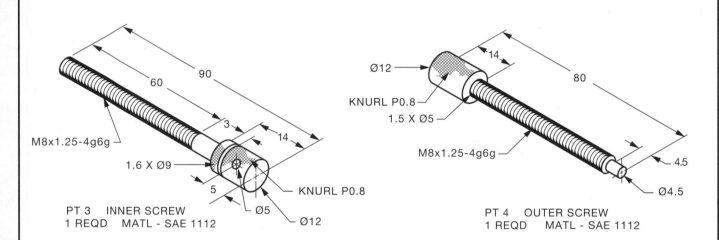

90

60

3

14

M8x1.25-4g6g

1.6 X Ø9

5

Ø5

KNURL P0.8

Ø12

PT 3   INNER SCREW
1 REQD   MATL - SAE 1112

Ø12

KNURL P0.8

1.5 X Ø5

14

80

M8x1.25-4g6g

4.5

Ø4.5

PT 4   OUTER SCREW
1 REQD   MATL - SAE 1112

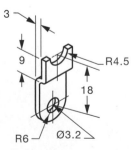

3

9

R4.5

18

R6

Ø3.2

PT 5   CLIP   1 REQD
MATL - 16 USS GA (1.52) STL

PT 6   CAP SCREW
1 REQD   M3x8 LG RD HD

ASSIGNMENT:
ON A CENTIMETRE GRID SHEET (1 mm SQUARES), MAKE A ONE-VIEW
ASSEMBLY DRAWING OF THE PARALLEL CLAMPS. USE SIMPLIFIED
THREAD CONVENTIONS (UNIT 16). PREPARE A BILL OF MATERIAL
SIMILAR TO THE ONE SHOWN IN FIGURE 33-2 CALLING FOR ALL
THE PARTS. IDENTIFY THE PARTS ON THE ASSEMBLY. THE ONLY
DIMENSION REQUIRED IS THE MAXIMUM OPENING OF THE JAWS.
SCALE 1:1.

METRIC
DIMENSIONS ARE IN MILLIMETRES

NOTE:
THREAD CONTROLLING ORGANIZATION
AND STANDARD—ASME B1.13M-2001

**PARALLEL CLAMP
ASSEMBLY**

A-90M

UNIT 34

STRUCTURAL STEEL SHAPES

Structural steel is widely used in the metal trades for fabricating machine parts because the many standard shapes lend themselves to many types of construction. Assignments A-91, A-94, and A-96 show three assemblies using components made from structural steel shapes.

Steel produced at the rolling mills and shipped to the fabricating shop comes in a variety of shapes and forms (about 600). At this stage, it is called *plain material*.

The great bulk of this material can be designated as Figure 34–1.

Abbreviations

When structural steel shapes are designated on drawings, a standard method of abbreviating should be followed that will identify the group of shapes without reference to the manufacturer and without the use of inches and pounds per foot. Therefore, it

SYMBOL	WWF	W	M	S	C	MC
SHAPE	I	I	I	I	[	[
NAME	WELDED WIDE FLANGE SHAPES	WIDE FLANGE SHAPES	MISCELLANEOUS SHAPES	STANDARD BEAMS	STANDARD CHANNELS	MISC. CHANNELS

SYMBOL	WWT	WT OR MT	L	
SHAPE	T	T	L	L
NAME	STRUCTURAL TEES		EQUAL LEG	UNEQUAL LEG
			ANGLES	

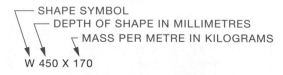

SHAPE SYMBOL
DEPTH OF SHAPE IN INCHES
WEIGHT IN POUNDS PER FOOT

W 18 X 114

(A) INCH DESIGNATION

SHAPE SYMBOL
DEPTH OF SHAPE IN MILLIMETRES
MASS PER METRE IN KILOGRAMS

W 450 X 170

(B) METRIC DESIGNATION

FIGURE 34–1 ■ Common structural steel shapes and drawing callout

is recommended that structural steel be abbreviated as listed in Figure 34–2.

The abbreviations are intended only for use on design drawings. When lists of materials are being prepared for ordering from the mills, the requirements of the respective mills from which the material is to be ordered should be observed.

S-shaped beams and all standard and miscellaneous channels have a slope on the inside flange of 16.67 percent (16.67 percent slope is equivalent to 9° 28′ or a bevel of 1:6). All other beams have parallel face flanges.

PHANTOM OUTLINES

A part or mechanism not included in the actual detail or assembly drawing is sometimes shown to clarify how the mechanism will connect with or operate from an adjacent part. This part is shown by drawing thin dashed lines (one long line and two short dashes) in the operating position. Such a drawing of the extra part is known as a *phantom drawing* or view drawn in *phantom*, Figure 34–3.

On the drawing of the four-wheel trolley (Assignment A-91), the track on which the wheels

SHAPE	U.S. CUSTOMARY EXAMPLES SEE NOTE 1		METRIC SIZE EXAMPLES SEE NOTE 2
	NEW DESIGNATION	OLD DESIGNATION	
Welded Wide Flange Shapes (WWF Shapes)			
- Beam	WWF48 X 230	48WWF320	WWF1000 X 244
- Columns			WWF350 X 315
Wide Flange Shapes (W Shapes)	W24 X 76	24WF76	W600 X 114
	W14 X 26	14B26	W160 X 18
Miscellaneous Shapes (M Shapes)	M8 X 18.5	8M18.5	M200 X 56
	M10 X 9	10JR9.0	M160 X 30
Standard Beams (S Shapes)	S24 X 100	24I100	S380 X 64
Standard Channels (C Shapes)	C12 X 20.7	12C20.7	C250 X 23
Structural Tees			
- cut from WWF Shapes	WWT24 X 160	ST24WWF160	WWT280 X 210
- cut from W Shapes	WT12 X 38	ST122F38	WT130 X 16
- cut from M Shapes	MT4 X 9.25	ST4M9.25	MT100 X 14
Bearing Piles (HP Shapes)	HP14 X 73	14BP73	HP350 X 109
Angles (L Shapes)	L6 X 6 X .75	L6 X 6 X 3/4	L75 X 75 X 6
(leg dimensions X thickness)	L6 X 4 X .62	L6 X 4 X 5/8	L150 X 100 X 13
Plates (width X thickness)	20 X .50	20 X 1/2	500 X 12
Square Bar (side)	◻1.00	BAR 1 ◻	◻25
Round Bar (diameter)	Ø1.25	BAR 1-1/4 Ø	Ø30
Flat Bar (width X thickness)	250 X .25	BAR 2-1/2 X 1/4	60 X 6
Round Pipe (type of pipe X OD X wall thickness)	12.75 OD X .375	12-3/4 X 3/8	XS 102 OD X 8
Square and Rectangular Hollow Structural Sections (outside dimensions X wall thickness)	HSS4 X 4 X .375 HSS8 X 4 X .375	4X 4RT X 3/8 8 X 4RT X 3/8	HSS102 X 102 X 8
Steel Pipe Piles (OD X wall thickness)			320 OD X 6

NOTE 1 - VALUES SHOWN ARE NOMINAL DEPTH (INCHES) X WEIGHT PER FOOT LENGTH (POUNDS).

NOTE 2 - VALUES SHOWN ARE NOMINAL DEPTH (MILLIMETRES) X MASS PER METRE LENGTH (KILOGRAMS).

NOTE 3 - METRIC SIZE EXAMPLES SHOWN ARE NOT NECESSARILY THE EQUIVALENTS OF THE INCH SIZE EXAMPLES SHOWN.

FIGURE 34–2 ■ Abbreviations for shapes, plates, bars, and tubes

run is an S beam. The wheels are set at an angle to the vertical plane in order to ride upon the sloping bottom flange of the S beam. The outline of the beam is shown by dashed lines, and, while not an integral part of the trolley, the outline or phantom view of the S beam shows clearly how the trolley operates.

CONICAL WASHERS

Conical washers, Figure 34–4, are available in several sizes to accommodate the slopes on structural steel shapes. A typical application can be found on the four-wheel trolley, parts **D** and **W** (Assignment A-91).

REFERENCES

Canadian Institute of Steel Construction
American Institute of Steel Construction
CAN3-B78.1-M83 Technical Drawings—General
    Principles

INTERNET RESOURCES

American Institute of Steel Construction For information on structural steel design and construction with links to an online library, training CD-ROM, directories, and job postings, see: http://www.aisc.org

American Iron and Steel Institute For news and information about the use of iron and steel in manufacturing and construction, see: http://www.steel.org

IDS Development—Nebraska Education For information on line types used on engineering drawings, see: http://idsdev.mccneb.edu/djackson/lineintro.htm

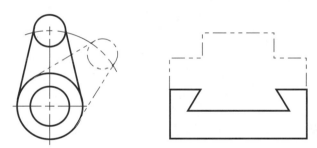

FIGURE 34–3 ■ Phantom lines

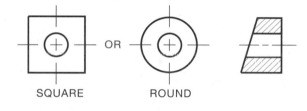

FIGURE 34–4 ■ Conical washers

QUESTIONS:

1. What does hidden line (E) indicate?

2. Which cutting plane in the primary auxiliary view indicates (A) where the section to the left of line N-N is taken, (B) where the section to the right of line N-N is taken?

3. What is the slope of angle (J)?

4. Locate part (K) in the section view.

5. What is the wheel diameter of the trolley?

6. Locate parts (2), (3), (5), (A), (C), (V), and (Z) in the primary auxiliary view.

7. What are the names of parts (T), (U), (V), (W), (X), and (Y) ?

8. What is the diameter of the bearing rollers?

9. Determine distance (L).

10. How many not-to-scale dimensions are shown?

11. What type of line is used to show the S beam?

S BEAM - S10 X 35

WHEEL - Ø8.00

SHAFT - Ø1.374

BEARING - 2.835 OD

      -ROLLERS - Ø.562

NOTE:
THREAD CONTROLLING ORGANIZATION
AND STANDARD-ASME B1.1-2003

REVISIONS	1	21/04/04	F. NEWMAN
		2.50 WAS 2.60	

ASSIGNMENTS:

1. ON A 1.00-INCH GRID SHEET (.10 IN. SQUARES), SKETCH THE SECONDARY AUXILIARY VIEW DRAWING OF PART A. THE SECONDARY AUXILIARY VIEW IS POSITIONED ABOVE AND PROJECTED FROM THE PRIMARY AUXILIARY VIEW. THE WIDTH OF THE PART IS TO BE SHORTENED BY USING CONVENTIONAL BREAKS AND ITS WIDTH TO BE DETERMINED BY THE STUDENT. ADD DIMENSIONS. SCALE 1:2.

2. ON 1.00-INCH GRID SHEETS (.10 IN. SQUARES), SKETCH WORKING DRAWINGS OF PARTS C AND D. SCALE 1:1.

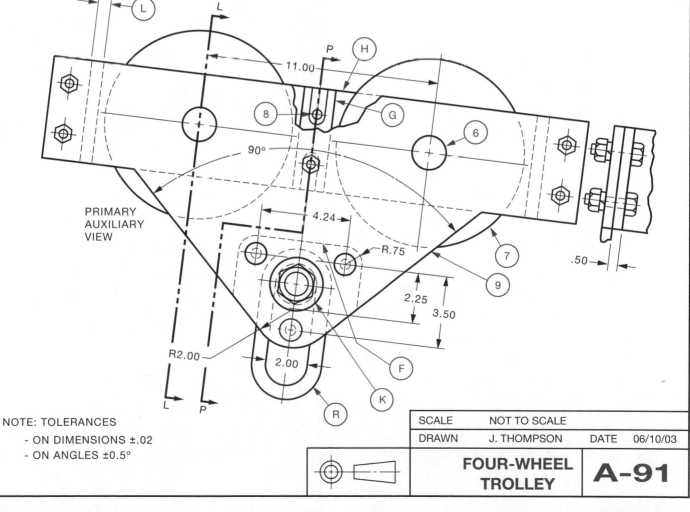

PRIMARY AUXILIARY VIEW

NOTE: TOLERANCES
- ON DIMENSIONS ±.02
- ON ANGLES ±0.5°

SCALE	NOT TO SCALE		
DRAWN	J. THOMPSON	DATE	06/10/03

FOUR-WHEEL TROLLEY | **A-91**

WELDING DRAWINGS

The primary purpose of welding is to unite pieces of metal so they will operate properly as a unit to support the loads to be carried. To design and build such a structure to be economical and efficient, a basic knowledge of welding is essential. Figure 35–1 illustrates basic welding terms.

The use of welding symbols on a drawing enables the designer to clearly specify the type and size of weld to meet the design requirements. Basic welding joints are shown in Figure 35–2. Points that must be made clear are the type of weld, the joint penetration, the weld size, and the root opening (if any). These points can be clearly indicated on the drawing by the welding symbol.

WELDING SYMBOLS

Welding symbols are a shorthand language. They save time and money and, if used correctly, ensure understanding and accuracy. Welding symbols should be a universal language; for this reason, the symbols of the American Welding Society have been adopted.

A distinction between the terms *weld symbol* and *welding symbol* should be understood. The weld symbol indicates the type of weld. The welding

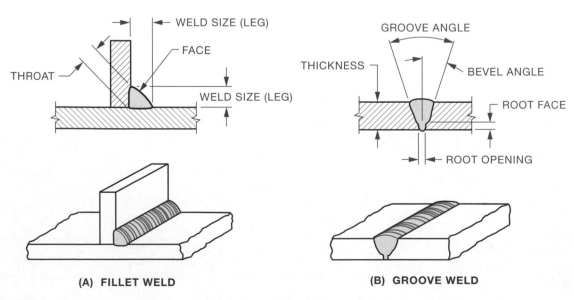

(A) FILLET WELD **(B) GROOVE WELD**

FIGURE 35–1 ■ Basic welding terms

symbol is a method of representing the weld on drawings. It includes supplementary information and consists of the following eight elements. Not all elements need be used unless required for clarity.

1. reference line
2. arrow
3. basic weld symbol
4. dimensions and other data
5. supplementary symbols
6. finish symbols
7. tail
8. specification, process, or other reference

The size and spacing of welds shown on the welding symbol are given in inches or millimetres.

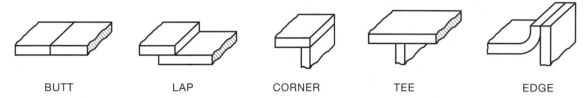

BUTT   LAP   CORNER   TEE   EDGE

FIGURE 35–2 ■ Basic welding joints

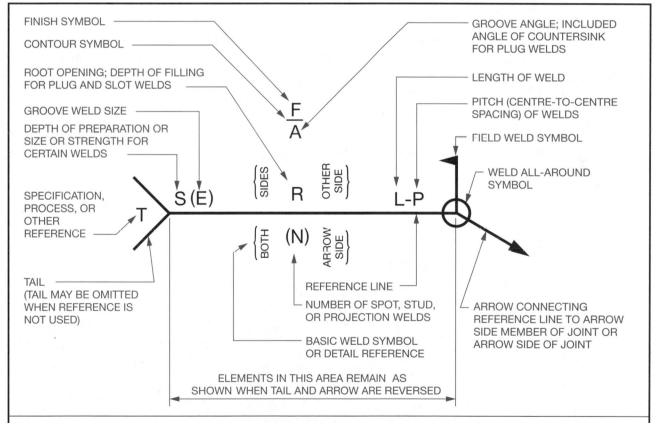

FINISH SYMBOL

CONTOUR SYMBOL

ROOT OPENING; DEPTH OF FILLING FOR PLUG AND SLOT WELDS

GROOVE WELD SIZE

DEPTH OF PREPARATION OR SIZE OR STRENGTH FOR CERTAIN WELDS

SPECIFICATION, PROCESS, OR OTHER REFERENCE

TAIL (TAIL MAY BE OMITTED WHEN REFERENCE IS NOT USED)

GROOVE ANGLE; INCLUDED ANGLE OF COUNTERSINK FOR PLUG WELDS

LENGTH OF WELD

PITCH (CENTRE-TO-CENTRE SPACING) OF WELDS

FIELD WELD SYMBOL

WELD ALL-AROUND SYMBOL

ARROW CONNECTING REFERENCE LINE TO ARROW SIDE MEMBER OF JOINT OR ARROW SIDE OF JOINT

REFERENCE LINE

NUMBER OF SPOT, STUD, OR PROJECTION WELDS

BASIC WELD SYMBOL OR DETAIL REFERENCE

ELEMENTS IN THIS AREA REMAIN AS SHOWN WHEN TAIL AND ARROW ARE REVERSED

NOTE: SIZE, WELD SYMBOL, LENGTH OF WELD, AND SPACING MUST READ IN THAT ORDER FROM LEFT TO RIGHT ALONG THE REFERENCE LINE. NEITHER ORIENTATION NOR REFERENCE LINE NOR LOCATION ALTER THIS RULE.
THE PERPENDICULAR LEG OF △, ∨, ∨, ⌐ WELD SYMBOLS MUST BE AT LEFT.
ARROW AND OTHER SIDE WELDS ARE OF THE SAME SIZE UNLESS OTHERWISE SHOWN.
SYMBOLS APPLY BETWEEN ABRUPT CHANGES IN DIRECTION OF WELDING UNLESS GOVERNED BY THE "ALL AROUND" SYMBOL OR OTHERWISE DIMENSIONED.

FIGURE 35–3 ■ Standard location of elements of a welding symbol

Figure 35–3 illustrates the position of the weld symbols and other information in relation to the welding symbol. The various weld symbols that may be applied to the basic welding symbol are shown in Figure 35–4. Figure 35–5 shows the actual shape of many of the weld types symbolized in Figure 35–4.

Supplementary symbols may also be added to the welding symbol. The supplementary symbols are illustrated in Figure 35–6.

Any welding joint indicated by a symbol will always have one arrow side and one other side. The terms arrow side, other side, and both sides are used accordingly to locate the weld with respect to the joint.

Tail of Welding Symbol

The welding symbol and allied process to be used may be specified by placing the appropriate letter designations from Figure 35–7 in the tail of the welding symbol, Figure 35–8.

Codes, specifications, or any other applicable documents may be specified by placing the reference in the tail of the welding symbol. Information contained in the referenced document need not be repeated in the welding symbol.

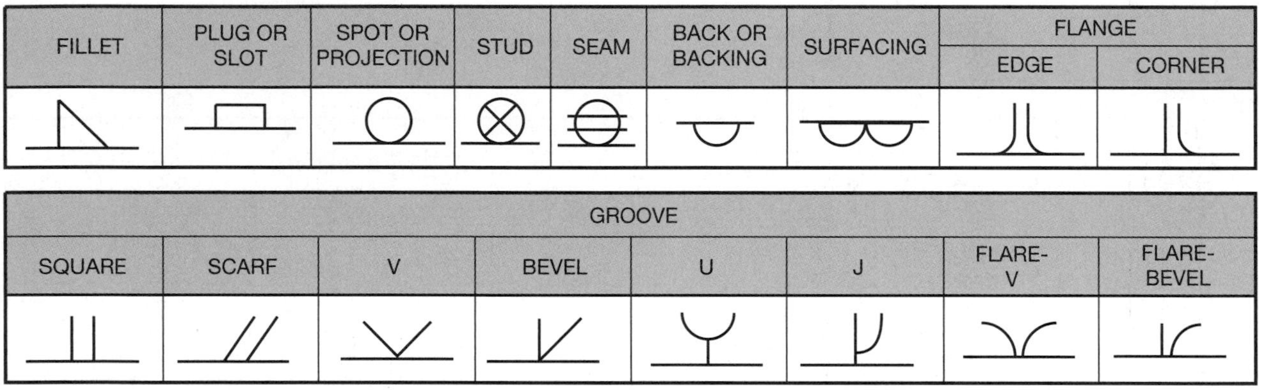

FIGURE 35–4 ■ Basic weld symbols shown on the reference line

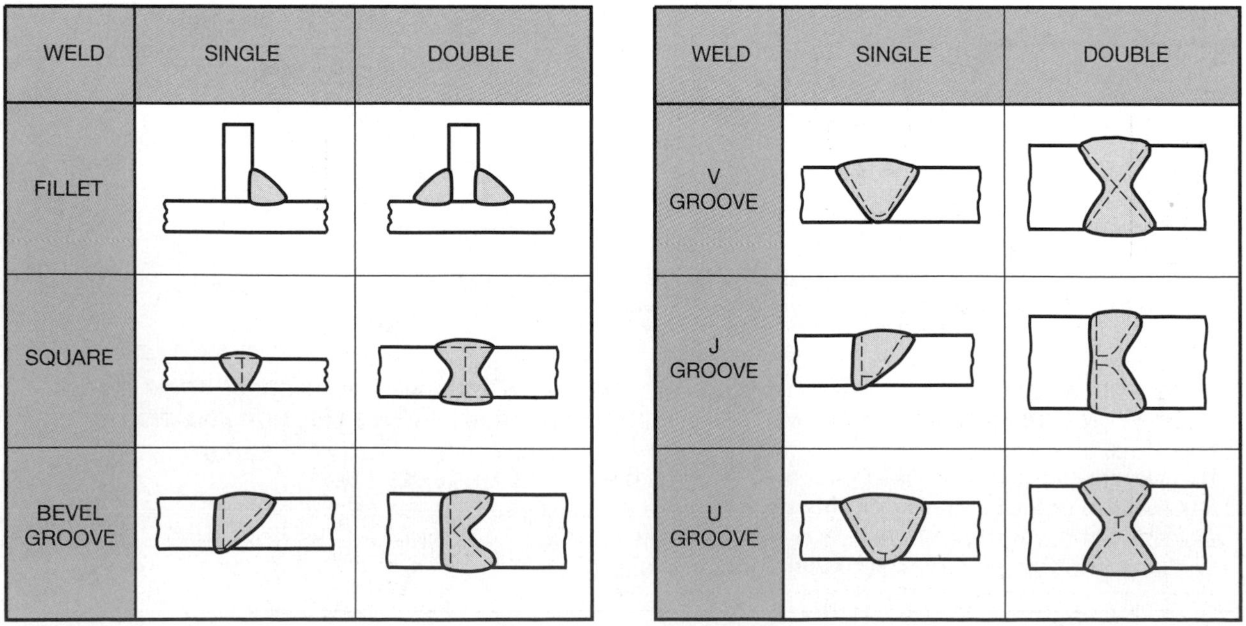

FIGURE 35–5 ■ Types of welds

WELD ALL AROUND	FIELD WELD	MELT-THRU	BACKING OR SPACER MATERIAL	CONSUMABLE INSERT	CONTOUR		
					FLUSH	CONVEX	CONCAVE

FIGURE 35–6 ■ Supplementary symbols

Welding Process	Welding Process (Specific)	Letter Designation
Brazing (B)	Infrared Brazing	IRB
	Torch Brazing	TB
	Furnace Brazing	FB
	Induction Brazing	IB
	Resistance Brazing	RB
	Dip Brazing	DB
Oxyfuel Gas Welding (OFW)	Oxyacetylene Welding	OAW
	Oxyhydrogen Welding	OHW
	Pressure Gas Welding	PGW
Resistance Welding (RW)	Resistance-Spot Welding	RSW
	Resistance-Seam Welding	RSEW
	Projection Welding	PW
	Flash Welding	FW
	Upset Welding	UW
	Percussion Welding	PEW
Arc Welding (AW)	Stud Arc Welding	SW
	Plasma-Arc Welding	PAW
	Submerged Arc Welding	SAW
	Gas Tungsten-Arc Welding	GTAW
	Gas Metal-Arc Welding	GMAW
	Flux Cored Arc Welding	FCAW
	Shielded Metal-Arc Welding	SMAW
	Carbon-Arc Welding	CAW
Other Processes	Thermit Welding	TW
	Laser Beam Welding	LBW
	Induction Welding	IW
	Electroslag Welding	ESW
	Electron Beam Welding	EBW
Solid State Welding (SSW)	Ultrasonic Welding	USW
	Friction Welding	FRW
	Forge Welding	FOW
	Explosion Welding	EXW
	Diffusion Welding	DFW
	Cold Welding	CW

Cutting Method	Letter Designation
Arc Cutting	AC
Air Carbon-Arc Cutting	AAC
Carbon-Arc Cutting	CAC
Metal-Arc Cutting	MAC
Plasma-Arc Cutting	PAC
Oxygen Cutting	OC
Chemical Flux Cutting	FOC
Metal Powder Cutting	POC
Oxygen-Arc Cutting	AOC

FIGURE 35–7 ■ Designation of welding process by letters

Multiple Reference Lines

Two or more reference lines may be used to describe a sequence of operations. The first operation is specified on the reference line nearest the arrow. Subsequent operations are specified sequentially on other reference lines, Figure 35–9.

Weld Locations on Symbol

Welds on the arrow side of the joint are shown by placing the weld symbol on the bottom side of the reference line. Welds on the other side of the joint are shown by placing the weld symbol on the top side of the reference line. Welds on both sides of the joint are shown by placing the weld symbol on both sides of the reference line. A weld extending completely around a joint is indicated by means of a weld-all-around symbol placed at the intersection of the reference line and the arrow.

Field welds (welds not made in the shop or at the initial place of construction) are indicated with the field weld symbol placed at the intersection of the reference line and the arrow.

All weld dimensions on a drawing may be subject to a general note, such as: ALL FILLET WELDS .25 UNLESS OTHERWISE NOTED.

Only the basic fillet welds will be discussed in this unit. Figure 35–10 illustrates several typical fillet welding symbols and the resulting welds.

FILLET WELDS

Fillet welds are triangular in shape and are used to join surfaces that are perpendicular to each other: lapped, tee, and corner joints. No preparation of the surfaces of the metal is required.

1. Fillet weld symbols are drawn with the perpendicular leg always to the left.

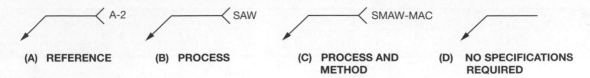

(A) REFERENCE   **(B) PROCESS**   **(C) PROCESS AND METHOD**   **(D) NO SPECIFICATIONS REQUIRED**

FIGURE 35–8 ■ Location of specifications, processes, and other references on welding symbols

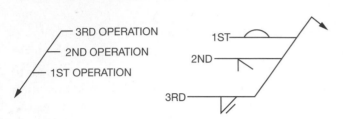

FIGURE 35–9 ■ Multiple reference lines

2. Dimensions of fillet welds are shown on the same side of the reference line and to the left of the weld symbol.

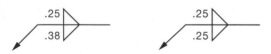

3. The dimensions of fillet welds on both sides of a joint are shown whether the dimensions are identical or different.

4. The dimension does not need to be shown when a general note is placed on the drawing to specify the dimension of fillet welds.

NOTE: SIZE OF FILLET WELDS .25 UNLESS OTHERWISE SPECIFIED.

5. The *length* of a fillet weld, when indicated on the welding symbol, is shown to the right of the weld symbol.

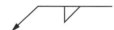

6. The *pitch* (centre-to-centre spacing) of an intermittent fillet weld is shown as the distance between centres of increments on one side of the joint. It is shown to the right of the length dimension following a hyphen.

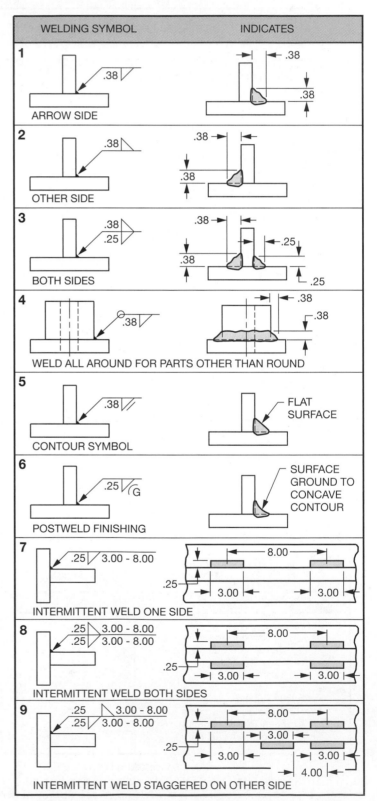

WELDING SYMBOL	INDICATES
1 ARROW SIDE	
2 OTHER SIDE	
3 BOTH SIDES	
4 WELD ALL AROUND FOR PARTS OTHER THAN ROUND	
5 CONTOUR SYMBOL	
6 POSTWELD FINISHING	
7 INTERMITTENT WELD ONE SIDE	
8 INTERMITTENT WELD BOTH SIDES	
9 INTERMITTENT WELD STAGGERED ON OTHER SIDE	

FIGURE 35–10 ■ Typical fillet welds

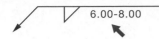

7. Staggered intermittent fillet welds are illustrated by staggering the weld symbols.

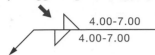

8. Fillet welds that are to be welded with approximately flat, convex, or concave faces without postweld finishing are specified by adding the flat, convex, or concave contour symbol to the weld symbol.

9. Fillet welds whose faces are to be finished approximately flat, convex, or concave by postweld finishing are specified by adding both the appropriate contour and finishing symbol to the weld symbol.

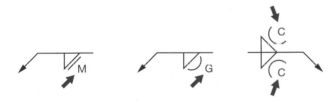

The following finishing symbols may be used to specify the method of finishing, but not the degree of finish:

C — Chipping
G — Grinding
H — Hammering
M — Machining
R — Rolling

10. A weld with a length less than the available joint length whose location is significant is specified on the drawing in a manner similar to that in Figure 35–11.

11. A continuous weld extending around a series of connected joints may be specified by the addition of the weld-all-around symbol at the junction of the arrow and reference line. The series of joints may involve different directions and may lie on more than one plane, Figure 35–12(A).

Welds extending around a pipe or circular or oval holes do not require the weld-all-around symbol, Figure 35–12(B).

REFERENCES

Canadian Welding Bureau
American Welding Society

INTERNET RESOURCES

American Welding Society For information on all aspects of welding with links to related organizations and materials, see: http://www.aws.org

eFunda For information on types of welding, see: http://www.efunda.com/home.cfm

Machine Design For information on welding and weld joints, see *Machine Design, Fastening/Joining Reference* at: http://www.machinedesign.com

Unified Engineering, Inc. For information on fillet welds, see: http://www.unified-eng.com/scitech/weld/weld.html

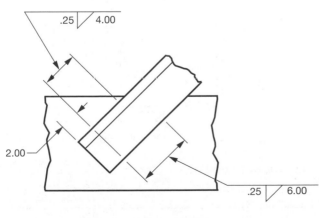

(A) DRAWING CALLOUT

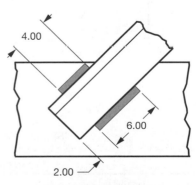

(B) INTERPRETATION

FIGURE 35–11 ■ Weld lengths located by dimensions

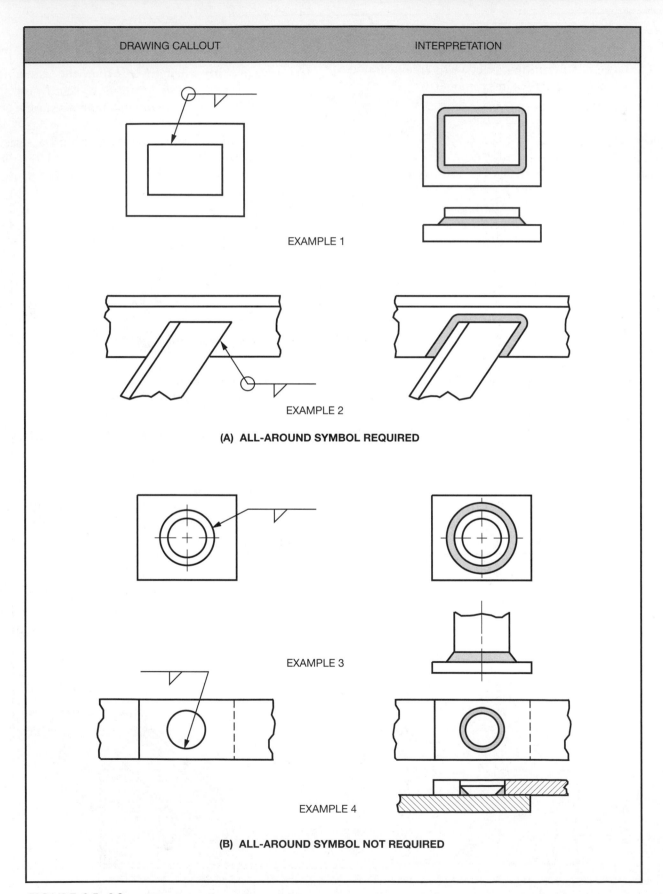

DRAWING CALLOUT	INTERPRETATION

EXAMPLE 1

EXAMPLE 2

(A) ALL-AROUND SYMBOL REQUIRED

EXAMPLE 3

EXAMPLE 4

(B) ALL-AROUND SYMBOL NOT REQUIRED

FIGURE 35–12 ■ The use of all-around symbols

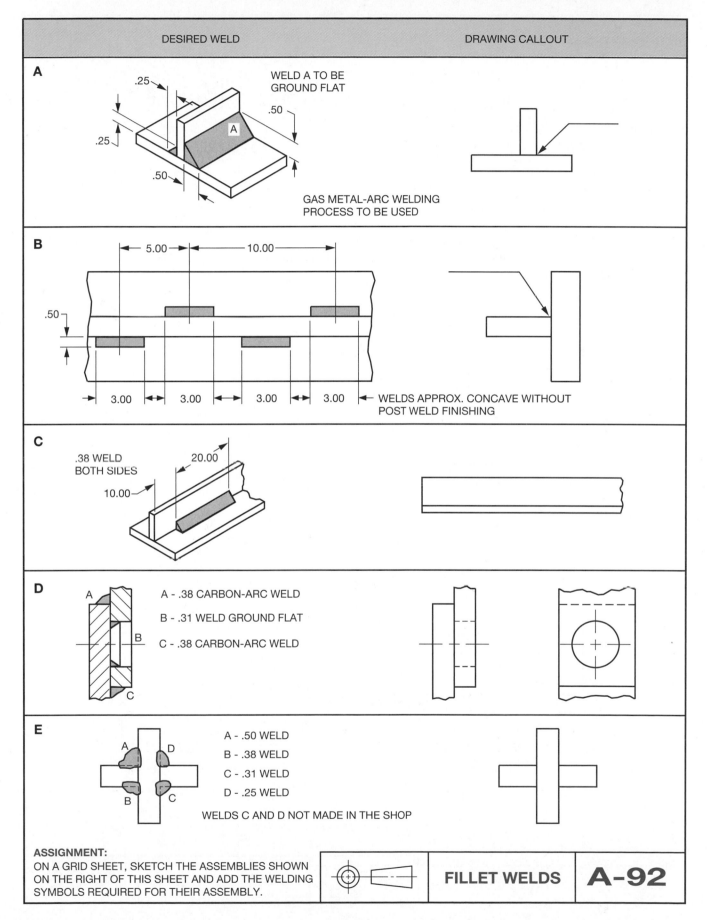

DESIRED WELD	DRAWING CALLOUT

A

.25 .25 .50 .50

WELD A TO BE GROUND FLAT

GAS METAL-ARC WELDING PROCESS TO BE USED

B

5.00 10.00 .50 3.00 3.00 3.00 3.00

WELDS APPROX. CONCAVE WITHOUT POST WELD FINISHING

C

.38 WELD BOTH SIDES 20.00 10.00

D

A - .38 CARBON-ARC WELD

B - .31 WELD GROUND FLAT

C - .38 CARBON-ARC WELD

E

A - .50 WELD

B - .38 WELD

C - .31 WELD

D - .25 WELD

WELDS C AND D NOT MADE IN THE SHOP

ASSIGNMENT:
ON A GRID SHEET, SKETCH THE ASSEMBLIES SHOWN ON THE RIGHT OF THIS SHEET AND ADD THE WELDING SYMBOLS REQUIRED FOR THEIR ASSEMBLY.

FILLET WELDS | **A-92**

ASSIGNMENT:

ON A ONE-INCH GRID SHEET (.10 IN. SQUARES), SKETCH THE ASSEMBLY SHOWN BELOW TO THE SCALE OF 1:2. ADD THE FOLLOWING WELD INFORMATION TO THE DRAWING:

- HORIZONTAL SHAFTS WELDED BOTH SIDES TO ARMS WITH .25 FILLET WELDS ON OUTSIDE AND .19 FILLET WELDS INSIDE. WELD TO BE FLAT WITHOUT POSTWELD FINISHING.
- ARMS WELDED BOTH SIDES TO BASE WITH .19 FILLET WELDS.
- RIBS WELDED BOTH SIDES TO BASE AND ARMS WITH .12 FILLET WELDS.
- VERTICAL SHAFT WELDED TO BASE WITH .25 FILLET WELD GROUND CONCAVE.

PROCESS - CARBON ARC WELDING.

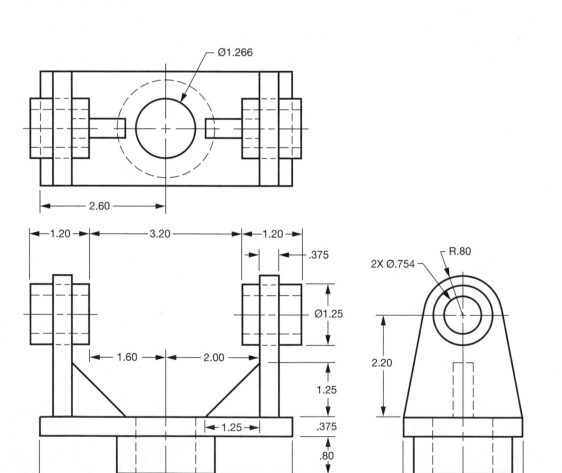

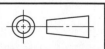

SHAFT SUPPORT | **A-93**

GROOVE WELDS

Groove welds are used to join butted parts. The welds are classified as *square, scarf, bevel, v, u, j, flare-v,* and *flare bevel.* One or more beads (welding passes) may be used to produce the desired weld. For most of these welds, preparation of the metal surfaces before welding is required.

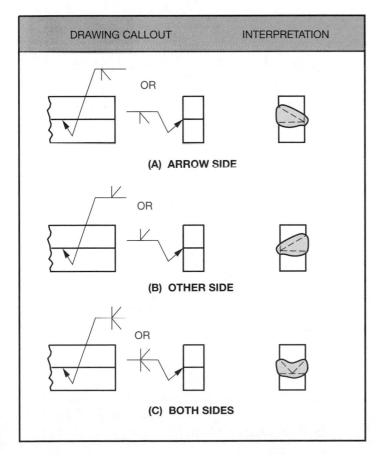

DRAWING CALLOUT	INTERPRETATION

(A) ARROW SIDE

(B) OTHER SIDE

(C) BOTH SIDES

FIGURE 36–1 ■ Application of break in arrow on welding symbol

1. Bevel-groove, J-groove, and flare bevel-groove weld symbols are always drawn with the perpendicular leg to the left.
2. Dimensions of single-groove welds are shown on the same side of the reference line as the weld symbol.
3. Each groove of a double-groove joint is dimensioned; however, the root opening need appear only once.
4. For bevel-groove and J-groove welds, a broken arrow is used, when necessary, to identify the member to be prepared, Figure 36–1.
5. The depth of groove preparation "S" and size (E) of a groove weld, when specified, are placed to the left of the weld symbol. Either or both may be shown. Except for square-groove welds, the groove weld size (E) in relation to the depth of the groove preparation "S" is shown as "S(E)," Figure 36–2.
6. Only the groove weld size is shown for square-groove welds.

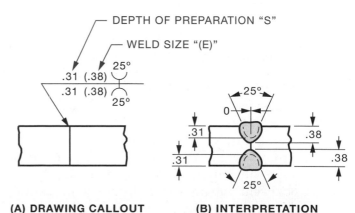

(A) DRAWING CALLOUT **(B) INTERPRETATION**

FIGURE 36–2 ■ Groove weld symbol showing use of combined dimensions

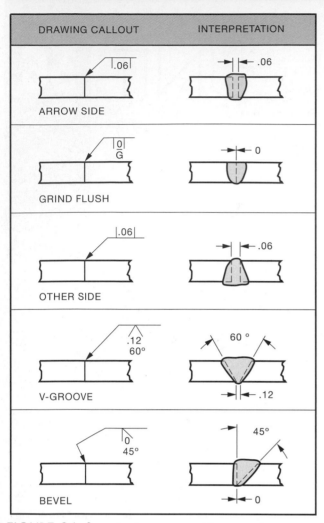

FIGURE 36–3 ■ Single-groove welds—complete joint penetration

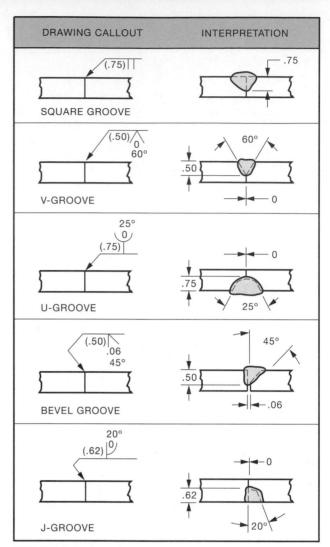

FIGURE 36–4 ■ Single-groove welds—partial penetration

7. When no depth of groove preparation and no groove weld size are specified on the welding symbol for single-groove and symmetrical double-groove welds, complete joint penetration is required, Figure 36–3.

8. When the groove welds extend only partly through the member being joined, the size of the weld is shown on the weld symbol, Figures 34–4, 34–5, and 34–6.

9. A dimension not in parentheses placed to the left of a bevel, V-, J-, or U-groove weld symbol indicates only the depth of penetration.

10. Groove welds that are to be welded with approximately flush or convex faces without postweld finishing are specified by adding the flush or convex contour symbol to the weld symbol.

11. Groove welds whose faces are to be finished flush or convex by postweld finishing are specified by adding both the appropriate contour and the finishing symbol to the weld symbol. Standard finishing symbols are:

   C — Chipping
   G — Grinding
   H — Hammering
   M — Machining
   R — Rolling

12. The size of flare-groove welds when no weld size is given is considered as extending only to the tangent points indicated by dimension "S," Figure 36–7. For application of flare-groove welds with partial joint penetration, see Figure 36–7.

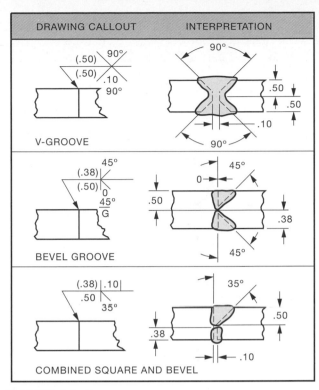

FIGURE 36–5 ■ Double-groove welds

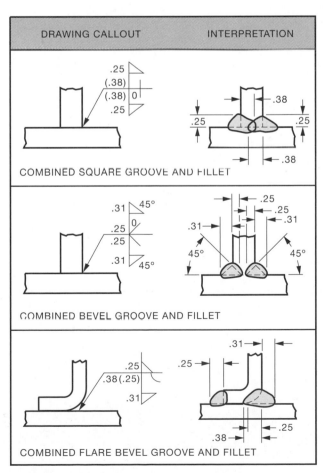

FIGURE 36–6 ■ Combined groove and fillet welds

SUPPLEMENTARY SYMBOLS

Back and Backing Welds

The back or backing weld symbol is used to indicate bead-type back or backing welds of single-groove welds.

The back and backing weld symbols are identical. The sequence of welding determines which designation applies. The back weld is made after the groove weld and the backing weld is made before the groove weld.

1. The back weld symbol is placed on the side of the reference line opposite a groove weld symbol. When a single reference line is used, BACK WELD is specified in the tail of the symbol. Alternatively, if a multiple reference line is used, the back weld symbol is placed on a reference line subsequent to the reference line specifying the groove weld, Figure 36–8(A).

2. The backing weld symbol is placed on the side of the reference line opposite the groove weld

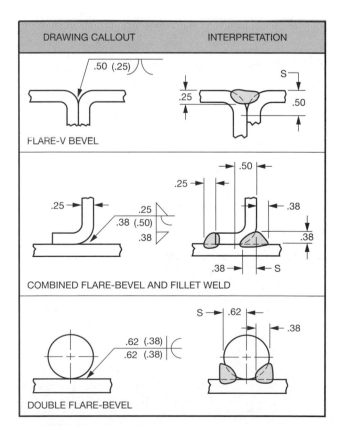

FIGURE 36–7 ■ Flare-V and flare bevel groove welds with partial joint penetration

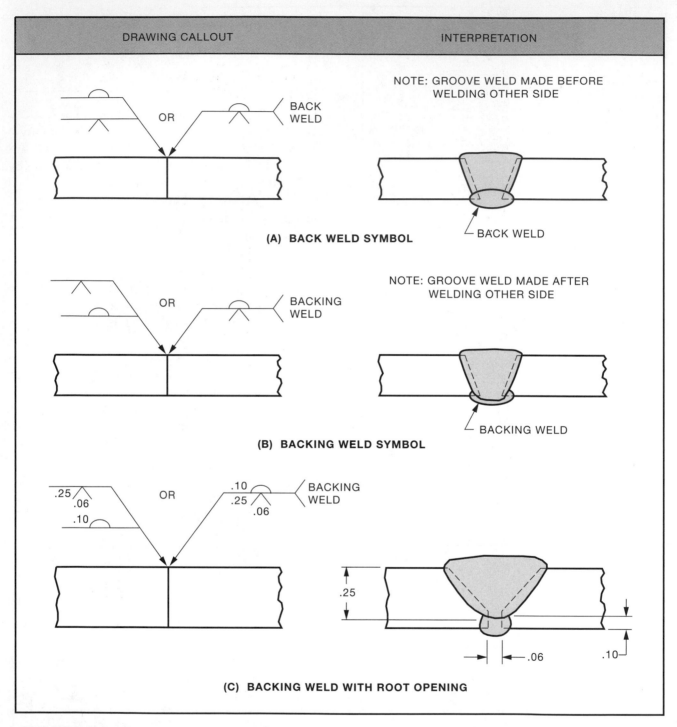

DRAWING CALLOUT · INTERPRETATION

OR — BACK WELD

NOTE: GROOVE WELD MADE BEFORE WELDING OTHER SIDE

BACK WELD

(A) BACK WELD SYMBOL

OR — BACKING WELD

NOTE: GROOVE WELD MADE AFTER WELDING OTHER SIDE

BACKING WELD

(B) BACKING WELD SYMBOL

.25 .06 .10 OR .10 .25 .06 BACKING WELD

.25 .06 .10

(C) BACKING WELD WITH ROOT OPENING

FIGURE 36–8 ■ Application of back and backing weld symbol

symbol. When a single reference line is used, BACKING WELD is specified in the tail of the arrow. If a multiple reference line is used, the backing weld symbol is placed on a reference line before that specifying the groove weld, Figures 36–8(B) and (C).

Melt-Through Symbol

The melt-through symbol is used only when complete root penetration plus visible root reinforcement is required in welds made from one side.

The melt-through symbol is placed on the side of the reference line opposite the weld symbol, Figure 36–9.

The height of root reinforcement may be specified by placing the required dimension to the left of the melt-through symbol. The height of root reinforcement may be unspecified.

REFERENCES

Canadian Welding Bureau
ANSI/AWS A2.4-86 Standard Symbols for Welding, Brazing, and NDE

INTERNET RESOURCES

eFunda For information on types of welding, see: http://www.efunda.com/home.cfm

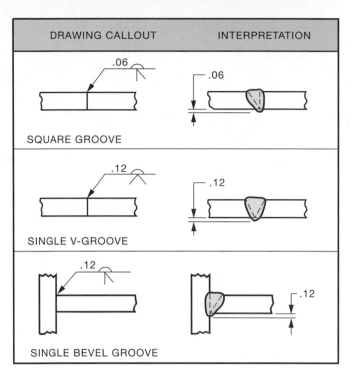

FIGURE 36–9 ■ Application of melt-through groove weld symbols

Machine Design For information on welding and weld joints, see *Machine Design, Fastening/Joining Reference* at: http://www.machinedesign.com

Unified Engineering, Inc. For information on groove welds, see: http://www.unified-eng.com/scitech/weld/weld.html

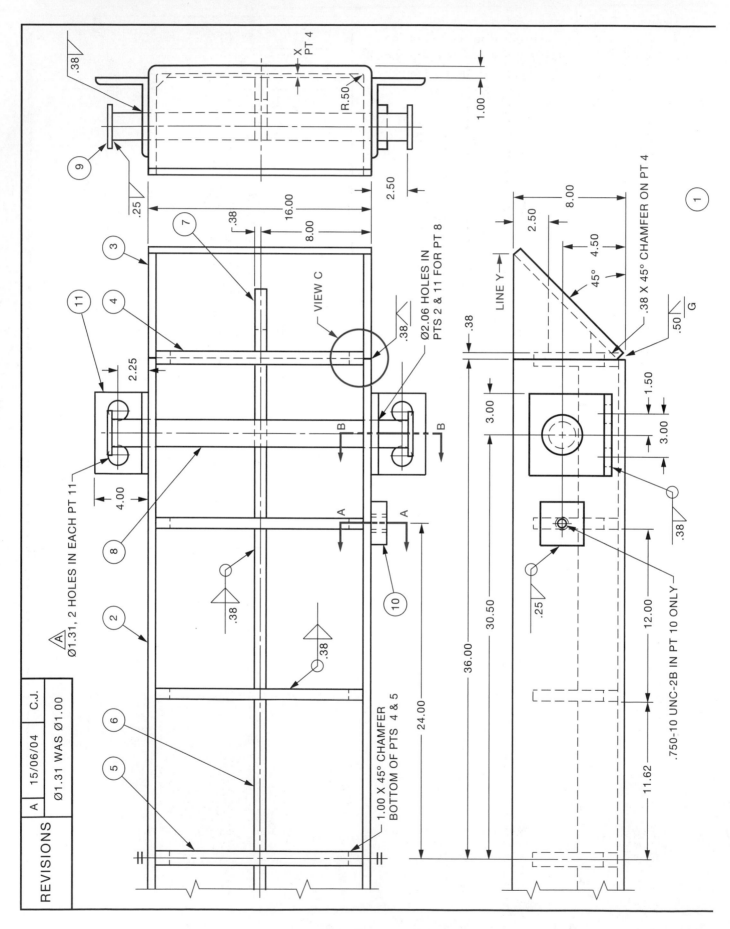

REVISIONS

A	15/06/04	C.J.
	Ø1.31 WAS Ø1.00	

⚠️A Ø1.31, 2 HOLES IN EACH PT 11

.38

9

.25

7

3

11

4

2.25

4.00

8

2

6

5

VIEW C

Ø2.06 HOLES IN
PTS 2 & 11 FOR PT 8

.38

10

1.00 X 45° CHAMFER
BOTTOM OF PTS 4 & 5

16.00

8.00

.38

24.00

B

B

A

A

.38

.38

X
PT 4

R.50

1.00

2.50

8.00

2.50

LINE Y

.38

3.00

4.50

45°

.38 X 45° CHAMFER ON PT 4

.50 / G

1.50

3.00

.38

1

30.50

36.00

11.62

12.00

.25

.750-10 UNC-2B IN PT 10 ONLY

NEL

QUESTIONS:

1. How many Ø1.31 holes are there in the complete assembly?

2. How deep is the .750 tapped hole?

3. What was the original size of the Ø1.31 hole?

4. What is the overall height of the assembly?

5. Determine the distance from line Y to the centre line of the assemb y.

6. Determine distance X.

7. What is the developed width of part 2? Use inside dimensions of channel.

8. What is the clearance for fitting on the length of pt. 5?

9. What is the length of (A) pt. 2, (B) pt. 4, allow .25 for clearance, (C) pt. 7, (D) pt. 8?

10. What is the difference be־ween pt. 4 and pt. 5?

11. How many 1.00 chamfers are needed?

12. What does G mean on .50 bevel weld symbols?

13. What type of weld is used to fasten pt. 11 to pt. 2?

14. What type of weld is used to join pt. 3 to pt. 2 at the sides?

15. How many parts make up the assembly?

16. Complete the missing sizes in the bill of material. Use inside travel for calculating part 2.

NOTE:

UNLESS OTHERWISE SPECIFIED:
- TOLERANCE ON LINEAR DIMENSIONS ±.04
- TOLERANCE ON ANGLES ±0.5°
- TOLERANCE ON HOLES ±.004
- THREAD CONTROLLING ORGANIZATION AND STANDARD-ASME B1.1-2003

ASSIGNMENT:
ON A ONE-INCH GRID SHEET (.10 IN. SQUARES), SKETCH SECTION VIEWS AT A-A AND B-B AND AN ENLARGED VIEW AT VIEW C. SHOW ONLY THE PAR־S AND THE WELDS. SCALE 1:1.

| TOP VIEW | RIGHT-SIDE VIEW |
| FRONT VIEW | |

ARRANGEMENT OF VIEWS

QTY	ITEM	MATL	DESCRIPTION	PT NO.
4	LOCATING ANGLE	STL	6.00 X 4.00 X .50 X 6.00 LG	11
2	GROUND BAR	STL BAR	1.00 X 3.00 X 3.00	10
4	RETAINER	STL PL	.50 X Ø3.00	9
2	DRAW BAR	STL RD	Ø2.00 X	8
2	GUSSET	STL BAR	.75 X 3.00 X	7
6	GUSSET	STL BAR	.75 X 3.00 X 11.25	6
5	GUSSET	STL BAR	.75 X 6.00 X 14.75	5
2	GUSSET	STL BAR	.75 X 6.00 X	4
2	END PLATE	STL PL	.50 X 10.62 X 25.90	3
1	BASE	STL PL	.50 X X	2
1	SKID ASS'Y			1

BASE SKID

A-94

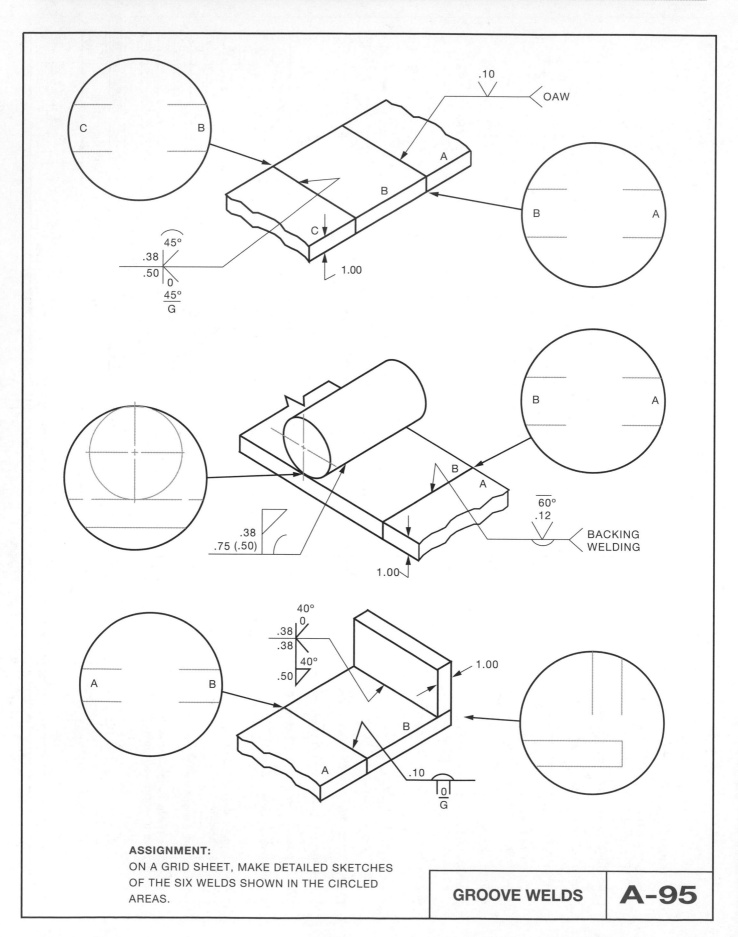

.10

OAW

C   B

B   A

.45°
.38
.50   0
.45°
G

A

B

C

1.00

B   A

.38
.75 (.50)

60°
.12

BACKING
WELDING

B
A

1.00

40°
0
.38
.38
40°
.50

1.00

A   B

B

A

.10
0
G

ASSIGNMENT:
ON A GRID SHEET, MAKE DETAILED SKETCHES
OF THE SIX WELDS SHOWN IN THE CIRCLED
AREAS.

GROOVE WELDS   **A-95**

UNIT **37**

OTHER BASIC WELDS

Plug and Slot Welds

Plug and slot welds are used to join overlapping parts. The top member contains holes (round for plug welds) or slots (elongated for slot welds). Weld metal, deposited in the holes, fuses the two parts together. The holes may be completely or partially filled with weld metal.

Plug Welds (Figure 37–1)

1. Holes in the arrow-side member of a joint for plug welding are specified by placing the weld symbol below the reference line.

2. Holes in the other-side member of a joint for plug welding are indicated by placing the weld symbol above the reference line.

3. The size of a plug weld is shown on the same side and to the left of the weld symbol.

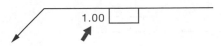

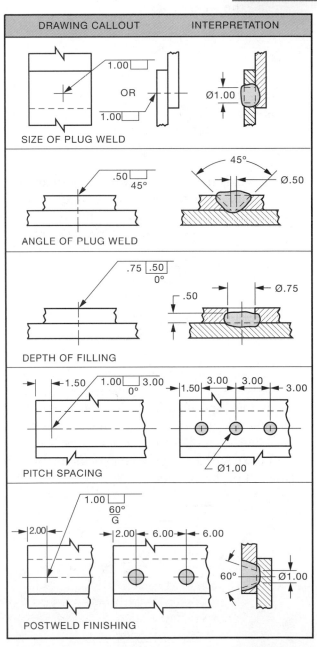

FIGURE 37–1 ■ Plug welds

4. The included angle of countersink of plug welds is the user's standard, unless otherwise specified. Included angle, when not the user's standard, is shown.

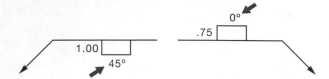

5. The depth of filling of plug welds is complete unless otherwise indicated. When the depth of filling is less than complete, the depth of filling, in inches or millimetres, is shown inside the weld symbol.

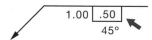

6. Pitch (centre-to-centre spacing) of plug welds is shown to the right of the weld symbol.

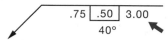

7. Plug welds that are to be welded with approximately flush or convex faces without postweld finishing are specified by adding the flush or convex contour symbol to the weld symbol.

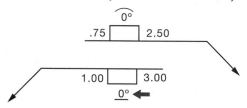

8. Plug welds whose faces are to be finished approximately flush or convex by postweld finishing are specified by adding both the appropriate contour and the finishing symbol to the welding symbol. Welds that require a flat but not flush surface require an explanatory note in the tail of the symbol.

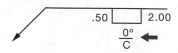

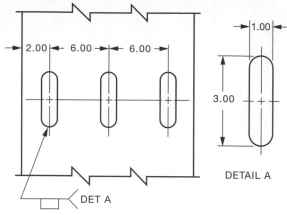

EXAMPLE 1 SLOTS PERPENDICULAR TO LINE OF WELD

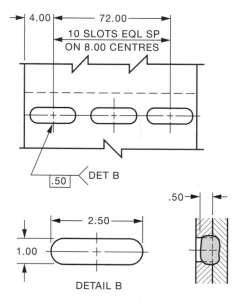

EXAMPLE 2 SLOTS PARALLEL TO LINE OF WELD

FIGURE 37–2 ■ Slot welds

Slot Welds (Figure 37–2)

1. Slots in the arrow-side member of a joint for slot welding are specified by placing the weld symbol below the reference line. Slot orientation must be shown on the drawing.

2. Slots in the other-side member of a joint for slot welding are specified by placing the weld symbol above the reference line.

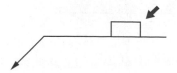

3. Depth of filling of slot welds is complete unless otherwise specified. When the depth of filling is less than complete, the depth of filling, in inches or millimetres, is shown inside the welding symbol.

4. Length, width, spacing, included angle of countersink, orientation, and location of slot welds cannot be specified on the welding symbol. These data are to be specified on the drawing or by a detail with reference to it on the welding symbol.

5. Slot welds that are to be welded with approximately flush or convex faces without postweld finishing are specified by adding the flush or convex contour symbol to the weld symbol.

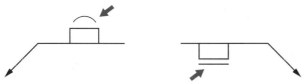

6. Slot welds whose faces are to be finished approximately flush or convex by postweld finishing are specified by adding both the appropriate contour and finishing symbol to the welding symbol. Welds that require a flat but not flush surface require an explanation note in the tail of the symbol.

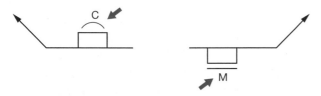

Spot Welds (Figure 37–3)

Spot welding is the most common type of resistance welding used to join sheet metal parts. The parts to be joined are placed under pressure between two electrodes. An electrical charge is then passed between the two electrodes at controlled interval spacing.

1. The symbol for all spot or projection welds is a circle, regardless of the welding process used.

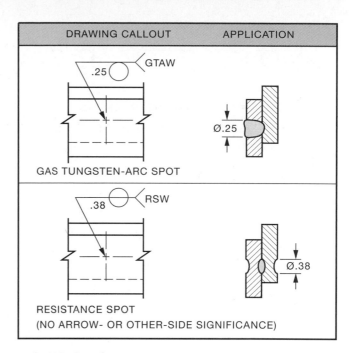

FIGURE 37–3 ■ Spot welds

There is no attempt to provide symbols for different ways of making a spot weld, such as resistance, arc, or electron beam welding. The symbol for a spot weld is a circle placed: below the reference line, indicating arrow side; above the reference line, indicating other side; on the reference line, indicating that there is no arrow or other side.

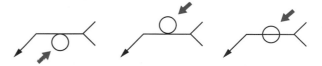

2. Dimensions of spot welds are shown on the same side of the reference line as the weld symbol, or on either side when the symbol is located astride the reference line and has no arrow-side or other-side significance. They are dimensioned by either the size or the strength. The size is designated as the diameter of the weld and is shown to the left of the weld symbol. The strength of the spot weld is designated in pounds (or newtons) per spot and is shown to the left of the weld symbol.

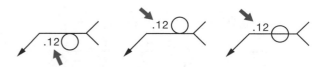

SPECIFYING DIAMETER OF SPOT

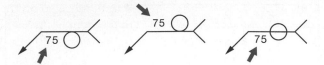

SPECIFYING STRENGTH OF SPOT

3. The process reference is specified in the tail of the welding symbol.

4. When projection welding is used, the spot weld symbol is used, and the projection welding process is referenced in the tail of the symbol. The spot weld symbol is located above or below (not on) the reference line to designate on which member the embossment is placed.

5. The pitch (centre-to-centre spacing) is shown to the right of the weld symbol.

6. When spot welding extends less than the distance between abrupt changes in the direction of the welding or less than the full length of the joint, the extent is dimensioned.

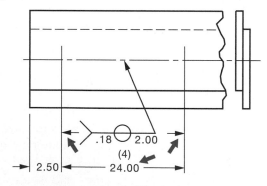

(A) DRAWING CALLOUT

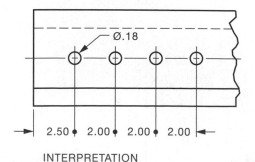

INTERPRETATION

7. Where the exposed surface of either member of a spot welded joint is to be welded with approximately flush or convex faces without postweld finishing, the surface is specified by adding the flush or convex contour symbol to the weld symbol.

8. Spot welds whose faces are to be finished approximately flush or convex by postweld finishing are specified by adding both the appropriate contour and finishing symbol to the welding symbol. Welds that require a flat but not flush surface require an explanatory note in the tail of the symbol.

MACHINE FLAT

Seam Welds (Figure 37–4)

Seam welding is similar to spot welding except that the charges between electrodes are more closely spaced, which produces a continuous-type weld. Seam wells can be continuous or intermittent.

1. The symbol for all seam welds is a circle traversed by two horizontal parallel lines. This symbol is used for all seam welds regardless of the way they are made. The seam weld symbol is placed below the reference line to indicate arrow side; above the reference line to indicate other side; on the reference line to indicate that there is no arrow or other side significance.

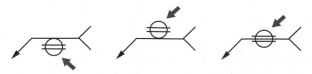

2. Dimensions of seam welds are shown on the same side of the reference line as the weld symbol or on either side when the symbol is centred on the reference line. They are dimensioned by either size or strength. The size of the seam welds is designated as the width of the

DRAWING CALLOUT	INTERPRETATION

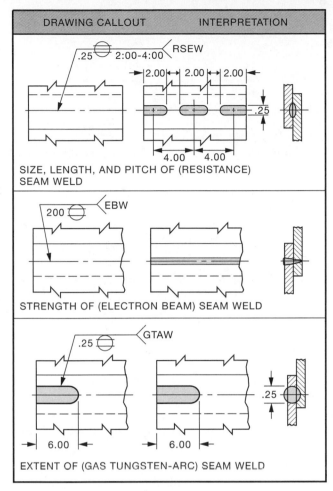

SIZE, LENGTH, AND PITCH OF (RESISTANCE) SEAM WELD

STRENGTH OF (ELECTRON BEAM) SEAM WELD

EXTENT OF (GAS TUNGSTEN-ARC) SEAM WELD

FIGURE 37–4 ■ Seam welds

weld at the faying (fitted) surfaces and is shown to the left of the weld symbol. The strength of seam welds is designated in pounds per linear inch (lb/in.) or newtons per millimetre (N/mm) and is shown to the left of the weld symbols.

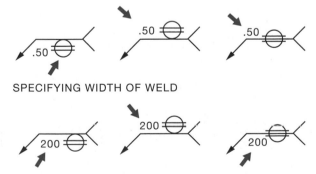

SPECIFYING WIDTH OF WELD

SPECIFYING STRENGTH OF WELD

3. The process reference is specified in the tail of the welding symbol.

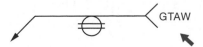

4. The length of a seam weld, when indicated on the welding symbol, is shown to the right of the weld symbol. When seam welding extends for the full distance between abrupt changes in the direction of the welding, no length dimension needs to be shown on the welding symbol. When a seam weld extends less than the full length of the joint, the extent of the weld should be shown.

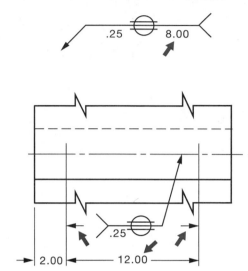

5. The pitch of an intermittent seam weld is shown as the distance between centres of the weld increments. The pitch is shown to the right of the length dimension.

6. When the exposed surface of either member of a seam-welded joint is to be welded with approximately flush or convex faces without postweld finishing, that surface is specified by adding the flush or convex contour symbol to the weld symbol.

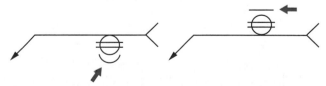

7. Seam welds with faces to be finished approximately flush or convex are specified by adding both the appropriate contour and finish symbol to the welding symbol.

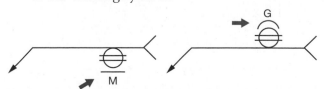

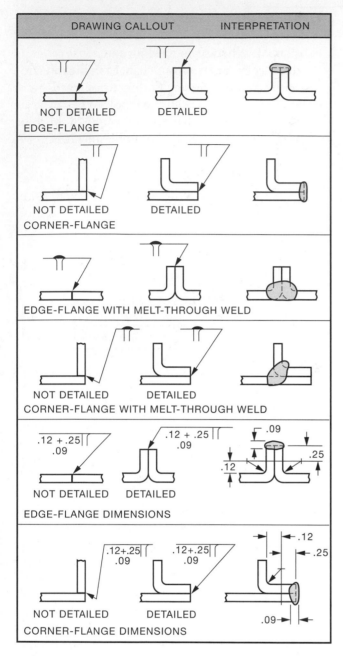

FIGURE 37–5 ■ Flange welds

Flange Welds (Figure 37–5)

These welding symbols are intended to be used for light-gauge metal joints involving the flaring or flanging of the edges to be joined.

1. Edge-flange welds are shown by the edge-flange-weld symbol.

2. Corner-flange welds on joints detailed on the drawing are specified by the corner-flange weld symbol. Weld symbols are always drawn with the perpendicular leg to the left.

3. Corner-flange welds on joints not detailed on the drawing are specified by the corner-flange weld symbol. A broken arrow points to the member being flanged.

4. Edge-flange welds requiring complete joint penetration are specified by the edge-flange weld symbol with the melt-through symbol placed on the opposite side of the reference line. The same welding symbol is used for joints either detailed or not detailed on the drawing.

5. Corner-flange welds requiring complete joint penetration are specified by the corner-flange weld symbol with the melt-through symbol placed on the opposite side of the reference line. A broken arrow points to the member to be flanged where the joint is not detailed.

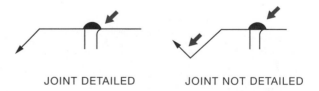

JOINT DETAILED   JOINT NOT DETAILED

6. Dimensions of flange welds are shown on the same side of the reference line as the weld symbol. The radius and the height, separated by a plus (+), are placed to the left of the weld symbol. The radius and the height read in that order from left to right along the reference line.

WHERE T = WELD THICKNESS
H = HEIGHT OF FLANGE
R = RADIUS OF FLANGE

7. The size (thickness) of flange welds is specified by a dimension placed above or below the flange dimensions.

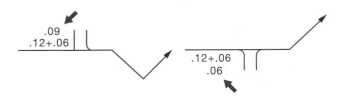

REFERENCES

Canadian Welding Bureau
ANSI/AWS A2.4-86 Standard Symbols for Welding, Brazing, and NDE

INTERNET RESOURCES

American Welding Society For information on all aspects of welding with links to related organizations and materials, see: http://www.aws.org

eFunda For information on types of welding, see: http://www.efunda.com/home.cfm

Machine Design For information on welding and weld joints, see *Machine Design, Fastening/Joining Reference* at: http://www.machinedesign.com

Unified Engineering, Inc. For information on plug, spot, and flange welds, see:http://www .unified-eng.com/scitech/weld/weld.html

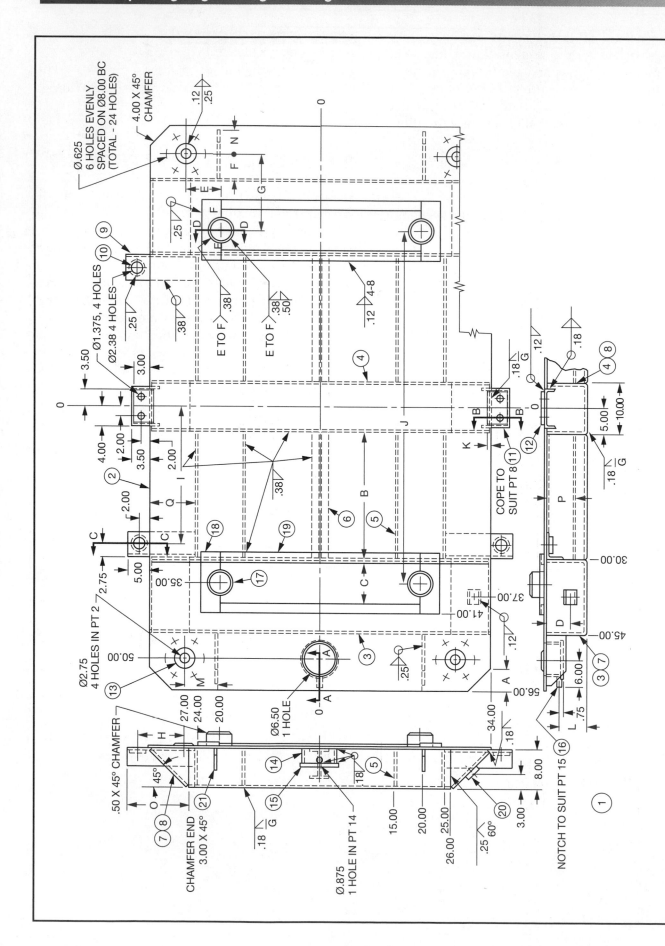

A-96 — BASE ASSEMBLY

DRAWN N. FREY
SCALE NONE

QTY	ITEM	MATL	DESCRIPTION	PT NO.
4	RIB	STL	.50 X 4.00 X 10.00	21
1	GROUND PAD		DWG A - 4158	20
4	RETAINER	STL	.50 X 2.00 X LG	19
4	RETAINER	STL	.75 X 4.00 X LG	18
4	BUMPER PIN	STL	Ø5.00 X 3.00	17
1	NIPPLE	STL	.50 IPS X 3.00 LG	16
1	END PLATE	STL	Ø7.20 X .50 THK	15
1	SUMP	STL	6.00 IPS X 3.00 LG	14
4	FLANGE	STL	WELD - IN FLANGE 3.00 IPS	13
2	SUPPORT	STL	.75 X 4.50 X 7.00	12
2	SUPPORT	STL	C8 X 11.5 X LG	11
4	SUPPORT	STL	Ø3.50 X 1.00 THK	10
4	DRAW BAR	STL	L5.00 X 3.50 X .50 X LG	9
2	RIB END	STL	.50 X 10.50 X 20.00 LG	8
4	RIB END	STL	.50 X 10.50 X 25.00	7
2	RIB	STL	S6 X 12.5 X 25.00 LG	6
4	RIB	STL - PL	.50 X X LG	5
1	RIB	STL - PL	.50 X X LG	4
2	RIB	STL - PL	.50 X X LG	3
1	BASE PLATE	STL - PL	.50 X X LG	2
1	BASE ASSEMBLY			1

NOTES:

UNLESS OTHERWISE SPECIFIED:
- DIMENSIONS SYMMETRICAL AROUND CENTRE LINE
- TOLERANCE ON DIMENSIONS ±.06 EXCEPT HOLES
- TOLERANCE ON HOLES ±.02
- TOLERANCE ON ANGLES ±1.0°

- DIMENSIONS ARE TO CENTRE LINES UNLESS OTHERWISE SHOWN.

- REFER TO HANDBOOK FOR STRUCTURAL SIZES AND SHAPES.

ARRANGEMENT OF VIEWS

LEFT-SIDE VIEW	TOP VIEW
	FRONT VIEW

QUESTIONS:

1. Determine dimensions A to P.

2. What is the overall (A) width, (B) depth, and (C) height of the complete assembly?

3. Assuming the formed channels have square inside corners, what is the width of the material used to make pt. 3? (Use inside travel.)

4. What is the (A) width, and (B) depth of the base plate, pt. 2?

5. If steel weighs .281 pound per cu. in., what is the weight of the base plate, pt. 2? (Disregard holes and chamfers.)

6. Determine the distance from the bottom of the S-beam (pt. 6) to the bottom of the base assembly.

7. What is the angle between the Ø.62 holes?

8. If 6-in. pipe has an OD of 6.62 in., what distance does pt. 15 project beyond pt. 14?

9. Determine the distance pt. 16 projects into pt. 14.

10. What is the area enclosed by parts 17, 18, and 19?

11. Complete the bill of material.

ASSIGNMENTS:

1. ON ONE-INCH GRID SHEETS (.10 IN. SQUARES), SKETCH SECTION VIEWS AT A-A, B-B, C-C, AND D-D. SHOW ONLY THE PARTS AND THE WELDS. SCALE 1:1.

2. ON A ONE-CENTIMETRE (1 MM SQUARES) GRID SHEET, SKETCH ONE-HALF THE DEVELOPMENT OF PT. 8, SHOWING DIMENSIONS AND INDICATING BEND LINES. USE INSIDE TRAVEL. LET THE ONE-CENTIMETRE GRID EQUAL ONE INCH.

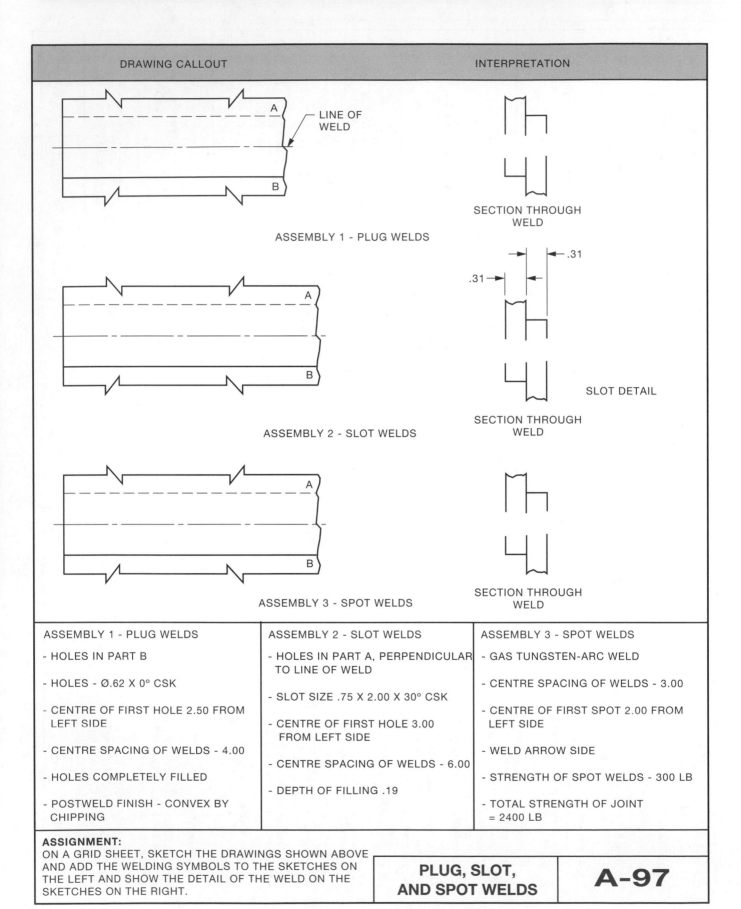

DRAWING CALLOUT	INTERPRETATION

ASSEMBLY 1 - PLUG WELDS

LINE OF WELD

SECTION THROUGH WELD

ASSEMBLY 2 - SLOT WELDS

.31

.31

SLOT DETAIL

SECTION THROUGH WELD

ASSEMBLY 3 - SPOT WELDS

SECTION THROUGH WELD

ASSEMBLY 1 - PLUG WELDS

- HOLES IN PART B

- HOLES - Ø.62 X 0° CSK

- CENTRE OF FIRST HOLE 2.50 FROM LEFT SIDE

- CENTRE SPACING OF WELDS - 4.00

- HOLES COMPLETELY FILLED

- POSTWELD FINISH - CONVEX BY CHIPPING

ASSEMBLY 2 - SLOT WELDS

- HOLES IN PART A, PERPENDICULAR TO LINE OF WELD

- SLOT SIZE .75 X 2.00 X 30° CSK

- CENTRE OF FIRST HOLE 3.00 FROM LEFT SIDE

- CENTRE SPACING OF WELDS - 6.00

- DEPTH OF FILLING .19

ASSEMBLY 3 - SPOT WELDS

- GAS TUNGSTEN-ARC WELD

- CENTRE SPACING OF WELDS - 3.00

- CENTRE OF FIRST SPOT 2.00 FROM LEFT SIDE

- WELD ARROW SIDE

- STRENGTH OF SPOT WELDS - 300 LB

- TOTAL STRENGTH OF JOINT = 2400 LB

ASSIGNMENT:
ON A GRID SHEET, SKETCH THE DRAWINGS SHOWN ABOVE AND ADD THE WELDING SYMBOLS TO THE SKETCHES ON THE LEFT AND SHOW THE DETAIL OF THE WELD ON THE SKETCHES ON THE RIGHT.

PLUG, SLOT, AND SPOT WELDS

A-97

DRAWING CALLOUT	INTERPRETATION

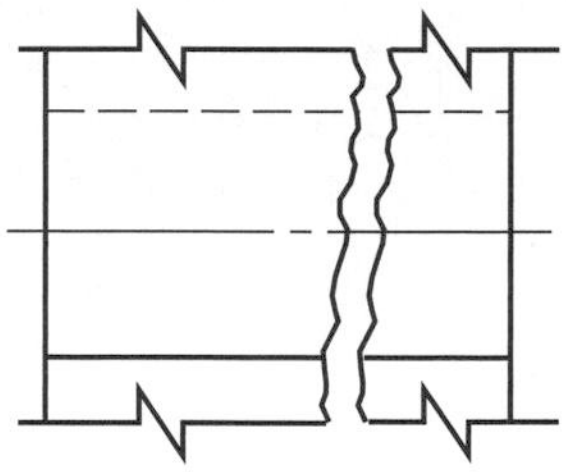

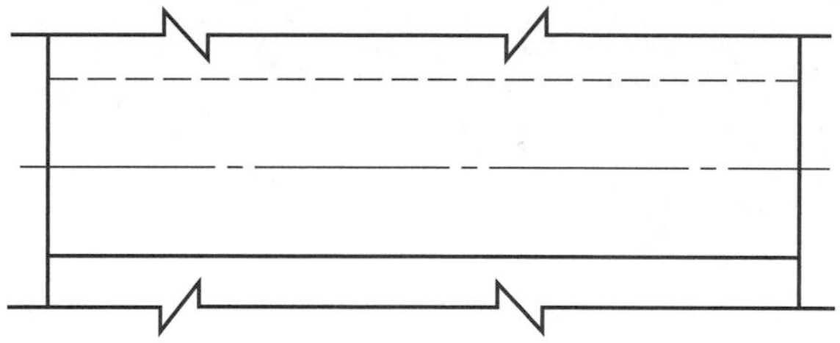

ASSEMBLY 1 - SEAM WELD

SECTION THROUGH WELD

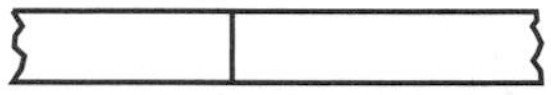

EDGE-FLANGE WELD ONLY

COMPLETE JOINT PREPARATION

EDGE-FLANGE WELD ONLY

COMPLETE JOINT PREPARATION

ASSEMBLY 2 - EDGE-FLANGE WELD

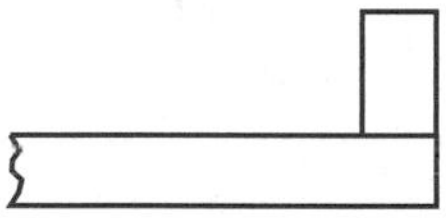

CORNER-FLANGE WELD ONLY

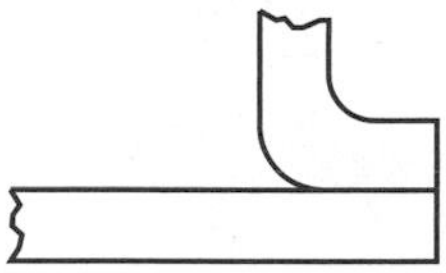

COMPLETE JOINT PREPARATION

CORNER-FLANGE WELD ONLY

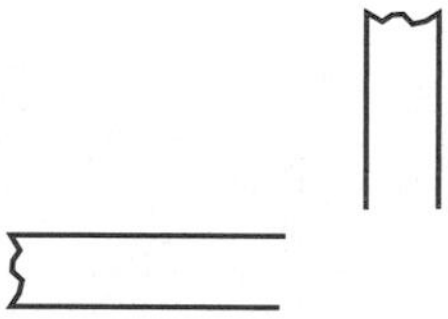

COMPLETE JOINT PREPARATION

ASSEMBLY 3 - CORNER-FLANGE WELD

ASSEMBLY 1 RESISTANCE SEAM WELD	ASSEMBLY 2 EDGE-FLANGE WELD	ASSEMBLY 3 CORNER-FLANGE WELD
- SIZE .25	- WELD THICKNESS .10	- WELD THICKNESS .20
- LENGTH 1.50	- HEIGHT OF FLANGE .20	- HEIGHT OF FLANGE .16
- PITCH 3.00	- RADIUS OF FLANGE .30	- RADIUS OF FLANGE .24

ASSIGNMENT:
ON A GRID SHEET, SKETCH THE DRAWINGS SHOWN ABOVE AND ADD THE WELDING SYMBOLS TO THE SKETCHES ON THE LEFT AND SHOW THE DETAIL OF THE WELD ON THE SKETCHES ON THE RIGHT.

SEAM AND FLANGE WELDS

A-98

GEARS

The function of a gear is to transmit rotary or reciprocating motion from one machine part to another. *Gears* are often used to reduce or increase the r/min (rotations per minute) of a shaft. Gears are rolling cylinders or cones. They have teeth on their contact surfaces to ensure the transfer of motion, Figure 38–1.

The many kinds of gears may be grouped according to the position of the shafts they connect. *Spur gears* connect parallel shafts, *bevel gears* connect shafts having intersecting axes, and *worm gears* connect shafts having axes that do not intersect. A spur gear with a rack converts rotary motion to reciprocating or linear motion. The smaller of two gears is the *pinion*.

FIGURE 38–1 ■ Gears

A simple gear drive consists of a toothed driving wheel meshing with a similar driving wheel. Tooth forms are designed to ensure uniform angular rotation of the driven wheel during tooth engagement.

SPUR GEARS

Spur gears, which are used for drives between parallel shafts, have teeth on the rim of the wheel, Figure 38–2. A pair of spur gears operates as though

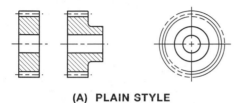

(A) PLAIN STYLE

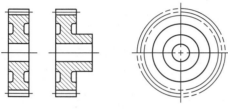

(B) WEBBED STYLE

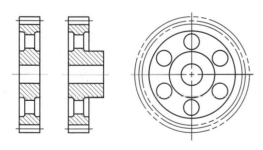

(C) WEBBED WITH CORED HOLES

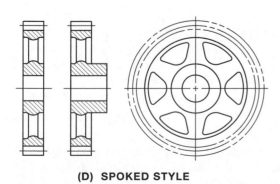

(D) SPOKED STYLE

FIGURE 38–2 ■ Stock spur gear styles

it consists of two cylindrical surfaces with formed teeth that maintain constant speed ratio between the driving and the driven wheel.

Gear design is complex, dealing with such problems as strength, wear, noise, and material selection. Usually, a designer selects a gear from a catalogue. Most gears are made of cast iron or steel. However, brass, bronze, and plastic are used when such factors as wear or noise must be considered.

Theoretically, the teeth of a spur gear are built around the original frictional cylindrical surface called the *pitch circle,* Figure 38–3.

The angle between the direction of pressure between contacting teeth and a line tangent to the pitch circle is the *pressure angle.* The 14.5° pressure angle has been used for many years and remains useful for duplicate or replacement gearing. The 20° and 25° pressure angles have become the standard for new gearing because of their smoother and quieter operation and greater load-carrying ability.

Gear Terms

The following terms, shown in Figures 38–4 and 38–5, are used in spur gear train calculations.

Pitch Diameter (PD)

The diameter of an imaginary circle on which the gear tooth is designed.

Number of Teeth (N)

The number of teeth on the gear.

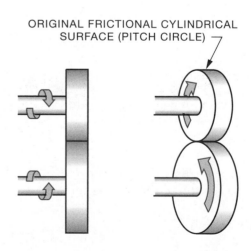

ORIGINAL FRICTIONAL CYLINDRICAL SURFACE (PITCH CIRCLE)

FIGURE 38–3 ■ The pitch circle of spur gears

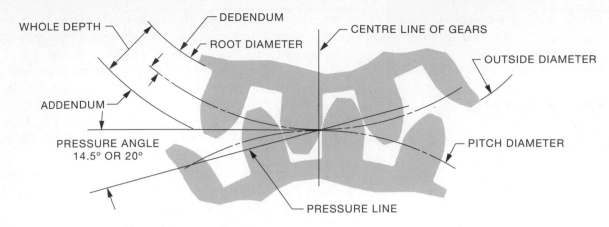

FIGURE 38–4 ∎ Gear teeth terms

Diametral Pitch (DP)

The diametral pitch is a ratio of the number of teeth (N) to a unit length of pitch diameter, DP = N/PD.

Outside Diameter (OD)

The overall gear diameter.

Root Diameter (RD)

The diameter at the bottom of the tooth.

Addendum (ADD)

The radial distance from the pitch circle to the top of the tooth.

Dedendum (DED)

The radial distance from the pitch circle to the bottom of the tooth.

Whole Depth (WD)

The overall height of the tooth.

Clearance (CL)

The radial distance between the bottom of one tooth and the top of the mating tooth.

Circular Pitch (CP)

The distance measured from the point of one tooth to the corresponding point on the adjacent tooth on the circumference of the pitch diameter.

Circular Thickness (T)

The thickness of a tooth or space measured on the circumference of the pitch diameter.

Chordal Thickness (Tc)

The thickness of a tooth or space measured along a chord on the circumference of the pitch diameter.

Chordal Addendum (ADDc)

Also known as corrected addendum, the perpendicular distance from the chord to the outside circumference of the gear.

Chordal Thickness and Corrected Addendum

After the gear teeth have been milled or generated, the width of the tooth space and the thickness of the tooth, measured on the pitch circle, should be equal.

Instead of measuring the curved length of line *(circular thickness of tooth),* it is more convenient to measure the length of the straight line *(chordal thickness),* which connects the ends of that arc.

The *corrected* or *chordal addendum* is the radial distance extending from the addendum circle to the chord.

A gear tooth vernier caliper may be used to measure accurately the thickness of a gear tooth at the pitch line. To use the gear tooth vernier, which measures only a straight line or chordal distance, it is necessary to set the tongue to the computed chordal addendum and then measure the chordal thickness.

TERM AND SYMBOL	FORMULA	
	INCH GEARS	METRIC GEARS
Pitch Diameter - PD	PD = N ÷ DP	PD = MDL X N
Number of Teeth - N	N = PD X DP	N = PD ÷ MDL
Module - MDL		MDL = PD ÷ N
Diametral Pitch - DP	DP = N ÷ PD	
Addendum - ADD	14.5° or 20° ADD = 1 ÷ DP 20° stub ADD = 0.8 ÷ DP	14.5° or 20° ADD = MDL 20° stub ADD = 0.8 X MDL
Dedendum - DED	14.5° or 20° DED = 1.157 ÷ DP 20° stub DED = 1 ÷ DP	14.5° or 20° DED = 1.157 X MDL 20° stub DED = MDL
Whole Depth - WD	14.5° or 20° WD = 2.157 ÷ DP 20 stub WD = 1.8 ÷ DP	14.5° or 20° WD = 2.157 X MDL 20° stub WD = 1.8 X MDL
Clearance - CL	14.5° or 20° CL = 0.157 ÷ DP 20° stub CL = 0.2 ÷ DP	14.5° or 20° CL = 0.157 X MDL 20° stub CL = 0.2 X MDL
Outside Diameter - OD	14.5° or 20° OD = PD + 2 ADD = (N + 2) ÷ DP	14.5° or 20° OD = PD + 2 ADD = PD + 2 MDL

	20° stub
	OD = PD + 2 ADD
	   = (N + 1.6) ÷ DP
		20° stub
		OD = PD + 2 ADD
		   = PD + 1.6 MDL

Root Diameter - RD	14.5° or 20°
	RD = PD - 2 DED
	   = (N - 2.314) ÷ DP
		14.5° or 20°
		RD = PD - 2 DED
		   = PD - 2.314 MDL

	20° stub
	RD = PD - 2 DED
	   = (N - 2) ÷ DP
		20° stub
		RD − PD - 2 DED
		   = PD - 2 MDL

Base Circle - BC	BC = PD Cos PA	BC = PD Cos PA
Pressure Angle - PA	14.5° or 20°	14.5° or 20°

Circular Pitch - CP	CP = 3.1416 PD ÷ N
	   = 3.1416 ÷ DP
		CP = 3.1416 PD ÷ N
		   = 3.1416 MDL

Circular Thickness - T	T = 3.1416 PD ÷ 2 N
	   = 1.5708 ÷ DP
		T = 3.1416 PD ÷ 2N
		   = 1.5708 PD ÷ N
		   = 1.5708 MDL

Chordal Thickness - Tc	TC = PD sin (90° ÷ N)	Tc = PD sin (90° ÷ N)
Chordal Addendum - ADDc	ADDc = ADD + (T^2 ÷ 4 PD)	ADDc = ADD + (T^2 ÷ 4PD)
Working Depth - WKG DP	WKG DP = 2 ADD	WKG DP = 2 ADD

FIGURE 38–5 ■ Spur gear abbreviations and formulas

Working Drawings of Spur Gears

The working drawings of gears, which are normally cut from blanks, are not complicated. A sectional view is sufficient unless a front view is required to show web or arm details. Because the teeth are cut to shape by cutters, they need not be shown in the front view, Assignment A-99.

ANSI recommends the use of phantom lines for the outside and root circles. In the section view, the root and outside circles are shown as solid lines.

The dimensioning for the gear is divided into two groups because the finishing of the gear blank and the cutting of the teeth are separate operations in the shop. The gear blank dimensions are shown on the drawing, whereas the gear tooth information is given in a table.

The only differences in terminology between inch-size and metric-size gear drawings are the terms *diametral pitch* and *module.*

For inch-size gears, the term *diametral pitch* is used instead of the term *module.* The diametral pitch is a ratio of the number of teeth to unit length of pitch diameter.

$$\text{diametral pitch} = DP = \frac{N}{PD}$$

Module, the term used on metric gears, is the length of pitch diameter per tooth measured in millimetres.

$$\text{module} = MDL = \frac{PD}{N}$$

The module is equal to the reciprocal of the diametral pitch and thus is not its metric dimensional equivalent. If the diametral pitch is known, the module can be obtained.

$$\text{module} = 25.4 \div \text{diametral pitch}$$

Gears used in North America are designed in the inch system and have a standard diametral pitch instead of a preferred standard module. Therefore, the diametral pitch should be referenced beneath the module when gears designed with standard inch pitches are used. For gears designed with standard modules, the diametral pitch need not be referenced on the gear drawing. The standard modules for metric gears are 0.8, 1, 1.25, 2.25, 3, 4, 6, 7, 8, 9, 10, 12, and 16. See Figure 38–6 for gear teeth sizes.

Examples of Spur Gear Calculations

The pitch diameter of a gear can easily be found if the number of teeth and diametral pitch are known. The outside diameter is equal to the pitch diameter plus two addendums. The addendum for a 14.5 or 20° spur gear tooth is equal to 1 ÷ DP. *Examples*

1. A 14.5° spur gear has a DP of 4 and 34 teeth.

$$\text{pitch diameter} = \frac{N}{DP} = 34 \div 4 = 8.500 \text{ in.}$$

$$OD = PD + 2\ ADD = 8.500 + 2(1/4) = 9.000 \text{ in.}$$

2. The outside diameter of a 14.5° spur gear is 6.500 in. The gear has 24 teeth.

$$OD = \frac{N+2}{DP} = \frac{24+2}{DP}$$

$$DP = \frac{26}{6.500} = 4$$

$$\text{addendum} = \frac{1}{DP} = 1/4 = .250 \text{ in.}$$

$$\begin{aligned}\text{pitch diameter} &= OD - 2\ ADD \\ &= 6.500 - 2(.250) \\ &= 6.000 \text{ in.}\end{aligned}$$

3. A 14.5° spur gear has a module of 6.35 and 38 teeth.

$$\begin{aligned}\text{pitch diameter} &= N \times MDL \\ &= 38 \times 6.35 \\ &= 241.3 \text{ mm}\end{aligned}$$

$$\begin{aligned}OD = \\ PD + 2\ ADD &= 241.3 + 2(6.35) \\ &= 254 \text{ mm}\end{aligned}$$

REFERENCE

ASME Y14.7.1-1971 (R2003) Gear Drawing Standards— Part 1

INTERNET RESOURCES

Animated Worksheets For information on spur gears, see: http://www.animatedworksheets.co.uk (mechanisms)

Boston Gear For information on all aspects of gears and gear drives, see: http://www.bostongear.com

Flying Pig For information on mechanisms and movements, see: http://www.flying-pig.co.uk (mechanisms and movements)

Howstuffworks For information on all types of gears and gear applications, see: http://science.howstuffworks.com/gear3.htm

Machine Design For information on spur gears and related topics, see *Machine Design, Mechanical Reference* at: http://www.machinedesign.com

TechStudent.Com For information, including illustrations, on spur gears and simple gear trains, see: http://www.technologystudent.com (Gears and Pulleys)

MODULE	DIAMETRAL PITCH	PRESSURE ANGLE	
FOR METRIC-SIZE GEARS	FOR INCH-SIZE GEARS	14.5°	20°
6.35	4	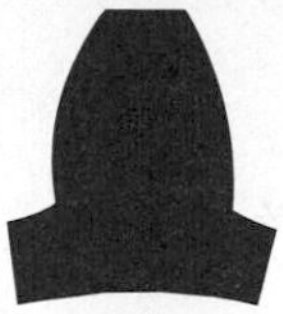	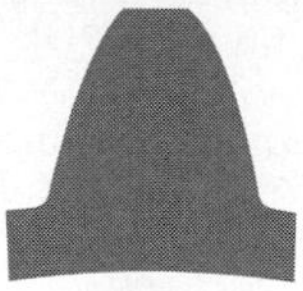
5.08	5	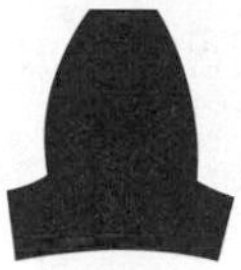	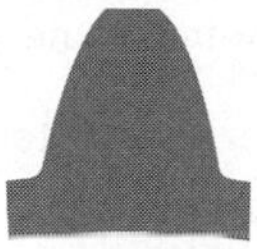
4.23	6	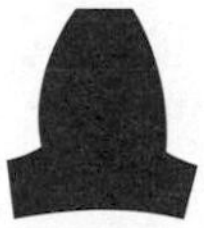	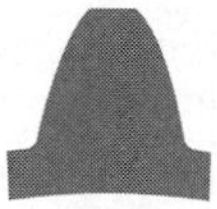
3.18	8	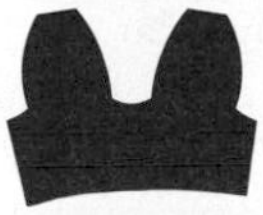	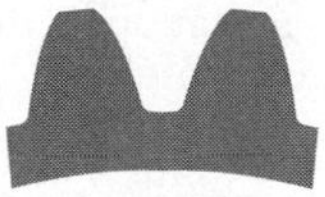
2.54	10	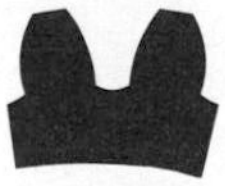	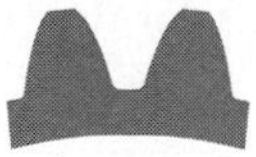
2.17	12	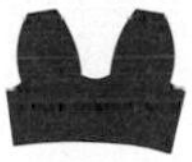	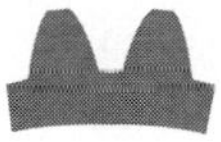
1.59	16	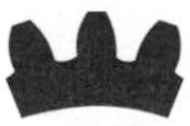	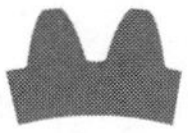
1.27	20	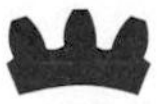	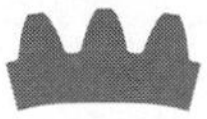
1.06	24	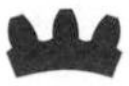	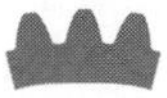

NOTE: MODULE SIZES SHOWN
ARE CONVERTED INCH SIZES

FIGURE 38–6 ■ Gear teeth sizes

QUESTIONS:

1. What is the hub diameter?

2. What is the maximum thickness of the spokes?

3. What is the average width of the spokes?

4. How many surfaces indicate that allowance must be added to pattern for finishing?

5. Determine distance Ⓙ for the pattern. Assume .10 is allowed on pattern for each surface to be finished.

6. Determine distance Ⓚ for the pattern.

7. What is the outside diameter of the pattern?

8. Determine distance Ⓛ for the pattern.

9. What is the width of the pattern?

10. What is the diametral pitch of the gear?

11. Calculate the centre-to-centre distance if this gear were to mesh with a pinion having (A) 24 teeth, (B) 36 teeth, (C) 32 teeth.

12. Calculate distances Ⓔ through Ⓗ.

13. Calculate the following: addendum, dedendum, circular pitch, and root diameter.

14. Complete the missing information in the cutting data table.

15. Calculate the limits of size for the Ø1.500 hole if an LT3 fit between the hole and shaft is required.

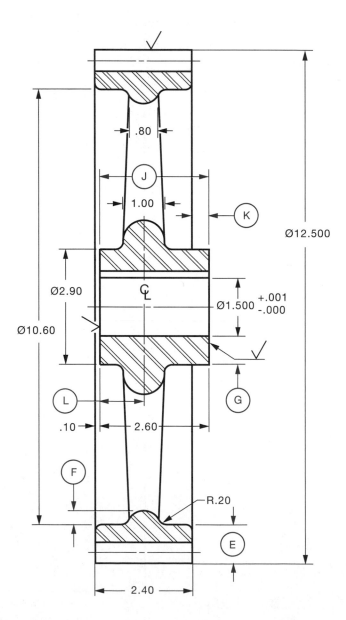

NOTE: ROUNDS & FILLETS TO BE R.10
UNLESS OTHERWISE SPECIFIED
∨ SHOWN TO BE .10 ∇ 63/

CUTTING DATA	
NUMBER OF TEETH	48
PITCH DIAMETER	
DIAMETRAL PITCH	4
PRESSURE ANGLE	20°
WHOLE DEPTH	
CHORDAL ADDENDUM	
CHORDAL THICKNESS	

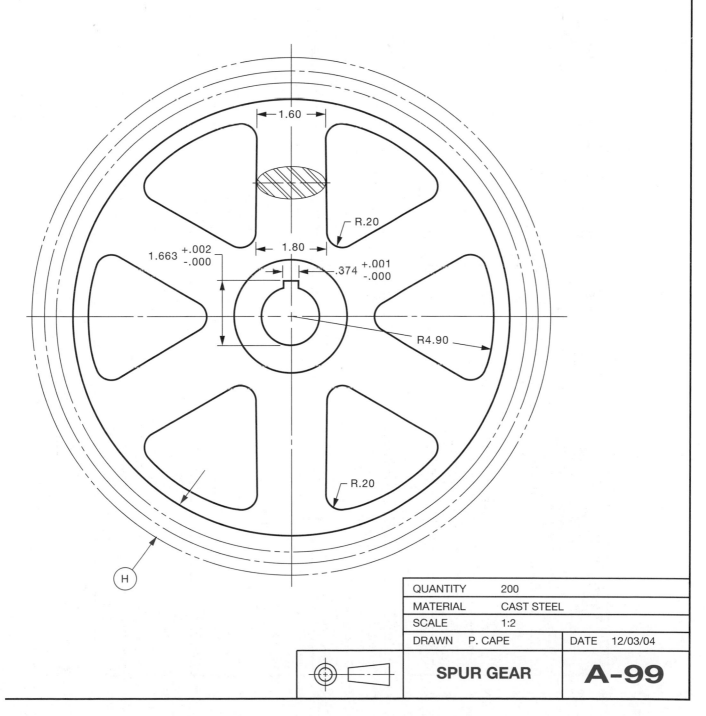

QUANTITY	200	
MATERIAL	CAST STEEL	
SCALE	1:2	
DRAWN P. CAPE		DATE 12/03/04

SPUR GEAR   **A-99**

INCH GEARS

TERM AND SYMBOL	GEAR 1	GEAR 2	GEAR 3	GEAR 4
Pitch Diameter - PD		5.000	3.000	2.250
Number of Teeth - N	24	40		36
Diametral Pitch - DP	6		10	
Addendum - ADD				
Dedendum - DED				
Whole Depth - WD				
Clearance - CL				
Outside Diameter - OD				
Root Diameter - RD				
Pressure Angle - PA	20°	14.5°	20°	14.5°
Circular Pitch - CP				
Circular Thickness - T				
Chordal Thickness - Tc				
Chordal Addendum - ADDc				

METRIC GEARS

TERM AND SYMBOL	GEAR 5	GEAR 6	GEAR 7	GEAR 8
Pitch Diameter - PD		89.04		
Number of Teeth - N	40			30
Module - MDL	5.08	3.18	6	4
Addendum - ADD				
Dedendum - DED				
Whole Depth - WD				
Clearance - CL				
Outside Diameter - OD			228	
Root Diameter - RD				
Pressure Angle - PA	14.5°	20°	20°	14.5°
Circular Pitch - CP				
Circular Thickness - T				
Chordal Thickness - Tc				
Chordal Addendum - ADDc				

SPUR GEAR CALCULATIONS | **A-100**

ASSIGNMENT:
SKETCH CHARTS SIMILAR TO THOSE SHOWN ABOVE AND
COMPLETE THE MISSING INFORMATION.

UNIT 39

BEVEL GEARS

With a working knowledge of the parts, principles, and formulas underlying spur gears, one can easily interpret and understand drawings of bevel gears.

Spur gears transmit motion by or through shafts that are parallel and in the same plane, whereas bevel gears transmit motion between shafts that are in the same plane but whose axes

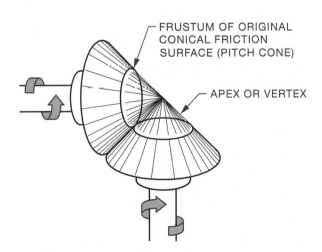

FIGURE 39–1 ■ Principle as applied to bevel gears

would meet if extended. Theoretically, the teeth of a spur gear may be said to be built about the original frictional cylindrical surface known as *pitch circle,* whereas the teeth of a bevel gear are formed about the frustum of the original conical friction surface called *pitch cone,* Figure 39–1.

One common type of bevel gear is the mitre gear—a pair of bevel gears of the same size that transmit motion at right angles.

Whereas any two spur gears of the same diametral pitch will mesh, this is not true of bevel gears except for mitre gears. On each pair of mating bevel gears, the diameters of the gears determine the angles at which the teeth are cut.

Working drawings of bevel gears, like spur gears, give only the dimensions of the bevel gear blank. Cutting data for the teeth are given in a note or table. A single section view is used unless a second view is required to show such details as spokes. Sometimes both the bevel gear and pinion are drawn together. Dimensions and cutting data depend on the method used in cutting the teeth, but the information in Figures 39–2 and 39–3 is commonly used.

REFERENCE

ASME Y14.7.2-1978 (R2004) Gear and Spline Drawing Standards—Part 2

INTERNET RESOURCES

Animated Worksheets For information on bevel gears, see: http://www.animatedworksheets.co .uk (mechanisms)

Boston Gear For information on all aspects of gears and gear drives, see: http://www.bostongear .com

Howstuffworks For information on all types of gears and gear applications, see: http://science .howstuffworks.com/gear3.htm

Machine Design For information on mitre gears and gear applications, see *Machine Design, Mechanical Reference* at: http://www .machinedesign.com

TechStudent.Com For information, including illustrations, on bevel gears, see: http://www .technologystudent.com (Gears and Pulleys)

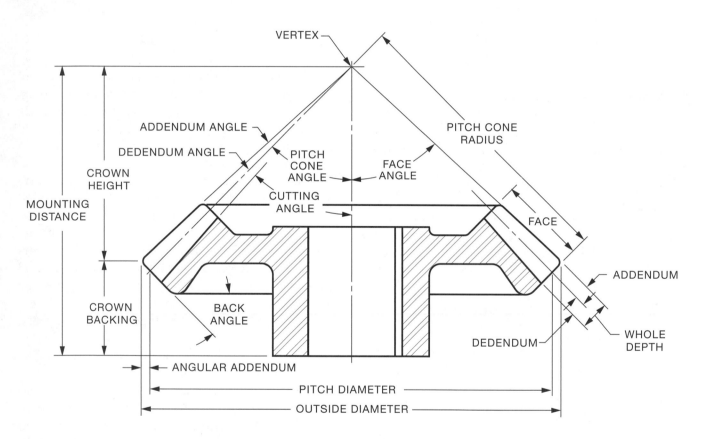

FIGURE 39–2 ■ Bevel gear nomenclature

TERM	FORMULA
Addendum, dedendum, whole depth, pitch diameter, diametral pitch, number of teeth, circular pitch, chordal thickness, circular thickness	Same as for spur gears. Refer to Figure 38-5.
Pitch Cone Angle (Pitch Angle)	$\text{Tan pitch angle} = \dfrac{\text{PD of gear}}{\text{PD of pinion}} = \dfrac{\text{N of gear}}{\text{N of pinion}}$
Pitch Cone Radius	$\dfrac{\text{PD}}{2 \times \text{sin of pitch angle}}$
Addendum Angle	$\text{Tan addendum angle} = \dfrac{\text{Addendum}}{\text{Pitch cone radius}}$
Dedendum Angle	$\text{Tan dedendum angle} = \dfrac{\text{Dedendum}}{\text{Pitch cone radius}}$
Face Angle	Pitch cone angle plus addendum angle
Cutting Angle	Pitch cone angle minus dedendum angle
Back Angle	Same as pitch cone angle
Angular Addendum	Cosine of pitch cone angle x addendum
Outside Diameter	Pitch diameter plus two angular addendums
Crown Height	Divide $\frac{1}{2}$ the outside diameter by the tangent of the face angle
Face Width	$1\frac{1}{2}$ to $2\frac{1}{2}$ times the circular pitch
Chordal Addendum	$\text{Addendum} + \dfrac{\text{Circular thickness}^2 \times \text{Cos pitch cone angle}}{4\,\text{PD}}$

FIGURE 39–3 ■ Bevel gear formulas

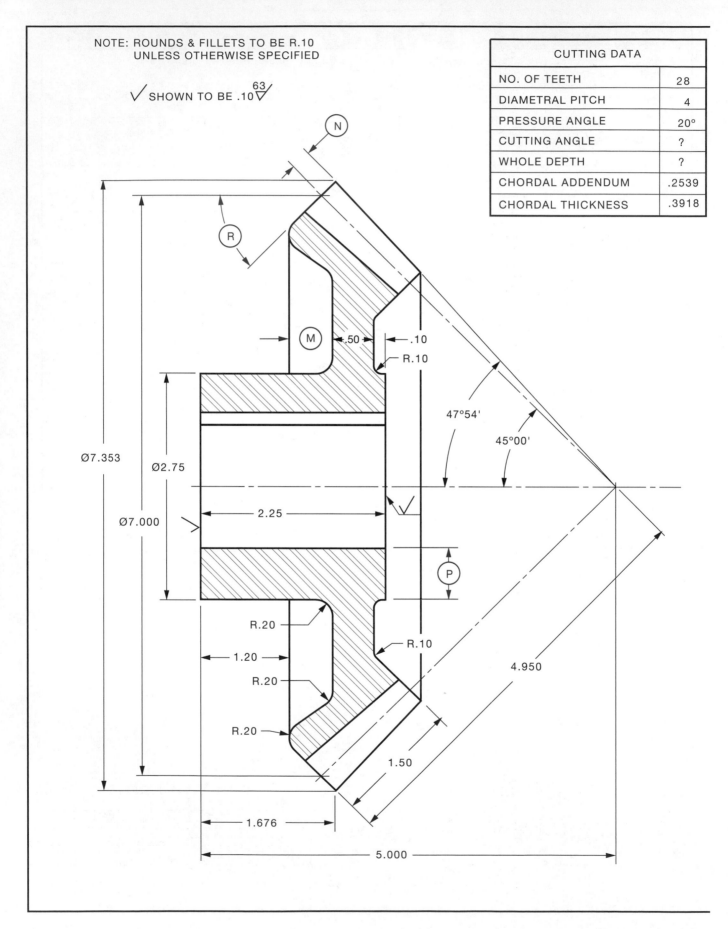

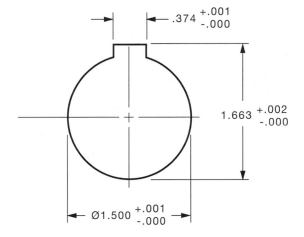

.374 $^{+.001}_{-.000}$

1.663 $^{+.002}_{-.000}$

Ø1.500 $^{+.001}_{-.000}$

QUESTIONS:

1. The drafter neglected to put on angle dimension (R). What should it be?

2. How many finished surfaces are indicated?

3. List those dimensions shown on the drawing that are not used by the pattern maker. Assume that the hole will not be cored.

4. What is the pitch cone angle?

5. What is the pitch diameter?

6. List those dimensions on the drawing that the machinist would use to machine the blank before the teeth are cut.

7. What is the depth of teeth at the large end?

8. What is the pitch cone radius?

9. What is the face angle?

10. What is the addendum angle?

11. What is the cutting angle?

12. What is the mounting distance?

13. What is the crown height?

14. What is the angular addendum?

15. What is the face width?

16. Determine dimensions (M), (N), and (P).

17. Calculate the limits of size for the Ø1.500 hole if an LT2 fit between the hole and shaft is required.

18. What was the width of the cast hub before finishing?

QUANTITY		50
MATERIAL		CAST STEEL
SCALE		1:1
DRAWN	J. HELSEL	DATE 23/10/04
MITRE GEAR		**A-101**

GEAR TRAINS

Centre Distance

The centre distance between the two shaft centres is determined by adding the pitch diameter of the two gears together and dividing the sum by 2.

Examples

1. An 8DP, 24-tooth pinion mates with a 96-tooth gear. Find the centre distance.

   pitch diameter of pinion =
   N ÷ DP = 24 ÷ 8 = 3.000 in.

   pitch diameter of gear =
   N ÷ DP = 96 ÷ 8 = 12.000 in.

   sum of the two pitch diameters =
   3.000 + 12.000 = 15.000 in.

   centre distance =
   1/2 sum of the two pitch diameters =

   $$\frac{15.000}{2} = 7.500 \text{ in.}$$

2. A 2.54 module, 28-tooth pinion mates with an 84-tooth gear. Find the centre distance.

   pitch diameter (PD) =
   number of teeth × module =
   28 × 2.54 = 71.12 (pinion) =
   84 × 2.54 = 213.36 (gear)

   sum of the two pitch diameters =
   71.12 + 213.36 = 284.48 mm

   centre distance =
   1/2 sum of the two pitch diameters =
   284.48 ÷ 2 = 142.24 mm

Ratio

The ratio of gears is a relationship between any of the following:

■ the r/min (revolutions per minute) of the gears

■ the number of teeth on the gears

■ the pitch diameter of the gears

The ratio is obtained by dividing the larger value of any of the three by the corresponding smaller value.

3. A gear rotates at 90 r/min and the pinion at 360 r/min.

   $$\text{ratio} = \frac{360}{90} = 4 \text{ or ratio} = 4{:}1$$

4. A gear has 72 teeth; the pinion 18 teeth.

   $$\text{ratio} = \frac{72}{18} = 4 \text{ or ratio} = 4{:}1$$

5. A gear with a pitch diameter of 8.500 in. meshes with a pinion having a pitch diameter of 2.125 in.

$$ratio = \frac{PD\ of\ gear}{PD\ of\ pinion} = \frac{8.500}{2.125} = 4$$

$$or\ ratio = 4:1$$

Figure 40–1 illustrates how this type of information would be shown on an engineering sketch.

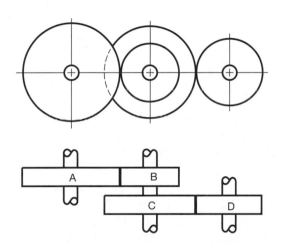

GEAR	PD	N	DP	R/MIN	CENTRE DISTANCE
A	6.000	24	4	300	
					4.500
B	3.000	12	4	600	
C	6.000	48	8	600	
					4.000
D	2.000	16	8	1800	

FIGURE 40–1 ■ Gear train data

Motor Drive

Assignment A-102 shows a motor drive similar to the type used to operate a load-ratio control switch on a power transformer.

The load-ratio control switch is operated by a small motor with a speed of 1080 r/min. The shaft speed at the switch is reduced to 9 r/min by a series of spur and mitre gears.

When the operator pushes a button, the motor is activated until the circuit breaker pointer rotates 90° and one of the arms depresses the roller and breaks the circuit. During this time the load-ratio control switch shaft will rotate 360°, moving the contactor in the load-ratio control switch one position. This will be shown on the dial by the position indicator.

To simplify the assembly, only the pitch diameters of the gears are shown and much of the hardware has been omitted.

When referring to the direction in which the gears rotate, the terms clockwise (CWISE or CW) and anticlockwise (AWISE), or counterclockwise (CCW), are used.

INTERNET RESOURCES

Boston Gear For information on all aspects of gears and gear drives, see: http://www.bostongear.com

Howstuffworks For information on all types of gears and gear applications, see: http://science.howstuffworks.com/gear3.htm

Machine Design For information on gear trains and gear applications, see *Machine Design, Mechanical Reference* at: http://www.machinedesign.com

TechStudent.Com For information on gear trains, see: http://www.technologystudent.com/gears/geardex1.htm

GEAR DATA				
GEAR	NO. OF TEETH	PITCH DIAMETER	DP	R/MIN
G_1	24		20	
G_2		4.800		270
G_3	20	1.000		
G_4	100			
G_5		1.000	20	
G_6		6.000		
G_7	18			
G_8		7.200		
G_9	72		20	
G_{10}		2.500	10	
G_{11}	25			

SHAFT DATA			
SHAFT	GEARS ON SHAFT	R/MIN	SHAFT * ROTATION
S_1			
S_2			
S_3			
S_4			
S_5			
S_6			
S_7			C-C WISE

* AS VIEWED FROM FRONT OR BOTTOM OF MOTOR DRIVE ASSEMBLY

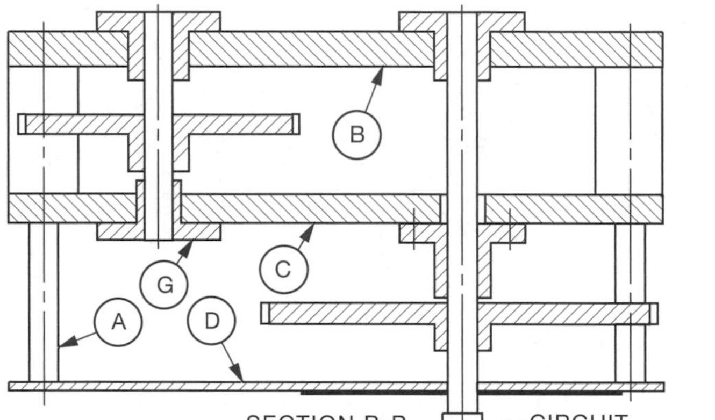
B
C
G
A
D

SECTION B-B

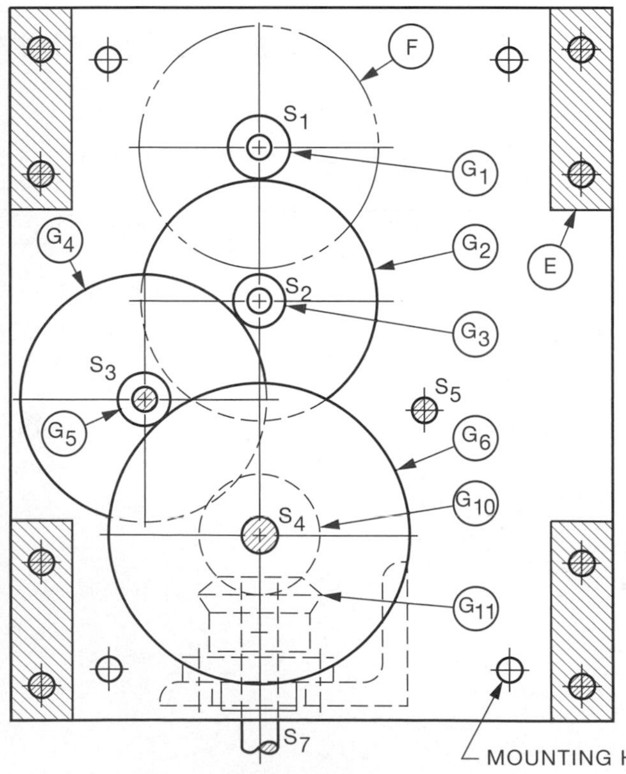
F
S_1
G_1
G_2
G_3
G_4
S_2
S_3
G_5
S_5
G_6
G_{10}
S_4
G_{11}
E
S_7
MOUNTING HOLES

SECTION A-A

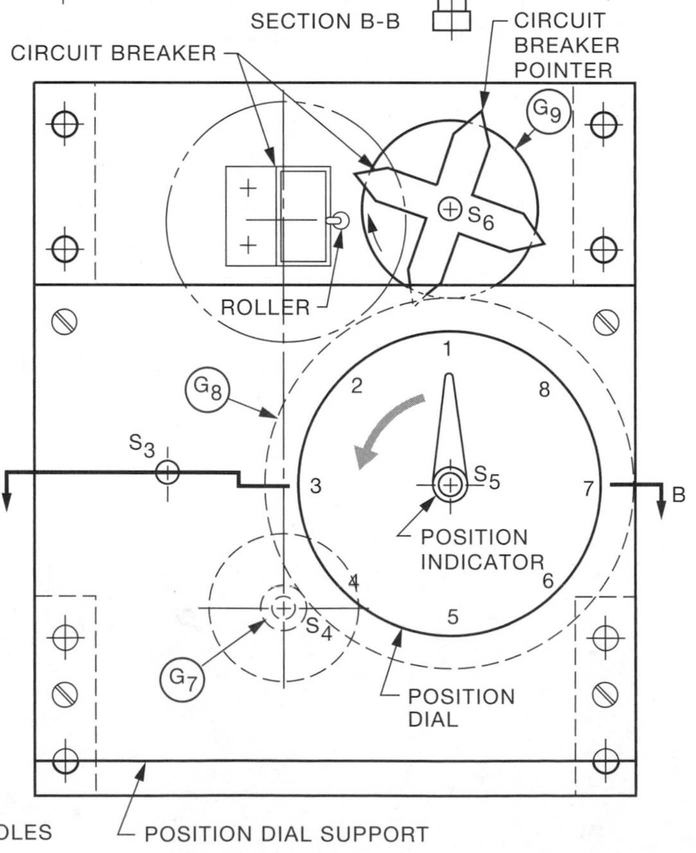
CIRCUIT BREAKER
CIRCUIT BREAKER POINTER
G_9
S_6
ROLLER
G_8
S_3
B
B
1
2
3
4
5
6
7
8
S_5
POSITION INDICATOR
S_4
G_7
POSITION DIAL
POSITION DIAL SUPPORT

ASSIGNMENT:

ON A GRID SHEET, SKETCH THE GEAR AND SHAFT DATA
CHARTS SHOWN AND COMPLETE THE MISSING INFORMATION.

QUESTIONS:

1. What are the names of parts (A) to (G)?

2. How many spur gears are shown?

3. How many mitre gears are shown?

4. How many gear shafts are there?

5. What is the ratio between the following gears? (A) G_1 and G_2, (B) G_3 and G_4, (C) G_5 and G_6, (D) G_7 and G_8, (E) G_8 and G_9, (F) G_{10} and G_{11}.

6. What is the centre-to-centre distance between the following shafts? (A) S_1 and S_2, (B) S_2 and S_3, (C) S_3 and S_4, (D) S_4 and S_5, (E) S_5 and S_6.

7. How many seconds does it take to turn the load-ratio control switch one position?

8. How many seconds does it take the position indicator to move continuously from position 4 to position 7?

9. What is the r/min ratio between the motor and the switch?

10. If the shaft S_7 rotates 180°, how many degrees does the position indicator rotate?

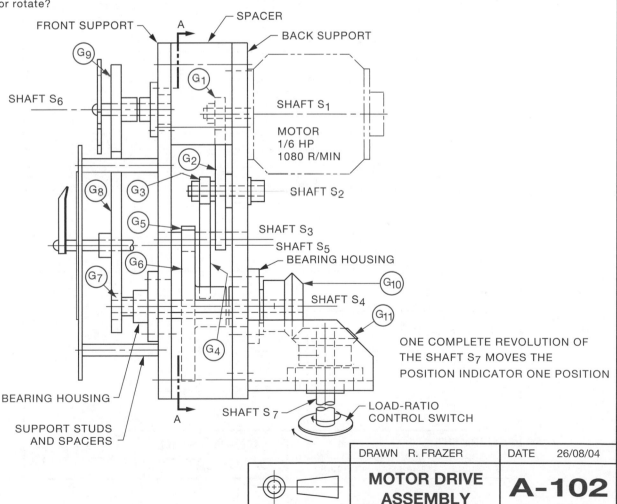

FRONT SUPPORT — SPACER — BACK SUPPORT

G_9

G_1

SHAFT S_6

SHAFT S_1

MOTOR
1/6 HP
1080 R/MIN

G_2

SHAFT S_2

G_8 G_3

G_5

SHAFT S_3

SHAFT S_5

BEARING HOUSING

G_6

G_7

G_{10}

SHAFT S_4

G_{11}

G_4

ONE COMPLETE REVOLUTION OF
THE SHAFT S_7 MOVES THE
POSITION INDICATOR ONE POSITION

BEARING HOUSING

SHAFT S_7 — LOAD-RATIO
CONTROL SWITCH

SUPPORT STUDS
AND SPACERS

DRAWN R. FRAZER	DATE 26/08/04
MOTOR DRIVE ASSEMBLY	**A-102**

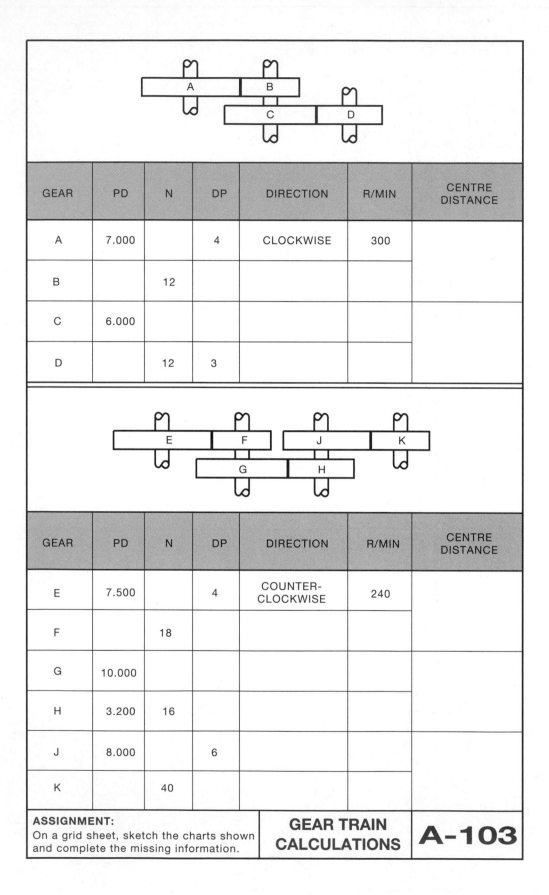

GEAR	PD	N	DP	DIRECTION	R/MIN	CENTRE DISTANCE
A	7.000		4	CLOCKWISE	300	
B		12				
C	6.000					
D		12	3			

GEAR	PD	N	DP	DIRECTION	R/MIN	CENTRE DISTANCE
E	7.500		4	COUNTER-CLOCKWISE	240	
F		18				
G	10.000					
H	3.200	16				
J	8.000		6			
K		40				

ASSIGNMENT:
On a grid sheet, sketch the charts shown and complete the missing information.

GEAR TRAIN CALCULATIONS

A-103

CAMS

The cam is invaluable in the design of automatic machinery, making it possible to impart any desired motion to another mechanism.

A *cam* is a rotating, oscillating, or reciprocating machine element with a surface or groove formed to impart special or irregular motion to a second part called a *follower*. The follower rides against the curved surface of the cam. The distance that the follower rises and falls in a defined period of time is determined by the cam profile.

Types of Cams

The type and shape of cam used is dictated by the required relationship of the parts and their motions, Figure 41–1. The cams that are generally used are either radial or cylindrical. The follower of a radial, or face, cam moves in a plane perpendicular to the axis of the cam; in the cylindrical type of cam, the movement of the follower is parallel to the cam axis.

A simple OD (outside diameter), or plate, cam is shown in Figure 41–2. The hole in the plate is bored off-centre, causing the follower to move up and down as it revolves. The follower can be any type that will roll or slide on the surface of the cam. The one used with this cam is called a flat face follower.

The cam shown in Figure 41–3 is a drum, or barrel-type, cam. It transmits motion transversely to a lever connected to a conical follower, which rides in the groove as the cam revolves.

Cam Displacement Diagrams

In preparing cam drawings, a cam displacement diagram is drawn first to plot the motion of the follower. The curve on the drawing represents the path of the follower, not the face of the cam. The diagram may be any convenient length, but it is often drawn equal to the circumference of the base circle of the cam, and the height is drawn equal to the follower displacement. The lines drawn on the motion diagram are shown as radial lines on the cam drawing; the sizes are transferred from the motion diagram to the cam drawing.

Figure 41–4 shows a cam displacement diagram having a modified uniform type of motion plus two dwell periods. Most cam displacement diagrams have 360° cam displacement angles. For drum or cylindrical grooved cams, the displacement diagram is often replaced by the developed surface of the cam.

The cylindrical feeder cam (Assignment A-104) is a drum, or barrel, cam. In addition to the working views of the cam, a development of the contour of the grooves is shown. This development aids the machinist in scribing and laying out the contour of the cam action lobes on the surface of the cam blank preparatory to machining grooves.

Regardless of the type of cam or follower, the purpose of all cams is to impart motion to other mechanisms in various directions to actuate mechanisms to do specific jobs.

INTERNET RESOURCES

Animated Worksheets For information on cams and cam followers, see: http://www.animatedworksheets .co.uk (mechanisms)

Industrial Motion Control For information on cams and cam followers, see: http://www.camcoindex .com

Design & Technology Online For information on cams and cam followers, see: http://www .dtonline.org/ (mechanisms)

Flying Pig For information on mechanisms and movements, see: http://www.flying-pig.co.uk (mechanisms and movements)

Saltire For information on cams and cam followers, see: http://www.saltire.com/cams.html

TechStudent.Com For information, including illustrations, on cams, see: http://www.technologystudent .com (Mechanisms)

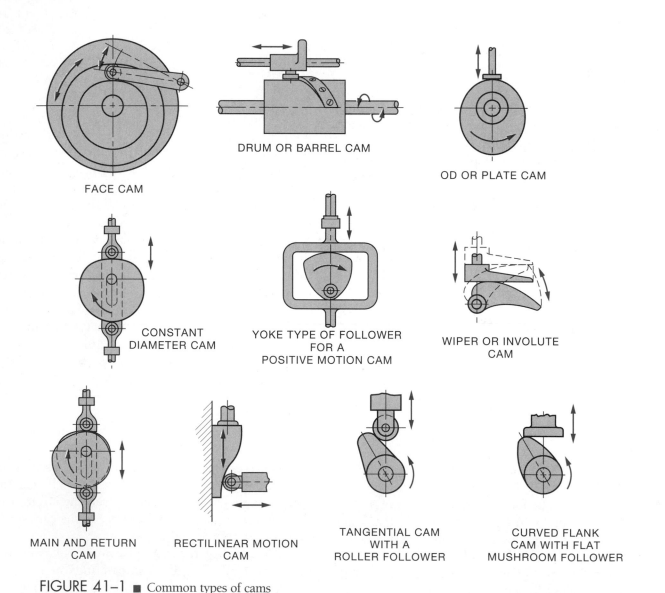

FACE CAM

DRUM OR BARREL CAM

OD OR PLATE CAM

CONSTANT DIAMETER CAM

YOKE TYPE OF FOLLOWER FOR A POSITIVE MOTION CAM

WIPER OR INVOLUTE CAM

MAIN AND RETURN CAM

RECTILINEAR MOTION CAM

TANGENTIAL CAM WITH A ROLLER FOLLOWER

CURVED FLANK CAM WITH FLAT MUSHROOM FOLLOWER

FIGURE 41–1 ■ Common types of cams

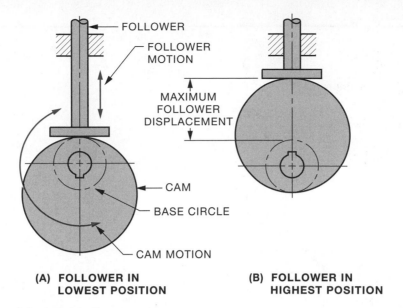

FIGURE 41–2 ■ Eccentric plate cam

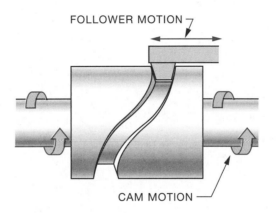

FIGURE 41–3 ■ Drum, or barrel-type, cam

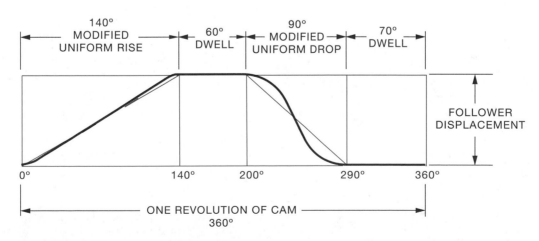

FIGURE 41–4 ■ Cam displacement diagram

QUESTIONS:

1. Through what thickness of metal is hole ② drilled?

2. Locate line ④ in the right-side view.

3. Locate line ⑪ in the cam displacement diagram.

4. Locate line ⑥ in another view.

5. Locate line ⑨ in the right-side view.

6. What is the maximum permissible diameter of hole ②?

7. Locate line ⑧ in the front view.

8. Locate line ㊸ in the right-side view.

9. What is the total follower displacement of the cam follower for (A) the finishing cut and (B) the roughing cut?

10. Assuming a .06 in. allowance for machining, what would be the outside diameter of the cam before finishing?

11. What is the total number of through holes?

12. Locate lines ㉚ to ㊶ on other view.

13. Determine distances Ⓐ to Ⓜ.

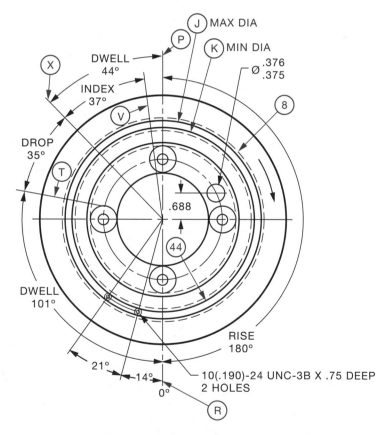

ROUGHING CAM DATA
LEFT-SIDE VIEW

THREAD CONTROLLING ORGANIZATION
AND STANDARD-ASME B1.1-2003

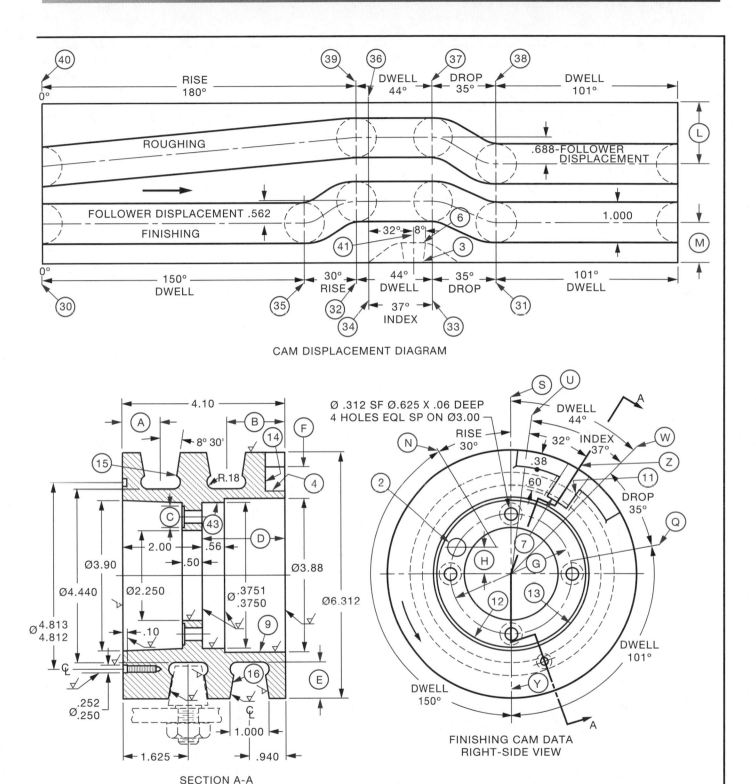

CAM DISPLACEMENT DIAGRAM

SECTION A-A

FINISHING CAM DATA
RIGHT-SIDE VIEW

Ø .312 SF Ø.625 X .06 DEEP
4 HOLES EQL SP ON Ø3.00

NOTE: UNLESS OTHERWISE SHOWN:

- TOLERANCE ON TWO-PLACE DECIMAL DIMENSIONS ±.02
- TOLERANCE ON THREE-PLACE DECIMAL DIMENSIONS ±.005
- TOLERANCE ON ANGLES ±30'
- SURFACES ✓ TO BE 63/

MATERIAL		
SCALE	NOT TO SCALE	
DRAWN	D. SMITH	DATE 06/11/04

CYLINDRICAL FEEDER CAM   **A-104**

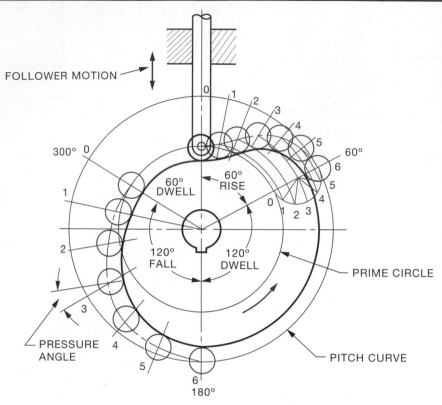

FOLLOWER MOTION

300° 0

60° DWELL   60° RISE

120° FALL   120° DWELL

PRESSURE ANGLE

60°

PRIME CIRCLE

PITCH CURVE

180°

CAM DRAWING

POSITION IN DEGREES	DISTANCE FROM THE PRIME CIRCLE
0	
10	
20	
30	
40	
50	
60	
180	
200	
220	
240	
260	
280	
300	
360	

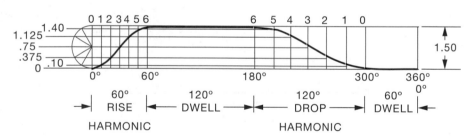

1.125
.75
.375
0

1.40

.10

1.50

0°   60°   180°   300°   360°
                              0°

60° RISE   120° DWELL   120° DROP   60° DWELL

HARMONIC   HARMONIC

DISPLACEMENT DIAGRAM

ASSIGNMENT:

1. ON A GRID SHEET, SKETCH THE CHART SHOWN ABOVE. USING 0° AS THE STARTING POSITION AND THE POSITIONS SHOWN, CALCULATE THE DISTANCE FROM THE CENTRE OF THE FOLLOWER TO THE PRIME CIRCLE.

2. IF THE CAM SHAFT ROTATES ONE COMPLETE REVOLUTION EVERY THREE MINUTES, HOW MUCH TIME DOES IT TAKE FOR THE CAM FOLLOWER TO (A) RISE 1.50 INCHES AND (B) DROP 1.50 INCHES?

3. HOW MUCH TIME DOES THE CAM FOLLOWER REMAIN AT DWELL WHEN (A) IN THE UPPER POSITION, AND (B) IN THE LOWER POSITION?

PLATE CAM   **A-105**

UNIT 42

ANTIFRICTION BEARINGS

Antifriction bearings, also known as *roller-element* bearings, use a type of rolling element between the loaded members. Relative motion is accommodated by rotation of the elements. Roller-element bearings are usually housed in bearing races conforming to the element shapes. A cage or separator is often used to locate the elements within the bearings. These bearings are usually categorized by the form of the rolling element and in some instances by the load type they carry,

Figure 42–1 and 42–2. Roller-element bearings are generally classified as either ball or roller.

Ball Bearings

Ball bearings can be divided into three categories: radial, angular contact, and thrust. Radial-contact ball bearings are designed for applications in which the load is primarily radial with only low-magnitude thrust loads. Angular-contact bearings are used where loads are combined radial and high thrust, and where precise shaft location is required. Thrust bearings handle loads that are primarily thrust.

SINGLE ROW, DEEP GROOVE BALL BEARINGS

The *single row, deep groove ball bearing* will sustain, in addition to radial load, a substantial thrust load in either direction, even at very high speeds. This advantage results from the intimate contact existing between the balls and the deep, continuous groove in each ring. When using this type of bearing, careful alignment between the shaft and housing is essential. This bearing is also available with seals, which exclude dirt and retain lubricant.

ANGULAR CONTACT BALL BEARINGS

The *angular contact ball bearing* supports a heavy thrust load in one direction, sometimes combined with a moderate radial load. A steep contact angle, assuring the highest thrust capacity and axial rigidity, is obtained by a high thrust supporting shoulder on the inner ring and a similar high shoulder on the opposite side of the outer ring. These bearings can be mounted singly or, when the sides are flush ground, in tandem for constant thrust in one direction; mounted in pairs, also when sides are flush ground, for a combined load, either face-to-face or back-to-back.

FIGURE 42–1A ■ Roller-element bearings

CYLINDRICAL ROLLER BEARINGS

The *cylindrical roller bearing* has high radial capacity and provides accurate guiding of the rollers, resulting in a close approach to true rolling. Consequent low friction permits operation at high speed. Those with flanges on one ring only allow a limited free axial movement of the shaft in relation to the housing. They are easy to dismount even when both rings are mounted with a tight fit. The double-row type assures maximum radial rigidity and is particularly suitable for machine tool spindles.

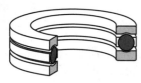

BALL THRUST BEARINGS

The *ball thrust bearing* is designed for thrust load in one direction only. The load line through the balls is parallel to the axis of the shaft, resulting in high thrust capacity and minimum axial deflection. Flat seats are preferred, particularly where the load is heavy or where close axial positioning of the shaft is essential, for example, in machine tool spindles.

SPHERICAL ROLLER THRUST BEARINGS

The *spherical roller thrust bearing* is designed to carry heavy thrust loads, or combined loads that are predominantly thrust. This bearing has a single row of rollers that roll on a spherical outer race with full self-alignment. The cage, centred by an inner ring sleeve, is constructed so that lubricant is pumped directly against the inner ring's unusually high guide flange. This ensures good lubrication between the roller ends and the guide flange. The spherical roller thrust bearing operates best with relatively heavy oil lubrication.

FIGURE 42–1B ■ Roller-element bearings, continued

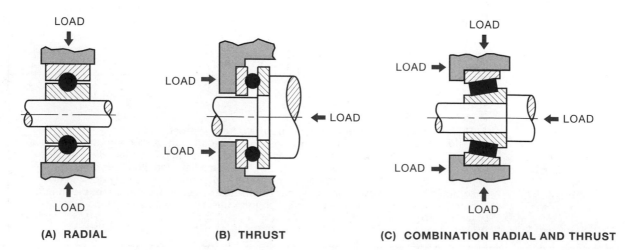

(A) RADIAL **(B) THRUST** **(C) COMBINATION RADIAL AND THRUST**

FIGURE 42–2 ■ Types of bearing loads

Roller Bearings

Roller bearings have higher load capacities than ball bearings for a given envelope size. They are widely used in moderate-speed, heavy-duty applications. The four principal types of roller bearings are: cylindrical, needle, tapered, and spherical. Cylindrical roller bearings use cylinders with approximate length-diameter ratios ranging from 1:1 to 1:3 as rolling elements. Needle roller bearings use cylinders or needles of greater length-diameter ratios. Tapered and spherical roller bearings are capable of supporting combined radial and thrust loads.

The rolling elements of tapered roller bearings are truncated cones. Spherical roller bearings are available with both barrel and hourglass roller shapes. The primary advantage of spherical roller bearings is their self-aligning capability. Plain bearings are covered in Unit 25.

RETAINING RINGS

Retaining rings, or snap rings, are designed to provide a removable shoulder to locate, retain, or lock components accurately on shafts and in bores and housings, Figure 42–3 and 42–4. They are easily installed and removed. Because they are usually made of spring steel, retaining rings have a high shear strength and impact capacity. In addition to fastening and positioning, a number of rings are designed for taking up end play caused by accumulated tolerances or wear in the parts being retained.

O-RING SEALS

O-rings are used as an axial mechanical seal (a seal that forms a running seal between a moving shaft and a housing) or a static seal (no moving parts). The advantage of using an O-ring as a gasket-type seal, Figure 42–5, over conventional gaskets is that the nuts need not be tightened uniformly and sealing compounds are not required. A rectangular groove is the most common type of groove used for O-rings.

CLUTCHES

Clutches are used to start and stop machines or rotating elements without starting or stopping the prime mover. They are also used for automatic disconnection and quick starts and stops and to permit shaft rotation in one direction only, such as the *overrunning* clutch shown in Figure 42–6. A full complement of sprags between concentric inner and outer races transmits power from one race to the other by wedging action of the sprags when either race is rotated in the driving direction. Rotation in the opposite direction frees the sprags, and the clutch is disengaged or overruns, Figure 42–6. This type of clutch is used in the power drive, Assignment A-106.

BELT DRIVES

A *belt drive* consists of an endless flexible belt connecting two wheels or pulleys. Belt drives depend on friction between belt and pulley surfaces for transmission of power.

In a V-belt drive, the belt has a trapezoidal cross section and runs in V-shaped grooves on the pulleys. These belts are made of cords or cables, impregnated and covered with rubber or other organic compound. The covering is formed to produce the required cross section. V-belts are usually manufactured as endless belts, although open-end and link types are available.

With V-belts, the friction for the transmission of the driving force is increased by the wedging of the belt into the grooves on the pulley.

AXIAL ASSEMBLY RINGS

INTERNAL   EXTERNAL

BASIC TYPES: Designed for axial assembly. Internal ring is compressed for insertion into bore or housing; external ring is expanded for assembly over shaft. Both rings seat in deep grooves and are secure against heavy thrust loads and high rotational speeds.

INTERNAL   EXTERNAL

INVERTED RINGS: Same tapered construction as basic types, with lugs inverted to abut bottom of groove. Section height increased to provide higher shoulder. Uniformly concentric with housing or shaft. Rings provide better clearance and more attractive appearance than basic types.

END PLAY RINGS

INTERNAL   EXTERNAL

BOWED RINGS: For assemblies in which accumulated tolerances cause objectionable end play between ring and retained part. Bowed construction permits rings to provide resilient end-play takeup in axial direction while maintaining tight grip against groove bottom.

EXTERNAL   INTERNAL

RADIAL RINGS: Bowed E-rings are used for providing resilient end-play takeup in an assembly.

SELF-LOCKING RINGS

EXTERNAL   EXTERNAL

CIRCULAR EXTERNAL RINGS: Push-on type of fastener with inclined prongs that bend from their initial position to grip the shaft. Ring at left has arched rim for increased strength and thrust load capacity; extra-long prongs accommodate wide shaft tolerances. Ring at right has flat rim, shorter locking prongs, and smaller OD.

INTERNAL

CIRCULAR INTERNAL RINGS: Designed for use in bores and housings. Function in same manner as external types except that locking prongs are on the outside rim.

RADIAL LOCKING RINGS

EXTERNAL

CRESCENT RINGS: Have a tapered section similar to the basic axial types. Remain circular after installation on a shaft and provide a tight grip against the groove bottom.

EXTERNAL

E-RINGS: Provide a large bearing shoulder on small-diameter shafts and are often used as a spring retainer. Three heavy prongs, spaced about 120° apart, provide contact surface with groove bottom.

FIGURE 42–3 ■ Stamped retaining rings

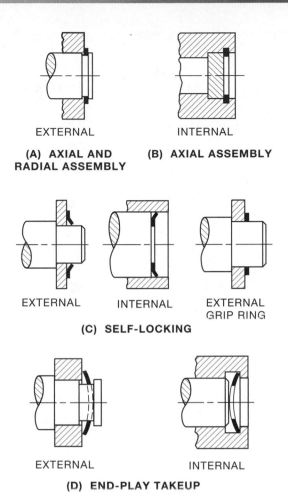

(A) AXIAL AND RADIAL ASSEMBLY

EXTERNAL

(B) AXIAL ASSEMBLY

INTERNAL

EXTERNAL

INTERNAL

EXTERNAL GRIP RING

(C) SELF-LOCKING

EXTERNAL

INTERNAL

(D) END-PLAY TAKEUP

FIGURE 42–4 ■ Retaining ring application

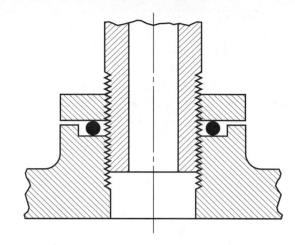

FIGURE 42–5 ■ O-ring seal

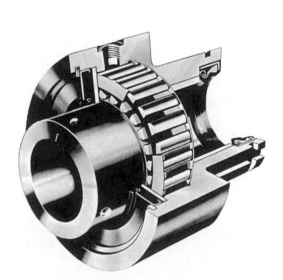

FIGURE 42–6 ■ Overrunning clutch

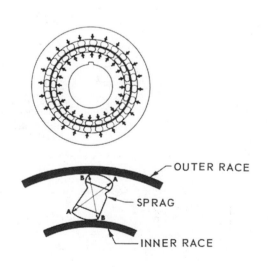

OUTER RACE

SPRAG

INNER RACE

V-Belt Sizes

To facilitate interchangeability and ensure uniformity, V-belt manufacturers have developed industrial standards for the various types of V-belts, Figure 42–7. Industrial V-belts are made in two types: heavy duty (conventional and narrow) and light duty. Conventional belts are available in A, B, C, D, and E sections. Narrow belts are made in 3V, 5V, and 8V sections. Light-duty belts come in 3L, 4L, and 5L sections.

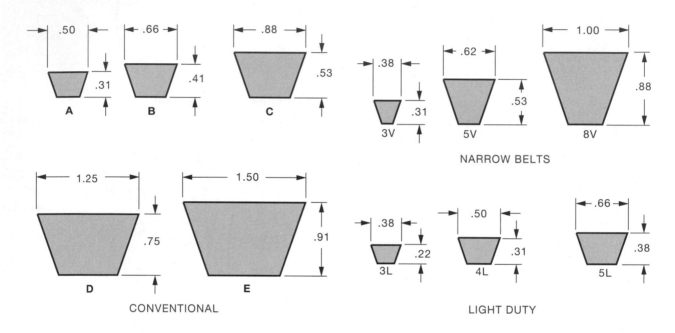

NARROW BELTS

CONVENTIONAL

LIGHT DUTY

FIGURE 42–7 ■ V-belt sizes

FIGURE 42–8 ■ V-belt sheave and bushing

Sheaves and Bushings

Sheaves, the grooved wheels of pulleys, are sometimes equipped with tapered bushings for ease of installation and removal, Figure 42–8. They have extreme holding power, providing the equivalent of a shrink fit. Both sheave and bushing used in the power drive, Assignment A-106, have a six-hole drilling arrangement, making it possible to insert the cap screw from either side. This is especially advantageous for applications where space is at a premium.

REFERENCES

A. O. Dehayt, "Basic Bearing Types," *Machine Design* 40, No. 14

ASME B27.7-1977 (R1999) General Purpose Tapered and Reduced Cross Section Retaining Rings, Metric

SKF Co. Ltd.

ASME B-18.27-1998 Tapered and Reduced Cross Section Retaining Rings (Inch Series)

INTERNET RESOURCES

Design and Technology Online For information on pulleys, see: http://www.dtonline.org/ (mechanisms)

eFunda For information on O-rings, see http://www.efunda.com/home.cfm

Gates Rubber Company For information on automotive and industrial drive belts, see: http://www.gates.com

Machine Design For information on bearings, retaining rings, clutches, seals, and belt drives, see *Machine Design, Mechanical Reference* at: http://www.machinedesign.com

TechStudent.Com For information on pulleys and pulley systems, see: http://www.technologystudent.com (gears and pulleys)

The Timken Company For information on precision bearings and motion control components and assemblies, see: http://www.torrington.com/

Truarc Retaining Rings For information on retaining rings, see: http://www.truarc.com

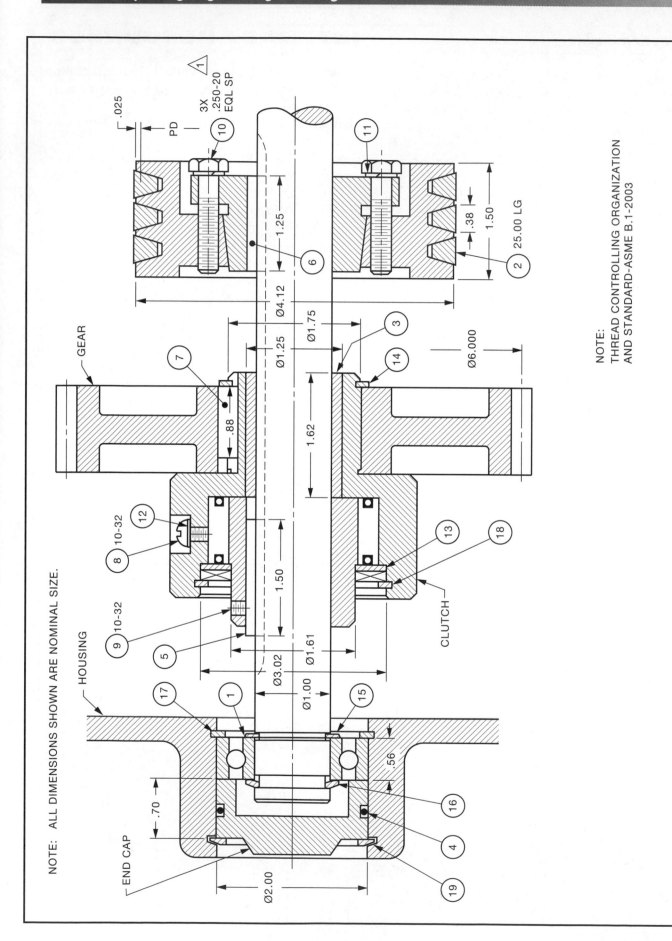

NOTE:
THREAD CONTROLLING ORGANIZATION
AND STANDARD-ASME B.1-2003

NOTE: ALL DIMENSIONS SHOWN ARE NOMINAL SIZE.

QUESTIONS:

1. How many cap screws fasten the sheave to the bushing?

2. List five parts or methods that are used to lock or join parts together on this assembly.

3. How many V-belts are used?

4. How many keys are there?

5. How many retaining rings are used?

6. What type of bearing is part (3) ?

7. What type of bearing is part (1) ?

8. What prevents the oil from leaking out between the housing and the end cap?

9. Can the gear be driven in both directions?

10. If the pitch on the gear is 8, what is the number of teeth?

11. What is the pitch diameter of the sheave?

12. What size V-belt is required?

ASSIGNMENTS:

1. ON A 1.00-INCH GRID SHEET (.10 IN. SQUARES), MAKE A ONE-VIEW DETAILED SKETCH OF THE END CAP, USING DIMENSIONS TAKEN FROM O-RING CATALOGUES. USE AN RC7 FIT BETWEEN THE OUTSIDE DIAMETER OF THE CAP AND THE HOUSING. USE YOUR JUDGMENT FOR DIMENSIONS NOT SHOWN. SCALE 1:1.

2. ON A GRID SHEET, SKETCH A BILL OF MATERIAL SIMILAR TO THE ONE SHOWN IN FIGURE 33-2, SHOWING PARTS 1 TO 19. REFER TO MANUFACTURERS' CATALOGUES AND APPENDIX TABLES.

REVISIONS			
⚠	10/12/04	CAMPBELL	
	PT 10 WAS 12-24 UNC		

SCALE	NOT TO SCALE	
DRAWN	R. BROWN	DATE 15/10/04
POWER DRIVE		**A-106**

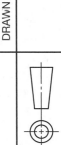

43 UNIT

RATCHET WHEELS

Ratchet wheels are used to transform reciprocating or oscillating motion into intermittent motion, to transmit motion in one direction only, or to serve as an indexing device.

Common forms of ratchets and pawls are shown in Figure 43–1. The teeth in the ratchet engage with the teeth in the pawl, permitting rotation in one direction only.

When a ratchet wheel and pawl are designed, points A, B, and C, as shown in Figure 43–1(A), are positioned on the same circle to ensure that the smallest forces are acting on the system.

Mechanical Advantage

Mechanical advantage is gained when a weight in one place lifts a heavier weight in another place, or a force applied at one point on a lever produces a

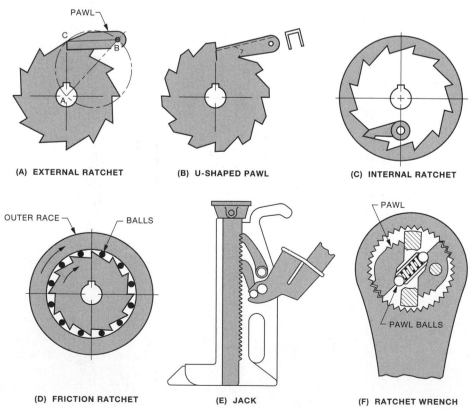

(A) EXTERNAL RATCHET   **(B) U-SHAPED PAWL**   **(C) INTERNAL RATCHET**

(D) FRICTION RATCHET   **(E) JACK**   **(F) RATCHET WRENCH**

FIGURE 43–1 ■ Ratchets and pawls

greater force at another point. Examples of mechanical advantage are the teeter-totter, winch, and gears, Figure 43–2.

The teeter-totter shows how mechanical advantage can be applied. If a 10-pound (lb.) weight is placed on one end of a teeter-totter and a 20 lb. weight is placed on the other end, the 10 lb. weight goes up and the 20 lb. weight goes down when placed equidistant from the fulcrum. If the fulcrum is moved to make the distance from the 10 lb. weight to the fulcrum twice that of the distance between the fulcrum and the 20 lb. weight, the weights become balanced. If the fulcrum is moved still closer to the 20 lb. weight, the 10 lb. weight goes down and the 20 lb. weight goes up.

As the distance between the fulcrum and the 10 lb. weight increases, the distance that the 10 lb. weight moves must increase proportionately to maintain the same movement of the 20 lb. weight. Thus, a light mass can move a heavier weight, but in doing so, the smaller weight must travel farther than the larger one.

Mechanical advantage is also gained when a winch or different-size gears are used. With the winch design shown in Figure 43–2(B), a mechanical advantage of 10 is obtained because the handle is 10 times the distance from the fulcrum compared with the centre of the rope.

The mechanical advantage of gears can easily be determined by obtaining the ratio between the number of teeth on the gears or the ratio between the pitch diameters, Figure 43–2(C).

In the winch, Assignment A-107, the centre of the handlebar is 10 in. from the centre of the shaft to which the pinion gear is attached. Half the pitch diameter of the pinion is .625 in. This produces a mechanical advantage of 10:.625 or 16:1. Further mechanical advantage is gained through the gear attached to the $\varnothing$1.00 shaft on which the rope revolves.

The hand will move about 63 in. when the handle is turned one complete revolution. This will turn the rope drum one-fifth of a revolution (50:10 teeth ratio), winding a $\varnothing$.25 rope up about 4 in. A greater force can be exerted using the winch than by simply pulling the rope.

INTERNET RESOURCES

Animated Worksheets For information on ratchets and actuators, see: http://www.animatedworksheets .co.uk (mechanisms)

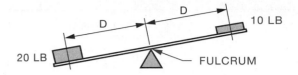

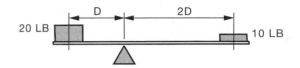

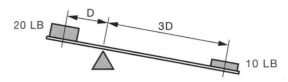

(A) TEETER-TOTTER

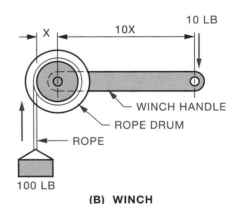

(B) WINCH

MECHANICAL ADVANTAGE = 3:1

MECHANICAL ADVANTAGE = 5:1

60 TEETH
20 TEETH

15 TEETH
75 TEETH

(C) GEARS

FIGURE 43–2 ■ Mechanical advantage

Globalspec For information on ratchets and related mechanisms and applications, see: http://www .globalspec.com/

TechStudent.Com For information, including illustrations, on ratchets and simple ratchet mechanisms and applications, see: http://www .technologystudent.com (Mechanisms)

ASSIGNMENT: ON ONE-INCH GRID SHEETS (.10 IN. SQUARES), MAKE A DETAILED SKETCH OF THE PARTS OF THE WINCH. DIMENSIONS SHOWN ARE NOMINAL SIZES AND ALLOWANCES AND TOLERANCES ARE TO BE DETERMINED. THE NUMBER OF VIEWS AND THE DRAWING SCALE TO BE SELECTED BY THE STUDENT.

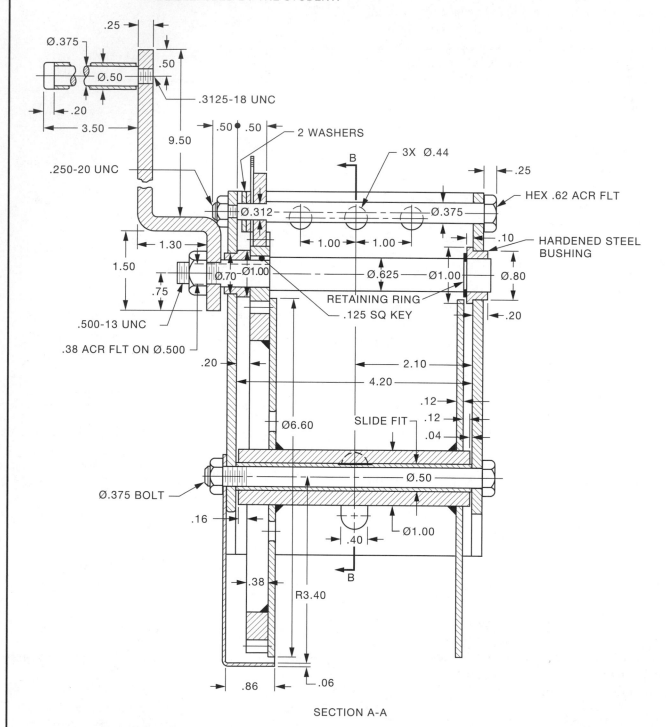

SECTION A-A

THREAD CONTROLLING ORGANIZATION
AND STANDARD–ASME B1.1-2003

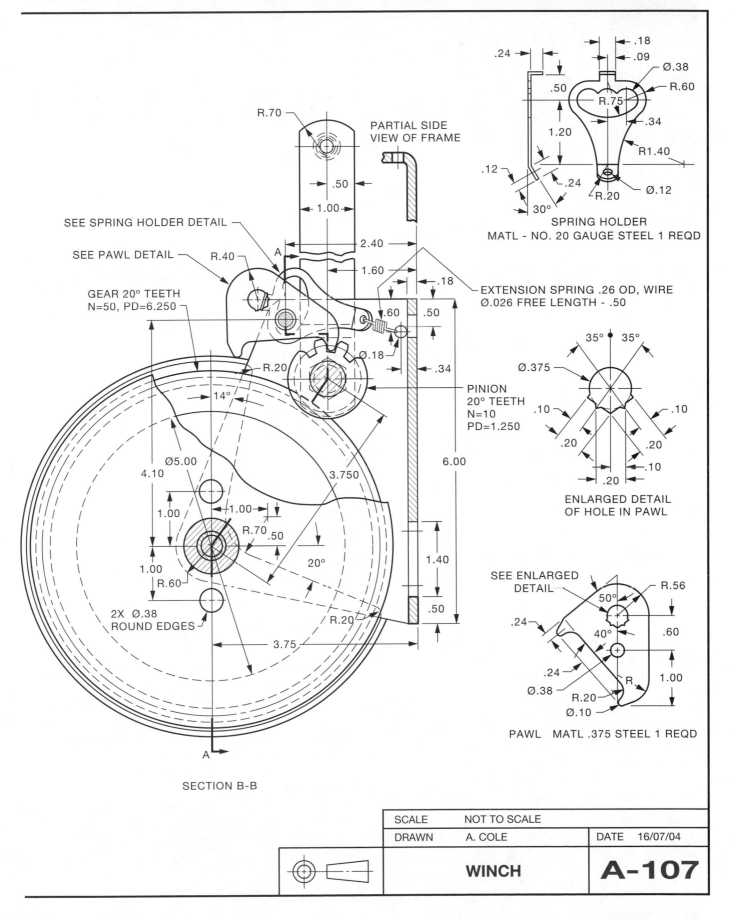

PARTIAL SIDE VIEW OF FRAME

R.70

.50
1.00

SEE SPRING HOLDER DETAIL

SEE PAWL DETAIL

R.40

A

2.40
1.60

.18

GEAR 20° TEETH
N=50, PD=6.250

R.20

14°

Ø5.00
4.10
1.00
1.00
R.70 .50
R.60

Ø.18

.60 .50

.34

EXTENSION SPRING .26 OD, WIRE
Ø.026 FREE LENGTH - .50

PINION
20° TEETH
N=10
PD=1.250

6.00

3.750

20°

3.75

R.20

2X Ø.38
ROUND EDGES

1.40

.50

SECTION B-B

A

SPRING HOLDER DETAIL (top right):
.24
.18
.09
Ø.38
R.60
.50
R.75
1.20
.34
R1.40
.12
.24
R.20
Ø.12
30°

SPRING HOLDER
MATL - NO. 20 GAUGE STEEL 1 REQD

ENLARGED DETAIL OF HOLE IN PAWL:
35° 35°
Ø.375
.10 .10
.20 .20
.10
.20

ENLARGED DETAIL
OF HOLE IN PAWL

PAWL DETAIL:
SEE ENLARGED
DETAIL
R.56
50°
.24
40°
.60
.24
Ø.38
R
R.20
1.00
Ø.10

PAWL   MATL .375 STEEL 1 REQD

SCALE	NOT TO SCALE	
DRAWN	A. COLE	DATE 16/07/04

WINCH

A-107

MODERN ENGINEERING TOLERANCING

An engineering drawing of a manufactured part conveys information from the designer to the manufacturer and inspector. It must contain all information necessary for the part to be correctly manufactured. It must also enable the inspector to determine precisely whether the finished parts are acceptable.

Therefore, each drawing must convey three items of information: the material to be used, the size or dimensions of the part, and the shape or geometric characteristics of the part. The drawing must also specify the permissible variation of size and form.

The actual size of a feature must be within the size limits specified on the drawing. Each measurement made at any cross section of the feature must not be greater than the maximum limit of size, nor smaller than the minimum limit of size, Figure 44–1. Although each part is within the prescribed tolerance zones, the parts may not be usable because of their deviation from their geometric form.

To meet functional requirements, it is often necessary to control errors of form, including squareness, roundness, and flatness, as well as deviation from true size. With mating parts, such as holes and shafts, it is usually necessary to ensure that they do not cross the boundary of perfect form at their maximum material size (the smallest hole or the largest shaft) because of being bent or otherwise deformed, Figure 44–2.

The system of *geometric tolerancing* offers a precise interpretation of drawing requirements. Geometric tolerancing controls geometric characteristics of parts, including:

- straightness
- flatness
- circularity
- cylindricity
- angularity
- parallelism
- perpendicularity
- runout
- profile
- position

Other techniques, such as datum systems, datum targets, and projected tolerance zones, were developed to facilitate this precise interpretation.

Geometric tolerances need not be used for every feature of a part. Generally, if each feature meets all dimensional tolerances, form variations will be adequately controlled by the accuracy of the manufacturing process and equipment used. A geometric tolerance is used when geometric errors must be limited more closely than might ordinarily be expected from the manufacturing process. A geometric tolerance is also used to state functional or interchangeability requirements.

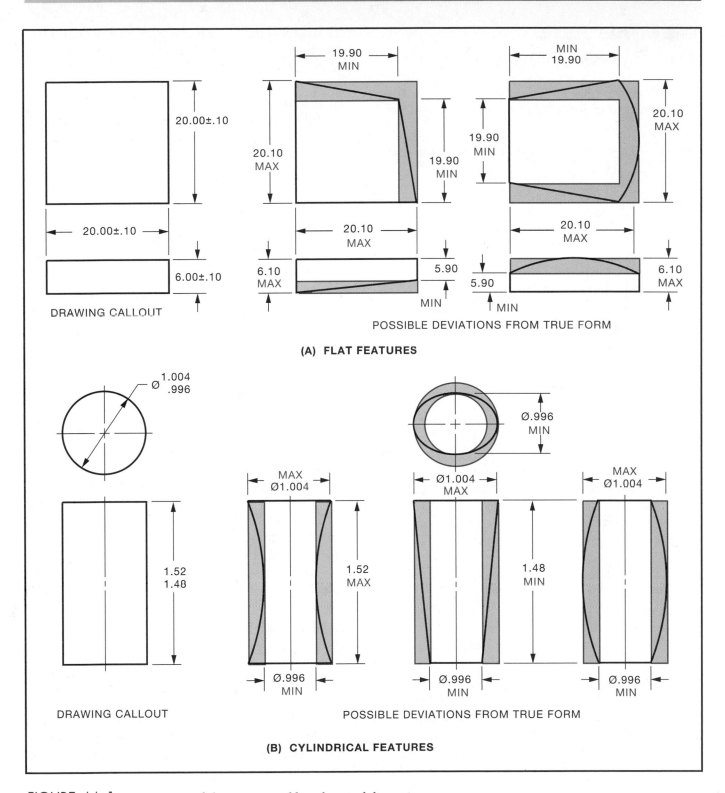

(A) FLAT FEATURES

(B) CYLINDRICAL FEATURES

FIGURE 44–1 ■ Deviations of shape permitted by toleranced dimensions

GEOMETRIC TOLERANCING

A geometric tolerance is the maximum permissible variation of form, orientation, or location of a feature from that indicated or specified on a drawing. The tolerance value represents the width or diameter of the tolerance zone within which the point, line, or surface of the feature should lie.

From this definition, it follows that a feature would be permitted to have any variation of form or take up any position within the specified geometric tolerance zone. For example, a line controlled in a single plane by a straightness tolerance of .008 in. must be contained within tolerance zone .008 in. wide, Figure 44–3.

Points, Lines, and Surfaces

The production and measurement of engineering parts deal, in most cases, with surfaces of objects. These surfaces may be flat, cylindrical, conical, or spherical, or have some irregular shape or contour.

Measurement, however, usually has to be made at specific points. A line or surface is evaluated dimensionally by making a series of measurements at various points along its length.

Surfaces are considered to be composed of a series of line elements running in two or more directions.

Points have position but no size; therefore, position is the only characteristic that requires control. Lines and surfaces have to be controlled for form, orientation, and location, and geometric tolerances provide for control of these characteristics, Figure 44–4. (Symbols will be introduced as required, but all are shown in the figure for reference.)

FEATURE CONTROL FRAME

Some geometric tolerances have been used for many years in the form of such notes as PARALLEL WITH SURFACE "A" WITHIN .001" and STRAIGHT WITHIN .12". Although such notes are now obsolete, the reader must recognize them on older drawings.

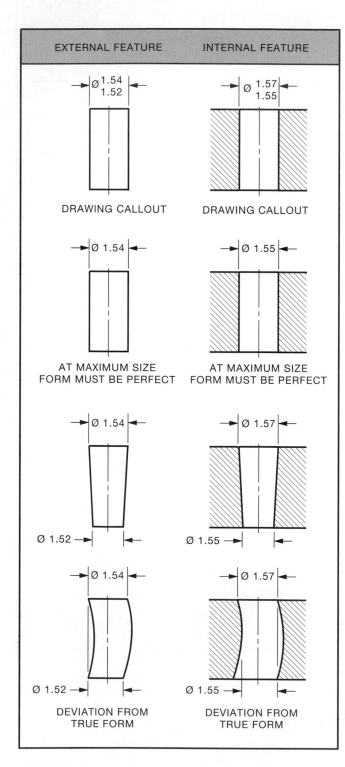

FIGURE 44–2 ■ Examples of deviation of form when perfect form at the maximum material size is required

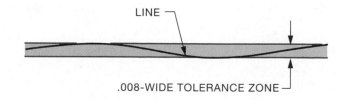

FIGURE 44–3 ■ Tolerance zone for straightness of a line

The current method is to specify geometric tolerances with a feature control frame, Figure 44–5. This consists of a rectangular frame divided into two or more compartments. The first compartment (starting from the left) contains the geometric characteristic; the second the allowable tolerance. Where applicable, the tolerance is preceded by the diameter symbol and followed by a modifying symbol. Other compartments are added when datums must be specified.

The feature control frame is related to the feature by one of the following methods.

Application to Surfaces (Figure 44–6A)

The arrowhead of the leader from the feature control frame should touch the surface of the feature

FEATURE	TYPE OF TOLERANCE	CHARACTERISTIC	SYMBOL	SEE UNIT
INDIVIDUAL FEATURES	FORM	STRAIGHTNESS	—	44 & 45
		FLATNESS	▱	46
		CIRCULARITY (ROUNDNESS)	○	46
		CYLINDRICITY	⌭	46
INDIVIDUAL OR RELATED FEATURES	PROFILE	PROFILE OF A LINE	⌒	53
		PROFILE OF A SURFACE	⌓	53
RELATED FEATURES	ORIENTATION	ANGULARITY	∠	48 & 49
		PERPENDICULARITY	⊥	
		PARALLELISM	//	
	LOCATION	POSITION	⊕	51
		CONCENTRICITY	◎	
		SYMMETRY	⩵	
	RUNOUT	CIRCULAR RUNOUT	*↗	54
		TOTAL RUNOUT	*⌰↗	
SUPPLEMENTARY SYMBOLS		AT MAXIMUM MATERIAL CONDITION	Ⓜ	45 & 51
		AT LEAST MATERIAL CONDITION	Ⓛ	
		PROJECTED TOLERANCE ZONE	Ⓟ	
		BASIC DIMENSION	XX	47
		DATUM FEATURE	▷─A	47
		DATUM TARGET	Ø.50 / A2	50

* MAY BE FILLED IN

FIGURE 44–4 ■ Geometric characteristic symbols

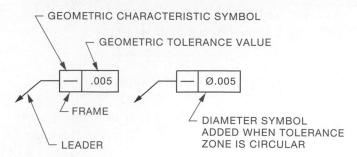

FIGURE 44–5 ■ Feature control frame

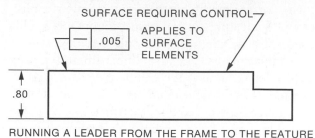

RUNNING A LEADER FROM THE FRAME TO THE FEATURE

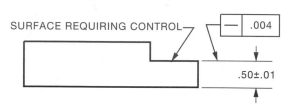

ATTACHED TO AN EXTENSION LINE USING A LEADER

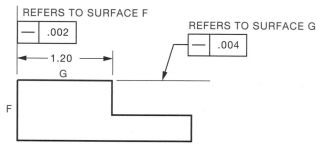

ATTACHED DIRECTLY TO AN EXTENSION LINE

(A) CONTROL OF SURFACE OR SURFACE ELEMENTS

or the extension line of the surface, but not in line with the dimension.

■ Attaching a side or end of the frame to an extension line extending from a place surface feature.

The leader from the feature control frame should be directed at the feature in its characteristic profile. Thus, in Figure 44–7, the straightness tolerance is directed to the side view, and the circularity tolerance to the end view. This may not always be possible; a tolerance connected to an alternative view, such as circularity tolerance connected to a side view, is acceptable. When it is more convenient or when space is limited, the arrowhead may be directed to an extension line, but not in line with the dimension line.

Applications to Features of Size (Figure 44–6B)

■ Locating the frame below or attached to the leader directed to the callout or dimension pertaining to the feature. (See Unit 45.)

■ Locating the frame below the size dimension to control the centre line, axis, or centre plane of the feature. (See Unit 45.)

When two or more feature control frames apply to the same feature, they are drawn together with a single leader and arrowhead, Figure 44–8.

ATTACHED TO THE DIMENSION LINE

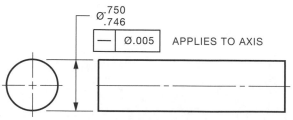

LOCATED BELOW DIMENSION CALLOUT

(B) CONTROL OF FEATURE OF SIZE

FIGURE 44–6 ■ Application of feature control frame

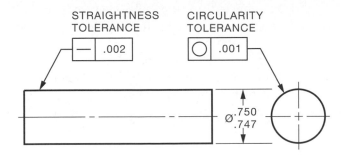

FIGURE 44-7 ■ Preferred location of feature control symbol when referring to a surface

FORM TOLERANCES

Form tolerances control straightness, flatness, circularity, and cylindricity. Orientation tolerances control angularity, parallelism, and perpendicularity.

Form tolerances are applicable to single (individual) features or elements of single features, so do not require locating dimensions. The form tolerance must be less than the size tolerance.

Form and orientation tolerances critical to function and interchangeability are specified where the tolerances of size and location do not provide sufficient control. A tolerance of form or orientation may be specified where no tolerance of size is given, for example, the control of flatness.

STRAIGHTNESS

Straightness is a condition in which the elements of a surface or its axes are in a straight line. The geometric characteristic symbol for straightness is a horizontal line, Figure 44-9. A straightness tolerance specifies a tolerance zone within which the considered element of the surface or axis must lie. A straightness tolerance is applied to the view where the elements to be controlled are represented by a straight line.

STRAIGHTNESS CONTROLLING SURFACE ELEMENTS

Straightness is fundamentally a characteristic of a line, such as the edge of a part or a line scribed on a surface. A straightness tolerance is specified on a drawing with a feature control frame, which is directed by a leader to the line requiring control,

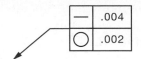

FIGURE 44-8 ■ Combined feature control frames directed to one surface

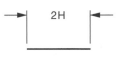

H = LETTER HEIGHT

FIGURE 44-9 ■ Straightness symbol

Figure 44-10. It states in symbolic form that the line shall be straight within .006 in. This means that the line shall be contained within a tolerance zone consisting of the area between two parallel straight lines in the same plane separated by the specified tolerance.

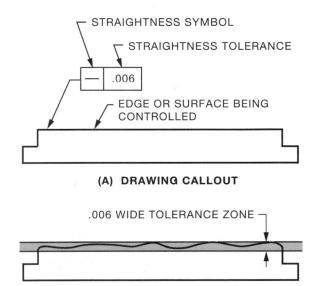

(A) DRAWING CALLOUT

(B) STRAIGHTNESS TOLERANCE ZONE

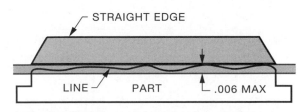

(C) CHECKING WITH A STRAIGHTEDGE

FIGURE 44-10 ■ Straightness tolerance applied to a flat surface

Theoretically, straightness could be measured by bringing a straightedge into contact with the line and determining that any space between the straightedge and the line does not exceed the specified tolerance.

For cylindrical parts or curved surfaces that are straight in one direction, the feature control frame should be directed to the side view, where line elements appear as a straight line, Figures 44–11 and 44–12.

A straightness tolerance thus applied to the surface controls surface elements only. Therefore, it would control bending or a wavy condition of the surface or a barrel-shaped part, but it would not

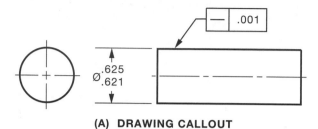

(A) DRAWING CALLOUT

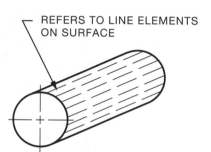

REFERS TO LINE ELEMENTS ON SURFACE

(B) REFERS TO LINE ELEMENTS ON SURFACE

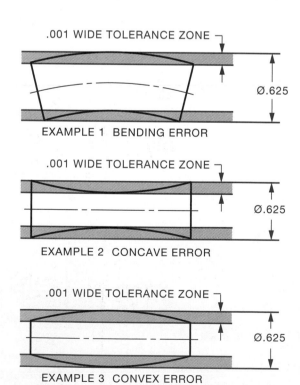

.001 WIDE TOLERANCE ZONE

Ø.625

EXAMPLE 1 BENDING ERROR

.001 WIDE TOLERANCE ZONE

Ø.625

EXAMPLE 2 CONCAVE ERROR

.001 WIDE TOLERANCE ZONE

Ø.625

EXAMPLE 3 CONVEX ERROR

(C) POSSIBLE VARIATIONS OF FORM

FIGURE 44–11 ■ Specifying the straightness of line elements of a cylindrical surface necessarily control the straightness of the axes or the conicity of the cylinder.

Straightness of a cylindrical surface is interpreted to mean that each line element of the surface shall be contained within a tolerance zone consisting of the space between two parallel lines, separated by the width of the specified tolerance, when the part is rolled along one of the planes. Theoretically, this could be measured by rolling the part on a flat surface and measuring the space between the part and the plate to ensure that it does not exceed the specified tolerance, Figure 44–13.

A straightness tolerance can be applied to a conical surface in the same manner as for a cylindrical surface, Figure 44–14, and will ensure that the rate of taper is uniform. The actual rate of taper, or the taper angle, must be separately toleranced.

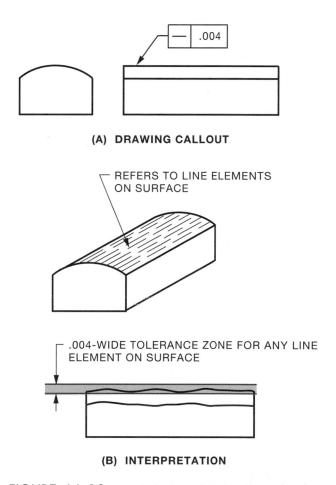

FIGURE 44–12 ■ Straightness of surface line elements

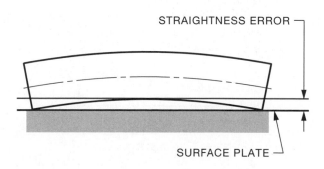

FIGURE 44–13 ■ Measuring the straightness of a cylindrical surface

A straightness tolerance applied to a flat surface indicates straightness control in one direction only and must be directed to the line on the drawing representing the surface to be controlled and the direction in which control is required, Figure 44–15. It is then interpreted to mean that each line element on the surface in the indicated direction shall lie within a tolerance zone.

Different straightness tolerances may be specified in two or more directions when required. However, if the same straightness tolerance is required in two coordinate directions on the same surface, a flatness tolerance rather than a straightness tolerance is used.

If it is not otherwise necessary to draw all three views, the straightness tolerances may all be shown on a single view by indicating the direction with short lines terminated by arrowheads, Figure 44–15C.

REFERENCES

CAN/CSA-B78.2-M91 Dimensioning and Tolerancing of Technical Drawings
ASME Y14.5M-1994 (R2004), Dimensioning and Tolerancing

INTERNET RESOURCES

Drafting Zone For information on geometric dimensioning and tolerancing, see: http://www .draftingzone.com

Effective Training Inc. For information on dimensioning and tolerancing, see: http://etinews .com/eti_solutions.html

eFunda For information on geometric dimensioning and tolerancing, see: http://www.efunda.com/ home.cfm

International Organization for Standardization For information and publications on international standards for dimensioning and tolerancing, see: http://www.iso.org

Wikipedia, the Free Encyclopedia For information on geometric dimensioning and tolerancing, see: http://en.wikipedia.org/wiki/ Engineering_drawing

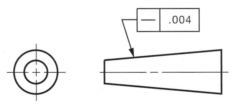

(A) DRAWING CALLOUT

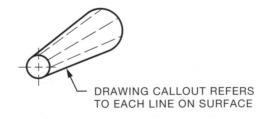

DRAWING CALLOUT REFERS
TO EACH LINE ON SURFACE

.004 WIDE TOLERANCE ZONE FOR
ANY LINE ELEMENT ON SURFACE

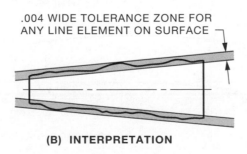

(B) INTERPRETATION

FIGURE 44–14 ■ Straightness of line elements on a conical surface

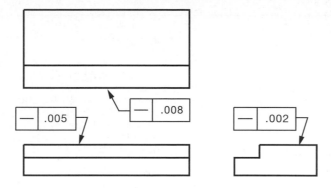

(A) DRAWING CALLOUT

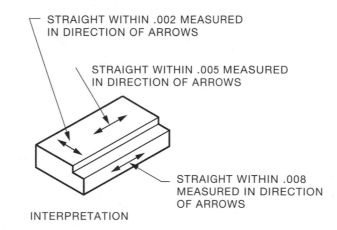

STRAIGHT WITHIN .002 MEASURED
IN DIRECTION OF ARROWS

STRAIGHT WITHIN .005 MEASURED
IN DIRECTION OF ARROWS

STRAIGHT WITHIN .008
MEASURED IN DIRECTION
OF ARROWS

INTERPRETATION

(B) STRAIGHTNESS TOLERANCES IN SEVERAL DIRECTIONS

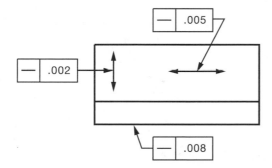

(C) THREE STRAIGHTNESS TOLERANCES ON ONE VIEW

FIGURE 44–15 ■ Straightness tolerances in several directions

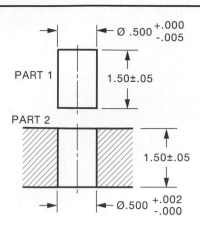

FIGURE 1

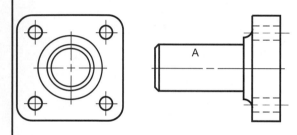

FIGURE 2

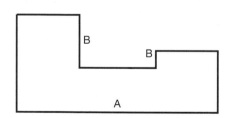

FIGURE 3

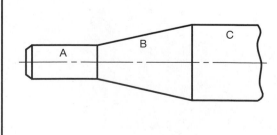

FIGURE 4

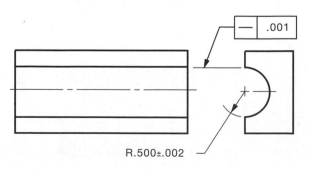

FIGURE 5

ASSIGNMENTS:

NOTE: ALL ASSIGNMENTS TO BE SKETCHED ON
ONE-INCH GRID SHEETS (.10 IN. SQUARES)

1. SKETCH THREE DIFFERENT PERMISSIBLE
DEVIATIONS (SIMILAR TO FIGURE 44-2) FOR A
RECTANGULAR PART .80 X 1.50 X .50 IN. THICK
HAVING A TOLERANCE OF ±.05 IN. ON THESE
DIMENSIONS. ADD DIMENSIONS.

2. MAKE A SKETCH, COMPLETE WITH DIMENSIONS,
OF A RING AND PLUG GAUGE TO CHECK THE HOLE
AND SHAFT IN FIGURE 1.

3. MAKE A SKETCH OF THE PART SHOWN IN FIGURE 2.
APPLY TWO FEATURE CONTROL FRAMES TO THE PART:
ONE FRAME TO DENOTE CIRCULARITY (ROUNDNESS),
THE OTHER FRAME TO CONTROL STRAIGHTNESS OF
THE SURFACE OF THE CYLINDRICAL FEATURE.

4. MAKE A SKETCH OF THE PART SHOWN IN FIGURE 3.
ADD THE FOLLOWING GEOMETRIC TOLERANCES TO
THE SKETCH:
(A) SURFACE "A" IS TO BE STRAIGHT WITHIN .005 IN.
(B) SURFACES "B" AND "C" ARE TO BE STRAIGHT
WITHIN .008 IN. USE A MINIMUM OF TWO METHODS
OF CONNECTING THE FEATURE CONTROL FRAME
TO THE SURFACES REQUIRING STRAIGHTNESS
CONTROL.

5. MAKE A SKETCH OF THE PART SHOWN IN FIGURE 4.
ADD THE FOLLOWING STRAIGHTNESS TOLERANCES
TO THE SURFACES:
(A) SURFACE "A"—.002 IN.
(B) SURFACE "B"—.001 IN.
(C) SURFACE "C"—.005 IN.

6. WHAT IS THE MAXIMUM PERMISSIBLE DEVIATION
FROM STRAIGHTNESS IF THE RADIUS IN FIGURE 5 IS:
(A) .498 IN.
(B) .500 IN.
(C) .502 IN.

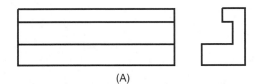

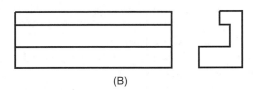

ASSIGNMENTS (CONT'D):

7. MAKE A SKETCH OF THE PART SHOWN IN FIGURE 6(B) AND PLACE THE FEATURE CONTROL FRAMES SHOWN IN FIGURE 6(A) ON THIS SKETCH.

8. IS THE PART SHOWN IN FIGURE 7 ACCEPTABLE? STATE YOUR REASON.

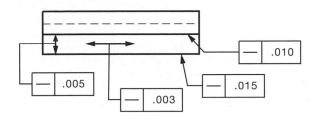

(A)

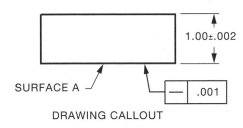

(B)

FIGURE 6

1.00±.002

SURFACE A

DRAWING CALLOUT

— .001

FIGURE 7

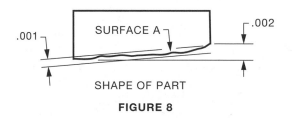

.001

SURFACE A

.002

SHAPE OF PART

FIGURE 8

STRAIGHTNESS TOLERANCE CONTROLLING SURFACE ELEMENTS

A-108

STRAIGHTNESS OF A FEATURE OF SIZE

Features of Size

Geometric tolerances so far considered concern only lines, line elements, and single surfaces. These are features having no diameter or thickness, and the form tolerances applied to them cannot be affected by feature size. In these examples, the feature control frame leader was directed to the surface or extension line of the surface but not to the size dimension. The straightness tolerance had to be less than the size tolerance.

Features of size have diameter or thickness. These may be cylinders, such as shafts and holes. They may be slots, tabs, or rectangular or flat parts where two parallel flat surfaces are considered to form a single feature. When applying a geometric tolerance to a feature of size, the feature control frame is associated with the size dimension or attached to an extension of the dimension line.

Circular Tolerance Zones

When the resulting tolerance zone is cylindrical, such as when straightness of the axis of a cylindrical feature is specified, a diameter symbol precedes the tolerance value in the feature control frame. The feature control frame is located below the dimension pertaining to the feature, Figure 45–1.

FEATURE OF SIZE DEFINITIONS

Before proceeding with examples of features of size, it is essential to understand certain terms.

Maximum Material Condition (MMC)

When a feature or part is at the limit of size where it contains the maximum amount of material, it is said to be at MMC. Thus, it is the maximum limit of size for an external feature, such as a shaft,

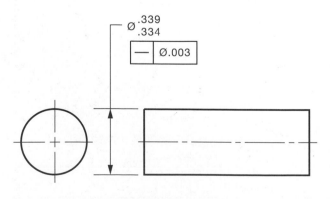

(A) DRAWING CALLOUT

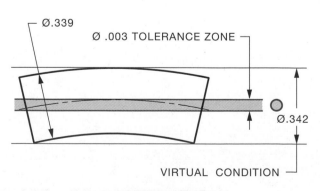

(B) INTERPRETATION

FIGURE 45–1 ■ Circular tolerance zone

or the minimum limit of size for an internal feature, such as a hole, Figure 45–2.

Virtual Condition

Virtual condition is the overall envelope of perfect form within which the feature would just fit. It is the boundary formed by the MMC limit of size of a feature plus or minus the applied geometric tolerances. For an external feature, such as a shaft, it is the maximum material size plus the effect of permissible form variations, such as straightness, flatness, roundness, cylindricity, and orientation tolerances. For an internal feature, such as a hole, it is the maximum material size minus the effect of such form variations, Figures 45–2 and 45–3.

Parts are generally toleranced so they will assemble when mating features are at MMC. Addi-

tional tolerance on form or location is permitted when features depart from their MMC size.

Least Material Condition (LMC)

This term refers to that size of a feature that results in the part containing the minimum amount of material. Thus it is the minimum limit of size for an external feature, such as a shaft, and the maximum limit of size for an internal feature, such as a hole, Figure 45–3.

Regardless of Feature Size (RFS)

This term means that the size of the geometric tolerance remains the same for any feature lying within its limits of size.

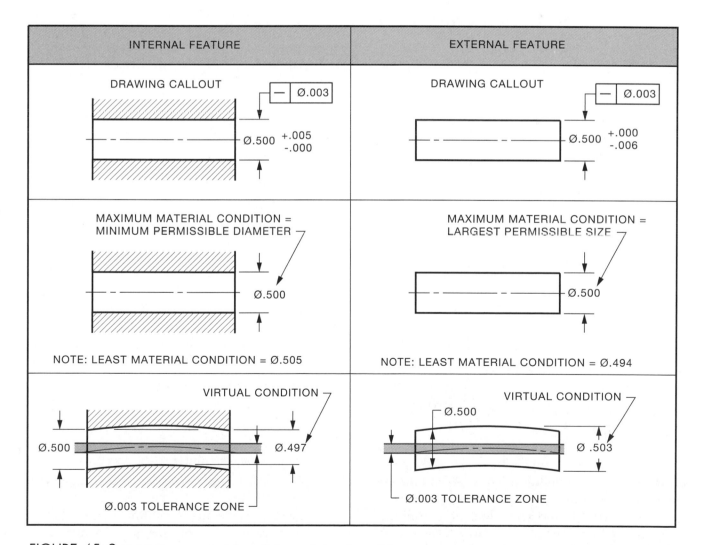

INTERNAL FEATURE	EXTERNAL FEATURE
DRAWING CALLOUT — Ø.003 Ø.500 +.005 / -.000	DRAWING CALLOUT — Ø.003 Ø.500 +.000 / -.006
MAXIMUM MATERIAL CONDITION = MINIMUM PERMISSIBLE DIAMETER Ø.500 NOTE: LEAST MATERIAL CONDITION = Ø.505	MAXIMUM MATERIAL CONDITION = LARGEST PERMISSIBLE SIZE Ø.500 NOTE: LEAST MATERIAL CONDITION = Ø.494
VIRTUAL CONDITION Ø.500 Ø.497 Ø.003 TOLERANCE ZONE	VIRTUAL CONDITION Ø.500 Ø .503 Ø.003 TOLERANCE ZONE

FIGURE 45–2 ■ Maximum material condition and virtual condition

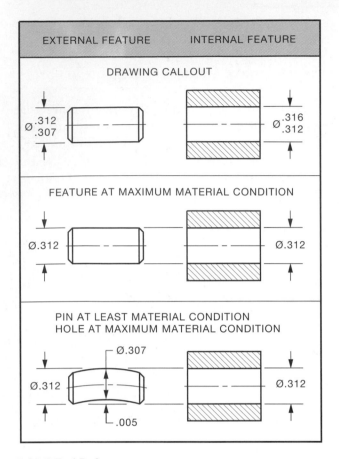

FIGURE 45–3 ■ Effect of form variation when only feature of size is specified

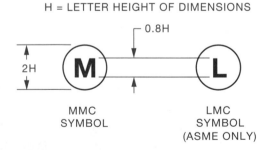

FIGURE 45–4 ■ Modifying symbols

MATERIAL CONDITION SYMBOLS

The modifying symbols used to indicate "at maximum material condition," and "at least material condition" are shown in Figure 45–4. The use of these symbols in local or general notes is prohibited.

Prior to 1994, the ANSI Y14.5 Dimensioning and Tolerancing Standard used a material condition symbol for "regardless of feature size," but this practice is now discontinued.

Applicability of RFS, MMC, and LMC

Applicability of RFS, MMC, and LMC is limited to features subject to variations in size. They may be datum features or other features whose axes or centre planes are controlled by geometric tolerances. In such cases, the following practices apply: RFS applies, with respect to the individual tolerance, datum reference, or both, where no modifying symbol is shown. (See Figures 45–2 and 45–11.) MMC or LMC must be specified on the drawing where it is required. (See Figure 45–12.)

EXAMPLES

If freedom of assembly of mating parts is the chief criterion for establishing a geometric tolerance for a feature of size, the least favourable assembly condition exists when the parts are made to the maximum material condition, that is, the largest diameter pin allowed entering a hole produced to the smallest allowable size. Further geometric variations can then be permitted, without jeopardizing assembly, as the features approach their LMC.

Example 1

The effect of a form tolerance is shown in Figure 45–3, where a cylindrical pin of ∅.307–.312 in. is intended to assemble into a round hole of ∅.312–.316 in. If both parts are at their maximum material condition of ∅.312 in., both would have to be perfectly round and straight to assemble. However, if the pin is at its least material condition of ∅.307 in., it can be bent up to .005 in. and still assemble in the smallest permissible hole.

Example 2

Another example, based on the location of features, is shown in Figure 45–5. This shows a part with two projecting pins required to assemble into a mating part having two holes at the same centre distance.

The worst assembly condition exists when the pins and holes are at their maximum material condition, which is ∅.250 in. Theoretically, these parts

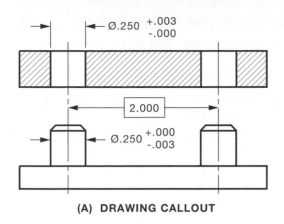

(A) DRAWING CALLOUT

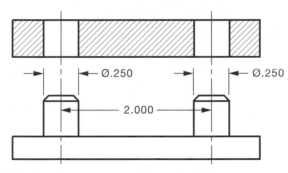

CENTRE DISTANCE MUST BE PERFECT
TO ASSEMBLE

(B) PINS AND HOLES AT MAXIMUM MATERIAL CONDITION

EACH CENTRE DISTANCE MAY BE INCREASED
OR DECREASED BY .003

(C) PINS AND HOLES AT LEAST MATERIAL CONDITION

FIGURE 45–5 ■ Effect of location

would just assemble if their form, orientation (squareness to the surface), and centre distances were perfect. However, if the pins and holes were at their least material condition of ∅.247 in. and ∅253 in., respectively, one centre distance could be increased and the other decreased by .003 in. without jeopardizing the assembly condition.

MAXIMUM MATERIAL CONDITION (MMC)

The symbol for MMC is shown in Figure 45–6. The symbol dimensions are based on percentages of the recommended letter height of dimensions.

If a geometric tolerance is required to be modified on an MMC basis, it is specified on the drawing by including the symbol Ⓜ immediately after the tolerance value in the feature control frame, Figure 45–6.

A form tolerance modified in this way can be applied only to a feature of size; it cannot be applied

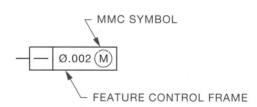

FIGURE 45–6 ■ Application of MMC symbol

to a single surface. It controls the boundary of the feature, such as a complete cylindrical surface or two parallel surfaces of a flat feature. This permits the feature surface or surfaces to cross the maximum material boundary by the amount of the form tolerance. If design requirements dictate that the virtual condition be kept within the maximum material boundary, the form tolerance must be specified as zero at MMC, Figure 45–7.

Application of MMC to geometric symbols is shown in Figure 45–8.

(A) FOR U.S. CUSTOMARY INCH DRAWINGS	(B) FOR METRIC DRAWINGS

FIGURE 45–7 ■ MMC symbol with zero tolerances

CHARACTERISTIC TOLERANCE		FEATURE BEING CONTROLLED
STRAIGHTNESS	——	NO FOR A PLANE SURFACE OR A LINE ON A SURFACE
PARALLELISM	//	
PERPENDICULARITY	⊥	YES FOR A FEATURE OF SIZE OF WHICH IS SPECIFIED BY A TOLERANCED DIMENSION, SUCH AS A HOLE, SHAFT, OR SLOT
ANGULARITY	∠	
POSITION	⊕	
FLATNESS	▱	NO FOR ALL FEATURES
CIRCULARITY (ROUNDNESS)	○	
CYLINDRICITY	⌭	
CONCENTRICITY	◎	
PROFILE OF A LINE	⌒	
PROFILE OF A SURFACE	⌓	
CIRCULAR RUNOUT	↗	
TOTAL RUNOUT	↗↗	
SYMMETRY	⩵	

FIGURE 45–8 ■ Application of MMC to geometric symbols

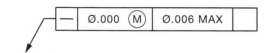

FIGURE 45–9 ■ Tolerance with a maximum value

Application with Maximum Value

It is sometimes necessary to ensure that the geometric tolerance does not vary over the full range permitted by the size variations. For such applications, a maximum limit may be set to the geometric tolerance, and this is shown in addition to that permitted at MMC, Figure 45–9.

REGARDLESS OF FEATURE SIZE (RFS)

When MMC or LMC is not specified with a geometric tolerance for a feature of size, no relationship is intended to exist between the feature size and the geometric tolerance. In other words, the tolerance applies regardless of feature size.

In this case, the geometric tolerance controls the form, orientation, or location of the axis, or median plane of the feature.

LEAST MATERIAL CONDITION (LMC)

The symbol for LMC is shown in Figure 45–4. It is the condition in which a feature of size contains the least amount of material within the stated limits of size.

Specifying LMC is limited to positional tolerance applications where MMC does not provide the desired control and RFS is too restrictive. LMC is used to maintain a desired relationship between the surface of a feature and its true position at tolerance extremes. It is used only with a tolerance of position. (See Unit 51.)

CSA will probably include the LMC symbol in its next revision of CAN/CSA-B78.2-M91 *Dimensioning and Tolerancing of Technical Drawings*.

STRAIGHTNESS OF A FEATURE OF SIZE

Figures 45–10 and 45–11 show examples of cylindrical parts where all circular elements of the surface are to be within the specified size tolerance; however, the boundary of perfect form at MMC may be violated. This violation is permissible when the feature control frame is associated with the size dimensions or attached to an extension of the dimension line. In these two figures, a diameter symbol precedes the tolerance value and the tolerance is applied on an RFS and an MMC basis, respectively. Normally, the straightness tolerance is smaller than the size tolerance, but a specific design may allow the situation depicted in the figures. The collective effect of size and form variations can produce a virtual condition equal to the MMC size plus the straightness tolerance, Figure 45–11. The derived median plane of the feature must lie within a cylindrical tolerance zone as specified.

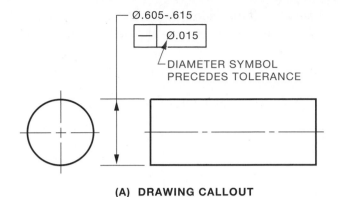

Ø.605-.615

— Ø.015

DIAMETER SYMBOL
PRECEDES TOLERANCE

(A) DRAWING CALLOUT

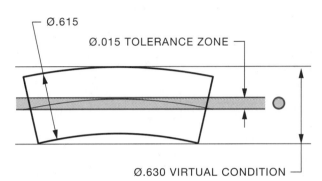

Ø.615

Ø.015 TOLERANCE ZONE

Ø.630 VIRTUAL CONDITION

FEATURE SIZE	DIAMETER TOLERANCE ZONE ALLOWED
.615	.015
.614	.015
.613	.015
↓	↓
.606	.015
.605	.015

(B) INTERPRETATION

FIGURE 45–10 ■ Specifying straightness—RFS

Straightness—RFS

When applied on an RFS basis, Figure 45–10, the maximum permissible deviation from straightness is .015 in., regardless of the feature size. Note that the absence of a modifying symbol indicates that RFS applies.

Straightness—MMC

If the straightness tolerance of .015 in. is required only at MMC, further straightness error can be permitted without jeopardizing assembly, as the feature approaches its least material size, Figure 45–11. The maximum straightness tolerance is the specified tolerance plus the amount the feature departs from its MMC size. The median line of the actual feature must lie within the derived cylindrical tolerance zone, as given in the table of Figure 45–11.

Straightness—Zero MMC

It is quite permissible to specify a geometric tolerance of zero MMC, which means that the virtual condition coincides with the maximum material size, Figure 45–12. Therefore, if a feature is at its maximum material limit everywhere, no errors of straightness are permitted.

Straightness on the MMC basis can be applied to any part or feature having straight line elements in a plane that includes the diameter or thickness. This also includes parts toleranced on an RFS basis. However, it should not be used for features that do not have a uniform cross section.

Straightness with a Maximum Value

If it is desired to ensure that the straightness error does not become too great when the part approaches the LMC, a maximum value may be added, Figure 45–13. The maximum overall tolerance follows in a separate compartment in the feature control frame.

Straightness per Unit Length

Straightness may be applied on a unit-length basis to prevent an abrupt surface variation within a relatively short length of the feature, Figure 45–14. Caution should be exercised when using unit control without specifying a maximum limit for the total length because of the relatively large variations that may result if no such restriction is applied. If the feature has a uniformly continuous bow throughout its length that just conforms to the tolerance applicable to the unit length, then the overall tolerance may result in an unsatisfactory part. Figure 45–15 illustrates the possible condition if the straightness per unit length given in Figure 45–14 is used alone, that is, if straightness for the total length is not specified.

REFERENCES

CAN/CSA B78.2-M91 Dimensioning of Tolerancing of Technical Drawings

ASME Y14.5M-1994 (R2004) Dimensioning and Tolerancing

INTERNET RESOURCES

Drafting Zone For information on geometric dimensioning and tolerancing, see: http://www .draftingzone.com

Effective Training Inc. For information on dimensioning and tolerancing, see: http://etinews.com/ eti_solutions.html

eFunda For information on geometric dimensioning and tolerancing, see: http://www.efunda.com/ home.cfm

Engineers Edge For information on geometric tolerancing and dimensioning, see: http://www .engineersedge.com/gdt.htm

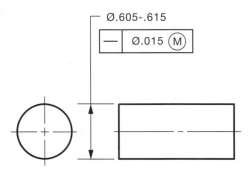

NOTE: PERFECT FORM AT MMC NOT REQ'D

(A) DRAWING CALLOUT

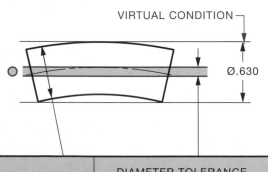

FEATURE SIZE	DIAMETER TOLERANCE ZONE ALLOWED
.615	.015
.614	.016
.613	.017
↓	↓
.606	.025
.605	.024

(B) INTERPRETATION

FIGURE 45–11 ■ Specifying straightness—MMC

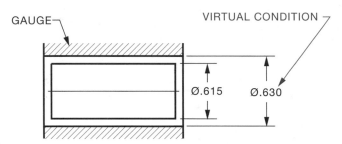

MAXIMUM DIAMETER OF THE PIN WITH PERFECT FORM IN A GAUGE

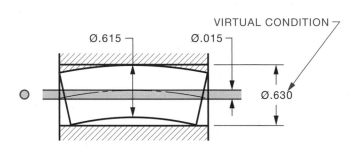

WITH PIN AT MAXIMUM DIAMETER (.615 IN), THE GAUGE WILL ACCEPT THE PIN WITH UP TO .015 IN. VARIATION IN STRAIGHTNESS

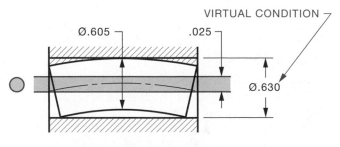

WITH PIN AT MINIMUM DIAMETER (.605 IN.), THE GAUGE WILL ACCEPT THE PIN WITH UP TO .025 IN. VARIATION IN STRAIGHTNESS

(C) ACCEPTANCE BOUNDARY

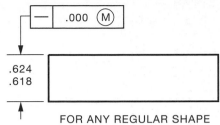

.624
.618

FOR ANY REGULAR SHAPE

DIAMETER SYMBOL ADDED IF TOLERANCE
ZONE IS CYLINDRICAL

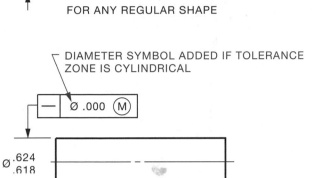

Ø .624
.618

FOR CYLINDRICAL SHAPES

(A) DRAWING CALLOUT

TOLERANCE ZONE FOR STRAIGHTNESS ERROR

VIRTUAL
CONDITION   FEATURE SIZE

.624

FEATURE SIZE	PERMISSIBLE STRAIGHTNESS ERROR
.624	.000
.623	.001
.622	.002
.621	.003
.620	.004
.619	.005
.618	.006

(B) PERMISSIBLE VARIATIONS

FIGURE 45–12 ■ Straightness—zero MMC

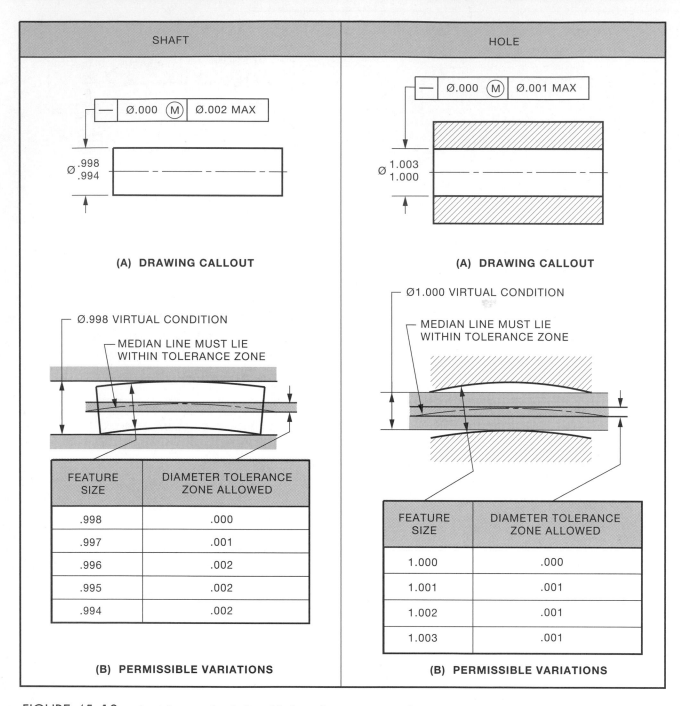

SHAFT	HOLE

(A) DRAWING CALLOUT

(A) DRAWING CALLOUT

FEATURE SIZE	DIAMETER TOLERANCE ZONE ALLOWED
.998	.000
.997	.001
.996	.002
.995	.002
.994	.002

(B) PERMISSIBLE VARIATIONS

FEATURE SIZE	DIAMETER TOLERANCE ZONE ALLOWED
1.000	.000
1.001	.001
1.002	.001
1.003	.001

(B) PERMISSIBLE VARIATIONS

FIGURE 45–13 ■ Straightness of a shaft and hole with a maximum value—MMC

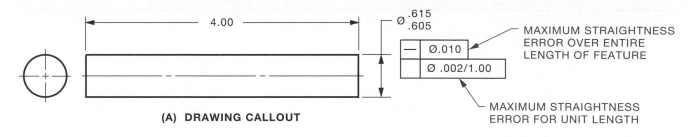

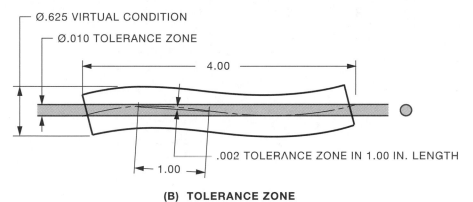

(A) DRAWING CALLOUT

MAXIMUM STRAIGHTNESS ERROR OVER ENTIRE LENGTH OF FEATURE

MAXIMUM STRAIGHTNESS ERROR FOR UNIT LENGTH

Ø.625 VIRTUAL CONDITION

Ø.010 TOLERANCE ZONE

.002 TOLERANCE ZONE IN 1.00 IN. LENGTH

(B) TOLERANCE ZONE

FIGURE 45–14 ■ Specifying straightness per unit length with specified total straightness, both RFS

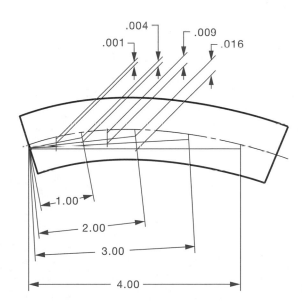

FIGURE 45–15 ■ Possible results of specifying straightness per unit length RFS with no maximum value

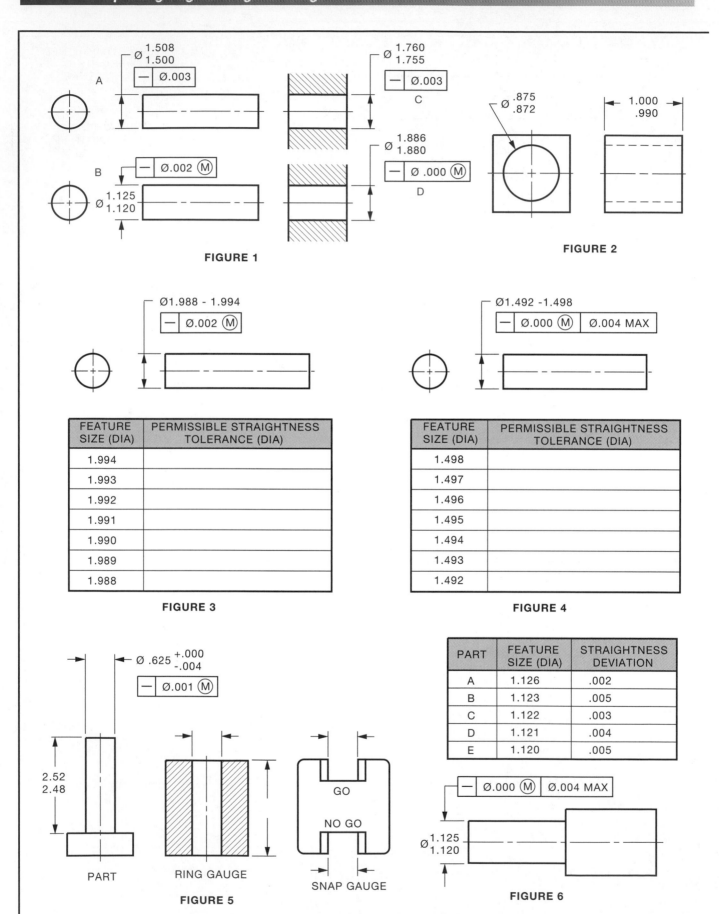

Ø 1.508
1.500

⊟ | Ø.003

A

Ø 1.760
1.755

⊟ | Ø.003

C

B

⊟ | Ø.002 Ⓜ

Ø 1.125
1.120

Ø 1.886
1.880

⊟ | Ø .000 Ⓜ

D

FIGURE 1

Ø .875
.872

1.000
.990

FIGURE 2

Ø1.988 - 1.994

⊟ | Ø.002 Ⓜ

Ø1.492 -1.498

⊟ | Ø.000 Ⓜ | Ø.004 MAX

FEATURE SIZE (DIA)	PERMISSIBLE STRAIGHTNESS TOLERANCE (DIA)
1.994	
1.993	
1.992	
1.991	
1.990	
1.989	
1.988	

FIGURE 3

FEATURE SIZE (DIA)	PERMISSIBLE STRAIGHTNESS TOLERANCE (DIA)
1.498	
1.497	
1.496	
1.495	
1.494	
1.493	
1.492	

FIGURE 4

Ø .625 +.000
-.004

⊟ | Ø.001 Ⓜ

2.52
2.48

PART

RING GAUGE

GO

NO GO

SNAP GAUGE

FIGURE 5

PART	FEATURE SIZE (DIA)	STRAIGHTNESS DEVIATION
A	1.126	.002
B	1.123	.005
C	1.122	.003
D	1.121	.004
E	1.120	.005

⊟ | Ø.000 Ⓜ | Ø.004 MAX

Ø 1.125
1.120

FIGURE 6

ASSIGNMENTS:

NOTE:
ALL ASSIGNMENTS TO BE SKETCHED ON
ONE-INCH GRID SHEETS (.10 IN. SQUARES)

1. WHAT IS THE VIRTUAL CONDITION FOR EACH
   PART SHOWN IN FIGURE 1?

2. THE HOLE IN FIGURE 2 DOES NOT HAVE A
   STRAIGHTNESS TOLERANCE. WHAT IS THE
   MAXIMUM PERMISSIBLE DEVIATION FROM
   STRAIGHTNESS IF PERFECT FORM AT THE
   MAXIMUM MATERIAL SIZE IS REQUIRED?

3. SKETCH CHARTS SIMILAR TO THOSE SHOWN
   IN FIGURES 3 AND 4. COMPLETE THE CHARTS,
   SHOWING THE LARGEST PERMISSIBLE
   STRAIGHTNESS FOR THE FEATURE SIZES SHOWN.

4. SKETCH AND DIMENSION THE RING AND SNAP
   GAUGE SHOWN IN FIGURE 5 TO CHECK THE PIN
   SHOWN. THE RING GAUGE SHOULD BE SUCH A SIZE
   AS TO CHECK THE LENGTH OF THE ENTIRE PIN.
   THE TWO OPEN ENDS OF THE SNAP GAUGE SHOULD
   MEASURE THE MINIMUM AND MAXIMUM
   ACCEPTABLE PIN DIAMETERS.

5. WITH REFERENCE TO FIGURE 6, ARE PARTS A TO E
   ACCEPTABLE? STATE YOUR REASONS IF THE PART
   IS NOT ACCEPTABLE.

6. SKETCH THE SHAFT SHOWN IN FIGURE 7 AND
   ADD A STRAIGHTNESS TOLERANCE OF .004 IN.
   REGARDLESS OF FEATURE SIZE TO THE SHAFT.

7. WITH REFERENCE TO FIGURE 8, WHAT IS THE
   MAXIMUM DEVIATION PERMITTED FROM
   STRAIGHTNESS FOR THE CYLINDRICAL SURFACE
   IF IT WAS (A) AT MMC? (B) AT LMC? (C) Ø.623?

8. SKETCH THE SHAFT SHOWN IN FIGURE 9 AND ADD
   A MAXIMUM STRAIGHTNESS TOLERANCE OF .002 IN.
   FOR ANY 1.00 INCH OF ITS LENGTH, BUT HAVE A
   MAXIMUM STRAIGHTNESS TOLERANCE OF .005 IN.
   OVER THE ENTIRE LENGTH. APPLY THE APPROPRIATE
   STRAIGHTNESS TOLERANCE TO THE SKETCH.

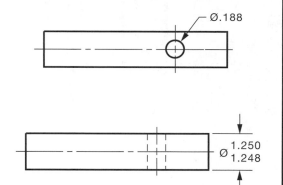

FIGURE 7

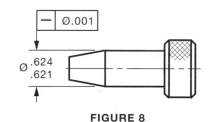

FIGURE 8

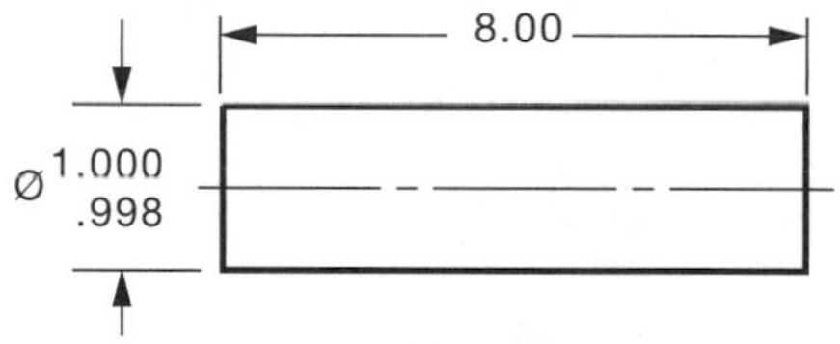

FIGURE 9

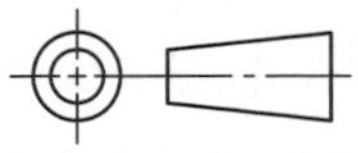

**STRAIGHTNESS OF
A FEATURE OF SIZE**

A-109

46 UNIT

FORM TOLERANCES

Form tolerances are used to control straightness, flatness, circularity, and cylindricity.

FLATNESS

Flatness of a surface is a condition in which all surface elements are in one plane. The symbol for flatness is a parallelogram with angles of 60°, Figure 46–1.

A flatness tolerance is applied to a line representing the surface of a part with a feature control frame, Figure 46–1.

A flatness tolerance means that all points on the surface shall be contained within a tolerance zone consisting of the space between two parallel planes that are separated by the specified tolerance. These two parallel planes must lie within the limits of size. These planes may be oriented in any manner to contain the surface; that is, they are not necessarily parallel to the base.

The flatness tolerance must be less than the size tolerance and contained within the limits of size.

When flatness tolerances are applied to opposite surfaces of a part and size tolerances are also shown, Figure 46–2, the flatness tolerance must be less than the size tolerance and lie within the limits of size.

Flatness per Unit Area

Flatness may be applied, as in the case of straightness, on a unit basis to prevent an abrupt surface variation within a relatively small area of the feature. The unit variation is used either in combination with a specified total variation or alone. Caution should be exercised when using unit control alone for the same reason given for straightness in Unit 45.

Because flatness involves surface area, the size of the unit area—for example, 1.00 × 1.00 in.—is specified to the right of the flatness tolerance, separated by a slash line, Figure 46–3.

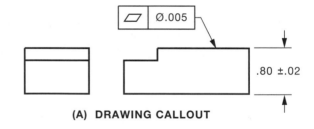

(A) DRAWING CALLOUT

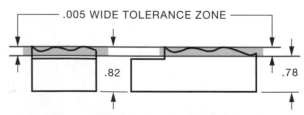

THE SURFACE MUST LIE BETWEEN TWO PARALLEL PLANES .005 IN. APART. ADDITIONALLY, THE SURFACE MUST BE LOCATED WITHIN THE SPECIFIED LIMITS OF SIZE.

(B) TOLERANCE ZONE

FIGURE 46–1 ■ Specifying flatness for a surface

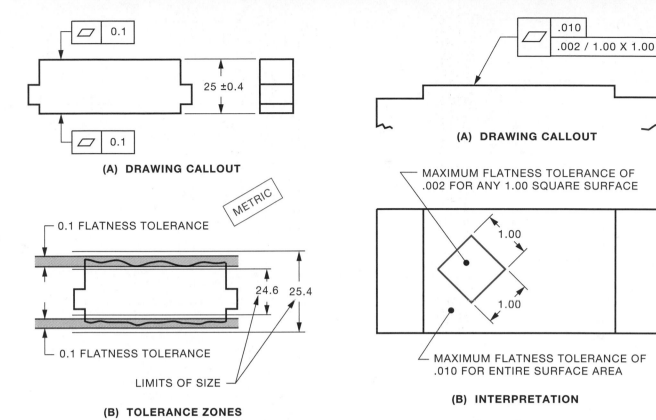

(A) DRAWING CALLOUT

METRIC

0.1 FLATNESS TOLERANCE

24.6 25.4

0.1 FLATNESS TOLERANCE

LIMITS OF SIZE

(B) TOLERANCE ZONES

FIGURE 46–2 ■ Location of flatness tolerance within limits of size

(A) DRAWING CALLOUT

MAXIMUM FLATNESS TOLERANCE OF .002 FOR ANY 1.00 SQUARE SURFACE

1.00

1.00

MAXIMUM FLATNESS TOLERANCE OF .010 FOR ENTIRE SURFACE AREA

(B) INTERPRETATION

FIGURE 46–3 ■ Overall flatness tolerance combined with a flatness for a unit area

CIRCULARITY

Circularity refers to a condition of a circular line or the surface of a circular feature where all points on the line or on the circumference of a plane cross section of the feature are the same distance from a common axis or centre point. It is similar to straightness except that it is wrapped around a circular cross section. Examples of circular features are disks, spheres, cylinders, and cones.

Errors of circularity (out-of-roundness) of a circular line on the periphery of a cross section of a circular feature may occur as (1) ovality, where differences appear between the major and minor axes; (2) lobing, where in some instances the diametral values may be constant or nearly so; or (3) random irregularities from a true circle, Figure 46–4.

The geometric characteristic symbol for circularity is a circle. A circularity tolerance may be specified by using this symbol in the feature control frame, Figure 46–5.

A circularity tolerance is measured radially and specifies the width between two concentric circular rings for a particular cross section within which the circular line or the circumference of the feature in that plane shall lie, Figure 46–6. Each

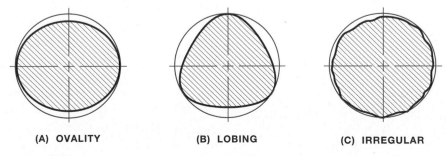

(A) OVALITY   **(B) LOBING**   **(C) IRREGULAR**

FIGURE 46–4 ■ Common types of circularity errors

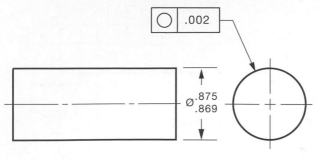

(A) DRAWING CALLOUT

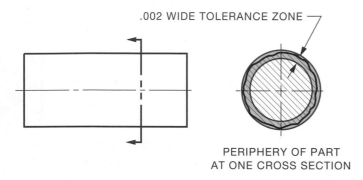

.002 WIDE TOLERANCE ZONE

PERIPHERY OF PART
AT ONE CROSS SECTION

PERIPHERY OF PART MUST LIE WITHIN LIMITS OF SIZE

(B) TOLERANCE ZONE

FIGURE 46–5 ■ Circularity tolerance applied to a cylindrical feature

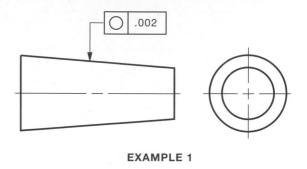

EXAMPLE 1

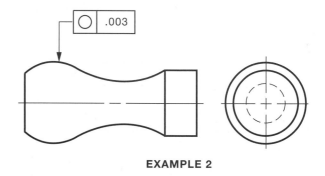

EXAMPLE 2

FIGURE 46–6 ■ Circularity tolerance applied to non-cylindrical features

CYLINDRICITY

Cylindricity is a condition of a surface in which all points of the surface are the same distance from a common axis. The cylindricity tolerance is a composite control of form that includes circularity, straightness, and parallelism of the surface element of a cylindrical feature. It is like a flatness tolerance wrapped around a cylinder.

The geometric characteristic symbol for cylindricity consists of a circle with two tangent lines at 60°.

A cylindricity tolerance that is measured radially specifies a tolerance zone bounded by two concentric cylinders within which the surface must lie. The cylindricity tolerance must be within the specified limits of size. Unlike that of circularity, cylindricity tolerance applies simultaneously to both circular and longitudinal elements of the surface, Figure 46–7. The leader from the feature control symbol may be directed to either view. The cylindricity tolerance must be less than half of the size tolerance.

Because each part is measured for form deviation, the total range of the specified cylindricity tolerance will not always be available.

The cylindricity tolerance zone is controlled by the measured size of the actual part. The part size is first determined; then the cylindricity tolerance

circular element of the surface must also be within the specified limits of size.

Because circularity is a form of tolerance, it is not related to datums.

A circularity tolerance may be specified with the circularity symbol in the feature control frame. It is expressed on an RFS basis. The absence of a modifying symbol in the feature control frame means that RFS applies to the circularity tolerance.

A circularity tolerance cannot be modified on an MMC basis because it controls surface elements only. The circularity tolerance must be less than half the size tolerance because it must lie in a space equal to half the size tolerance.

Circularity of Noncylindrical Parts

Noncylindrical parts are conical parts and other features that are circular in cross section but have variable diameters, such as those shown in Figure 46–6. Because many sizes of circles may be involved in the end view, it is usually best to direct the circularity tolerance to the longitudinal surfaces.

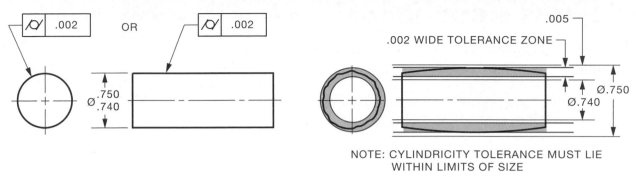

(A) DRAWING CALLOUT

(B) TOLERANCE ZONE

NOTE: CYLINDRICITY TOLERANCE MUST LIE WITHIN LIMITS OF SIZE

FIGURE 46–7 ■ Cylindrical tolerance directed to either view

is added as a refinement to the actual size of the part. In Figure 46–7, if the largest measurement of the produced part is ∅.748 in., which is near the high limit of size (.750 in.), the largest diameter of the two concentric cylinders for the cylindricity tolerance would be ∅.748 in. The smaller of the concentric cylinders would be .748 minus twice the cylindricity tolerance (2 × .002) = ∅.744 in. The cylindricity tolerance zone must also lie between the limits of size, and the entire cylindrical surface of the part must lie between these two concentric circles to be acceptable.

If, on the other hand, the largest diameter measured for a part was ∅.743 in., which is near the lower limit of size (.740 in.), the cylindricity deviation of that part cannot be greater than .0015 in., or it would exceed the lower limit of size.

Likewise, if the smallest measured diameter of a part was .748 in., which is near the high limit of size, the largest diameter of the two concentric cylinders for the cylindricity tolerance would be ∅.750 in., the maximum permissible diameter of the part. In this case, the cylindricity tolerance could not be greater than (.750–.748)/2 or .001 in.

Figure 46–8 shows some permissible form errors for the part shown in Figure 46–7.

Cylindricity tolerances can be applied only to cylindrical surfaces, such as round holes and shafts. No specific geometric tolerances have been devised for other circular forms, which require the use of several geometric tolerances. A conical surface, for example, must be controlled by a combination of tolerances for circularity, straightness, and angularity.

Because cylindricity is a form tolerance much like that of a flatness tolerance in that it controls surface elements only, it cannot be modified on an MMC basis. The absence of a modifying symbol in the feature control frame indicates that RFS applies.

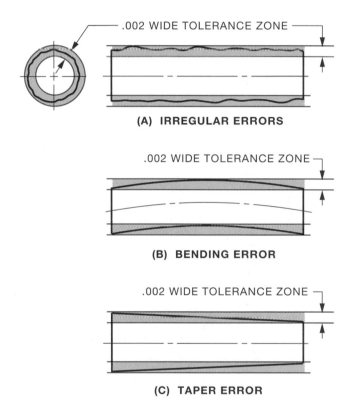

(A) IRREGULAR ERRORS

(B) BENDING ERROR

(C) TAPER ERROR

NOTE: CYLINDRICITY TOLERANCE MUST LIE WITHIN THE LIMITS OF SIZE

FIGURE 46–8 ■ Permissible form errors for part shown in Figure 46–7

REFERENCES

CAN/CSA-B78.2-M91 Dimensioning and Tolerancing of Technical Drawings

ASME Y14.5M-1994 (R2004) Dimensioning and Tolerancing

INTERNET RESOURCES

Drafting Zone For information on geometric dimensioning and tolerancing, see: http://www.draftingzone.com

Effective Training Inc. For information on dimensioning and tolerancing, see: http://etinews.com/eti_solutions.html

eFunda For information on geometric dimensioning and tolerancing, see: http://www.efunda.com/home.cfm

Engineers Edge For information on geometric tolerancing and dimensioning, see: http://www.engineersedge.com/gdt.htm

ASSIGNMENT

A-110 Form Tolerances

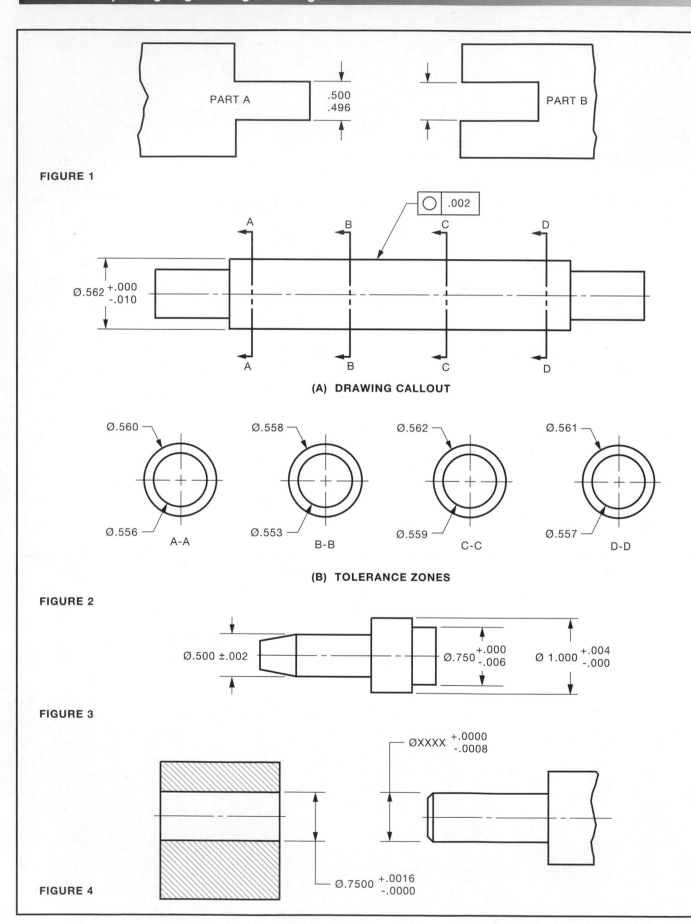

FIGURE 1

$\boxed{\bigcirc \;|\; .002}$

(A) DRAWING CALLOUT

Ø.562 $^{+.000}_{-.010}$

Ø.560 / Ø.556 A-A

Ø.558 / Ø.553 B-B

Ø.562 / Ø.559 C-C

Ø.561 / Ø.557 D-D

(B) TOLERANCE ZONES

FIGURE 2

Ø.500 ±.002   Ø.750 $^{+.000}_{-.006}$ Ø 1.000 $^{+.004}_{-.000}$

FIGURE 3

ØXXXX $^{+.0000}_{-.0008}$

Ø.7500 $^{+.0016}_{-.0000}$

FIGURE 4

ASSIGNMENTS:

NOTE: ALL ASSIGNMENTS TO BE SKETCHED ON
ONE-INCH GRID SHEETS (.10 IN. SQUARES).

1. SKETCH FIGURE 1. PART A MUST FIT INTO
PART B SO THAT THERE WILL NOT BE ANY
INTERFERENCE AND THE MAXIMUM CLEARANCE
WILL NEVER EXCEED .006 IN. ADD THE MAXIMUM
LIMITS OF SIZE TO PART B. FLATNESS TOLERANCES
OF .001 IN. ARE TO BE ADDED TO THE TWO SURFACES
OF EACH PART.

2. MEASUREMENTS FOR CIRCULARITY FOR THE PARTS
SHOWN IN FIGURE 2 WERE MADE AT THE CROSS
SECTIONS A-A TO D-D. ALL POINTS ON THE PERIPHERY
FELL WITHIN THE TWO RINGS. THE OUTER RING WAS
THE SMALLEST THAT COULD BE CIRCUMSCRIBED ABOUT
THE PROFILE AND THE INNER RING THE LARGEST THAT
COULD BE INSCRIBED WITHIN THE PROFILE. STATE
WHICH SECTIONS MEET DRAWING REQUIREMENTS.

3. SKETCH THE PART SHOWN IN FIGURE 3. ADD THE
LARGEST PERMISSIBLE CIRCULARITY TOLERANCE TO
EACH OF THE THREE DIAMETERS.

4. SKETCH THE PARTS SHOWN IN FIGURE 4.
   (A) THE PARTS MUST ASSEMBLE WITH A MINIMUM
   RADIAL CLEARANCE OF .0010 IN. (PER SIDE).
   DIMENSION THE SHAFT ACCORDINGLY.

   (B) WOULD ADDING A CYLINDRICITY TOLERANCE
   ALTER THE SIZE OF THE SHAFT?

   (C) WHAT IS THE LARGEST CYLINDRICITY TOLERANCE
   THAT COULD BE REALIZED FOR THE HOLE AND
   SHAFT IF THE FOLLOWING MEASUREMENTS WERE
   RECORDED: .7510 IN. IN HOLE, .7476 IN. IN SHAFT?

5. SKETCH THE PART SHOWN IN FIGURE 5. APPLY
CYLINDRICITY TOLERANCES TO THE THREE FEATURES.
THE CYLINDRICITY TOLERANCES ARE TO BE 25% OF THE
SIZE OF TOLERANCES SHOWN ON EACH SHAFT.

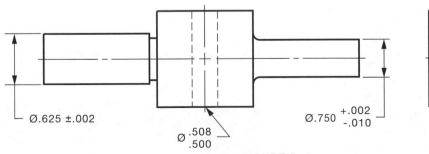

Ø.625 ±.002

Ø .508 / .500

Ø.750 +.002 / -.010

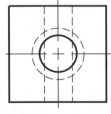

FIGURE 5

 FORM TOLERANCES | **A-110**

47 UNIT

DATUMS AND THE THREE-PLANE CONCEPT

A datum is a point, axis, or plane from which dimensions are measured or to which geometric tolerances are referenced. A datum has an exact form and represents an exact or fixed location for manufacture or measurement.

A datum feature is a feature of a part, such as an edge, surface, or hole, that forms the basis for a datum or is used to establish the location of a datum.

DATUMS FOR GEOMETRIC TOLERANCING

Datums are exact geometric points, axes, or surfaces, each based on one or more datum features of the part. Surfaces are usually flat or cylindrical, but other shapes are used when necessary. Because the datum features are physical surfaces of the part, they are subject to manufacturing errors and variations. For example, a flat surface of a part, if greatly magnified, will show some irregularity. If brought into contact with a perfect plane, this flat surface will touch only at the highest points, Figure 47–1. The true datums exist only in theory but are considered to be in the form of locating surfaces of machines, fixtures, and gauging equipment on which the part rests or with which it makes contact during manufacture and measurement.

THREE-PLANE SYSTEM

Geometric tolerances, such as straightness and flatness, refer to unrelated lines and surfaces and do not require the use of datums.

Orientation and locational tolerances refer to related features; that is, they control the relationship of features to one another or to a datum or datum system. Such datum features must be properly identified on the drawing.

Usually only one datum is required for orientation, but positional relationships may require a datum system consisting of two or three datums. These datums are designated as *primary, secondary,* and *tertiary.* When these datums are plane surfaces that are mutually perpendicular, they are commonly referred to as a three-plane datum system, or a datum reference frame.

Primary Datum

If the primary datum feature is a flat surface, it could be laid on a suitable plane surface, such as the

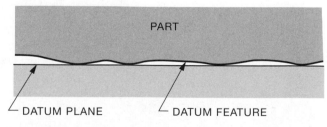

FIGURE 47–1 ■ Magnified section of a flat surface

surface of a gauge, which would then become a primary datum, Figure 47–2. Theoretically, there will be a minimum of three high spots on the flat surface coming in contact with the gauge surface.

Secondary Datum

If the part is brought into contact with a secondary plane while lying on the primary plane, it will theoretically touch at a minimum of two points, Figure 47–3.

Tertiary Datum

The part can be slid along while maintaining contact with both the primary and secondary planes until it contacts a third plane, Figure 47–4. This plane then becomes the tertiary datum and the part will, in theory, touch it at only one point.

These three planes constitute a datum system from which measurements can be taken. They will appear on the drawing as shown in Figure 47–5, except that the datum features should be identified in their correct sequence by the methods described later in the unit.

UNEVEN SURFACES

When establishing a datum plane from a datum-feature surface, it is assumed that the surface will be reasonably flat and that the part will rest on three high spots on the surface. If the surface has a tendency toward concavity, Figure 47–6, no particular problems will arise.

However, if the surface is somewhat convex, it will have a tendency to rock on one or two high spots. In such cases, the datum plane should lie in the direction where the rock is equalized as far as possible. This usually results in the least possible deviation of the actual surface from the datum plane. In Figure 47–7, the datum plane is plane B and not plane A, because this results in deviation Z, which is less than deviation X.

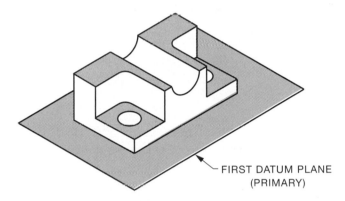

FIRST DATUM PLANE (PRIMARY)

PRIMARY DATUM FEATURE MUST TOUCH PRIMARY DATUM PLANE AT A MINIMUM OF THREE PLACES

FIGURE 47–2 ■ Primary datum

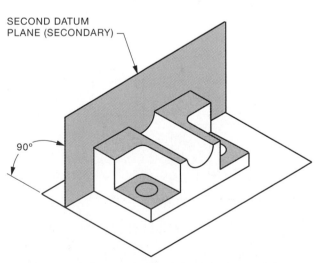

SECOND DATUM PLANE (SECONDARY)

90°

SECONDARY DATUM FEATURE MUST TOUCH SECONDARY DATUM PLANE AT A MINIMUM OF TWO PLACES WHILE RESTING ON DATUM PLANE A

FIGURE 47–3 ■ Secondary datum

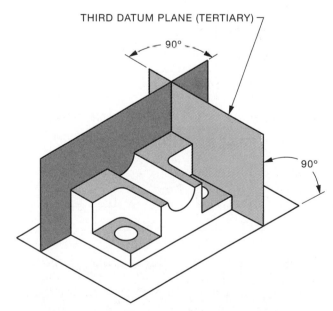

THIRD DATUM PLANE (TERTIARY)

90°

90°

TERTIARY DATUM FEATURE MUST TOUCH TERTIARY DATUM PLANE AT ONE PLACE

FIGURE 47–4 ■ Tertiary datum

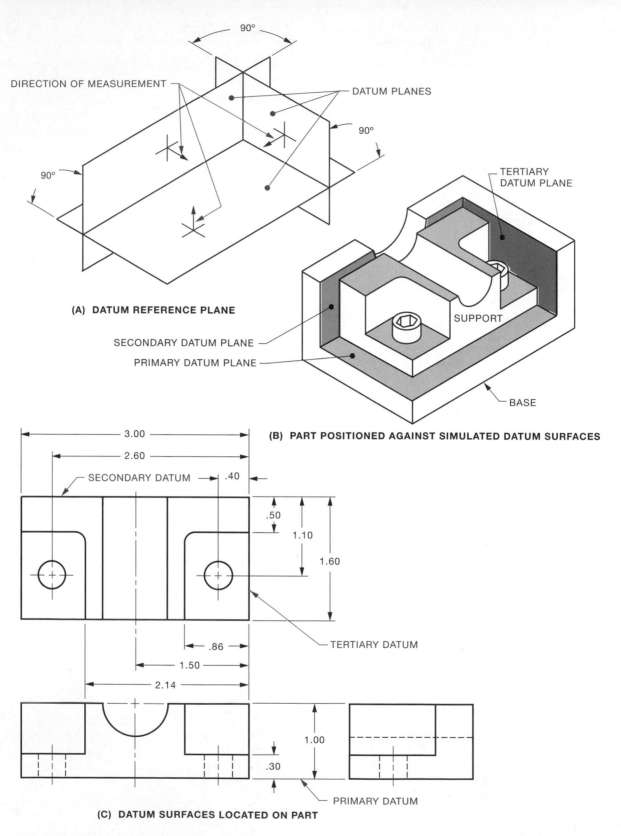

(A) **DATUM REFERENCE PLANE**

(B) **PART POSITIONED AGAINST SIMULATED DATUM SURFACES**

(C) **DATUM SURFACES LOCATED ON PART**

FIGURE *47–5* ■ Three-plane datum system

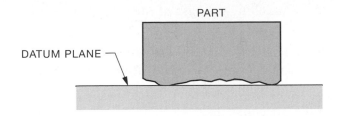

FIGURE 47–6 ■ Concave surface as a datum feature

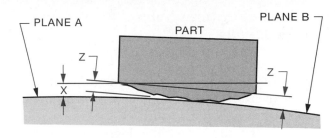

FIGURE 47–7 ■ Datum plane for convex feature

If such conditions are likely, the surface should be controlled by a flatness tolerance (and may have to be machined) or the datum-target method, explained in Unit 50, should be used.

DATUM FEATURE SYMBOL

Datum symbols have two functions. They indicate the datum surface or feature on the drawing and identify the datum feature so it can be easily referred to in other requirements.

The datum feature symbol is shown in Figure 47–8. The datum is identified by a capital letter placed horizontally in a square frame attached by a leader to a triangular base, which terminates at the datum feature. The only difference between the ASME and ISO datum feature symbols is the shape of the triangular base.

This identifying symbol may be directed to the datum feature in any of the following ways.

For Datum Features Not Subject to Size Variation

■ By attaching the base of the triangle to an extension line from the feature, providing it

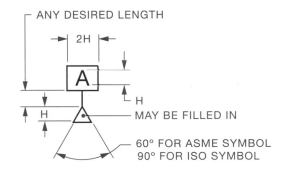

FIGURE 47–8 ■ Datum feature symbol

is a plane surface, but clearly separated from the dimension line, or to the surface itself, Figures 47–9(A) and (B).

■ When only a part of a surface is to be designated as a datum, by drawing a chain line parallel to the surface to indicate the portion of the surface acting as the datum, Figure 47–9 (C).

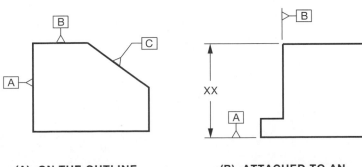

(A) ON THE OUTLINE OF A PART

(B) ATTACHED TO AN EXTENSION LINE

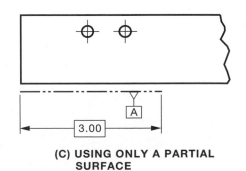

(C) USING ONLY A PARTIAL SURFACE

FIGURE 47–9 ■ Placement of datum feature symbols for features not subject to size variations

For Datum Features Subject to Size Variation

■ By attaching the base of the triangle to an extension of the dimension line pertaining to the feature of size when the datum is the axis or centre plane. The datum feature symbol may replace part of the dimension line, Figure 47–10(A).

■ By attaching the base of the triangle to a cylindrical feature, Figure 47–10(B).

■ By attaching the base of the triangle to the leader of a dimension where no feature control frame is used, Figure 47–10(C).

■ By attaching the base of the triangle above or below the feature control frame, Figure 47–10(D).

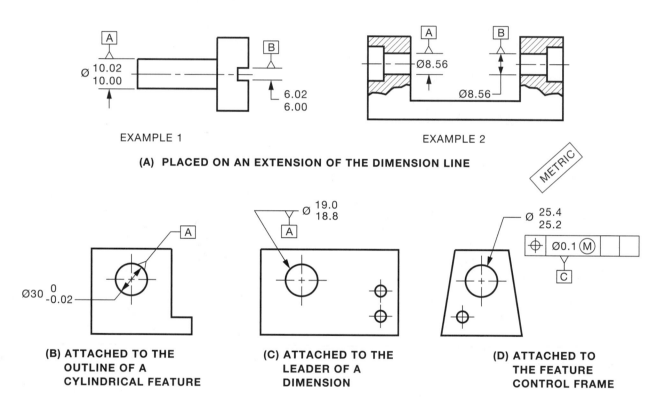

EXAMPLE 1

EXAMPLE 2

(A) PLACED ON AN EXTENSION OF THE DIMENSION LINE

(B) ATTACHED TO THE OUTLINE OF A CYLINDRICAL FEATURE

(C) ATTACHED TO THE LEADER OF A DIMENSION

(D) ATTACHED TO THE FEATURE CONTROL FRAME

FIGURE 47–10 ■ Placement of datum feature symbols for features subject to size variations

Former ANSI Datum Feature Symbol

Prior to 1994, the United States (ANSI) used the symbol shown in Figures 47–11 and 47–12 to identify the datum feature. Many drawings in existence show this symbol.

Association with Geometric Tolerances

The datum letter is placed in the feature control frame by adding an extra compartment for the datum reference, Figure 47–13.

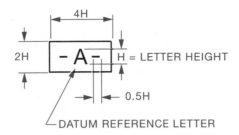

FIGURE 47–11 ■ Former ANSI datum feature symbol

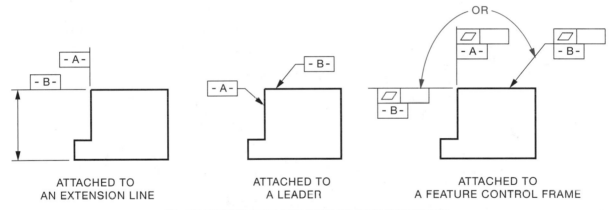

(A) FEATURES NOT SUBJECT TO SIZE VARIATION

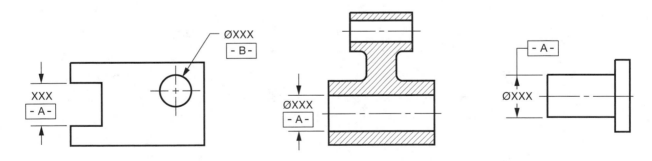

(B) FEATURES SUBJECT TO SIZE VARIATION

FIGURE 47–12 ■ Placement of former ANSI datum feature symbol

If two or more datum references are involved, additional frames are added and the datum references are placed in these frames in the correct order; that is, primary, secondary, and tertiary datums, Figure 47–14.

Multiple Datum Features

If a single datum is established by two datum features, such as two flat or cylindrical surfaces, Figure 47–15, the features are identified by separate letters. Both letters are then placed in the same compartment of the feature control frame, separated by a dash. The datum in this case is the common axis or plane between the two datum features.

Partial Surfaces as Datums

It is often desirable to specify only part of a surface, instead of the entire surface, to serve as a datum feature. This may be indicated with a thick chain line drawn parallel to the surface profile (dimension for length and location), Figure 47–16, or by a datum-target area as described in Unit 50. Figure 47–16 illustrates a long part where holes are located at only one end.

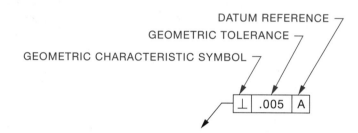

FIGURE 47–13 ■ Feature control symbol referenced to a datum

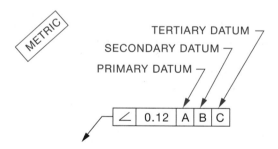

FIGURE 47–14 ■ Multiple datum references

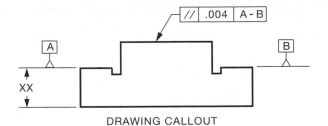

DRAWING CALLOUT

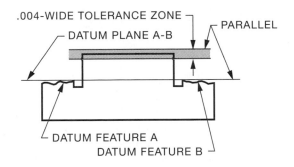

.004-WIDE TOLERANCE ZONE
PARALLEL
DATUM PLANE A-B

DATUM FEATURE A
DATUM FEATURE B

INTERPRETATION

(A) COPLANAR DATUM FEATURES

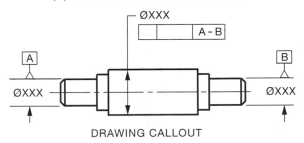

DRAWING CALLOUT

DATUM FEATURE B
DATUM FEATURE A
AXIS OF FEATURE BEING
CONTROLLED MUST LIE
WITHIN TOLERANCE
ZONE

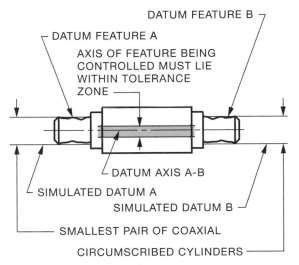

DATUM AXIS A-B
SIMULATED DATUM A
SIMULATED DATUM B
SMALLEST PAIR OF COAXIAL
CIRCUMSCRIBED CYLINDERS

INTERPRETATION

(B) COAXIAL DATUM FEATURES

FIGURE 47–15 ■ Two datum features for one datum

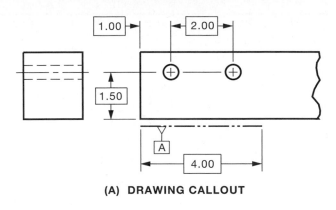

(A) DRAWING CALLOUT

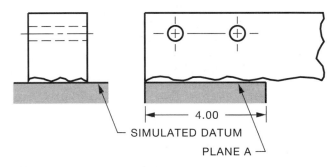

SIMULATED DATUM
PLANE A

(B) INTERPRETATION

FIGURE 47–16 ■ Partial datum

REFERENCES

CAN/CSA-B78.2-M91 Dimensioning and Tolerancing of Technical Drawings
ASME Y14.5M-1994 (R2004) Dimensioning and Tolerancing

INTERNET RESOURCES

Drafting Zone For information on geometric dimensioning and tolerancing, see: http://www.draftingzone.com

Effective Training Inc. For information on dimensioning and tolerancing, see: http://etinews.com/eti_solutions.html

eFunda For information on geometric dimensioning and tolerancing, see: http://www.efunda.com/home.cfm

Engineers Edge For information on geometric tolerancing and dimensioning, see: http://www.engineersedge.com/gdt.htm

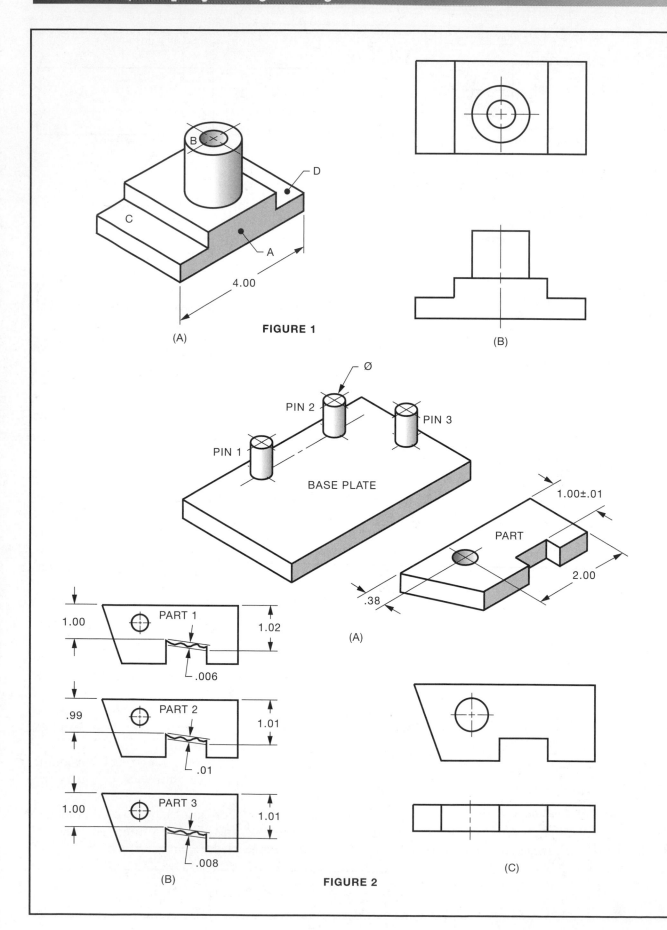

FIGURE 1

(A)

(B)

Ø

PIN 2

PIN 1

PIN 3

BASE PLATE

1.00±.01

PART

2.00

.38

(A)

1.00

PART 1

1.02

.006

.99

PART 2

1.01

.01

1.00

PART 3

1.01

.008

(B)

(C)

FIGURE 2

ASSIGNMENTS:

Note: Use one-inch grid sheets (.10 in. squares) for the sketching assignments.

1. Sketch the two views shown in Figure 1(B) and add the following information to the sketch:

   (A) Surface A is datum A and is to be straight within .008 in. for the 4.00 in. length, but the straightness error should not exceed .002 in. for any 1.00 in. length.
   (B) Surface B is datum B and is to be flat within .004 in.
   (C) The base is to be flat within .005 in.
   (D) Surfaces C and D are datum features C and D respectively which form a single datum.
   (E) The surface of the cylinder is to be straight within .003 in.

2. (A) Pins 1, 2, and 3 are used to establish the secondary and tertiary datums for the part shown in Figure 2. What is used for the primary datum?
   (B) Sketch the two views shown in Figure 2(C) and identify the primary, secondary, and tertiary datum planes as A, B, and C respectively.
   (C) How far is the centre of the hole from (1) tertiary datum? (2) secondary datum?
   (D) The back of the slot is to be flat within .008 in., and the secondary datum is to be flat within .004 in. Place these form tolerances on the sketch.
   (E) Are the parts shown in Figure 2(B) acceptable? If not, state your reasons.

3. (A) Sketch the part shown in Figure 3 and add the following to the sketch. The bottom surface is to be flat within .004 in. and is to be identified as datum B on the drawing.
   (B) What is the minimum height of the part?

4. What is the minimum number of contact points in a three-plane datum system for (A) primary datum? (B) secondary datum? (C) tertiary datum?

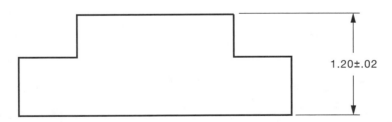

1.20±.02

FIGURE 3

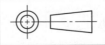 **DATUMS** **A-111**

ASSIGNMENTS:

Anyone involved with the use of technical drawings must be able to interpret drawings containing current and formerly used symbols and standards. On a centimetre grid sheet (1 mm squares), sketch the axle drawing twice, and add geometric tolerances and datums to these sketches. One drawing is to use current ASME drawing practices and symbols; the other to use former ANSI drawing practices and symbols.

Show the following information on both drawings:
1. Diameter M to be datum A
2. The end face of diameter N to be used as datum B
3. The width of the slot to be datum C
4. The end face to be flat within 0.25 mm
5. The axis of diameter M must be straight within 0.1 mm regardless of feature size
6. The surface of diameter L to be straight within 0.2 mm

QUESTIONS:

1. What are the names given to the planes of a three-plane datum system?

2. What is the name of the symbol that identifies a datum on a drawing?

3. What is the minimum number of contact points between the secondary datum feature and the datum plane?

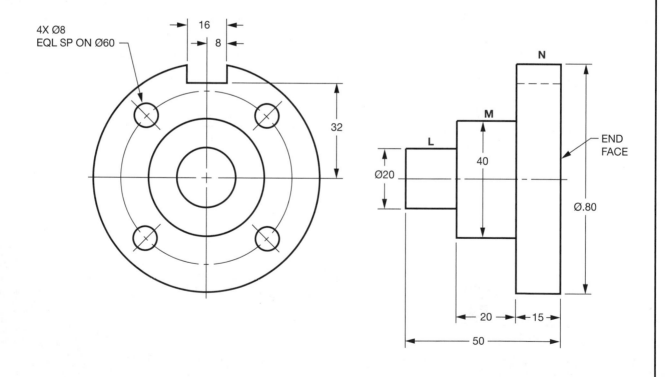

4X Ø8
EQL SP ON Ø60

16
8
32

N
M
L
Ø20
40
END FACE
Ø.80
20
15
50

METRIC
DIMENSIONS ARE IN MILLIMETRES

 AXLE A-112M

ORIENTATION TOLERANCES

Orientation is the angular relationship between two or more lines, surfaces, or other features. Orientation tolerances control angularity, parallelism, and perpendicularity. Because the limits of size generally control form and parallelism and tolerances of location control orientation (see Unit 51), the extent of this control should be considered before specifying form or orientation tolerances.

A tolerance of form or orientation may be specified where the tolerances of size and location do not provide sufficient control.

When applied to plane surfaces, orientation tolerances control flatness if a flatness tolerance is not specified.

The general geometric characteristic for orientation is termed *angularity*, which may describe angular relationships, of any angle, between straight lines or surfaces with straight line elements, such as flat or cylindrical surfaces. Special terms are used for two particular types of angularity: *perpendicularity,* or squareness, for features related to each other by a 90° angle; and *parallelism* for features related to one another by a 0° angle.

An orientation tolerance specifies a zone within which the considered feature, its line elements, its axis, or its centre plane must be contained.

Reference to a Datum

An orientation tolerance indicates a relationship between two or more features. Whenever possible, the feature to which the controlled feature is related should be designated as a datum. Sometimes this does not seem possible, for example, where two surfaces are equal and cannot be distinguished from one another. The geometric tolerance could theoretically be applied to both surfaces without a datum, but it is generally preferable to specify two similar requirements, using each surface in turn as the datum.

Angularity, parallelism, and perpendicularity are orientation tolerances applicable to related features. Relation to more than one datum feature should be considered if required to stabilize the tolerance zone in more than one direction.

There are three geometric symbols for orientation tolerances, Figure 48–1. The proportions are based on the height of the lettering used on the drawing.

Angularity Tolerance

Angularity is the condition of a surface or axis at a specified angle (other than 0° and 90°) from a datum plane or axis. An angularity tolerance for a flat surface specifies a tolerance zone, the width of which is defined by two parallel planes at a specified basic angle from a datum plane or axis. The surface of the considered feature must lie within this tolerance zone, Figure 48–2.

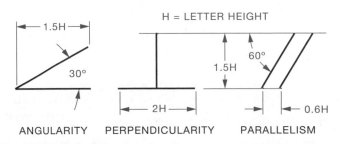

FIGURE 48–1 ■ Orientation symbols

For geometric tolerancing of angularity, the angle between the datum and the controlled feature should be stated as a basic angle. Therefore, it should be enclosed in a rectangular frame (basic dimension symbol), Figure 48–2, to indicate that the general tolerance note does not apply. The angle need not be stated for either perpendicularity (90°) or parallelism (0°).

Perpendicularity Tolerance

Perpendicularity is the condition of a surface at 90° to a datum plane or axis. A perpendicularity

tolerance for a flat surface specifies a tolerance zone defined by two parallel planes perpendicular to a datum plane or axis. The surface of the considered feature must lie within this, Figure 48–2.

Parallelism Tolerance

Parallelism is the condition of a surface equidistant at all points from a datum plane. A parallelism tolerance for a flat surface specifies a tolerance zone defined by two planes or lines parallel to a datum plane or axis. The line elements of the surface must be within this tolerance zone, Figure 48–2.

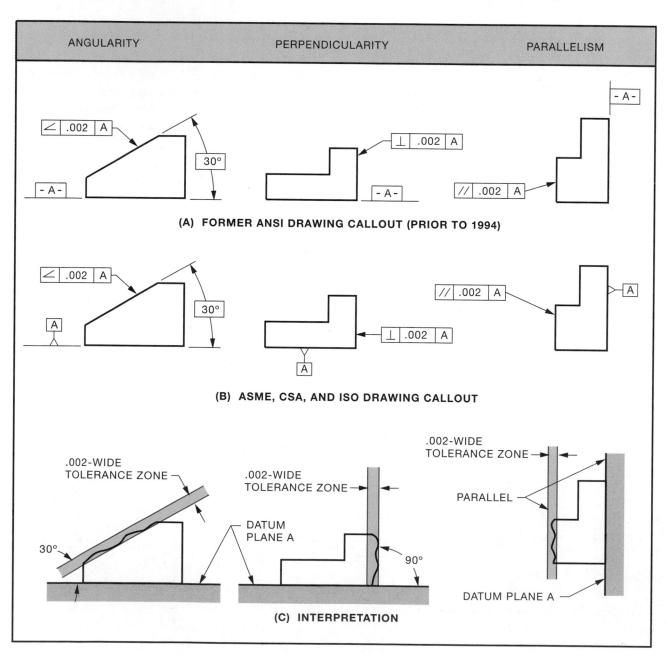

FIGURE 48–2 ■ Orientation tolerancing of flat surfaces

ORIENTATION TOLERANCING FOR FLAT SURFACES

Figure 48–2 shows three simple parts in which one flat surface is designated as a datum feature, and another flat surface is related to it by one of the orientation tolerances.

Each of these tolerances is interpreted to mean that the designated surface shall be contained within a tolerance zone consisting of the space between two parallel planes, separated by the specified tolerance (.002 in.) and related to the datum by the basic angle specified (30°, 90°, or 0°).

When orientation tolerances apply to a line or surface, a leader is attached to the feature control frame and is directed to the line or surface requiring control.

An orientation tolerance applied to a feature automatically ensures that the form of the feature is within the same tolerance.

Therefore, when an orientation tolerance is specified, there is no need to also specify a form tolerance for the same feature unless a smaller tolerance is necessary.

Control in Two Directions

The measuring principles for angularity indicate the method of aligning the part prior to making angularity measurements. Proper alignment ensures that line elements of the surface perpendicular to the angular line elements are parallel to the datum.

The part in Figure 48–3 will be aligned so that line elements running horizontally in the right-hand view will be parallel to datum A. However, these line elements will bear a proper relationship with the sides, ends, and top faces only if these surfaces are true and square with datum B.

Applying Form and Orientation Tolerances to a Single Feature

When both form and orientation tolerances are applied to a feature, the form tolerance must be less than the orientation tolerance. In Figure 48–4, the flatness of the surface must be controlled to a greater degree than its orientation. The flatness tolerance must lie within the angularity tolerance zone.

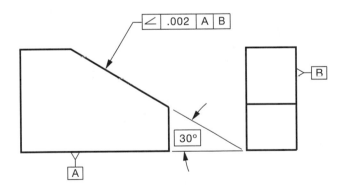

FIGURE 48–3 ■ Angularity referenced to a datum system

REFERENCES

CAN/CSA-B78.2-M91 Dimensioning and Tolerancing of Technical Drawings
ASME Y14.5M-1994 (R2004) Dimensioning and Tolerancing

INTERNET RESOURCES

Drafting Zone For information on geometric dimensioning and tolerancing, see: http://www.draftingzone.com

Effective Training Inc. For information on dimensioning and tolerancing, see: http://etinews.com/eti_solutions.html
eFunda For information on geometric dimensioning and tolerancing, see: http://www.efunda.com/home.cfm
Engineers Edge For information on geometric tolerancing and dimensioning, see: http://www.engineersedge.com/gdt.htm

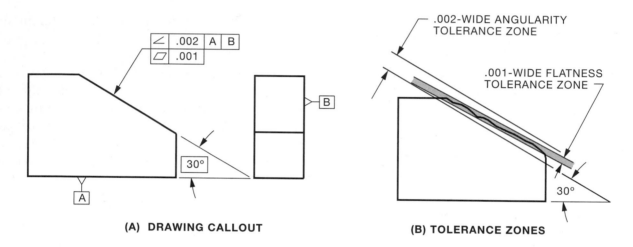

FIGURE 48–4 ■ Applying both an angularity and a flatness tolerance to a feature

ASSIGNMENT:

On a one-inch grid sheet (.10 in. squares), sketch three views of the stand shown below and add the following geometric tolerances to the drawing.

1. Surfaces A, B, and D are to be datums A, B, and D, respectively.

2. The back is to be perpendicular to the bottom within .01 in. and be flat within .006 in.

3. The top is to be parallel with the bottom within .005 in.

4. Surface C is to have an angularity tolerance of .008 in. with the bottom. Surface D is to be the secondary datum for this requirement.

5. The bottom is to be flat within .002 in.

6. The sides of the slot are to be parallel to each other within .002 in. and perpendicular to the back (datum B) within .004 in. One side of the slot is to be datum E.

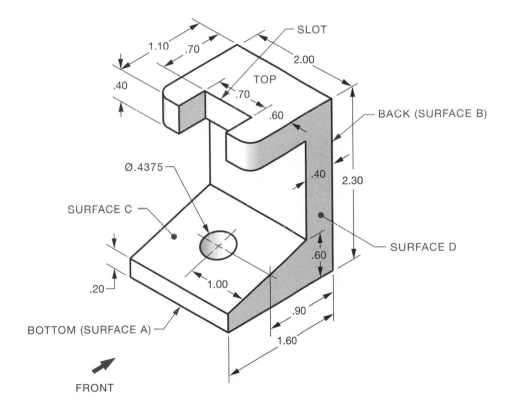

 **STAND** **A-113**

ASSIGNMENT:

On a centimetre grid sheet (1 mm squares), sketch the top, front, and left-side views of the cut-off stand to the scale of 1:2. From the information shown on the drawing below and the following, add the geometric tolerances and basic dimensions to the sketch.

1. Surfaces A, B, C, D, and E are to be datums A, B, C, D, and E, respectively.

2. Surface C is to have a flatness tolerance of 0.2 mm.

3. Surfaces F and G of the dovetail are to have an angularity tolerance of 0.05 mm with a single datum established by the two datum features D and E. These surfaces are to be flat within 0.02 mm.

4. Surface H is to be parallel to surface B within 0.05 mm.

5. Surface C is to be perpendicular to surfaces D and E within 0.04 mm.

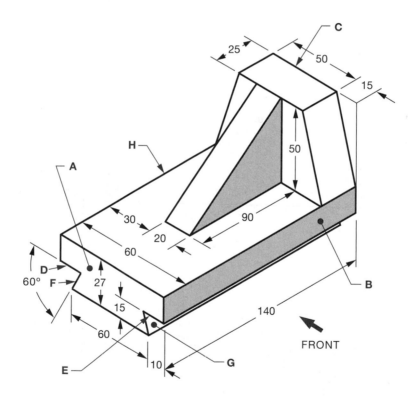

FRONT

METRIC
DIMENSIONS ARE IN MILLIMETRES

 CUT-OFF STAND | **A-114M**

NFL

ORIENTATION TOLERANCING FOR FEATURES OF SIZE

When orientation tolerances apply to the axis of cylindrical features or to the datum planes of two flat surfaces, the feature control frame is associated with the size dimension of the feature requiring control, Figure 49–1.

Tolerances intended to control orientation of the axis of a feature are applied to drawings, Figure 49–2. Although this unit deals mostly with cylindrical features, methods similar to those given here can be applied to noncircular features, such as square and hexagonal shapes.

The axis of the cylindrical feature must be contained within a tolerance zone consisting of the space between two parallel planes separated by the specified tolerance. The parallel planes are related to the datum by the basic angles of 45°, 90°, or 0° in Figure 49–2.

The absence of a modifying symbol in the tolerance compartment of the feature control frame indicates that RFS applies.

Angularity Tolerance

The tolerance zone is defined by two parallel planes at the specified basic angle from a datum plane or axis, within which the axis of the considered feature must lie. Figure 49–3 illustrates the tolerance zone for angularity.

Perpendicularity Tolerance

A perpendicularity tolerance specifies one of the following:

1. A cylindrical tolerance zone perpendicular to a datum plane or axis within which the centre line of the considered feature must lie, Figure 49–2.

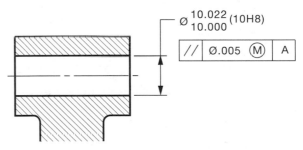

(A) ATTACHED TO A DIMENSION

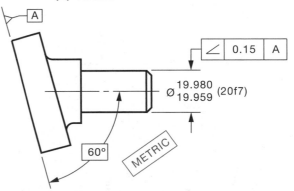

(B) ATTACHED TO THE EXTENSION OF THE DIMENSION LINE

FIGURE 49–1 ■ Feature control frame associated with size dimension

2. A tolerance zone defined by two parallel planes perpendicular to a datum axis within which the axis of the considered feature must lie, Figure 49–3.

When the tolerance is one of perpendicularity, the tolerance zone planes can be revolved around the feature axis without affecting the angle. The tolerance zone therefore becomes a cylinder. This cylindrical zone is perpendicular to the datum and has a diameter equal to the specified tolerance, Figure 49–4. A diameter symbol precedes the perpendicularity tolerance.

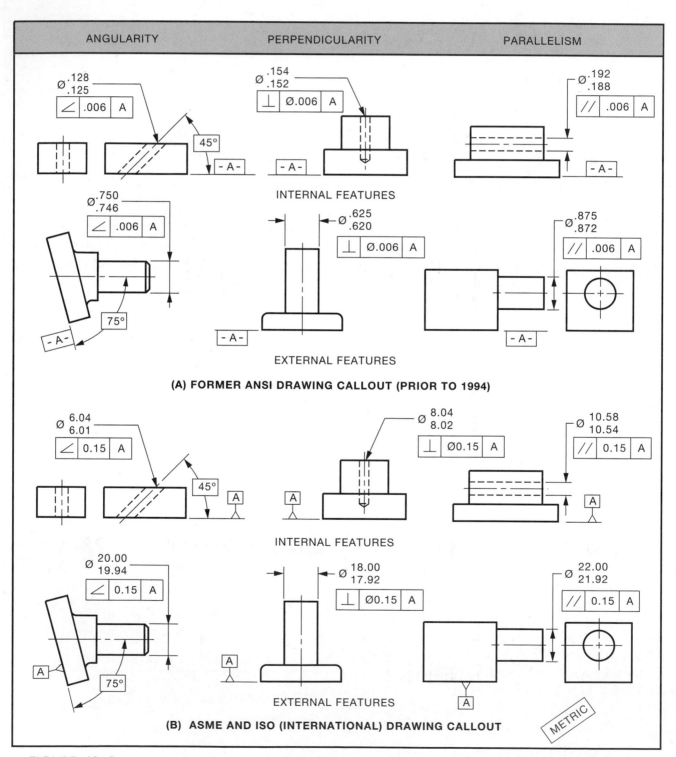

FIGURE 49–2 ■ Orientation tolerances for cylindrical features—RFS

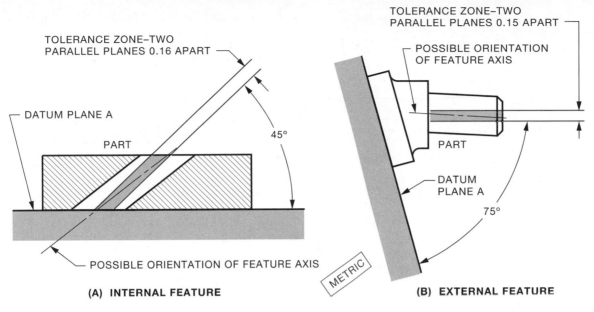

FIGURE 49–3 ■ Tolerance zone for angularity shown in Figure 49–2

Parallelism Tolerance

Parallelism is the condition of a surface equidistant at all points from a datum plane or an axis equidistant along its length from a datum axis or plane. A parallelism tolerance specifies a tolerance zone defined by two planes or lines parallel to a datum plane or axis, within which the axis of the considered feature must lie (Figure 49–5); or a cylindrical tolerance zone, the axis of which is parallel to the datum axis within which the axis of the considered feature must lie (Figure 49–10).

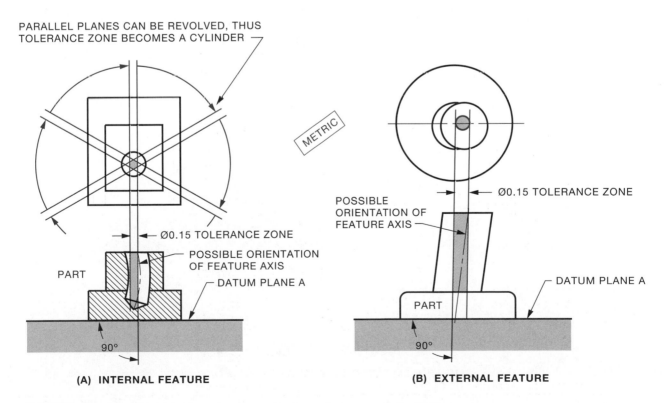

FIGURE 49–4 ■ Tolerance zone for perpendicularity shown in Figure 49–2

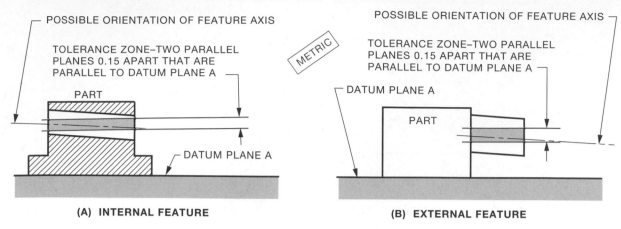

FIGURE 49–5 ■ Tolerance zones for parallelism shown in Figure 49–2

Control in Two Directions

The feature control frame for angularity shown in Figure 49–2 controls angularity with the base (datum A) only. If control with a side is also required, the side should be designated as the secondary datum, Figure 49–6. The centre line of the hole must lie within the two parallel planes.

Control on an MMC Basis

Example 1

As a hole is a feature of size, any of the tolerances shown in Figure 49–2 can be modified on an MMC basis. This is specified by adding the symbol Ⓜ after the tolerance; Figure 49–7 shows an example.

Examples 2 and 3

Because the cylindrical features represent features of size, orientation tolerances may be applied on an MMC basis. This is indicated by adding the modifying symbol after the tolerance, Figures 49–8 and 49–9.

INTERNAL CYLINDRICAL FEATURES

Figure 49–2 shows some simple parts in which the axis or centre line of a hole is related by an orientation tolerance to a flat surface. The flat surface is designated as the datum feature.

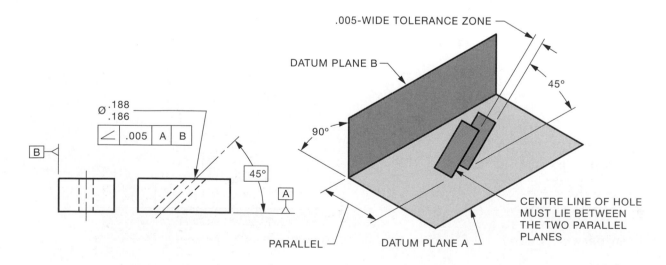

FIGURE 49–6 ■ Angularity tolerances referenced to two datums

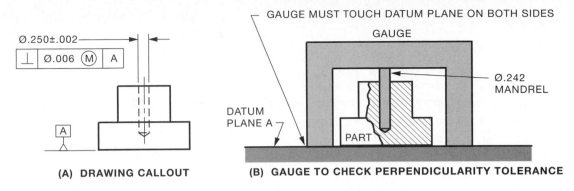

FIGURE 49–7 ■ Perpendicularity tolerance for a hole on an MMC basis

The axis of each hole must be contained within a tolerance zone consisting of the space between two parallel planes. These planes are separated by a specified tolerance of .006 in. for the parts shown in Figure 49–2(A), and by a specified tolerance of 0.15 mm for the parts shown in Figure 49–2(B).

Specifying Parallelism for an Axis

Figure 49–10 specifies parallelism for an axis when both the feature and the datum feature are shown on an RFS basis. Regardless of feature size, the feature axis must lie within a cylindrical tolerance zone of .002 in. diameter whose axis is parallel to datum axis A. The feature axis must be within any specified tolerance of location.

Figure 49–11 specifies parallelism for an axis when the feature is shown on an MMC basis and the datum feature is shown on an RFS basis. Where the feature is at the maximum material condition (.392 in.), the maximum parallelism tolerance is .002 in. diameter. Where the feature departs from

its MMC size, an increase in the parallelism tolerance is allowed equal to the amount of such departure. The feature axis must be within any specified tolerance of location.

Perpendicularity for a Median Plane

Regardless of feature size, the centre plane of the feature shown in Figure 49–12 must lie between two parallel planes, .005 in. apart, that are perpendicular to datum plane A. The feature centre plane must be within any specified tolerance of location.

Perpendicularity for an Axis (Both Feature and Datum RFS)

Regardless of feature size, the feature axis shown in Figure 49–13 must lie between two parallel planes, .005 in. apart, that are perpendicular to datum axis A. The feature axis must be within any specified tolerance of location.

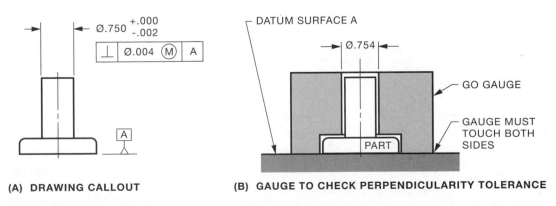

FIGURE 49–8 ■ Perpendicularity tolerance for a shaft on an MMC basis

Perpendicularity for an Axis (Tolerance at MMC)

Where the feature shown in Figure 49–14 is at the MMC (Ø2.000), its axis must be perpendicular within .002 in. to the datum plane A. Where the feature departs from MMC, an increase in the perpendicularity tolerance is allowed equal to the amount of such departure. The feature axis must be within the specified tolerance of location.

Perpendicularity for an Axis (Zero Tolerance at MMC)

Where the feature shown in Figure 49–15 is at the MMC (Ø2.000), its axis must be perpendicular to datum plane A. Where the feature departs from

MMC, an increase in the perpendicularity tolerance is allowed equal to the amount of such departure. The feature axis must be within any specified tolerance of location.

Perpendicularity with a Maximum Tolerance Specified

Where the feature shown in Figure 49–16 is at MMC (50.00 mm), its axis must be perpendicular to datum plane A. Where the feature departs from MMC, an increase in the perpendicularity tolerance is allowed equal to the amount of such departure, up to the 0.1 mm maximum. The feature axis must be within any specified tolerance of location.

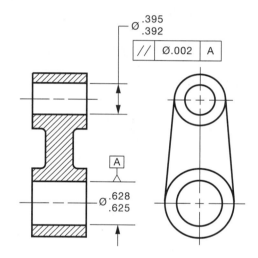

(A) DRAWING CALLOUT

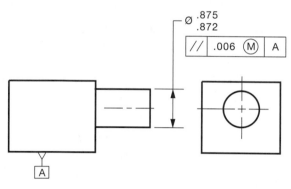

(A) DRAWING CALLOUT

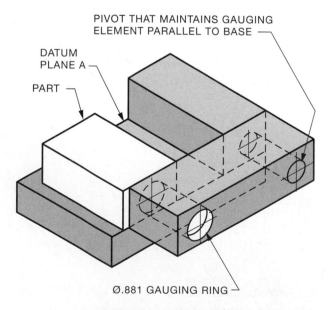

(B) GAUGE TO CHECK PARALLELISM TOLERANCE

FIGURE 49–9 ■ Parallelism tolerance for a shaft on an MMC basis

(B) TOLERANCE ZONE

FIGURE 49–10 ■ Specifying parallelism for an axis (both feature and datum feature RFS)

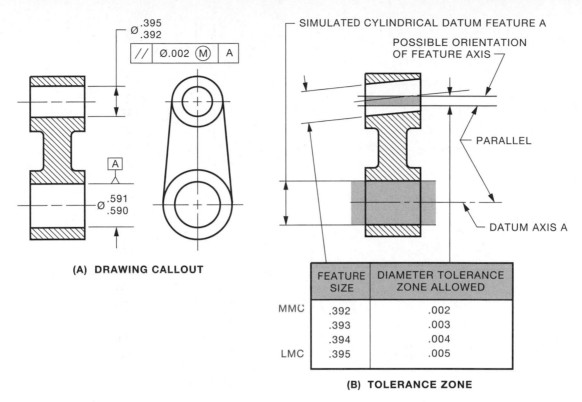

FIGURE 49-11 ■ Specifying parallelism for an axis (feature at MMC and datum feature RFS)

EXTERNAL CYLINDRICAL FEATURES

Perpendicularity for an Axis (Pin or Boss RFS)

Regardless of feature size, the feature axis shown in Figure 49–17 must lie within a cylindrical zone (.001 in. diameter) that is perpendicular to and projects from datum plane A for the feature height. The feature axis must be within any specified tolerance of location.

Perpendicularity for an Axis (Pin or Boss at MMC)

Where the feature shown in Figure 49–18 is at MMC (∅.625 in.), the maximum perpendicularity tolerance is .001 in. diameter. Where the feature

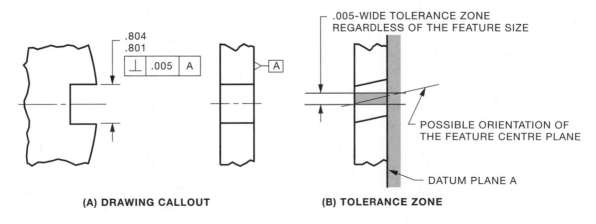

FIGURE 49–12 ■ Specifying perpendicularity for a median plane (feature RFS)

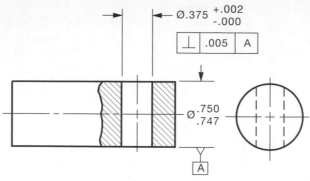

(A) DRAWING CALLOUT

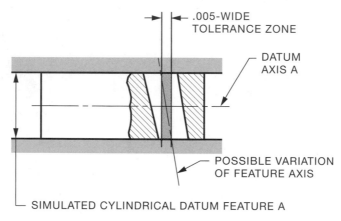

(B) TOLERANCE ZONE

FIGURE 49–13 ■ Specifying perpendicularity for an axis (both feature and datum feature RFS)

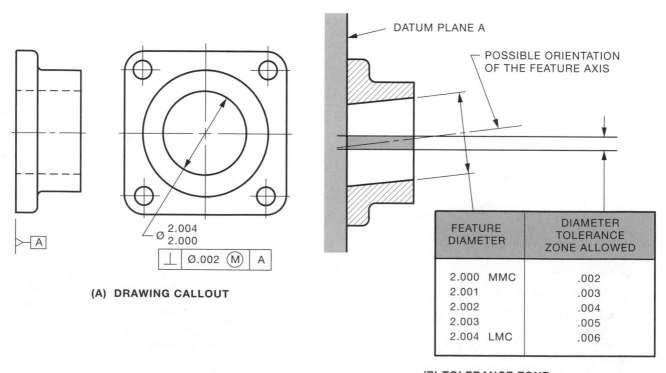

(A) DRAWING CALLOUT

FEATURE DIAMETER	DIAMETER TOLERANCE ZONE ALLOWED
2.000 MMC	.002
2.001	.003
2.002	.004
2.003	.005
2.004 LMC	.006

(B) TOLERANCE ZONE

FIGURE 49–14 ■ Specifying perpendicularity for an axis (tolerance at MMC)

departs from its MMC size, an increase in the perpendicularity tolerance is allowed equal to the amount of such departure. The feature axis must be within any specified tolerance of location.

REFERENCES

CAN/CSA B78.2-M91 Dimensioning and Tolerancing of Technical Drawings

ASME Y14.5M-1994 (R2004) Dimensioning and Tolerancing

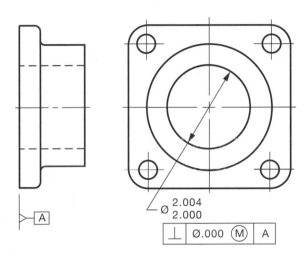

(A) DRAWING CALLOUT

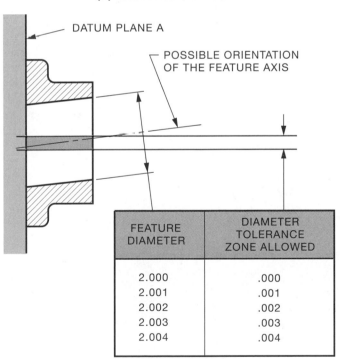

FEATURE DIAMETER	DIAMETER TOLERANCE ZONE ALLOWED
2.000	.000
2.001	.001
2.002	.002
2.003	.003
2.004	.004

(B) TOLERANCE ZONE

FIGURE 49–15 ■ Specifying perpendicularity for an axis (zero tolerance at MMC)

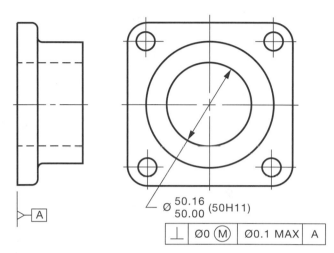

(A) DRAWING CALLOUT

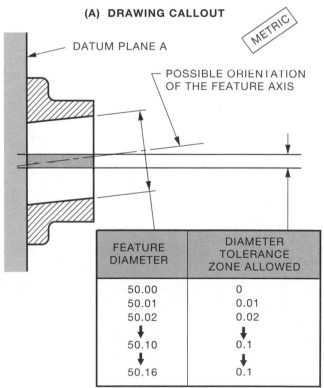

FEATURE DIAMETER	DIAMETER TOLERANCE ZONE ALLOWED
50.00	0
50.01	0.01
50.02	0.02
↓	↓
50.10	0.1
↓	↓
50.16	0.1

(B) TOLERANCE ZONE

FIGURE 49–16 ■ Specifying perpendicularity for an axis (zero tolerance at MMC with a maximum specified)

INTERNET RESOURCES

Drafting Zone For information on geometric dimensioning and tolerancing, see: http://www.draftingzone.com

Effective Training Inc. For information on dimensioning and tolerancing, see: http://etinews.com/eti_solutions.html

eFunda For information on geometric dimensioning and tolerancing, see: http://www.efunda.com/home.cfm

Engineers Edge For information on geometric tolerancing and dimensioning, see: http://www.engineersedge.com/gdt.htm

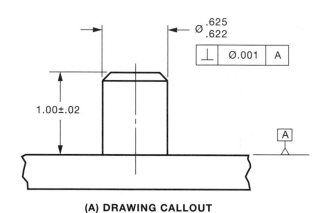

(A) DRAWING CALLOUT

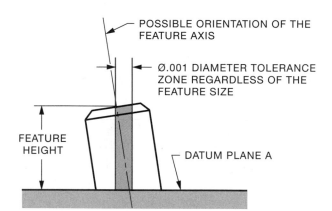

(B) TOLERANCE ZONE

FIGURE 49–17 ■ Specifying perpendicularity for an axis (pin or boss RFS)

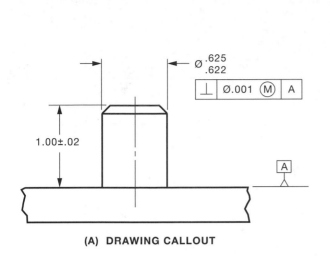

(A) DRAWING CALLOUT

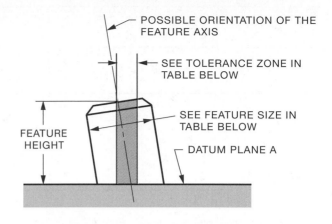

FEATURE SIZE	DIAMETER TOLERANCE ZONE ALLOWED
MMC .625	.001
.624	.002
.623	.003
LMC .622	.004

(B) TOLERANCE ZONE

FIGURE 49–18 ■ Specifying perpendicularity for an axis (pin or boss at MMC)

⊥	Ø.004	A

FEATURE SIZE Ø	DIAMETER TOLERANCE ZONE ALLOWED
2.000 2.001 ↕ 2.008 2.009	

⊥	Ø.000 Ⓜ	A

FEATURE SIZE Ø	DIAMETER TOLERANCE ZONE ALLOWED
2.000 2.001 ↕ 2.008 2.009	

⊥	Ø.000 Ⓜ	Ø.005 MAX	A

FEATURE SIZE Ø	DIAMETER TOLERANCE ZONE ALLOWED
2.000 2.001 ↕ 2.008 2.009	

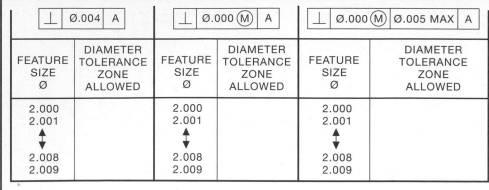

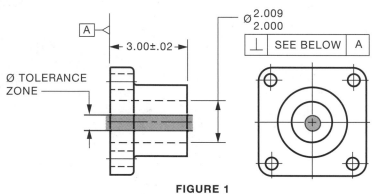

FIGURE 1

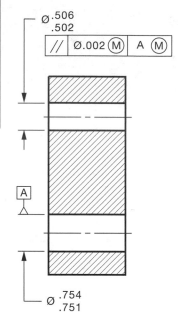

FIGURE 3

ASSIGNMENTS:

1. Sketch the tables shown in Figure 1 on a grid sheet and complete the tables showing the maximum permissible tolerance zones for the three callouts.

FIGURE 2

2. On an inch grid sheet (.10 in. squares), sketch the views shown in Figure 2, and add the following data to the sketch:

   (A) Surfaces marked A, B, and C are datums A, B, and C, respectively.

   (B) Surface A is perpendicular to surfaces B and C within .01 in.

   (C) Surface D is parallel to surface B within .004 in.

   (D) The slot is parallel to surface C within .002 in. and perpendicular to surface A within .001 in. at MMC.

   (E) The Ø1.750 hole has an RC7 fit (show the size of the hole as limits) and is perpendicular to surface A within .002 in. at MMC.

   (F) Surface E has an angularity tolerance of .010 in. with surface C.

   (G) Surface A is to be flat within .002 in. for any one-inch-square surface with a maximum flatness tolerance of .005 in.

   (H) Indicate which dimensions are basic.

3. Refer to Figure 3. What is the maximum diameter of the tolerance zone allowed when the Ø.506 in. hole is at (a) MMC? (b) LMC? (c) Ø.504 in.?

4. If the symbol M is removed from the tolerance in Figure 3, what is the maximum diameter of the tolerance zone allowed when the hole is at (a) MMC? (b) LMC? (c) Ø.504 in.?

ORIENTATION TOLERANCING FOR FEATURES OF SIZE	A-115

DATUM TARGETS

The full feature surface was used to establish a datum for the features so far designated as datum features. This may not always be practical.

1. The surface of a feature may be so large that a gauge designed to make contact with the full surface may be too expensive or too cumbersome to use.
2. Functional requirements of the part may necessitate the use of only a portion of a surface as a datum feature, for example, the portion that contacts a mating part in assembly.
3. A surface selected as a datum feature may not be sufficiently true, and a flat datum feature may rock when placed on a datum plane. Accurate and repeatable measurements from the surface would not be possible, as with surfaces of castings, forgings, weldments, and some sheet-metal and formed parts.

A useful technique to overcome such problems is the datum-target method. In this method, certain points, lines, or small areas on the surfaces are selected as the bases for establishing datums. For flat surfaces, this usually requires three target points or areas for a primary datum, two for a secondary datum, and one for a tertiary datum.

It is not necessary to use targets for all datums. It is quite logical, for example, to use targets for the primary datum and other surfaces or features for secondary and tertiary datums if required; or to use a flat surface of a part as the primary datum and to locate fixed points or lines on the edges as secondary and tertiary datums.

Datum targets should be spaced as far apart from each other as possible to provide maximum stability for making measurements. Dimensions locating target areas are basic and are shown enclosed in a rectangular frame.

DATUM-TARGET SYMBOL

Points, lines, and areas on datum features are designated on the drawing with a datum-target symbol, Figure 50–1. The symbol is placed outside the part outline with a leader directed to the target point (indicated by "X"), target line, or target area, as applicable, Figure 50–2. ISO and CSA standards use an arrow at the end of this leader. ASME does not show this arrow, Figure 50–4. The use of a solid leader line indicates that the datum target is on the near (visible) surface. The use of a dashed leader line (Figure 50–10B) indicates that the datum target is on the far (hidden) surface. The leader should not be shown in either a horizontal or vertical position. The datum feature itself is identified in the usual manner with a datum-feature symbol.

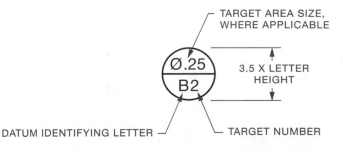

FIGURE 50–1 ■ Datum-target symbol

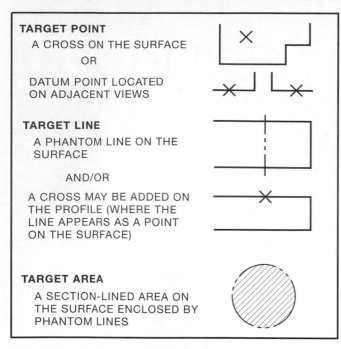

FIGURE 50–2 ■ Identification of datum targets

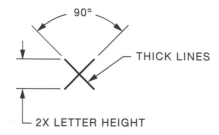

FIGURE 50–3 ■ Symbol for a datum-target point

The datum-target symbol is a circle with a diameter about 3.5 times the height of the lettering used on the drawing. The circle is divided horizontally into two halves. The lower half contains a letter identifying the associated datum, followed by the target number assigned sequentially starting with 1 for each datum. For example, in a three-plane, six-point datum system, if the datums are A, B, and C, the datums would be A1, A2, A3, B1, B2, and C1 (Figure 50–14). Where the datum target is an area, the area size may be entered in the upper half of the symbol; otherwise, the upper half is left blank.

Identification of Targets

Datum-Target Points

Each target point is shown in its desired location on the surface with a cross drawn at about 45° to the coordinate dimensions. The cross is twice the height of the lettering used, Figures 50–3 and 50–4(A). When the view that would show the location of the datum-target point is not drawn, its point location is dimensioned on the two adjacent views, Figure 50–4(B).

Target points may be represented on tools, fixtures, and gauges by spherically ended pins, Figure 50–5.

Datum-Target Lines

A datum-target line is indicated by the symbol ✕ on an edge view of a surface, a phantom line on the direct view, or both, Figure 50–6. When the length of the datum-target line must be controlled, its length and location are dimensioned.

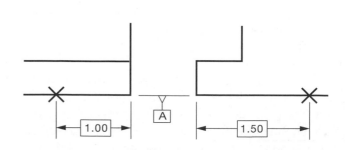

(A) DATUM POINTS SHOWN ON SURFACE

(B) DATUM POINTS LOCATED BY TWO VIEWS

FIGURE 50–4 ■ Datum target points

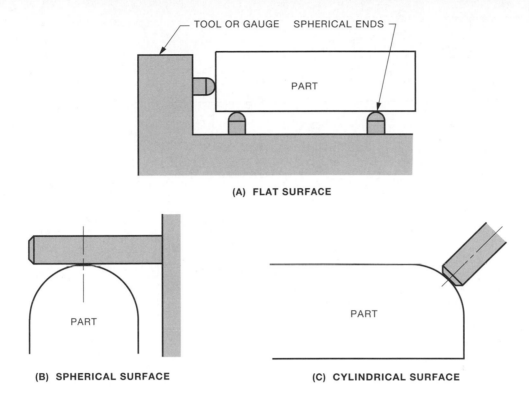

(A) FLAT SURFACE

(B) SPHERICAL SURFACE   (C) CYLINDRICAL SURFACE

FIGURE 50–5 ■ Location of part on datum-target points

Datum-target lines can be represented in tooling and gauging by the side of a round pin, Figure 50–7.

If a line is designated as a tertiary datum feature, it will touch the gauge pin theoretically at only one point. If it is a secondary datum feature, it will touch at two points.

The application and use of a surface and three lines as datum features are shown in Figures 50–8 and 50–9.

Datum-Target Areas

Where an area or areas of flat contact are necessary to ensure establishment of the datum (that is, where spherical or pointed pins would be inadequate), a target area of the desired shape is specified. The datum-target area is indicated by section lines inside a phantom outline of the desired shape, with controlling dimensions added. The diameter of circular areas is given in the upper half of the datum-target symbol, Figure 50–10(A). Where a circular target area is too small to be drawn to scale, the method shown in Figure 50–10(B) may be used.

Datum-target areas may have any desired shape, a few of which are shown in Figure 50–11. Target areas should be as small as possible but consistent with functional requirements.

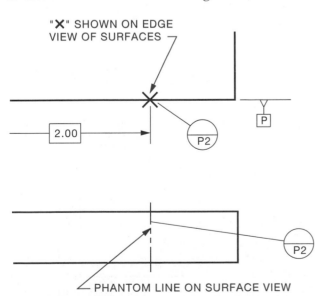

FIGURE 50–6 ■ Datum-target line

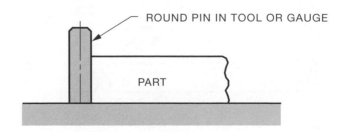

FIGURE 50–7 ■ Locating a datum line

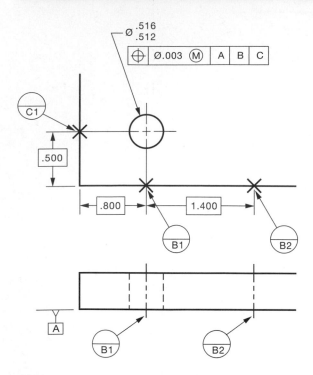

FIGURE 50–8 ■ Part with a surface and three target lines used as datum features

Targets Not in the Same Plane

In most applications, datum target points that form a single datum are all located on the same surface, Figure 50–4(A). However, they may be located on different surfaces to meet functional requirements, Figure 50–12. The datum plane may be located

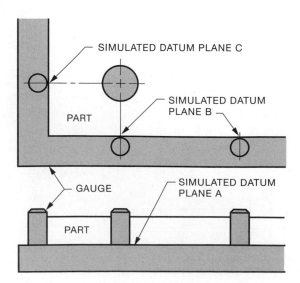

FIGURE 50–9 ■ Location of part in Figure 50–8 in a gauge

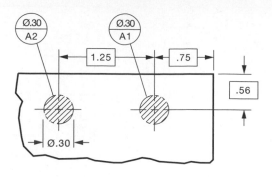

(A) TARGET AREAS ON NEAR SIDE

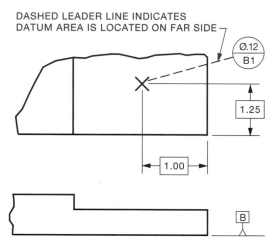

(B) TARGET AREAS ON FAR SIDE

FIGURE 50–10 ■ Datum-target areas

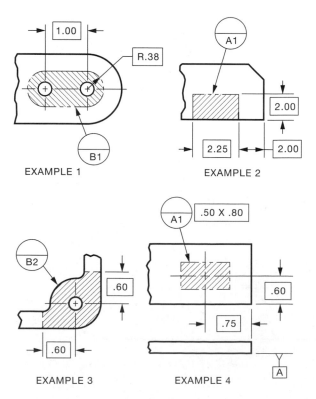

FIGURE 50–11 ■ Typical target areas

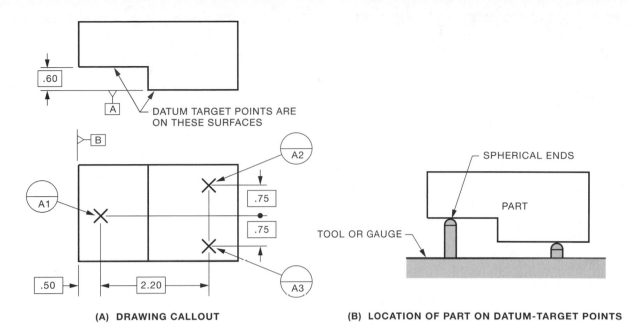

(A) DRAWING CALLOUT

(B) LOCATION OF PART ON DATUM-TARGET POINTS

FIGURE 50–12 ■ Datum-target points on different planes used as the primary datum

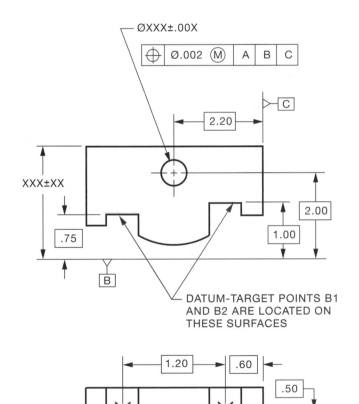

FIGURE 50–13 ■ Datum outside of part profile

in space, not actually touching the part, Figure 50–13. In such applications, the controlled features must be dimensioned from the specified datum, and the position of the datum from the datum targets must be shown with exact datum dimensions. For example, in Figure 50–13, datum B is positioned with datum dimensions .75 in., 1.00 in., and 2.00 in. The top surface is controlled from this datum with a toleranced dimension. The hole is positioned with the basic dimension 2.00 in. and a positional tolerance.

Dimensioning for Target Location

The location of datum targets is shown with basic dimensions. Each dimension is shown, without tolerances, enclosed in a rectangular frame, indicating that the general tolerance does not apply. Dimensions locating a set of datum targets should be dimensionally related or have a common origin.

Application of datum targets and datum dimensioning is shown in Figure 50–14.

REFERENCES

CAN/CSA-B78.2-M91 Dimensioning and Tolerancing of Technical Drawings
ASME Y14.5M-1994 (R2004) Dimensioning and Tolerancing

INTERNET RESOURCES

Drafting Zone For information on geometric dimensioning and tolerancing, see: http://www.draftingzone.com

Effective Training For information on dimensioning and tolerancing, see: http://etinews.com/eti_solutions.html

eFunda For information on geometric dimensioning and tolerancing, see: http://www.efunda.com/home.cfm

Engineers Edge For information on geometric tolerancing and dimensioning, see: http://www.engineersedge.com/gdt.htm

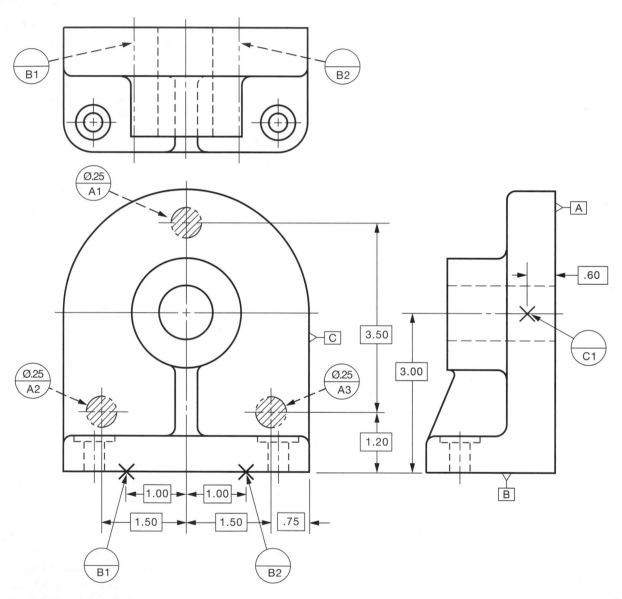

FIGURE 50–14 ■ Application of datum targets and dimensioning

ASSIGNMENT:

ON A ONE-INCH GRID SHEET (.10 IN. SQUARES) SKETCH THE BEARING HOUSING SHOWN BELOW AND ADD THE DATUM INFORMATION SHOWN IN THE TABLE AND THE DIMENSIONS RELATED TO THE DATUMS ON THIS SKETCH. SCALE 1:2.

QUESTIONS:

1. WHAT ARE THE THREE TYPES OF DATUM TARGETS?
2. WHAT TYPE OF DIMENSIONS ARE USED TO LOCATE DATUM TARGETS?
3. WHAT IS THE MINIMUM NUMBER OF CONTACT POINTS FOR A PRIMARY DATUM?
4. WHAT IS THE MINIMUM NUMBER OF CONTACT POINTS FOR A TERTIARY DATUM?
5. WHAT TYPE OF LEADER LINE INDICATES THAT THE DATUM TARGET IS ON THE FAR (HIDDEN) SURFACE?
6. WHAT INFORMATION IS CONTAINED IN THE LOWER HALF OF THE DATUM-TARGET SYMBOL?
7. HOW IS A TARGET POINT IDENTIFIED ON A DRAWING?
8. HOW IS A DATUM TARGET LINE IDENTIFIED ON THE EDGE VIEW OF A SURFACE?

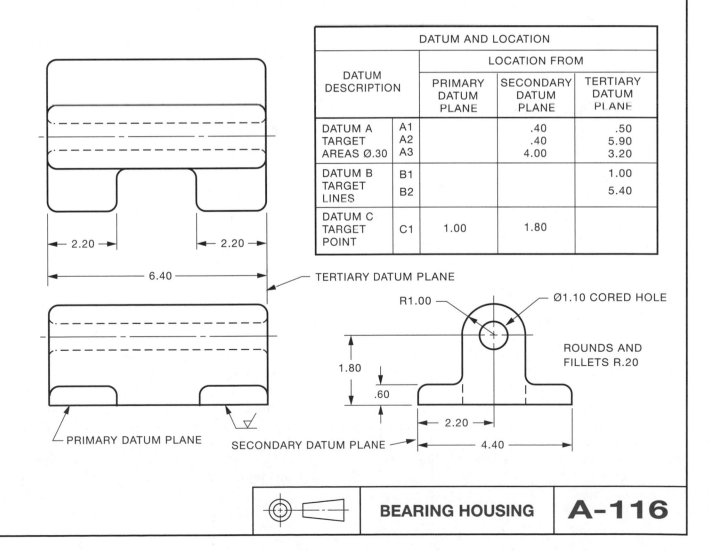

DATUM AND LOCATION				
DATUM DESCRIPTION		LOCATION FROM		
		PRIMARY DATUM PLANE	SECONDARY DATUM PLANE	TERTIARY DATUM PLANE
DATUM A TARGET AREAS Ø.30	A1		.40	.50
	A2		.40	5.90
	A3		4.00	3.20
DATUM B TARGET LINES	B1			1.00
	B2			5.40
DATUM C TARGET POINT	C1	1.00	1.80	

2.20

2.20

6.40

TERTIARY DATUM PLANE

PRIMARY DATUM PLANE

SECONDARY DATUM PLANE

R1.00

Ø1.10 CORED HOLE

ROUNDS AND FILLETS R.20

1.80

.60

2.20

4.40

BEARING HOUSING   **A-116**

51 UNIT

TOLERANCING OF FEATURES BY POSITION

The location of features is one of the most frequently used applications of dimensions on technical drawings. Tolerancing may be either by coordinate tolerances applied to the dimensions or by geometric (positional) tolerancing.

Positional tolerancing is especially useful when applied on an MMC basis to groups or patterns of holes or other small features in the mass production of parts. This method meets functional requirements in most cases and permits assessment with simple gauging procedures.

Most of the materials in this unit are devoted to the principles involved in the location of small round holes, because they represent the most common applications.

TOLERANCING METHODS

The location of a single hole is usually indicated by rectangular coordinate dimensions extending from suitable edges or other features of the part to the axis of the hole. Other dimensioning methods, such as polar coordinates, may be used.

There are two standard methods of tolerancing the location of holes: coordinate and positional tolerancing.

1. Coordinate tolerancing, Figure 51–1(A), refers to tolerances applied directly to the coordinate dimensions or to applicable tolerances specified in a general tolerance note.
2. Positional tolerancing, Figures 51–1(B) to 51–1(D), refers to a tolerance zone within which the centre line of the hole or shaft is permitted to vary from its true position. Positional tolerancing can be further classified according to the type of modifying associated with the tolerance:

- Positional tolerancing, regardless of feature size (RFS).

- Positional tolerancing, maximum material condition basis (MMC).

- Positional tolerancing, least material condition basis (LMC)—shown in ASME drawing standards and most likely to be adopted by CSA and ISO.

These positional tolerancing methods are part of the system of geometric tolerancing.

When the MMC or LMC modifying symbol is not shown in the feature control frame, it is understood that RFS applies.

Any of these tolerancing methods can be substituted for another, although with differing results. It is necessary, however, to first analyze the widely used method of coordinate tolerancing to explain and understand the advantages and disadvantages of the positional tolerancing methods.

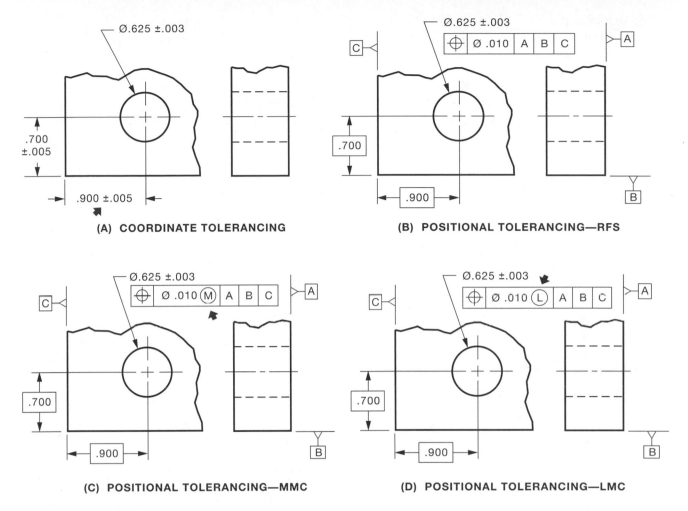

FIGURE 51–1 ■ Comparison of tolerancing methods

COORDINATE TOLERANCING

Coordinate dimensions and tolerances may be applied to the location of a single hole, Figure 51–2. They indicate the location of the hole axis and result in a rectangular or wedge-shaped tolerance zone within which the axis of the hole must lie.

If the two coordinate tolerances are equal, the tolerance zone formed will be a square; unequal tolerances result in a rectangular tolerance zone. Polar dimensioning, in which one of the locating dimensions is a radius, gives an annular segment tolerance zone. For simplicity, square tolerance zones are used in the analysis of most examples in this unit.

The tolerance zone extends for the full depth of the hole, or the whole length of the axis, Figure 51–3. In most of the illustrations, tolerances will be

analyzed as they apply at the surface of the part, where the axis is represented by a point.

Maximum Permissible Error

The actual position of the feature axis may be anywhere within the rectangular tolerance zone. For square tolerance zones, the maximum allowable variation from the desired position occurs in a direction of 45° from the direction of the coordinate dimensions, Figure 51–4.

For rectangular tolerance zones, this maximum tolerance is the square root of the sum of the squares of the individual tolerances. This is expressed mathematically as

$$\sqrt{X^2 + Y^2}$$

For the examples in Figure 51–2, the tolerance zones are shown in Figure 51–5. The maximum tolerance values are

Example 1

$$\sqrt{.010^2 + .010^2} = .014 \text{ in.}$$

Example 2

$$\sqrt{.010^2 + .020^2} = .022 \text{ in.}$$

For polar coordinates, the extreme variation is

$$\sqrt{A^2 + T^2}$$

Where: A = R tan a
            T = tolerance on radius
            R = mean radius
            a = angular tolerance

Example 3

$$\sqrt{(1.25 \times .017)^2 + .020^2} = .030 \text{ in.}$$

Note: Mathematically, the formula for Example 3 is incorrect, but the difference in results using the more complicated correct formula is insignificant for the tolerances normally used.

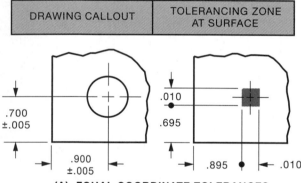

(A) EQUAL COORDINATE TOLERANCES

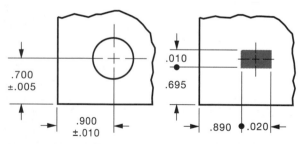

(B) UNEQUAL COORDINATE TOLERANCES

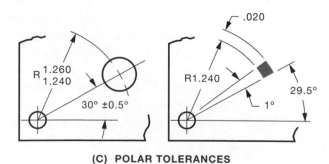

(C) POLAR TOLERANCES

FIGURE 51–2 ■ Tolerance zones for coordinate tolerances

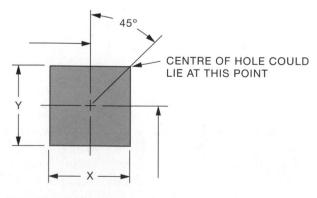

FIGURE 51–3 ■ Tolerance zone extending into a part

FIGURE 51–4 ■ Maximum permissible error for square tolerance zone

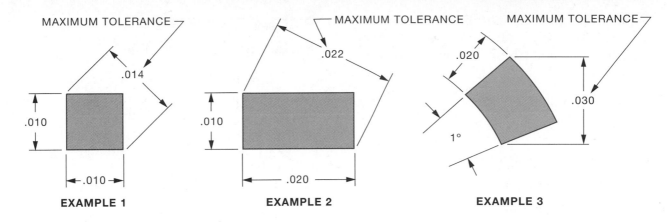

EXAMPLE 1   EXAMPLE 2   EXAMPLE 3

FIGURE 51–5 ■ Tolerance zones for examples shown in Figure 51–2

Some values of tan A for commonly used angular tolerances are as follows.

A	Tan a	A	Tan a	A	Tan a
0° 5'	.00145	0° 25'	.00727	0° 45'	.01309
0° 10'	.00291	0° 30'	.00873	0° 50'	.01454
0° 15'	.00436	0° 35'	.01018	0° 55'	.01600
0° 20'	.00582	0° 40'	.01164	1° 0'	.01745

Use of Chart

A quick and easy method for finding the maximum positional error permitted with coordinate tolerancing without having to calculate squares and square roots is to use a chart, Figure 51–6.

In the first example, Figure 51–2, the tolerance in both directions is .010 in. The extensions of the horizontal and vertical lines of .010 in the chart intersect at point A, which lies between the radii of .014 and .015 in. When rounded to three decimal places, this indicates a maximum permissible variation from true position of .014 in.

In the second example, Figure 51–2, the tolerances are .010 in. in one direction and .020 in.

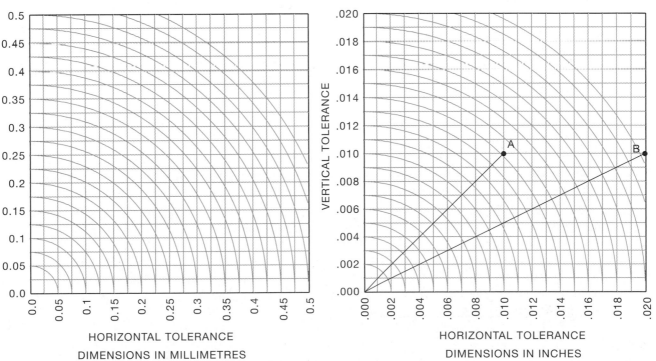

HORIZONTAL TOLERANCE

DIMENSIONS IN MILLIMETRES

HORIZONTAL TOLERANCE

DIMENSIONS IN INCHES

FIGURE 51–6 ■ Charts for calculating maximum tolerance using coordinate tolerancing

in the other. The extensions of the vertical and horizontal lines at .010 and .020 in., respectively, in the chart intersect at point B, which lies between the radii of .022 and .023 in. When rounded off to three decimal places, this indicates a maximum variation of position of .022 in. Figure 51–6 also shows a chart for use with tolerances in millimetres.

ADVANTAGES OF COORDINATE TOLERANCING

The advantages of direct coordinate tolerancing are:

1. It is simple and easily understood and, therefore, is a method commonly used.
2. It permits direct measurements to be made with standard instruments and does not require the use of special-purpose functional gauges or other calculations.

DISADVANTAGES OF COORDINATE TOLERANCING

There are a number of disadvantages to the direct tolerancing method:

1. It results in a square or rectangular tolerance zone within which the axis must lie. For a square zone, this permits a variation in a 45° direction of about 1.4 times the specified tolerance. This amount of variation may necessitate the specification of tolerances that are only 70 percent of those that are functionally acceptable.
2. It may result in an undesirable accumulation of tolerances when several features are involved, especially when chain dimensioning is used.
3. It is more difficult to assess clearances between mating features and components than when positional tolerancing is used, especially when a group or a pattern of features is involved.
4. It does not correspond to the control exercised by fixed functional GO gauges often desirable in mass production of parts.

POSITIONAL TOLERANCING

Positional tolerancing is part of the system of geometric tolerancing. It defines a zone within which the centre, axis, or centre plane of a feature of size is permitted to vary from true (theoretically exact) position. A positional tolerance is indicated by the position symbol, a tolerance, a material condition basis, and appropriate datum references placed in a feature control frame. Basic dimensions represent the exact values to which geometric positional tolerances are applied elsewhere, by symbols or notes on the drawing. They are enclosed in a rectangular frame (basic dimension symbol), Figure 51–7. Where the dimension represents a diameter, the symbol ∅ is included in the rectangular frame. General tolerance notes do not apply to basic dimensions. The frame size need not be any larger than necessary to enclose the dimension. It is necessary to identify features on the part to establish datums for dimensions locating true positions. The datum features are identified with datum-feature symbols, and the applicable datum references are included in the feature control frame.

The geometric characteristic symbol for position is a circle with two solid centre lines, Figure 51–7. This symbol is used in the feature control frame in the same manner as for other geometric tolerances.

MATERIAL CONDITION BASIS

Positional tolerancing is applied on an MMC, RFS, or LMC basis. When applied on an MMC or LMC basis, that appropriate symbol follows the specified tolerance and, where required, the applicable datum reference in the feature control frame. When no modifying symbol is shown after the tolerance, the RFS condition applies.

Because positional tolerance controls the position of the centre, axis, or centre plane of a feature of size, the feature control frame is normally attached to the size of the feature, Figure 51–8.

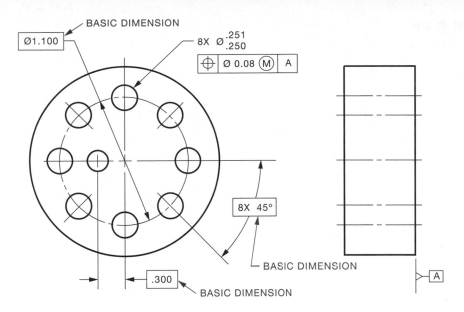

FIGURE 51–7 ■ Identifying basic dimensions

NOTE: NO MODIFYING SYMBOL SHOWN AFTER
TOLERANCE WHEN RFS APPLIES

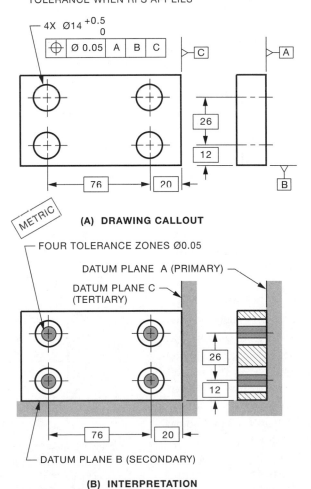

(A) DRAWING CALLOUT

(B) INTERPRETATION

FIGURE 51–8 ■ Positional tolerancing—RFS

POSITIONAL TOLERANCING FOR CIRCULAR FEATURES

The positional tolerance represents the diameter of a cylindrical tolerance zone, located at true position as determined by the basic dimensions on the drawing. The axis or centre line of the feature must lie within this cylindrical tolerance zone.

Except that the tolerance zone is circular instead of square, a positional tolerance on this basis has the same meaning as direct coordinate tolerancing but with equal tolerances in all directions.

With rectangular coordinate tolerancing, the maximum permissible error in location is not the value indicated by the horizontal and vertical tolerances, but rather is equivalent to the length of the diagonal between the two tolerances. For square tolerance zones, this is 1.4 times the specified tolerance values. The specified tolerance can therefore be increased to an amount equal to the diagonal of the coordinate tolerance zone without affecting the clearance between the hole and its mating part.

This does not affect the clearance between the hole and its mating part, yet it offers 57 percent more tolerance area, Figure 51–9. Such a change would most likely result in a reduction in the number of parts rejected for positional errors.

Positional Tolerancing—MMC

The positional tolerance and MMC of mating features are considered in relation to each other. MMC by itself means a feature of a finished product contains the maximum amount of material permitted by the toleranced size dimension of that feature. For holes, slots, and other internal features, maximum material is the condition in which these factors are at their minimum allowable sizes. For shafts, as well as for bosses, lugs, tabs, and other external features, maximum material is the condition in which these are at their maximum allowable sizes.

A positional tolerance applied on an MMC basis may be explained in either of these ways:

1. **In terms of the surface of a hole.** While maintaining the specified size limits of the hole, no element of the hole surface shall be inside a theoretical boundary having a diameter equal to the minimum limit of size (MMC) minus the positional tolerance located at true position, Figure 51–10.

2. **In terms of the axis of the hole.** Where a hole is at MMC (minimum diameter), its axis must fall within a cylindrical tolerance zone whose axis is located at true position. The diameter of this zone is equal to the positional tolerance, Figure 51–11, holes A and B. This tolerance zone also defines the limits of variation in the attitude of the axis of the hole in relation to the datum surface, Figure 51–11, hole C. It is only when the feature is at MMC that the specified positional tolerance applies. Where the actual size of the feature is larger than MMC, additional or bonus positional tolerance results. This increase of positional tolerance is equal to the difference between the specified maximum material limit of size (MMC) and the actual size of the feature.

The problems of tolerancing for the position of holes are simplified when positional tolerancing is applied on an MMC basis. Positional tolerancing simplifies measuring procedures of functional GO gauges. It also permits an increase in positional variations as the size departs from the maximum material size without jeopardizing free assembly of mating features.

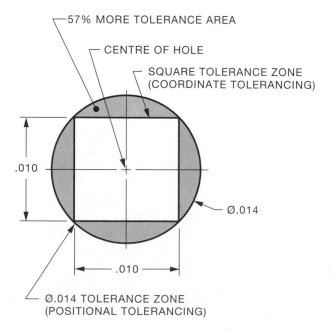

FIGURE 51–9 ■ Relationship of tolerance zones

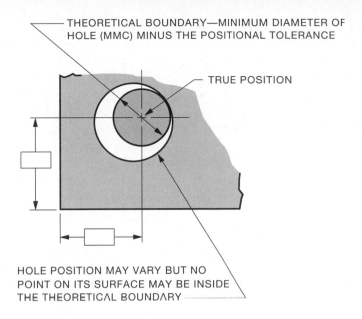

THEORETICAL BOUNDARY—MINIMUM DIAMETER OF
HOLE (MMC) MINUS THE POSITIONAL TOLERANCE

TRUE POSITION

HOLE POSITION MAY VARY BUT NO
POINT ON ITS SURFACE MAY BE INSIDE
THE THEORETICAL BOUNDARY

FIGURE 51-10 ■ Boundary for surface of a hole—MMC

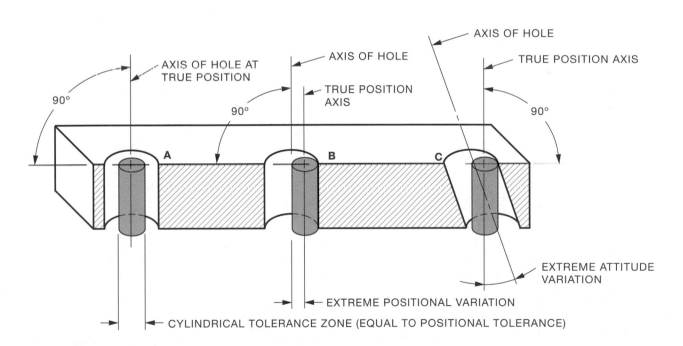

AXIS OF HOLE

AXIS OF HOLE AT
TRUE POSITION

AXIS OF HOLE

TRUE POSITION AXIS

90°

90°

TRUE POSITION
AXIS

90°

A

B

C

EXTREME ATTITUDE
VARIATION

EXTREME POSITIONAL VARIATION

CYLINDRICAL TOLERANCE ZONE (EQUAL TO POSITIONAL TOLERANCE)

HOLE A—AXIS OF HOLE IS COINCIDENT WITH TRUE POSITION AXIS

HOLE B—AXIS OF HOLE IS LOCATED AT EXTREME POSITION TO THE LEFT OF TRUE POSITION AXIS (BUT WITHIN
TOLERANCE ZONE)

HOLE C—AXIS OF HOLE IS INCLINED TO EXTREME ATTITUDE WITHIN TOLERANCE ZONE

NOTE: THE LENGTH OF THE TOLERANCE ZONE IS EQUAL TO THE LENGTH OF THE FEATURE UNLESS OTHERWISE
SPECIFIED ON THE DRAWING.

FIGURE 51-11 ■ Hole axes in relationship to positional tolerance zones

A positional tolerance on an MMC basis is specified on a drawing, on either the front or the side view, Figure 51–12. The MMC symbol Ⓜ is added in the feature control frame immediately after the tolerance.

A positional tolerance applied to a hole on an MMC basis means that the boundary of the hole must fall outside a perfect cylinder having a diameter equal to the minimum limit of size minus the positional tolerance. This cylinder is located with its axis at true position. The hole must meet its diameter limits.

The effect is illustrated in Figure 51–13, where the gauge cylinder is shown at true position, and the minimum and maximum diameter holes are drawn to show the extreme permissible variations in position in one direction.

Therefore, if a hole is at its MMC (minimum diameter), the position of its axis must lie within a circular tolerance zone having a diameter equal to the specified tolerance. If the hole is at its maximum diameter (LMC), the diameter of the tolerance zone for the axis is increased by the amount of the feature tolerance. The greatest deviation of the axis in one direction from true position is therefore

$$\frac{H + P}{2} = \frac{.006 + .028}{2} = .017 \text{ in.}$$

Where: H = hole diameter tolerance
P = positional tolerance

Positional tolerancing on an MMC basis is preferred when production quantities warrant the provision of functional GO gauges, because gauging is then limited to one simple operation, even when a group of holes is involved. This method also facilitates manufacture by permitting larger variations in position when the diameter departs from MMC. It cannot be used when it is essential that variations in location of the axis be observed RFS.

Positional Tolerancing at Zero MMC

The application of MMC permits the tolerance to exceed the value specified, provided features are within size limits and parts are acceptable. This is accomplished by adjusting the minimum size limit of a hole to the absolute minimum required for the insertion of an applicable fastener located precisely at true position and specifying a zero tolerance at MMC, Figure 51–14. In this case, the positional tolerance allowed is totally dependent on the actual size of the considered feature.

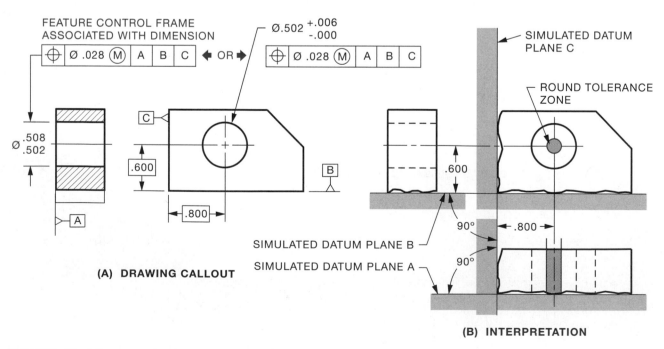

(A) DRAWING CALLOUT

(B) INTERPRETATION

FIGURE 51–12 ■ Positional tolerancing—MMC

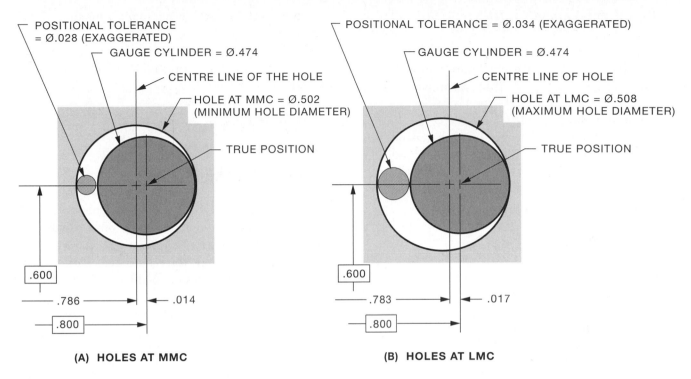

(A) **HOLES AT MMC**

(B) **HOLES AT LMC**

FIGURE 51–13 ■ Positional variations for tolerancing for Figure 51–12

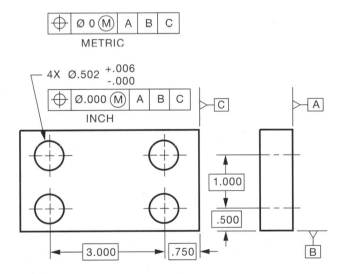

FIGURE 51–14 ■ Positional tolerancing—at zero MMC

Positional Tolerancing—RFS

In certain cases, the design or function of a part may require the positional tolerance or datum reference, or both, to be maintained regardless of actual feature sizes. RFS, where applied to the positional tolerance of circular features, requires the axis of each feature to be located within the specified positional tolerance regardless of the size of the feature, Figure 51–15. This requirement imposes a closer control of the features involved and introduces complexities in verification.

Positional Tolerancing—LMC

Where positional tolerancing at LMC is specified, the stated positional tolerance applies when the feature contains the least amount of material permitted by its toleranced size dimension, Figure 51–16. In this example, LMC is used to maintain a maximum wall thickness.

Specifying LMC is limited to applications where MMC does not provide the desired control and RFS is too restrictive.

ADVANTAGES OF POSITIONAL TOLERANCING

It is practical to replace coordinate tolerances with a positional tolerance having a value equal to the diagonal of the coordinate tolerance zone. This provides 57 percent more tolerance area, Figure 51–9, and would probably result in the rejection of fewer parts for positional errors.

A simple method for checking positional tolerance errors is to take coordinate measurements

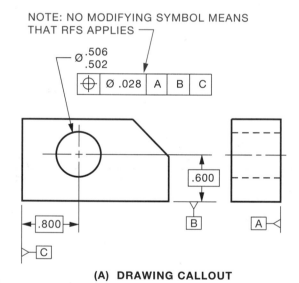

(A) DRAWING CALLOUT

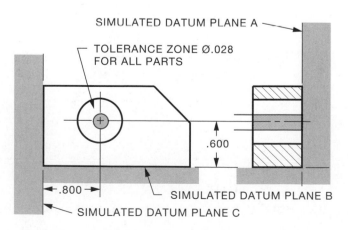

FIGURE 51–15 ■ Positional tolerancing—RFS

and evaluate them on a chart, Figure 51–17. For example, the four parts shown in Figure 51–18 were rejected when the coordinate tolerances were applied to them.

If the parts had been toleranced using the positional tolerance (RFS) method (Figure 51–15) and given a tolerance of ∅.028 in. (equal to the diagonal of the coordinate tolerance zone), three of the parts—A, B, and D—would not have been rejected.

If the parts shown in Figure 51–18 had been toleranced using the positional tolerance (MMC) method and given a tolerance of ∅.028 in. at MMC

(Figure 51–12), part C, which was rejected using the RFS tolerancing method (Figure 51–15), would not have been rejected if it had been straight. The positional tolerance can be increased to ∅.034 in. for a part having a diameter of .508 in. (LMC) without jeopardizing the function of the part (Figure 51–13).

REFERENCES

CAN/CSA B78.2-M91 Dimensioning and Tolerancing of Technical Drawings
ASME Y14.5M-1994 (R2004) Dimensioning and Tolerancing

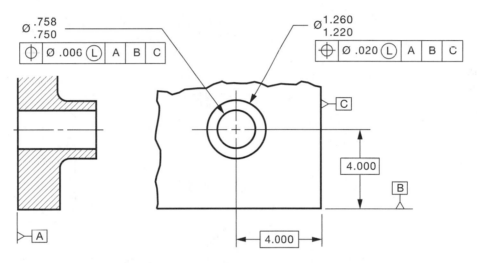

(A) DRAWING CALLOUT

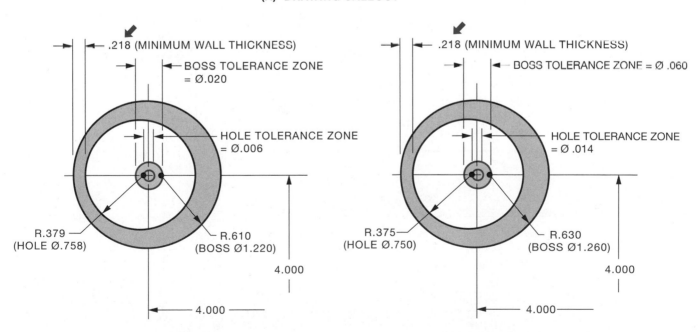

(B) TOLERANCE ZONES WHEN HOLE AT LMC

(C) TOLERANCE ZONES WHEN HOLE AT MMC

FIGURE 51–16 ■ LMC applied to a boss and hole

INTERNET RESOURCES

Drafting Zone For information on geometric dimensioning and tolerancing, see: http://www.draftingzone.com

Effective Training Inc. For information on dimensioning and tolerancing, see: http://etinews.com/eti_solutions.html

eFunda For information on geometric dimensioning and tolerancing, see: http://www.efunda.com/home.cfm

Engineers Edge For information on geometric tolerancing and dimensioning, see: http://www.engineersedge.com/gdt.htm

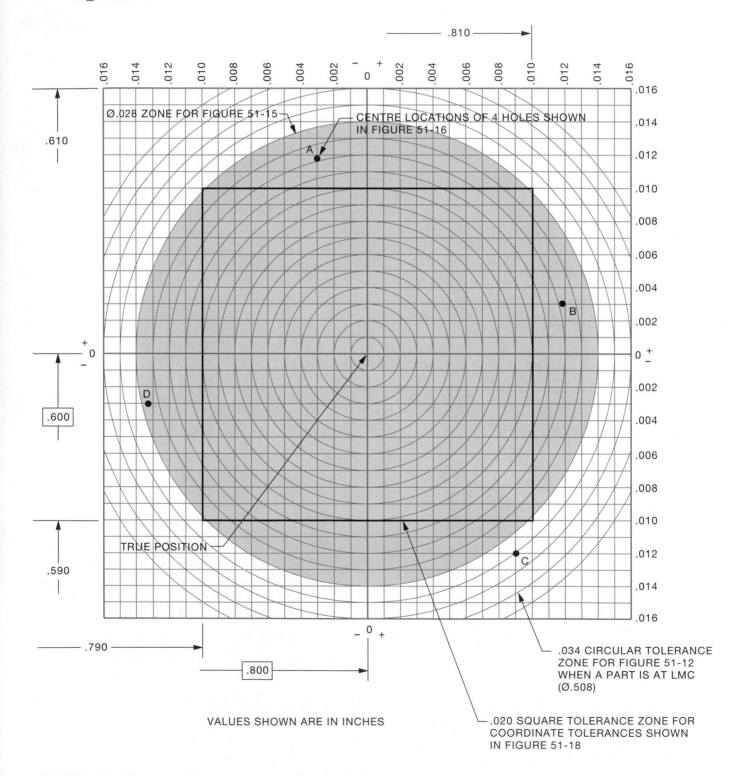

FIGURE 51–17 ■ Charts for evaluating positional tolerancing

NEL

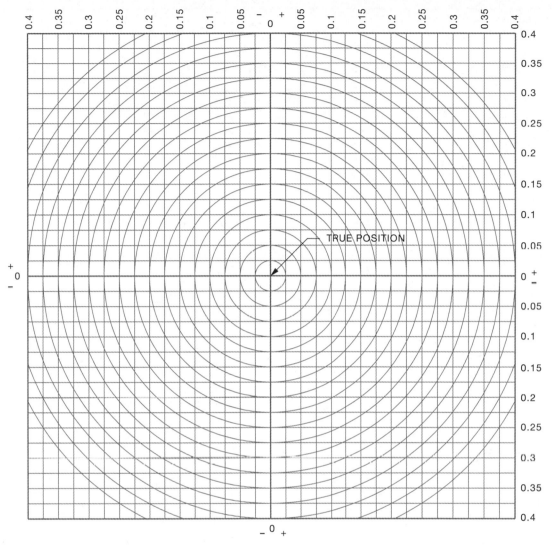

VALUES SHOWN ARE IN MILLIMETRES

FIGURE 51–17 CONT'D ■ Charts for evaluating positional tolerancing

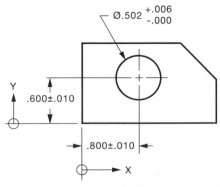

Ø.502 +.006 / -.000

.600±.010

.800±.010

(A) DRAWING CALLOUT

PART	HOLE DIA	HOLE LOCATION		COMMENT
		X	Y	
A	.503	.797	.612	REJECTED
B	.504	.812	.603	REJECTED
C	.508	.809	.588	REJECTED
D	.506	.787	.597	REJECTED

REFER TO FIGURE 51–17 FOR LOCATION ON CHART

(B) LOCATION AND SIZE OF REJECTED PARTS

FIGURE 51–18 ■ Parts A to D rejected because hole centres do not lie within coordinate tolerance zone

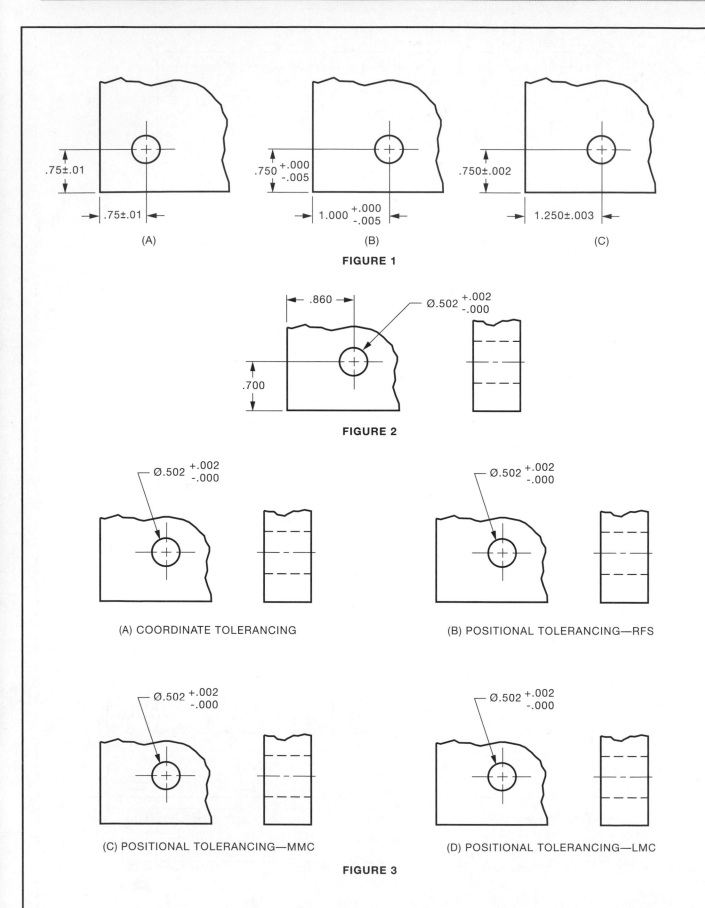

(A)

(B)

(C)

FIGURE 1

FIGURE 2

(A) COORDINATE TOLERANCING

(B) POSITIONAL TOLERANCING—RFS

(C) POSITIONAL TOLERANCING—MMC

(D) POSITIONAL TOLERANCING—LMC

FIGURE 3

ASSIGNMENT:

Make sketches as needed to answer the following questions.

QUESTIONS:

1. If coordinate tolerances in Figure 1 are given, what is the maximum distance between centres of mating holes for parts made to these drawing callouts?

2. To assemble correctly, the holes shown in Figure 2 must not vary more than .0014 in. in any direction from its true position when the hole is at its smallest size. Show suitable tolerancing, dimensioning, and datums where required on the drawings in Figure 2 to achieve this by using:
   (A) coordinate tolerancing
   (B) positional tolerancing—RFS
   (C) positional tolerancing—MMC
   (D) positional tolerancing—LMC

3. Refer to Question 2 and Figure 3. What would be the maximum permissible deviation from true position when the hole was at its largest size?

4. The part in Figure 4A is set on a revolving table, adjusted so that the part revolves about the true position centre of the Ø.316 in. hole. If the indicators give identical readings and the results in Figure 4B are obtained, which parts are acceptable?

5. What is the positional error for each part in Figure 4B?

6. If MMC instead of RFS had been shown in the feature control frame in Figure 4A, what is the diameter of the mandrel that would be required to check the parts?

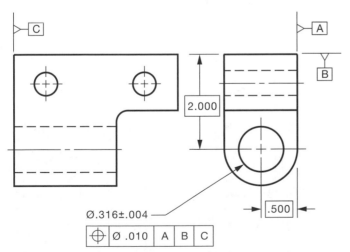

(A) DRAWING CALLOUT

PART NO.	SIZE OF MANDREL	HIGHEST READING	LOWEST READING
1	.316	.014	.008
2	.320	.008	-.004
3	.314	.026	.012
4	.312	.016	.006
5	.316	.015	.009
6	.318	.018	.010

(B) READINGS FOR PARTS

FIGURE 4

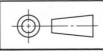 **POSITIONAL TOLERANCING** | **A-117**

52 UNIT

SELECTION OF DATUM FEATURES FOR POSITIONAL TOLERANCING

When selecting datums for positional tolerancing, the first consideration is to select the primary datum feature. It is usual to specify as the primary datum the surface in which the hole is produced. This will ensure that the true position of the axis is perpendicular to this surface or at a basic angle if other than 90°. This surface is resting on the gauging plane or surface plate for measuring. Secondary and tertiary datum features are then selected and identified, if required, Figures 52–1 and 52–2.

Positional tolerancing is also useful for parts having holes not perpendicular to the primary surface. This principle is illustrated in Figure 52–3.

LONG HOLES

It is not always essential to have the true position of a hole perpendicular to the face into which the hole is produced. It may be functionally more important, especially with long holes, to have it parallel to one of the sides, Figure 52–4. In this example, the sides are designated as primary and secondary datums. Gauging is facilitated

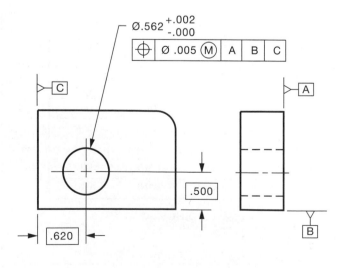

(A) DRAWING CALLOUT

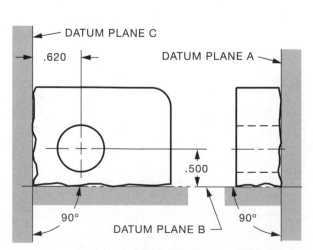

(B) INTERPRETATION OF TRUE POSITION

FIGURE 52–1 ■ Part with three datum features specified—MMC

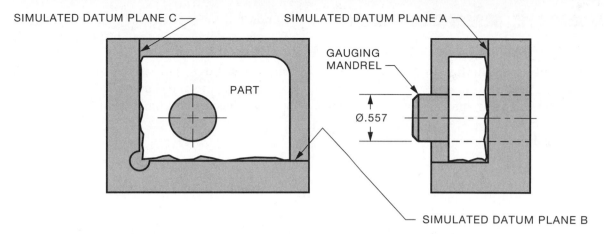

SIMULATED DATUM PLANE C

SIMULATED DATUM PLANE A

GAUGING MANDREL

PART

Ø.557

SIMULATED DATUM PLANE B

PART MUST SLIDE OVER Ø.557 GAUGE PIN, LIE FLAT ON BASE OF GAUGE (SIMULATED DATUM PLANE A), AND TOUCH SIMULATED DATUM PLANE B AT LEAST AT TWO POINTS ALONG ITS LENGTH, WHILE SIMULTANEOUSLY TOUCHING SIMULATED DATUM PLANE C AT LEAST AT ONE POINT.

FIGURE 52–2 ■ Gauge for the part in Figure 52–1

if the positional tolerance is specified on an MMC basis.

CIRCULAR DATUMS

Example 1

Circular features, such as holes or external cylindrical features, can be used as datums just as readily as flat surfaces. In the simple part shown in Figure 52–5, the true position of the small hole is established from the flat surface, datum A, and the large hole, datum D. Specifying datum hole D on an MMC basis facilitates gauging.

Example 2

In other cases, such as that in Figure 52–6, the datum could be either the axis of the hole or the axis of the outside cylindrical surface. In such applications, one must determine whether the true position should be established perpendicular to the surface as shown or parallel with the datum axis. In the latter case, datum A would not be specified, and it would not be necessary to ensure that the gauge made full contact with the surface.

In a group of holes, it may be desirable to indicate one of the holes as the datum from which all the other

holes are located (Unit 53 and 54). All circular datums of this type may be specified on an MMC basis when required, and this is preferred for ease of gauging.

MULTIPLE HOLE DATUMS

On an MMC basis, any number of holes or similar features that form a group or pattern may be specified as a single datum. All features forming such a datum must be related with a positional tolerance. The actual datum position is based on the virtual condition of all features in the group; that is, the collectible effect of the maximum material sizes of the features and the specified positional tolerance.

In Figure 52–7, the gauging element that locates the datum position would have four Ø.240 pins located at true position with respect to one another. This setup automatically checks the positional tolerance specified for these four holes.

REFERENCES

CAN/CSA B78.2-M91 Dimensioning and Tolerancing of Technical Drawings
ASME Y14.5M-1994 (R2004) Dimensioning and Tolerancing

INTERNET RESOURCES

Drafting Zone For information on geometric dimensioning and tolerancing, see: http://www.draftingzone.com

Effective Training Inc. For information on dimensioning and tolerancing, see: http://etinews.com/eti_solutions.html

eFunda For information on geometric dimensioning and tolerancing, see: http://www.efunda.com/home.cfm

Engineers Edge For information on geometric tolerancing and dimensioning, see: http://www.engineersedge.com/gdt.htm

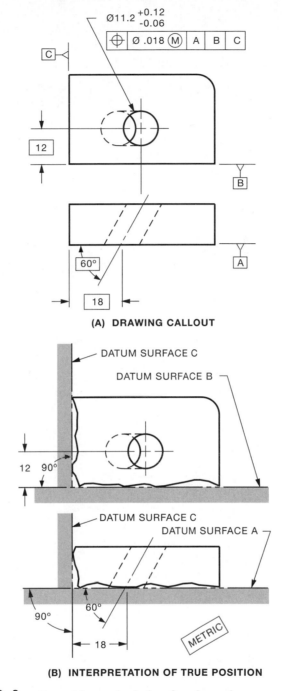

⌀11.2 $^{+0.12}_{-0.06}$

| ⌖ | ⌀ .018 Ⓜ | A | B | C |

(A) DRAWING CALLOUT

DATUM SURFACE C

DATUM SURFACE B

12 90°

DATUM SURFACE C

DATUM SURFACE A

90° 60°

18

METRIC

(B) INTERPRETATION OF TRUE POSITION

FIGURE 52–3 ■ Part with angular hole referred to a datum system—MMC

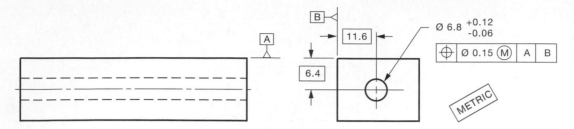

FIGURE 52–4 ■ Datum system for a long hole—MMC

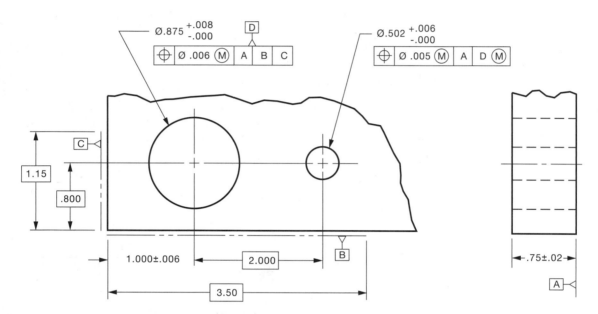

FIGURE 52–5 ■ Specifying an internal circular feature as a datum—MMC

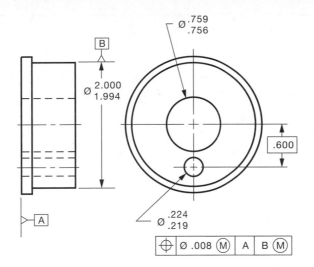

FIGURE 52–6 ■ Specifying an external circular feature as a datum—MMC

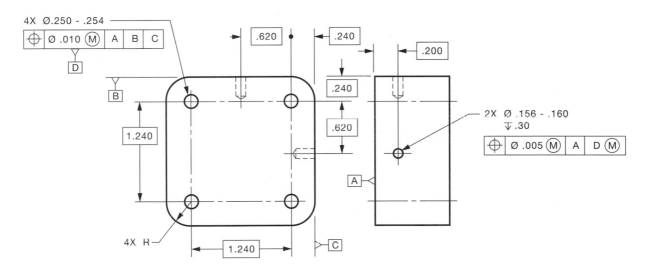

FIGURE 52–7 ■ Specifying a group of holes as a single datum—MMC

ASSIGNMENT:

On a one-inch grid sheet (.10 in. squares), sketch a suitable gauge
to check the positional tolerance for the Ø.750-.755 in. hole.

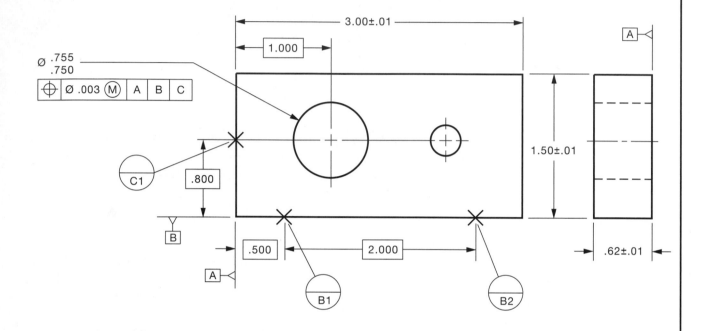

PROFILE TOLERANCES

A profile is the outline form or shape of a line or surface. A line profile may be the outline of a part or feature depicted in a view on a drawing. It may represent the edge of a part or refer to line elements of a surface in a single direction, such as the outline of cross sections through the part. A surface profile outlines the form or shape of a complete surface in three dimensions.

The elements of a line profile are straight lines, arcs, or other curved lines. The elements of a surface profile are flat surfaces, spherical surfaces, cylindrical surfaces, or surfaces composed of various line profiles in two or more directions.

A profile tolerance specifies a uniform boundary along the true profile within which the elements of the surface must lie. MMC is not applicable to profile tolerances. Where used as a refinement of size, the profile tolerance must be contained within the size limits.

Profile Symbols

There are two geometric characteristic symbols for profiles, one for lines and one for surfaces. Separate symbols are required because it is often necessary to distinguish between line elements of a surface and the complete surface itself.

The symbol for a line profile consists of a semicircle with a diameter equal to twice the lettering size used on the drawing. The symbol for profile of a surface is similar, but the semicircle is closed by a straight line at the bottom, Figure 53–1. All other geometric tolerances of form and orientation are merely special cases of profile tolerancing.

Profile tolerances are used to control the position of lines and surfaces that are neither flat nor cylindrical.

PROFILE OF A LINE

A profile-of-a-line tolerance may be directed to a line of any length or shape. Datums may be used in some circumstances but not when the only requirement is the profile shape taken cross section by cross section. Profile-of-a-line tolerancing is used where it is not desirable to control the entire surface of the feature as a single entity.

A profile-of-a-line tolerance is specified in the usual manner by including the symbol and tolerance in a feature control frame directed to the line to be controlled, Figure 53–2.

The tolerance zone established by the profile-of-a-line tolerance is two-dimensional, extending along the length of the considered feature.

If the line on the drawing to which the tolerance is directed represents a surface, the tolerance applies to all line elements of the surface parallel to the plane of the view on the drawing, unless otherwise specified.

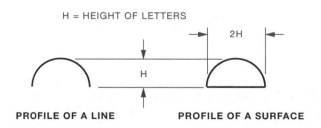

FIGURE 53–1 ■ Profile symbols

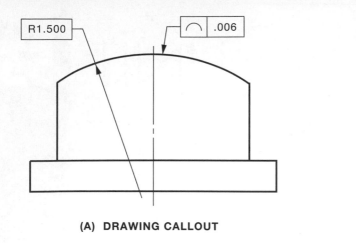

(A) DRAWING CALLOUT

(B) BILATERAL TOLERANCE ZONE

FIGURE 53–2 ■ Simple profile with a bilateral profile-of-a-line tolerance zone

The tolerance indicates a tolerance zone consisting of the area between two parallel lines, separated by the specified tolerance, which are themselves parallel to the basic form of the line being toleranced.

Bilateral and Unilateral Tolerances

The profile tolerance zone, unless otherwise specified, is equally disposed about the basic profile in a form known as a bilateral tolerance zone. The width of this zone is always measured perpendicular to the profile surface. The tolerance zone may be considered to be bounded by two lines enveloping a series of circles, each having a diameter equal to the specified profile tolerance. Their centres are on the theoretical basic profile, Figure 53–2.

Occasionally, it is desirable to have the tolerance zone wholly on one side of the basic profile instead of equally divided on both sides. This unilateral tolerance zone is specified by showing a phantom line drawn parallel and close to the profile surface. The tolerance is directed to this line, Figure 53–3. The zone line need extend only enough to make its application clear.

Specifying All-Around Profile Tolerancing

Where a profile tolerance applies all around the profile of a part, the symbol used to designate "all around" is placed on the leader from the feature control frame, Figure 53–4.

Method of Dimensioning

The true profile is established with basic dimensions, each of which is enclosed in a

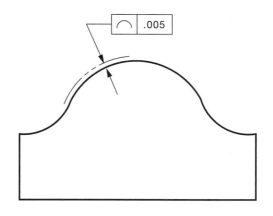

(A) TOLERANCE ZONE ON OUTSIDE OF TRUE PROFILE

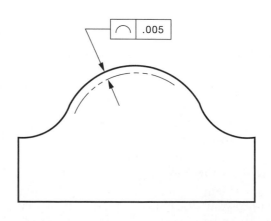

(B) TOLERANCE ZONE ON INSIDE OF TRUE PROFILE

FIGURE 53–3 ■ Unilateral tolerance zones

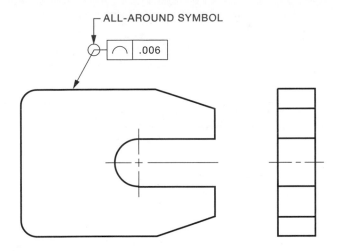

FIGURE 53–4 ■ Profile tolerance required for all around the surface

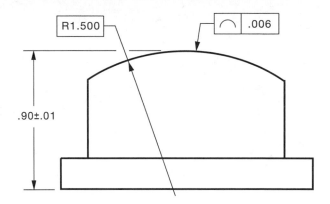

FIGURE 53–5 ■ Positional and form as separate requirements

rectangular frame to indicate that the tolerance in the general tolerance note does not apply.

When the profile tolerance is not intended to control the position of the profile, there must be clear distinction between dimensions that control the position of the profile and those that control its form or shape.

The simple part in Figure 53–5 shows a dimension of .90 ± .01 in. controlling the height of the profile. This dimension must be separately measured. The radius of 1.500 in. is a basic dimension and becomes part of the profile. Therefore, the profile tolerance zone has radii of 1.497 and 1.503 in., but is free to float in any direction within the limits of the positional tolerance zone to enclose the curved profile.

Figure 53–6 shows a more complex profile, which is located by a single toleranced dimension. There are, however, four basic dimensions defining the true profile. The tolerance on the height indicates a tolerance zone .06 in. wide, extending the full length of the profile, which is established by basic dimensions. No other dimension exists to affect the orientation or height. The profile tolerance specifies a .008 in. wide tolerance zone, which may lie anywhere within the .06 in. tolerance zone.

Extent of Controlled Profile

The profile is generally intended to extend to the first abrupt change or sharp corner unless otherwise specified. In Figure 53–6, it extends from the upper left- to the upper right-hand corners. If

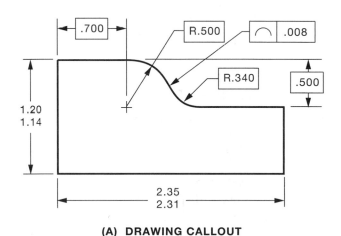

(A) DRAWING CALLOUT

(B) PROFILE TOLERANCE ZONE

FIGURE 53–6 ■ Profile defined by basic dimensions

the extent of the profile is not clearly identified by sharp corners or by basic profile dimensions, it must be indicated by a note under the feature control symbol, such as A ←→ B, meaning between points A and B, Figure 53–7.

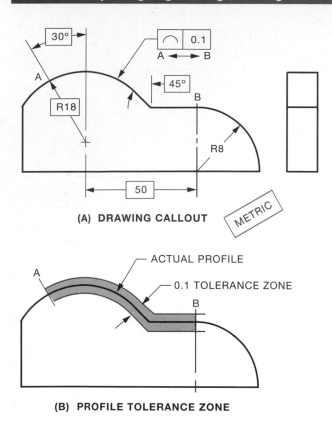

(A) DRAWING CALLOUT

METRIC

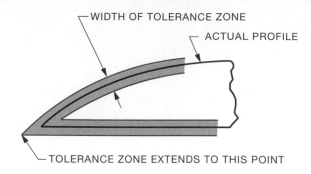

FIGURE 53–8 ■ Controlling the profile of a sharp corner

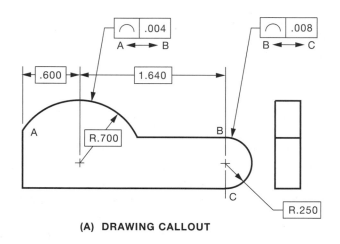

ACTUAL PROFILE

0.1 TOLERANCE ZONE

(B) PROFILE TOLERANCE ZONE

FIGURE 53–7 ■ Specifying extent of profile tolerance

If the controlled profile includes a sharp corner, the tolerance boundary is considered to extend to the intersection of the boundary lines, Figure 53–8. Because the intersecting surfaces may lie anywhere within the converging zone, the actual part contour could conceivably be round. If this is undesirable, the drawing must indicate the design requirements, possibly by specifying the maximum radius.

If different profile tolerances are required on different segments of a surface, the extent of each profile tolerance is indicated by reference letters identifying the extremities, Figure 53–9.

SURFACE PROFILE (PROFILE OF A SURFACE)

If the same tolerance is intended to apply over the whole surface, rather than lines or line elements in specific directions, the profile-of-a-surface symbol is used, Figure 53–10. Whereas the profile tolerance may be directed to the surface in either view, it is usually directed to the view showing the shape of the profile.

The profile-of-a-surface tolerance indicates a tolerance zone having the same form as the basic

(A) DRAWING CALLOUT

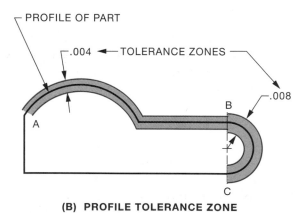

(B) PROFILE TOLERANCE ZONE

FIGURE 53–9 ■ Dual tolerance zones

surface, with a uniform width equal to the specified tolerance within which the entire surface must lie. It is used to control form or combinations of size, form, and orientation. Where used as a refinement of size, the profile tolerance must be contained within the size limits.

MMC is not applicable to profile tolerances.

The basic rules for profile-of-a-line tolerancing apply to profile-of-a-surface tolerancing except that profile-of-a-surface tolerance usually requires reference to datums to provide proper orientation of the profile. This is specified simply by indicating suitable datums. Figure 53–10 shows a simple part where two datums are designated.

The criterion that distinguishes between a profile tolerance applying to position and to orientation is whether the profile is related to the datum by a basic dimension or by a toleranced dimension.

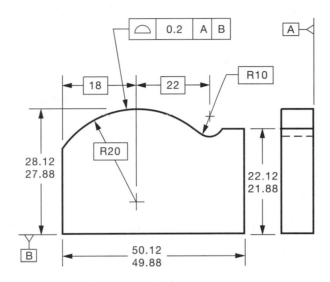

(A) PROFILE TOLERANCE CONTROLS FORM AND ORIENTATION OF A PROFILE

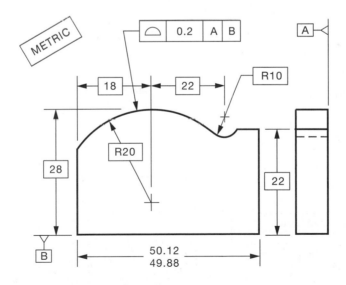

(B) PROFILE TOLERANCE CONTROLS FORM, ORIENTATION, AND POSITION OF A PROFILE

FIGURE 53–10 ■ Comparison of profile-of-a-surface tolerances

Profile tolerancing may be used to control the form and orientation of plane surfaces. In Figure 53–11, profile of a surface is used to control a plane surface inclined to a datum feature.

Where a profile-of-a-surface tolerance applies all around the profile of a part, the symbol used to designate "all around" is placed on the leader from the feature control frame, Figure 53–12.

REFERENCES

CAN/CSA-B78.2-M91 Dimensioning and Tolerancing of Technical Drawings

ASME Y14.5M-1994 (R2004) Dimensioning and Tolerancing

INTERNET RESOURCES

Drafting Zone For information on geometric dimensioning and tolerancing, see: http://www .draftingzone.com

Effective Training Inc. For information on dimensioning and tolerancing, see: http://etinews.com/ eti_solutions.html

eFunda For information on geometric dimensioning and tolerancing, see: http://www.efunda.com/ home.cfm

Engineers Edge For information on geometric tolerancing and dimensioning, see: http://www .engineersedge.com/gdt.htm

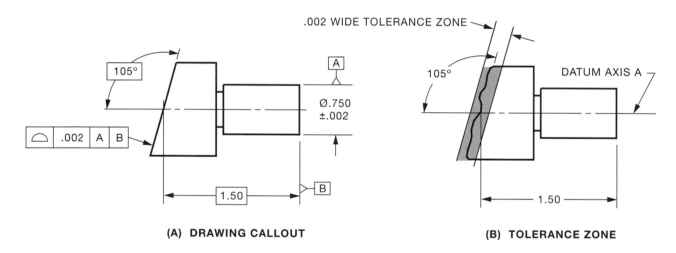

(A) DRAWING CALLOUT

(B) TOLERANCE ZONE

FIGURE 53–11 ■ Specifying profile-of-a-surface tolerance for a plane surface

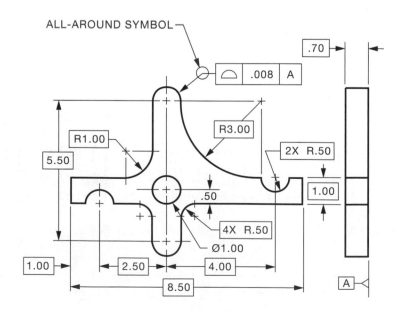

FIGURE 53–12 ■ Profile-of-a-surface tolerance required for all around the surface

ASSIGNMENTS:
Use one-inch grid sheets (.10 in. squares) for the sketching assignments below.

1. The profile form B to A (clockwise) shown in Figure 1 requires a profile-of-a-line tolerance .004 in. It is essential that the point between B and A remain sharp, having a maximum .010-in. radius. The remainder of the profile requires a profile-of-a-line tolerance of .020 in. Sketch Figure 1 showing the geometric tolerance and basic dimensions to meet these requirements.

2. The part shown in Figure 2 requires an all-around profile-of-a-line tolerance of .005 in. located on the outside of the true profile. Sketch Figure 2 showing the geometric tolerance and basic dimensions to meet these requirements.

3. With the information given below and that on Figure 3, make a sketch and add dimensions showing the geometric tolerances, datums, and basic dimensions. Profile-of-a-surface tolerances are to be applied to the part as follows:

   a) Between points A and B - .005 in.
   b) Between points B and C - .004 in.
   c) Between points C and D - .002 in.

   These tolerances are to be referenced to datum surfaces marked E and F, in that order.

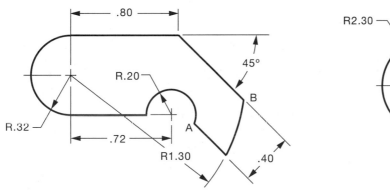

FIGURE 1

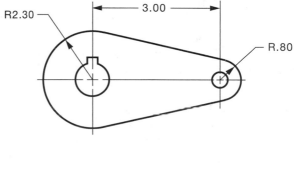

FIGURE 2

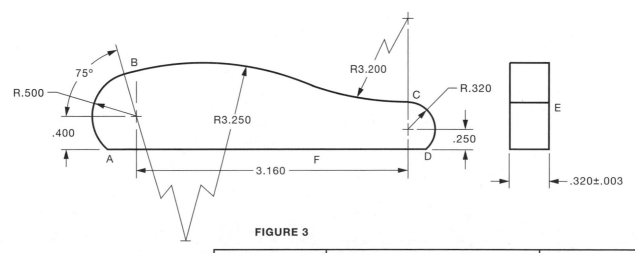

FIGURE 3

PROFILE TOLERANCING **A-119**

RUNOUT TOLERANCES

Runout is a composite tolerance used to control the functional relationship of one or more features of a part to a datum axis. The types of features controlled by runout tolerances include those surfaces constructed around a datum axis and those constructed at right angles to a datum axis, Figure 54–1.

Each feature must be within its runout tolerance when rotated about the datum axis.

The datum axis is established by a diameter of sufficient length, two diameters having sufficient axial separation, or a diameter and a face at right angles to it. Features used as datums for establishing axes should be functional, such as mounting features that establish an axis of rotation.

The tolerance specified for a controlled surface is the total tolerance, or full indicator movement (FIM) in inspection and international terminology. Both the tolerance and the datum feature apply only on an RFS basis.

There are two types of runout control: circular runout and total runout. The type depends on design requirements and manufacturing considerations. The geometric characteristic symbols for runout are shown in Figure 54–2.

CIRCULAR RUNOUT

Circular runout provides control of circular elements of a surface. The tolerance is applied independently at any cross section as the part is rotated 360°. Where applied to surfaces constructed around a datum axis, circular runout controls variations such as circularity and coaxiality. Where applied to surfaces constructed at right angles to the datum axis, circular runout controls wobble at all diametral positions.

In Figure 54–3, the surface is measured at several positions, as shown by the three indicator positions. At each position, the indicator movement during one revolution of the part must not exceed

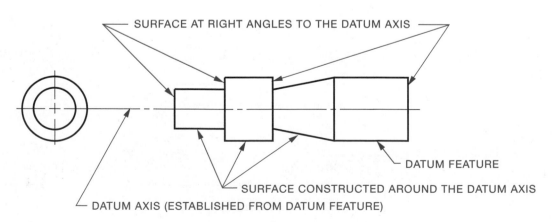

SURFACE AT RIGHT ANGLES TO THE DATUM AXIS

DATUM FEATURE

SURFACE CONSTRUCTED AROUND THE DATUM AXIS

DATUM AXIS (ESTABLISHED FROM DATUM FEATURE)

FIGURE 54–1 ■ Features applicable to runout tolerances

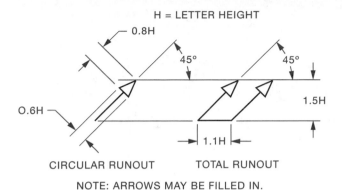

CIRCULAR RUNOUT   TOTAL RUNOUT

NOTE: ARROWS MAY BE FILLED IN.

FIGURE 54–2 ■ Runout symbols

the specified tolerance, in this case .005 inch. For a cylindrical feature, runout error is caused by eccentricity and errors of roundness. It is not affected by taper (conicity) or errors of straightness of the straight line elements, such as barrel shaping.

Figure 54–4 shows a part where the tolerance is applied to a surface that is at right angles to the axis. In this case, an error—generally referred to as wobble—will be shown if the surface is flat but not perpendicular to the axis, (B). No error will be

indicated if the surface is convex or concave but otherwise perfect, (C).

Circular runout can also be applied to curved surfaces. Unless otherwise specified, measurement is always made normal to the surface.

A runout tolerance directed to a surface applies to the full length of the surface up to an abrupt change in direction.

If a control is intended to apply to more than one portion of a surface, additional leaders and arrowheads may be used where the same tolerance applies. If different tolerance values are required, separate tolerances must be specified.

Where a runout tolerance applies to a specific portion of a surface, a thick chain line is drawn adjacent to the surface profile to show the desired length. Basic dimensions are used to define the extent of the portion indicated, Figure 54–5.

If only part of a surface or several consecutive portions require the same tolerance, the length to which the tolerance applies may be indicated, Figure 54–6.

Circular runout can be applied only on an RFS basis.

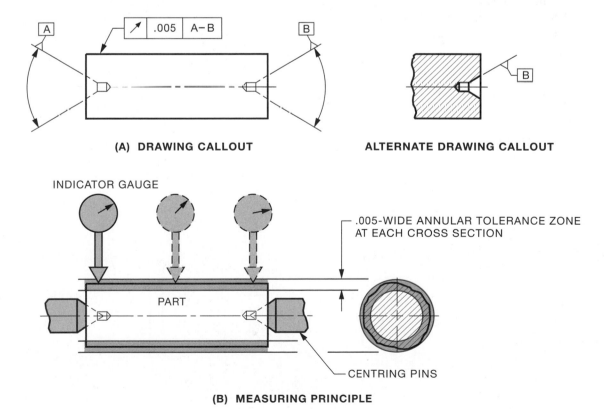

(A) DRAWING CALLOUT   **ALTERNATE DRAWING CALLOUT**

(B) MEASURING PRINCIPLE

FIGURE 54–3 ■ Circular runout for cylindrical features

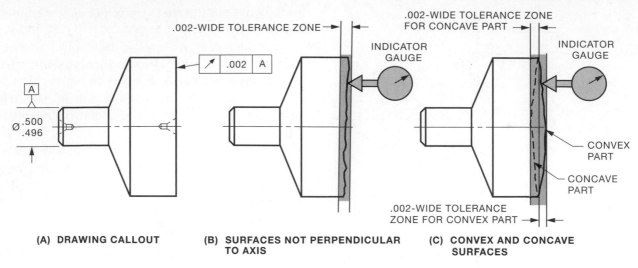

(A) DRAWING CALLOUT

(B) SURFACES NOT PERPENDICULAR TO AXIS

(C) CONVEX AND CONCAVE SURFACES

FIGURE 54–4 ■ Circular runout perpendicular to datum axis

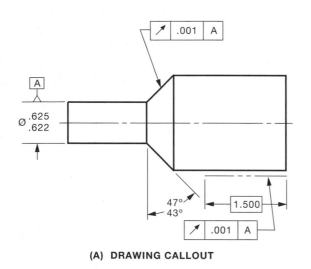

(A) DRAWING CALLOUT

(B) MEASURING PRINCIPLE

FIGURE 54–5 ■ Specifying circular runout relative to a datum diameter

TOTAL RUNOUT

Total runout concerns the runout of a complete surface, not merely the runout of each circular element. For measurement, the checking indicator must traverse the full length or extent of the surface while the part is revolved about its datum axis. Measurements are made over the whole surface without resetting the indicator. Total runout is the difference between the lowest indicator reading in any position on the same surface. In Figure 54–7, the tolerance zone is the space between two concentric cylinders separated by the specified tolerance and coaxial with the datum axis. In this case, the runout is affected not only by eccentricity and errors of roundness, but also by errors of straightness and conicity of the cylindrical surface.

A total runout may be applied to surfaces at various angles, as described for circular runout, and may

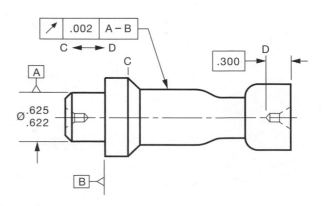

FIGURE 54–6 ■ Indication of length for a runout tolerance

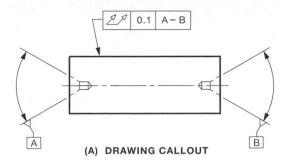

(A) DRAWING CALLOUT

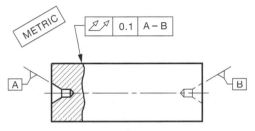

(B) ALTERNATIVE DRAWING CALLOUT FOR DATUM FEATURES

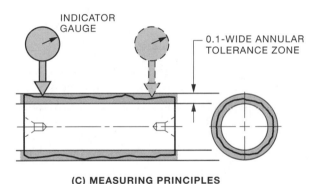

(C) MEASURING PRINCIPLES

FIGURE 54–7 ■ Tolerance zones for total runout

therefore control profile of the surface in addition to runout. However, for measurement, the indicator gauge must be capable of following the true profile direction of the surface. This is comparatively simple for straight surfaces, such as cylindrical surfaces and flat faces. For conical surfaces, the datum axis can be tilted to the taper angle so that the measured surface becomes parallel to a surface plate.

ESTABLISHING DATUMS

In many examples, the datum axis has been established from centres drilled in the two ends of the part, in which case the part is mounted between centres for measurement. This is an ideal method of mounting and revolving the part when such centres have been provided for manufacturing. When centres are not provided, any cylindrical or conical surface may be used to establish the datum axis if chosen on the basis of the functional requirements of the part. In some cases, a runout tolerance may also be applied to the datum feature. Some examples of suitable datum features and methods of establishing datum axes follow.

Measuring Principles

Example 1

Figure 54–8 shows a simple external cylindrical feature specified as the datum feature.

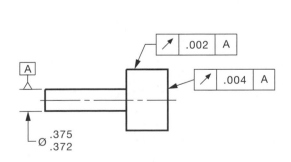

(A) DRAWING CALLOUT

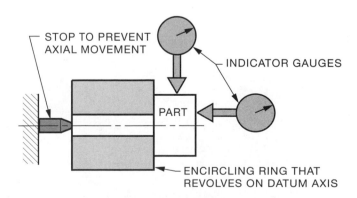

(B) MEASURING PRINCIPLE

FIGURE 54–8 ■ External cylindrical datum feature for runout tolerance

Measurement would require the datum feature to be held in an encircling ring capable of being revolved about the datum axis. Parts with these types of datum features are sometimes mounted in a V-block, although this practice permits precise measurements only if there are no significant roundness errors of the datum feature.

Example 2

The L-support method is particularly useful when two datum features are used, Figure 54–9. Measuring the part by using two L-supports is quite simple. It would be complicated were it necessary to fit the features into concentric encircling rings.

Example 3

Figure 54–10 illustrates the application of runout tolerances where two datum diameters act as a single datum axis to which the features are related. For measurement, the part may be mounted on a mandrel having a diameter equal to the maximum size of the hole.

When required, runout tolerances may be referenced to a datum system, usually consisting of two datum features perpendicular to each other. For measuring, the part is mounted on a flat surface capable of being rotated. Centring on the secondary datum requires some form of centralizing device, such as an expandable arbor.

Example 4

It may be necessary to control individual datum surface variations with respect to flatness, circularity, parallelism, straightness, or cylindricity. Where such control is required, the appropriate tolerances are specified. See Figure 54–11 for applying cylindricity to the datum.

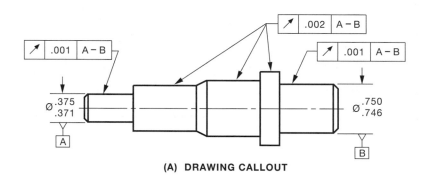

(A) DRAWING CALLOUT

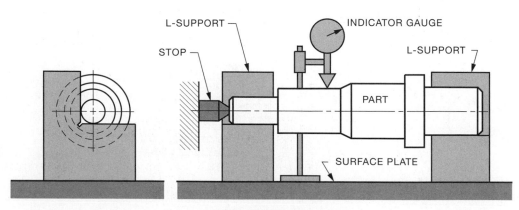

(B) MEASURING PRINCIPLE

FIGURE 54–9 ■ Runout tolerance with two datum features

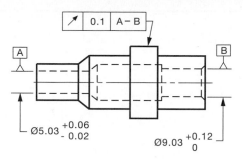

(A) DRAWING CALLOUT

Ø5.03 +0.06 / -0.02

Ø9.03 +0.12 / 0

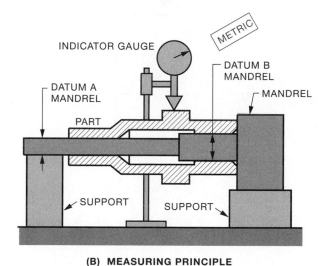

(B) MEASURING PRINCIPLE

FIGURE 54–10 ■ Specifying runout relative to two datum diameters

REFERENCES

CAN/CSA-B78.2-M91 Dimensioning and Tolerancing of Technical Drawings
ASME Y14.5M-1994 (R2004) Dimensioning and Tolerancing

INTERNET RESOURCES

Drafting Zone For information on geometric dimensioning and tolerancing, see: http://www.draftingzone.com

eFunda For information on geometric dimensioning and tolerancing, see: http://www.efunda.com/home.cfm

Engineers Edge For information on geometric tolerancing and dimensioning, see: http://www.engineersedge.com/gdt.htm

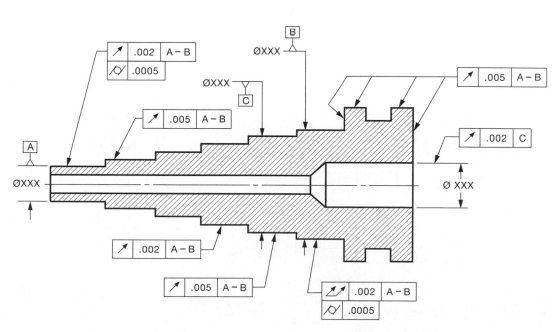

FIGURE 54–11 ■ Specifying runout relative to two datum diameters with form tolerances

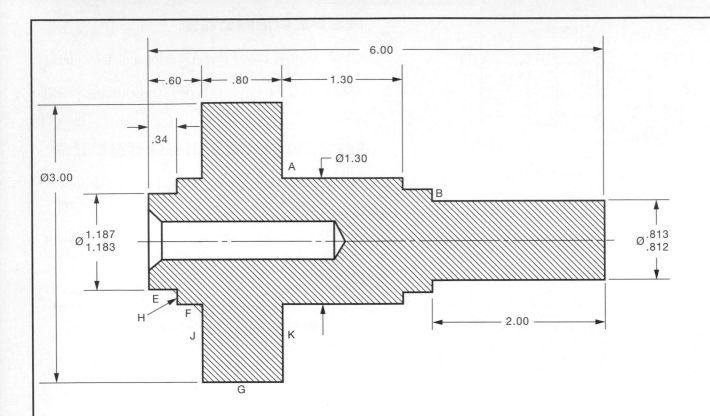

FIGURE 1

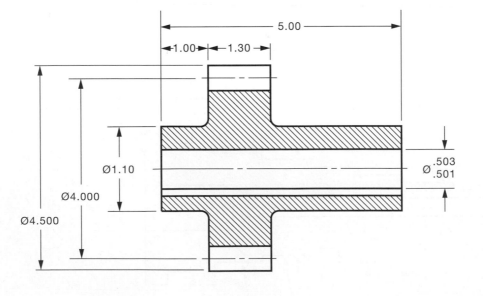

FIGURE 2

ASSIGNMENT:

Use inch grid sheets (.10 in. squares) for the sketching assignments below.

1. Sketch the part shown in Figure 1. Add the following runout tolerances and datums to the sketch:
   a) The Ø1.187 in. is to be datum C.
   b) A 1.20-in. length starting .40 in. from the right end of the part is to be datum D.
   c) Runout tolerances are related to the axis established by datums C and D.
   d) A total runout tolerance of .005 in. between positions A and B
   e) A circular runout tolerance of .002 in. for diameters E and F
   f) A circular runout tolerance of .005 in. for diameter G
   g) A circular runout tolerance of .004 in. for surface H
   h) A circular runout tolerance of .003 in. for surfaces J and K

2. Make a sketch of the gear shown in Figure 2. Add circular runout tolerances referenced to datum A. Both side faces of the gear portion require a tolerance of .015 in. The two hub portions require a tolerance of .010 in. The hole is to be datum A.

3. Make a sketch of the part shown in Figure 3. The part is intended to function by rotating with the two ends (datum diameters A and B) supported in bearings. These two datums collectively act as a coaxial datum for the larger diameters, which are required to have a total runout tolerance of .001 in. Add the above requirements to the drawing.

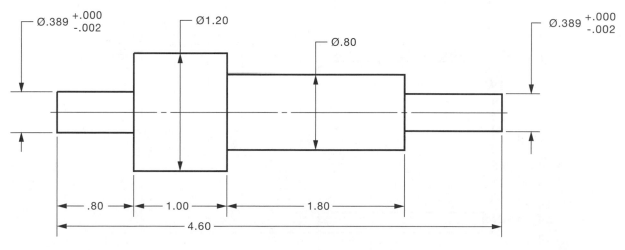

FIGURE 3

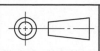 **RUNOUT TOLERANCES** | **A-120**

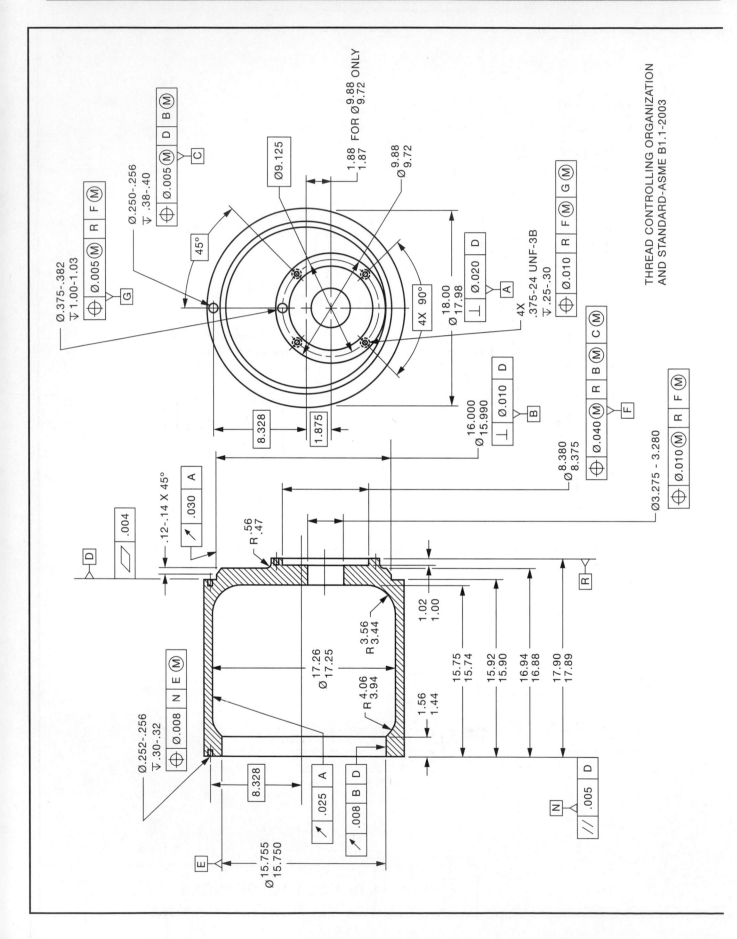

THREAD CONTROLLING ORGANIZATION
AND STANDARD-ASME B1.1-2003

A-121

HOUSING

ASSIGNMENT:

Use inch grid sheets (.10 in. squares) for the sketching assignments.

1. Sketch a suitable gauge to check the Ø3.275-3.280 hole.

2. Make a sketch of datum surface D showing the permissible tolerance zone.

3. Make a sketch of datum surface N showing the permissible tolerance zone.

QUESTIONS:

1. How many datum surfaces or points are indicated?

2. How many basic dimensions are indicated?

3. How many datum surfaces are flat?

4. How many datum surfaces are circular?

5. How many dimensions show positional tolerancing?

6. How many form tolerances are required?

7. How many orientation tolerances are shown?

8. How many features use datum A as a reference?

9. If the diameter of datum F was Ø8.375, what would be the maximum permissible positional tolerance?

10. What is the tertiary datum for the positional tolerance of the diameter shown as datum F?

11. The geometric tolerance placed on datum surface D controls _____.

12. With reference to the Ø3.275-3.280 hole, what is the maximum deviation permitted from true position when the hole is: (A) Ø3.275, (B) Ø3.280?

13. With reference to the Ø.252-.256 hole, what is the maximum deviation permitted from true position if the hole was (A) Ø.252, (B) Ø.256?

14. With reference to the Ø.250-.256 hole, what is the maximum deviation permitted from true position if the hole was: (A) Ø.250, (B) Ø.254, (C) Ø.256?

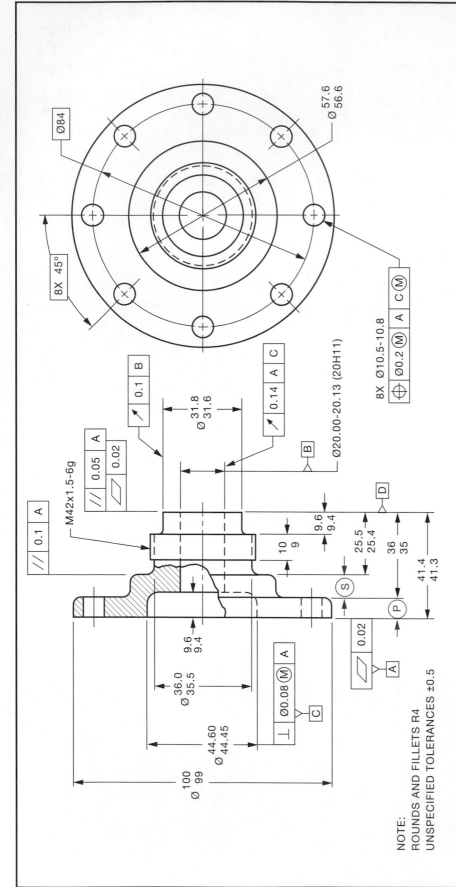

NOTE:
ROUNDS AND FILLETS R4
UNSPECIFIED TOLERANCES ±0.5

THREAD CONTROLLING ORGANIZATION
STANDARD-ASME B1.13M-2001

ASSIGNMENTS:

Use centimetre grid sheets (1mm squares) for the sketching assignments.

1. Sketch a suitable gauge to check the eight Ø10.5-10.8 holes.
2. Make a sketch of datum surface C showing the permissible tolerance zones.
3. Make a sketch of datum surface D showing the permissible tolerance zones.

QUESTIONS:

1. How many datum surfaces are indicated?
2. How many basic dimensions are shown?
3. How many different geometric tolerancing symbols are shown?
4. How many datum surfaces are circular?
5. How many features use datum A as a reference?
6. How many form tolerances are shown?
7. With reference to the Ø10.5-10.8 holes, what variation from the true position is permissible if the holes are (A) Ø10.5 (B) Ø10.8?
8. Calculate the following dimensions: (A) P min., (B) P max., (C) S min., (D) S max.

METRIC
DIMENSIONS ARE IN MILLIMETRES

END PLATE | **A-122M**

Fraction	Decimal-Inch		Millimetres	Fraction	Decimal-Inch		Millimetres
	Two Place	Three Place			Two Place	Three Place	
1/64	.02	.016	.04	33/64	.52	.516	13.1
1/32	.03	.031	.08	17/32	.53	.531	13.5
3/64	.05	.047	1.2	35/64	.55	.547	13.9
1/16	.06	.062	1.6	9/16	.56	.562	14.3
5/64	.08	.078	2	37/64	.58	.578	14.7
3/32	.09	.094	2.4	19/32	.59	.594	15.1
7/64	.11	.109	2.8	39/64	.61	.609	15.5
1/8	.12	.125	3.2	5/8	.62	.625	15.9
9/64	.14	.141	3.6	41/64	.64	.641	16.3
5/32	.16	.156	4	21/32	.66	.656	16.7
11/64	.17	.172	4.4	43/64	.67	.672	17.1
3/16	.19	.188	4.8	11/16	.69	.688	17.5
13/64	.20	.203	5.2	45/64	.70	.703	17.9
7/32	.22	.219	5.6	23/32	.72	.719	18.3
15/64	.23	.234	6	47/64	.73	.734	18.7
1/4	.25	.250	6.4	3/4	.75	.750	19.1
17/64	.27	.266	6.8	49/64	.77	.766	19.5
9/32	.28	.281	7.1	25/32	.78	.781	19.9
19/64	.30	.297	7.5	51/64	.80	.797	20.2
5/16	.31	.312	7.9	13/16	.81	.812	20.6
21/64	.33	.328	8.3	53/64	.83	.828	21
11/32	.34	.344	8.7	27/32	.84	.844	21.4
23/64	.36	.359	9.1	55/64	.86	.859	21.8
3/8	.38	.375	9.5	7/8	.88	.875	22.2
25/64	.39	.391	9.9	57/64	.89	.891	22.6
13/32	.41	.406	10.3	29/32	.91	.906	23
27/64	.42	.422	10.7	59/64	.92	.922	23.4
7/16	.44	.438	11.1	15/16	.94	.938	23.8
29/64	.45	.453	11.5	61/64	.95	.953	24.2
15/32	.47	.469	11.9	31/32	.97	.969	24.6
31/64	.48	.484	12.3	63/64	.98	.984	25
1/2	.50	.500	12.7	1	1.00	1.000	25.4

TABLE 1 ■ Chart for Converting Inch Dimensions to Millimetres

Across Flats	ACR FLT
And	&
Approximate	APPROX
Assembly	ASSY
Bill of Material	B/M
Bolt Circle	BC
Brass	BR
Bronze	BRZ
Brown and Sharpe Gauge	B & S GA
Carbon Steel	CS
Cast Iron	CI
Centimetre	cm
Centre Line	℄ or CL
Centre to Centre	C to C
Chamfer	CHAM
Circularity	CIR
Cold Rolled Steel	CRS
Concentric	CONC
Conical Taper	—▷
Copper	COP
Counterbore	CBORE or ⌴
Countersink	CSK or ⌵
Cubic Centimetre	cm^3
Cubic Metre	m^3
Datum	[A] △
Deep or Depth	↧
Degree (Angle)	°
Diametre	∅ or DIA
Diametral Pitch	DP
Dimension	DIM
Dimension Origin	⊙→
Drawing	DWG
Eccentric	ECC
Equally Spaced	EQL SP
Figure	FIG
Finish All Over	FAO
Gauge	GA
Grey Iron	GI
Head	HD
Heat Treat	HT TR
Hexagon	HEX
Inside Diametre	ID
International Organization for Standardization	ISO
International Pipe Standard	IPS
Kilogram	kg
Kilometre	km
Least Material Condition	LMC or Ⓛ
Left Hand	LH
Long	LG
Machined	✓
Machine Steel	MST
Malleable Iron	MI
Material	MATL
Maximum	MAX
Maximum Material Condition	MMC or Ⓜ
Metre	m
Metric Thread	M
Micrometre	µm
Mild Steel	MS
Millimetre	mm
Minimum	MIN
Minute (Angle)	(′)
Nominal	NOM
Not to Scale	<u>XXX</u>
Number	NO
Outside Diametre	OD
Parallel	PAR or //
Perpendicular	PERP or ⊥
Pitch Circle Diametre	PCD
Pitch Diametre	PD
Projected Tolerance Zone	Ⓟ
Radius	R
Reference or Reference Dimension	(xx)
Regardless of Feature Size	RFS
Revolutions per Minute	R/MIN
Right Hand	RH
Second (Arc)	″
Section	SECT
Slope	⊿
Spherical Radius	SR
Spotface	SF or ⌴
Square	SQ or □
Square Centimetre	cm^2
Square Metre	m^2
Steel	STL
Symmetrical	SYM or ⌯
Symmetry	⌯
Taper Pipe Thread	NPT
Thick	THK
Through	THRU
Undercut	UCUT
United States Gauge	USG
Wrought Iron	WI
Wrought Steel	WS

TABLE 2 ■ Abbreviations and Symbols Used on Technical Drawings

	Decimal-Inch and Millimetre Equivalents of Number Size Drills					Decimal-Inch and Millimetre Equivalents of Letter Size Drills		
No.	Decimal-Inch	mm	No.	Decimal-Inch	mm	Letter	Decimal-Inch	mm
1	.2280	5.8	31	.1200	3.0	A	.234	5.9
2	.2210	5.6	32	.1160	2.9	B	.238	6.0
3	.2130	5.4	33	.1130	2.9	C	.242	6.1
4	.2090	5.3	34	.1110	2.8	D	.246	6.2
5	.2055	5.2	35	.1100	2.8	E	.250	6.4
6	.2040	5.2	36	.1065	2.7	F	.257	6.5
7	.2010	5.1	37	.1040	2.6	G	.261	6.6
8	.1990	5.1	38	.1015	2.6	H	.266	6.8
9	.1960	5.0	39	.0995	2.5	I	.272	6.9
10	.1935	4.9	40	.0980	2.5	J	.277	7.0
11	.1910	4.9	41	.0960	2.4	K	.281	7.1
12	.1890	4.8	42	.0935	2.4	L	.290	7.4
13	.1850	4.7	43	.0890	2.3	M	.295	7.5
14	.1820	4.6	44	.0860	2.2	N	.302	7.7
15	.1800	4.6	45	.0820	2.1	O	.316	8.0
16	.1770	4.5	46	.0810	2.1	P	.323	8.2
17	.1730	4.4	47	.0785	2.0	Q	.332	8.4
18	.1695	4.3	48	.0760	1.9	R	.339	8.6
19	.1660	4.2	49	.0730	1.9	S	.348	8.8
20	.1610	4.1	50	.0700	1.8	T	.358	9.1
21	.1590	4.0	51	.0670	1.7	U	.368	9.3
22	.1570	4.0	52	.0635	1.6	V	.377	9.6
23	.1540	3.9	53	.0595	1.5	W	.386	9.8
24	.1520	3.9	54	.0550	1.4	X	.397	10.1
25	.1495	3.8	55	.0520	1.3	Y	.404	10.3
26	.1470	3.7	56	.0465	1.2	Z	.413	10.5
27	.1440	3.7	57	.0430	1.1			
28	.1405	3.6	58	.0420	1.1			
29	.1360	3.5	59	.0410	1.0			
30	.1285	3.3	60	.0400	1.0			

TABLE 3 ■ Number and Letter Size Drills

Metric Drill Sizes (mm)		Reference Decimal Equivalent (Inches)	Metric Drill Sizes (mm)		Reference Decimal Equivalent (Inches)	Metric Drill Sizes (mm)		Reference Decimal Equivalent (Inches)
Preferred	Available		Preferred	Available		Preferred	Available	
–	0.40	.0157	2.2	–	.0866	10	–	.3937
–	0.42	.0165	–	2.3	.0906	–	10.3	.4055
–	0.45	.0177	2.4	–	.0945	10.5	–	.4134
–	0.48	.0189	2.5	–	.0984	–	10.8	.4252
0.5	–	.0197	2.6	–	.1024	11	–	.4331
–	0.52	.0205	–	2.7	.1063	–	11.5	.4528
0.55	–	.0217	2.8	–	.1102	12	–	.4724
–	0.58	.0228	–	2.9	.1142	12.5	–	.4921
0.6	–	.0236	3	–	.1181	13	–	.5118
–	0.62	.0244	–	3.1	.1220	–	13.5	.5315
0.65	–	.0256	3.2	–	.1260	14	–	.5512
–	0.68	.0268	–	3.3	.1299	–	14.5	.5709
0.7	–	.0276	3.4	–	.1339	15	–	.5906
–	0.72	.0283	–	3.5	.1378	–	15.5	.6102
0.75	–	.0295	3.6	–	.1417	16	–	.6299
–	0.78	.0307	–	3.7	.1457	–	16.5	.6496
0.8	–	.0315	3.8	–	.1496	17	–	.6693
–	0.82	.0323	–	3.9	.1535	–	17.5	.6890
0.85	–	.0335	4	–	.1575	18	–	.7087
–	0.88	.0346	–	4.1	.1614	–	18.5	.7283
0.9	–	.0354	4.2	–	.1654	19	–	.7480
–	0.92	.0362	–	4.4	.1732	–	19.5	.7677
0.95	–	.0374	4.5	–	.1772	20	–	.7874
–	0.98	.0386	–	4.6	.1811	–	20.5	.8071
1	–	.0394	4.8	–	.1890	21	–	.8268
–	1.03	.0406	5	–	.1969	–	21.5	.8465
1.05	–	.0413	–	5.2	.2047	22	–	.8661
–	1.08	.0425	5.3	–	.2087	–	23	.9055
1.1	–	.0433	–	5.4	.2126	24	–	.9449
–	1.15	.0453	5.6	–	.2205	25	–	.9843
1.2	–	.0472	–	5.8	.2283	26	–	1.0236
1.25	–	.0492	6	–	.2362	–	27	1.0630
1.3	–	.0512	–	6.2	.2441	28	–	1.1024
–	1.35	.0531	6.3	–	.2480	–	29	1.1417
1.4	–	.0551	–	6.5	.2559	30	–	1.1811
–	1.45	.0571	6.7	–	.2638	–	31	1.2205
1.5	–	.0591	–	6.8	.2677	32	–	1.2598
–	1.55	.0610	–	6.9	.2717	–	33	1.2992
1.6	–	.0630	7.1	–	.2795	34	–	1.3386
–	1.65	.0650	–	7.3	.2874	–	35	1.3780
1.7	–	.0669	7.5	–	.2953	36	–	1.4173
–	1.75	.0689	–	7.8	.3071	–	37	1.4567
1.8	–	.0709	8	–	.3150	38	–	1.4961
–	1.85	.0728	–	8.2	.3228	–	39	1.5354
1.9	–	.0748	8.5	–	.3346	40	–	1.5748
–	1.95	.0768	–	8.8	.3465	–	41	1.6142
2	–	.0787	9	–	.3543	42	–	1.6535
–	2.05	.0807	–	9.2	.3622	–	43.5	1.7126
2.1	–	.0827	9.5	–	.3740	45	–	1.7717
–	2.15	.0846	–	9.8	.3858	–	46.5	1.8307

TABLE 4 ■ Metric Twist Drill Sizes

Size		Coarse Thread Series UNC & NC		Fine Thread Series UNF & NF		Extra Fine Series UNEF & NEF		8-Pitch Thread Series 8 N		12-Pitch Thread Series 12 N		16-Pitch Thread Series 16 N	
Number or Fraction	Decimal	Threads Per Inch	Tap Drill	Threads Per Inch	Tap Drill	Threads Per Inch	Tap Drill	Threads Per Inch	Tap Drill	Threads Per Inch	Tap Drill	Threads Per Inch	Tap Drill
	.060			80	3/64								
1	.073	64	No. 53	72	No. 53								
2	.086	56	No. 50	64	No. 50								
3	.099	48	No. 47	56	No. 45								
4	.112	40	No. 43	48	No. 42								
5	.125	40	No. 38	44	No. 37								
6	.138	32	No. 36	40	No. 33								
8	.164	32	No. 29	36	No. 29								
10	.190	24	No. 25	32	No. 21								
12	.216	24	No. 16	28	No. 14	32	No. 13						
1/4	.250	20	No. 7	28	No. 3	32	7/32						
5/16	.312	18	F	24	I	32	9/32						
3/8	.375	16	5/16	24	Q	32	11/32						
7/16	.438	14	U	20	25/64	28	13/32						
1/2	.500	13	27/64	20	29/64	28	15/32						
1/2	.500									12	27/64		
9/16	.562	12	31/64	18	33/64	24	33/64			12	31/64		
5/8	.625	11	17/32	18	37/64	24	37/64			12	35/64		
3/4	.750	10	21/32	16	11/16	20	45/64			12	43/64	16	11/16
7/8	.875	9	49/64	14	13/16	20	53/64			12	51/64	16	13/16
1	1.000	8	7/8	12	59/64	20	61/64	8	7/8	12	59/64	16	15/16
1 1/8	1.125	7	63/64	12	1 3/64	18	1 5/64	8	1	12	1 3/64	16	1 1/16
1 1/4	1.250	7	1 7/64	12	1 11/64	18	1 3/16	8	1 1/8	12	1 11/64	16	1 3/16
1 3/8	1.375	6	1 7/32	12	1 19/64	18	1 5/16	8	1 1/4	12	1 19/64	16	1 5/16
1 1/2	1.500	6	1 11/32	12	1 27/64	18	1 7/16	8	1 3/8	12	1 27/64	16	1 7/16
1 3/4	1.750	5	1 9/16			16	1 11/16	8	1 5/8	12	1 43/64	16	1 11/16
2	2.000	4 1/2	1 25/32			16	1 15/16	8	1 7/8	12	1 51/64	16	1 15/16
2 1/4	2.250	4 1/2	2 1/32					8	2 1/8	12	2 11/64	16	2 3/16
2 1/2	2.500	4	2 1/4					8	2 3/8	12	2 27/64	16	2 7/16
2 3/4	2.750	4	2 1/2					8	2 5/8	12	2 43/64	16	2 11/16
3	3.000	4	2 3/4					8	2 7/8	12	2 59/64	16	2 15/16

Colour shows unified thread

TABLE 5 ■ Unified and American (Inch) Threads

Nominal Size DIA (mm) Preferred	(other)	Series with Graded Pitches — Coarse Thread Pitch	Coarse Tap Drill Size	Fine Thread Pitch	Fine Tap Drill Size	Series with Constant Pitches — 4 Thread Pitch	4 Tap Drill Size	3 Thread Pitch	3 Tap Drill Size	2 Thread Pitch	2 Tap Drill Size	1.5 Thread Pitch	1.5 Tap Drill Size	1.25 Thread Pitch	1.25 Tap Drill Size	1 Thread Pitch	1 Tap Drill Size	0.75 Thread Pitch	0.75 Tap Drill Size	0.5 Thread Pitch	0.5 Tap Drill Size	0.35 Thread Pitch	0.35 Tap Drill Size
1.6		0.35	1.25																				
	1.8	0.35	1.45																				
2		0.4	1.6																				
	2.2	0.45	1.75																				
2.5		0.45	2.05																			0.35	2.15
3		0.5	2.5																			0.35	2.65
	3.5	0.6	2.9																			0.35	3.15
4		0.7	3.3																	0.5	3.5		
	4.5	0.75	3.7																	0.5	4		
5		0.8	4.2																	0.5	4.5		
*6		1	5															0.75	5.2				
**6.3		1	5.3																				
8		1.25	6.7	1	7											1	7	0.75	7.2				
10		1.5	8.5	1.25	8.7									1.25	8.7	1	9	0.75	9.2				
12		1.75	10.2	1.25	10.8							1.5	10.5	1.25	10.7	1	11						
	14	2	12	1.5	12.5							1.5	12.5	1.25	12.7	1	13						
16		2	14	1.5	14.5							1.5	14.5			1	15						
18		2.5	15.5	1.5	16.5					2	16	1.5	16.5			1	17						
	20	2.5	17.5	1.5	18.5					2	18	1.5	18.5			1	19						
22		2.5	19.5	1.5	20.5					2	20	1.5	20.5			1	21						
	24	3	21	2	22					2	22	1.5	22.5			1	23						
27		3	24	2	25					2	25	1.5	25.5			1	26						
30		3.5	26.5	2	28					2	28	1.5	28.5			1	29						
	33	3.5	29.5	2	31					2	31	1.5	31.5										
36		4	32	3	33					2	34	1.5	34.5										
	39	4	35	3	36					2	37	1.5	37.5										
42		4.5	37.5	3	39	4	38	3	39	2	40	1.5	40.5										
	45	4.5	39	3	42	4	41	3	42	2	43	1.5	43.5										
48		5	43	3	45	4	44	3	45	2	46	1.5	46.5										

* ISO thread size

** ASME thread size (to be discontinued)

TABLE 6 ■ Metric Threads

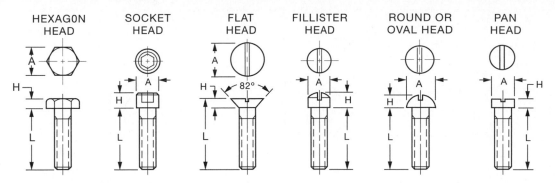

	Nominal Size		Hexagon Head		Socket Head		Flat Head		Fillister Head		Round or Oval Head	
	Fraction	Decimal	A	H	A	H	A	H	A	H	A	H
Imperial and U.S. Customary (Inches)	1/4	.250	.44	.17	.38	.25	.50	.14	.38	.22	.44	.19
	5/16	.312	.50	.22	.47	.31	.62	.18	.44	.25	.56	.25
	3/8	.375	.56	.25	.56	.38	.75	.21	.56	.31	.62	.27
	7/16	.438	.62	.30	.66	.44	.81	.21	.62	.36	.75	.33
	1/2	.500	.75	.34	.75	.50	.88	.21	.75	.41	.81	.35
	5/8	.625	.94	.42	.94	.62	1.12	.28	.88	.50	1.00	.44
	3/4	.750	1.12	.50	1.12	.75	1.38	.35	1.00	.59	1.25	.55
	7/8	.875	1.31	.58	1.31	.88	1.62	.42	1.12	.69		
	1	1.000	1.50	.67	1.50	1.00	1.88	.49	1.31	.78		
	1 1/8	1.125	1.69	.75	1.69	1.12	2.06	.53				
	1 1/4	1.250	1.88	.84	1.88	1.25	2.31	.60				
	1 1/2	1.500	2.25	1.00	2.25	1.50	2.81	.74				

	Nominal Size	Hexagon Head		Socket Head			Flat Head		Fillister Head		Pan Head	
		A	H	A	H	Key Size	A	H	A	H	A	H
Metric (Millimetres)	M3	5.5	2	5.5	3	2.5	5.6	1.6	6	2.4	5.6	1.9
	4	7	2.8	7	4	3	7.5	2.2	8	3.1	7.5	2.5
	5	8.5	3.5	9	5	4	9.2	2.5	10	3.8	9.2	3.1
	6	10	4	10	6	5	11	3	12	4.6	11	3.8
	8	13	5.5	13	8	6	14.5	4	16	6	14.5	5
	10	17	7	16	10	8	18	5	20	7.5	18	6.2
	12	19	8	18	12	10						
	14	22	9	22	14	12						
	16	24	10	24	16	14						
	18	27	12	27	18	14						
	20	30	13	30	20	17						
	22	36	15	33	22	17						
	24	36	15	36	24	19						
	27	41	17	40	27	19						
	30	46	19	45	30	22						

NOTE: Length sizes normally available in .25 inch and 10 mm increments

TABLE 7 ■ Common Cap Screws

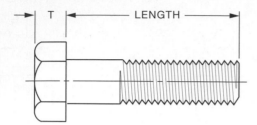

Imperial and U.S. Customary (Inches)			
Nominal Size Fraction	Decimal	Width across Flats	Thickness
1/4	.250	.44	.17
5/16	.312	.50	.22
3/8	.375	.56	.25
7/16	.438	.62	.30
1/2	.500	.75	.34
5/8	.625	.94	.42
3/4	.750	1.12	.50
7/8	.875	1.31	.58
1	1.000	1.50	.67
1 1/8	1.125	1.69	.75
1 1/4	1.250	1.88	.84
1 3/8	1.375	2.06	.91
1 1/2	1.500	2.25	1.00

Metric (Millimetres)		
Nominal Size (Millimetres)	Width across Flats	Thickness
4	7	2.8
5	8	3.5
6	10	4
8	13	5.5
10	17	7
12	19	8
14	22	9
16	24	10
18	27	12
20	30	13
22	32	14
24	36	15
27	41	17
30	46	19
33	50	21
36	55	23

NOTE: For bolt and cap screw sizes below 7/16 inch and 8 mm, length sizes normally available in .25 inch and 10 mm increments

TABLE 8 ∎ Hexagon-Head Bolts and Cap Screws

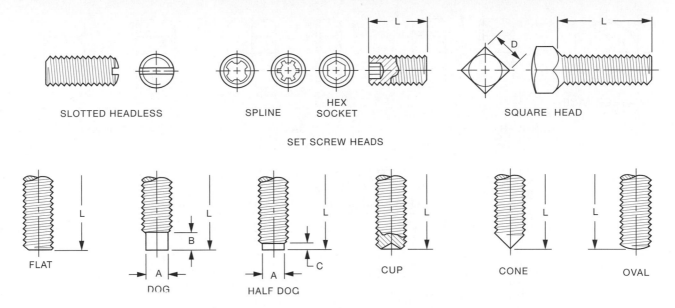

SET SCREW HEADS

SET SCREW POINTS

Imperial and U.S. Customary (Inches)			Metric (Millimetres)	
Nominal Size		Key Size	Nominal Size	Key Size
Number	Decimal			
4	.112	.050	M 1.4	0.7
5	.125	.062	2	0.9
6	.138	.062	3	1.5
8	.164	.078	4	2
10	.190	.094	5	2.5
12	.216	.109	6	3
1/4	.250	.125	8	4
5/16	.312	.156	10	5
3/8	.375	.188	12	6
1/2	.500	.250	16	8

TABLE 9 ■ Set screws

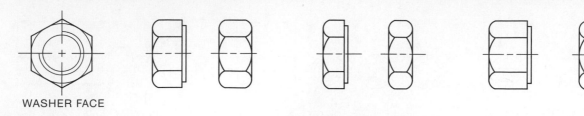

WASHER FACE

REGULAR JAM THICK

	Nominal Size		Distance across Flats	Thickness		
	Fraction	Decimal		Regular	Jam	Thick
Imperial and U.S. Customary (Inches)	1/4	.250	.44	.22	.16	.28
	5/16	.312	.50	.27	.19	.33
	3/8	.375	.56	.33	.22	.41
	7/16	.438	.69	.38	.25	.45
	1/2	.500	.75	.44	.31	.56
	9/16	.562	.88	.48	.31	.61
	5/8	.625	.94	.55	.38	.72
	3/4	.750	1.12	.64	.42	.81
	7/8	.875	1.31	.75	.48	.91
	1	1.000	1.50	.86	.55	1.00
	1 1/8	1.125	1.69	.97	.61	1.16
	1 1/4	1.250	1.88	1.06	.72	1.25
	1 3/8	1.375	2.06	1.17	.78	1.38
	1 1/2	1.500	2.25	1.28	.84	1.50

	Nominal Size (Millimetres)	Distance across Flats	Thickness		
			Regular	Jam	Thick
Metric (Millimetres)	4	7	3	2	5
	5	8	4	2.5	5
	6	10	5	3	6
	8	13	6.5	5	8
	10	17	8	6	10
	12	19	10	7	12
	14	22	11	8	14
	16	24	13	8	16
	18	27	15	9	18.5
	20	30	16	9	20
	22	32	18	10	22
	24	36	19	10	24
	27	41	22	12	27
	30	46	24	12	30
	33	50	26		
	36	55	29		
	39	60	31		

TABLE 10 ■ Hexagon-Head Nuts

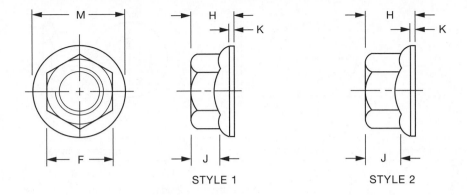

STYLE 1   STYLE 2

Metric (Millimetres)							
Nominal Nut Size and Thread Pitch	Width across Flats F	Style 1				Style 2	
		H	J	K	M	H	J
M6 × 1	10	5.8	3	1	14.2	6.7	3.7
M8 × 1.25	13	6.8	3.7	1.3	17.6	8	4.5
M10 × 1.5	15	9.6	5.5	1.5	21.5	11.2	6.7
M12 × 1.75	18	11.6	6.7	2	25.6	13.5	8.2
M14 × 2	21	13.4	7.8	2.3	29.6	15.7	9.6
M16 × 2	24	15.9	9.5	2.5	34.2	18.4	11.7
M20 × 2.5	30	19.2	11.1	2.8	42.3	22	12.6

TABLE 11 ■ Hex Flange Nuts

FLAT WASHER   LOCKWASHER

Nominal Screw Size		Flat Washer			Lockwasher		
Number or Fraction	Decimal	Inside Dia A	Outside Dia B	Thickness C	Inside Dia A	Outside Dia B	Thickness C
6	.138	.16	.38	.05	.14	.25	.03
8	.164	.19	.44	.05	.17	.29	.04
10	.190	.22	.50	.05	.19	.33	.05
12	.216	.25	.56	.07	.22	.38	.06
1/4	.250 N	.28	.63	.07	.26	.49	.06
1/4	.250 W	.31	.73	.07			
5/16	.312 N	.34	.69	.07	.32	.59	.08
5/16	.312 N	.38	.88	.08			
3/8	.375 W	.41	.81	.07	.38	.68	.09
3/8	.375 W	.44	1.00	.08			
7/16	.438 W	.47	.92	.07	.45	.78	.11
7/16	.438 W	.50	1.25	.08			
1/2	.500 N	.53	1.06	.10	.51	.87	.12
1/2	.500 W	.56	1.38	.11			
5/8	.625 N	.66	1.31	.10	.64	1.08	.16
5/8	.625 W	.69	1.75	.13			
3/4	.750 N	.81	1.47	.13	.76	1.27	.19
3/4	.750 W	.81	2.00	.15			
7/8	.875 N	.94	1.75	.13	.89	1.46	.22
7/8	.875 W	.94	2.25	.17			
1	1.000 N	1.06	2.00	.13	1.02	1.66	.25
1	1.000 W	1.06	2.50	.17			
1 1/8	1.125 N	1.25	2.25	.13	1.14	1.85	.28
1 1/8	1.125 W	1.25	2.75	.17			
1 1/4	1.250 N	1.38	2.50	.17	1.27	2.05	.31
1 1/4	1.250 W	1.38	3.00	.17			
1 3/8	1.375 N	1.50	2.75	.17	1.40	2.24	.34
1 3/8	1.375 W	1.50	3.25	.18			
1 1/2	1.500 N	1.62	3.00	.17	1.53	2.43	.38
1 1/2	1.500 W	1.62	3.50	.18			

N—SAE Sizes (Narrow)
W—Standard Plate (Wide)   INCH SIZES

TABLE 12 ■ Common Washer Sizes

FLAT WASHER   LOCKWASHER   SPRING LOCKWASHER

Bolt Size	Flat Washer			Lockwasher			Spring Lockwasher		
	ID	OD	Thickness	ID	OD	Thickness	ID	OD	Thickness
2	2.2	5.	0.5	2.1	3.3	0.5			
3	3.2	7	0.5	3.1	5.7	0.8			
4	4.3	9	0.8	4.1	7.1	0.9	4.2	8	0.3 0.4
5	5.3	11	1	5.1	8.7	1.2	5.2	10	0.4 0.5
6	6.4	12	1.5	6.1	11.1	1.6	6.2	12.5	0.5 0.7
7	7.4	14	1.5	7.1	12.1	1.6	7.2	14	0.5 0.8
8	8.4	17	2	8.2	14.2	2	8.2	16	0.6 0.9
10	10.5	21	2.5	10.2	17.2	2.2	10.2	20	0.8 1.1
12	13	24	2.5	12.3	20.2	2.5	12.2	25	0.9 1.5
14	15	28	2.5	14.2	23.2	3	14.2	28	1 1.5
16	17	30	3	16.2	26.2	3.5	16.3	31.5	1.2 1.7
18	19	34	3	18.2	28.2	3.5	18.3	35.5	1.2 2
20	21	36	3	20.2	32.2	4	20.4	40	1.5 2.25
22	23	39	4	22.5	34.5	4	22.4	45	1.75 2.5
24	25	44	4	24.5	38.5	5			
27	28	50	4	27.5	41.5	5			
30	31	56	4	30.5	46.5	6			

MILLIMETRE SIZES

TABLE 12 (CONT'D) ■ Common Washer Sizes

Imperial and U.S. Customary (Inches)					Metric (Millimetres)						
Diameter of Shaft	Square Key		Flat Key		Diameter of Shaft (mm)		Square Key		Flat Key		
	Nominal Size		Nominal Size				Nominal Size		Nominal Size		
Inclusive	W	H	W	H	Over	Up To	W	H	W	H	
.500 – .562	.125	.125	.125	.094	12	17	5	5			
.625 – .875	.188	.188	.188	.125	17	22	6	6			
.938 –1.250	.250	.250	.250	.188	22	30	7	7	8	7	
1.312 –1.375	.312	.312	.312	.250	30	38	8	8	10	8	
1.438 –1.750	.375	.375	.375	.250	38	44	9	9	12	8	
1.812 –2.250	.500	.500	.500	.375	44	50	10	10	14	9	
2.312 –2.750	.625	.625	.625	.438	50	58	12	12	16	10	

TABLE 13 ■ Square and Flat Stock Keys

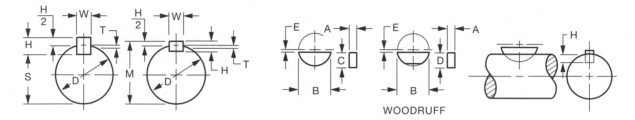

C = ALLOWANCE FOR PARALLEL KEYS = .005 IN. OR 0.12 MM

$$S = D - \frac{H}{2} - T = \frac{D - H + \sqrt{D^2 - W^2}}{2} \qquad T = \frac{D - \sqrt{D^2 - W^2}}{2}$$

W = NORMAL KEY WIDTH (INCHES OR MILLIMETRES)

$$M = D - T + \frac{H}{2} + C = \frac{D + H + \sqrt{D^2 - W^2} + C}{2}$$

Key No.	Nominal (A × B)		Imperial and U.S. Customary (Inches)				Metric (Millimetres)			
			Key			Keyseat	Key			Keyseat
	Millimetres	Inches	E	C	D	H	E	C	D	H
204	1.6 × 6.4	0.062 × 0.250	.05	.20	.19	.10	0.5	2.8	2.8	4.3
304	2.4 × 12.7	0.094 × 0.500	.05	.20	.19	.15	1.3	5.1	4.8	3.8
305	2.4 × 15.9	0.094 × 0.625	.06	.25	.24	.20	1.5	6.4	6.1	5.1
404	3.2 × 12.7	0.125 × 0.500	.05	.20	.19	.14	1.3	5.1	4.8	3.6
405	3.2 × 15.9	0.125 × 0.625	.06	.25	.24	.18	1.5	6.4	6.1	4.6
406	3.2 × 19.1	0.125 × 0.750	.06	.31	.30	.25	1.5	7.9	7.6	6.4
505	4.0 × 15.9	0.156 × 0.625	.06	.25	.24	.17	1.5	6.4	6.1	4.3
506	4.0 × 19.1	0.156 × 0.750	.06	.31	.30	.23	1.5	7.9	7.6	5.8
507	4.0 × 22.2	0.156 × 0.875	.06	.38	.36	.29	1.5	9.7	9.1	7.4
606	4.8 × 19.1	0.188 × 0.750	.06	.31	.30	.21	1.5	7.9	7.6	5.3
607	4.8 × 22.2	0.188 × 0.875	.06	.38	.36	.28	1.5	9.7	9.1	7.1
608	4.8 × 25.4	0.188 × 1.000	.06	.44	.43	.34	1.5	11.2	10.9	8.6
609	4.8 × 28.6	0.188 × 1.250	.08	.48	.47	.39	2.0	12.2	11.9	9.9
807	6.4 × 22.2	0.250 × 0.875	.06	.38	.36	.25	1.5	9.7	9.1	6.4
808	6.4 × 25.4	0.250 × 1.000	.06	.44	.43	.31	1.5	11.2	10.9	7.9

TABLE 14 ■ Woodruff Keys

Nominal Pipe Size	Imperial and U.S. Customary (Inches)					Metric (Millimetres)				
	Outside Diameter	Wall Thickness				Outside Diameter (Millimetres)	Wall Thickness (Millimetres)			
		Schedule 40 Pipe *	Schedule 80 Pipe **	Schedule 160 Pipe			Standard	Extra Strong	Double Extra Strong	
.125 (1/8)	.405	.068	.095	—		10.29	1.75	2.44	—
.250 (1/4)	.540	.088	.119	—		13.72	2.29	3.15	—
.375 (3/8)	.675	.091	.126	—		17.15	2.36	3.28	—
.500 (1/2)	.840	.109	.147	.188		21.34	2.82	3.84	7.80
.750 (3/4)	1.050	.113	.154	.219		26.67	2.92	3.99	8.08
1.00	1.315	.133	.179	.250		33.4	3.45	4.65	9.37
1.25	1.660	.140	.191	.250		42.4	3.63	4.98	9.98
1.50	1.900	.145	.200	.281		48.3	3.76	5.18	10.44
2.00	2.375	.154	.218	.344		60.3	3.99	5.66	11.35
2.50	2.875	.203	.276	.375		73.0	5.26	7.16	14.40
3.00	3.500	.216	.300	.438		88.9	5.61	7.77	15.62
3.50	4.000	.226	.318	—		101.6	5.87	8.25	16.54
4.00	4.500	.237	.337	.531		114.3	6.15	8.74	17.53
5.00	5.563	.258	.375	.625		141.3	6.68	9.73	19.51
6.00	6.625	.280	.432	.719		168.3	7.26	11.20	22.45
8.00	8.625	.322	.500	.906		219.1	8.36	12.98	22.73
10.00	10.750	.365	.594	1.125		273.1	9.45	12.95	—
12.00	12.750	.406	.688	1.312		323.9	—	12.95	—
14.00	14.000	.438	.750	1.406		355.6	9.73	12.95	—
16.00	16.000	.500	.844	1.594		406.4	9.73	12.95	—

*Standard
**Extra Strong Pipe

Nominal pipe sizes are specified in inches.
Outside diameter and wall thickness are specified in millimetres.

TABLE 15 ■ American Standard Wrought Steel Pipe

North American Gauges								European Gauges			
Ferrous Metals, such as Galvanized Steel, Tin Plate		Galvanized Steel, Tin Plate, Copper, Strip Steel and Steel, Copper, Aluminum Tubes						Nonferrous Metals, such as Copper, Brass, Aluminum		Nonferrous	
U.S. Standard (USS)		U.S. Standard (Revised)		Birmingham (BWG)		New Birmingham (BG)		Browne and Sharpe (B & S)		Imperial Standard (SWG)	
Gauge	In.	Gauge	In.	Gauge	In.	Gauge	In.	Gauge	In.	Gauge	In.
		3	.239					3	.229		
4	.234	4	.224	4	.238	4	.250	4	.204	4	.232
5	.219	5	.209	5	.220	5	.223	5	.182	5	.212
6	.203	6	.194	6	.203	6	.198	6	.162	6	.192
7	.188	7	.179	7	.180	7	.176	7	.144	7	.176
8	.172	8	.164	8	.165	8	.157	8	.129	8	.160
9	.156	9	.149	9	.148	9	.140	9	.114	9	.144
10	.141	10	.135	10	.134	10	.125	10	.102	10	.128
11	.125	11	.120	11	.120	11	.111	11	.091	11	.116
12	.109	12	.105	12	.109	12	.099	12	.081	12	.104
13	.094	13	.090	13	.095	13	.088	13	.072	13	.092
14	.078	14	.075	14	.083	14	.079	14	.064	14	.080
15	.070	15	.067	15	.072	15	.070	15	.057	15	.072
16	.063	16	.060	16	.065	16	.063	16	.051	16	.064
17	.056	17	.054	17	.058	17	.056	17	.045	17	.056
18	.050	18	.048	18	.049	18	.050	18	.040	18	.048
19	.044	19	.042	19	.042	19	.044	19	.036	19	.040
20	.038	20	.036	20	.035	20	.039	20	.032	20	.036
21	.034	21	.033	21	.032	21	.035	21	.029	21	.032
22	.031	22	.030	22	.028	22	.031	22	.025	22	.028
23	.028	23	.027	23	.025	23	.028	23	.023	23	.024
24	.025	24	.024	24	.022	24	.025	24	.020	24	.022
25	.022	25	.021	25	.020	25	.022	25	.018	25	.020
26	.019	26	.018	26	.018	26	.020	26	.016	26	.018
27	.017	27	.016	27	.016	27	.017	27	.014	27	.016
28	.016	28	.015	28	.014	28	.016	28	.013	28	.015
29	.014	29	.014	29	.013	29	.014	29	.011	29	.014
30	.012	30	.012	30	.012	30	.012	30	.010	30	.012
31	.011	31	.011	31	.010	31	.011	31	.009		
32	.010	32	.010	32	.009			32	.008	32	.011
33	.009	33	.009	33	.008	33	.009	33	.007	33	.010
34	.008	34	.008	34	.007	34	.008	34	.006	34	.009
				35	.005	35	.007			35	.008
36	.007	36	.007	36	.004	36	.006	36	.005		

TABLE 16 ■ Sheet Metal Gauges and Thicknesses

**EXAMPLE: RC2 SLIDING FIT FOR A
Ø1.50 NOMINAL HOLE DIAMETER**

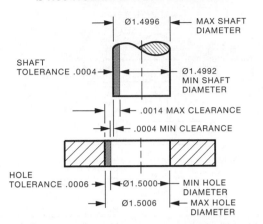

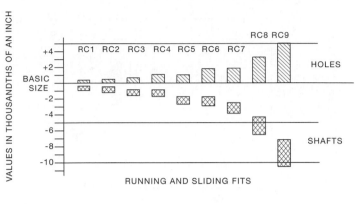

RUNNING AND SLIDING FITS

Nominal Size Range Inches		Class RC1 Precision Sliding			Class RC2 Sliding Fit			Class RC3 Precision Running			Class RC4 Close Running			Class RC5 Medium Running		
		Hole Tol. GR5	Minimum Clearance	Shaft Tol. GR4	Hole Tol. GR6	Minimum Clearance	Shaft Tol. GR5	Hole Tol. GR7	Minimum Clearance	Shaft Tol. GR6	Hole Tol. GR8	Minimum Clearance	Shaft Tol. GR7	Hole Tol. GR8	Minimum Clearance	Shaft Tol. GR7
Over	To	-0		+0	-0		+0	-0		+0	-0		+0	-0		+0
0	.12	+0.15	0.10	-0.12	+0.25	0.10	-0.15	+0.40	0.30	-0.25	+0.60	0.30	-0.40	+0.60	0.60	-0.40
.12	.24	+0.20	0.15	-0.15	+0.30	0.15	-0.20	+0.50	0.40	-0.30	+0.70	0.40	-0.50	+0.70	0.80	-0.50
.24	.40	+0.25	0.20	-0.15	+0.40	0.20	-0.25	+0.60	0.50	-0.40	+0.90	0.50	-0.60	+0.90	1.00	-0.60
.40	.71	+0.30	0.25	-0.20	+0.40	0.25	-0.30	+0.70	0.60	-0.40	+1.00	0.60	-0.70	+1.00	1.20	-0.70
.71	1.19	+0.40	0.30	-0.25	+0.50	0.30	-0.40	+0.80	0.80	-0.50	+1.20	0.80	-0.80	+1.20	1.60	-0.80
1.19	1.97	+0.40	0.40	-0.30	+0.60	0.40	-0.40	+1.00	1.00	-0.60	+1.60	1.00	-1.00	+1.60	2.00	-1.00
1.97	3.15	+0.50	0.40	-0.30	+0.70	0.40	-0.50	+1.20	1.20	-0.70	+1.80	1.20	-1.20	+1.80	2.50	-1.20
3.15	4.73	+0.60	0.50	-0.40	+0.90	0.50	-0.60	+1.40	1.40	-0.90	+2.20	1.40	-1.40	+2.20	3.00	-1.40
4.73	7.09	+0.70	0.60	-0.50	+1.00	0.60	-0.70	+1.60	1.60	-1.00	+2.50	1.60	-1.60	+2.50	3.50	-1.60
7.09	9.85	+0.80	0.60	-0.60	+1.20	0.60	-0.80	+1.80	2.00	-1.20	+2.80	2.00	-1.80	+2.80	4.50	-1.80
9.85	12.41	+0.90	0.80	-0.60	+1.20	0.80	-0.90	+2.00	2.50	-1.20	+3.00	2.50	-2.00	+3.00	5.00	-2.00
12.41	15.75	+1.00	1.00	-0.70	+1.40	1.00	-1.00	+2.20	3.00	-1.40	+3.50	3.00	-2.20	+3.50	6.00	-2.20

Nominal Size Range Inches		Class RC6 Precision Sliding			Class RC7 Sliding Fit			Class RC8 Precision Running			Class RC9 Close Running		
		Hole Tol. GR9	Minimum Clearance	Shaft Tol. GR8	Hole Tol. GR9	Minimum Clearance	Shaft Tol. GR8	Hole Tol. GR10	Minimum Clearance	Shaft Tol. GR9	Hole Tol. GR11	Minimum Clearance	Shaft Tol. GR10
Over	To	-0		+0	-0		+0	-0		+0	-0		+0
0	.12	+1.00	0.60	-0.60	+1.00	1.00	-0.60	+1.60	2.50	-1.00	+2.50	4.00	-1.60
.12	.24	+1.20	0.80	-0.70	+1.20	1.20	-0.70	+1.80	2.80	-1.20	+3.00	4.50	-1.80
.24	.40	+1.40	1.00	-0.90	+1.40	1.60	-0.90	+2.20	3.00	-1.40	+3.50	6.00	-2.20
.40	.71	+1.60	1.20	-1.00	+1.60	2.00	-1.00	+2.80	3.50	-1.60	+4.00	6.00	-2.80
.71	1.19	+2.00	1.60	-1.20	+2.00	2.50	-1.20	+3.50	4.50	-2.00	+5.00	7.00	-3.50
1.19	1.97	+2.50	2.00	-1.60	+2.50	3.00	-1.60	+4.00	5.00	-2.50	+6.00	8.00	-4.00
1.97	3.15	+3.00	2.50	-1.80	+3.00	4.00	-1.80	+4.50	6.00	-3.00	+7.00	9.00	-4.50
3.15	4.73	+3.50	3.00	-2.20	+3.50	5.00	-2.20	+5.00	7.00	-3.50	+9.00	10.00	-5.00
4.73	7.09	+4.00	3.50	-2.50	+4.00	6.00	-2.50	+6.00	8.00	-4.00	+10.00	12.00	-6.00
7.09	9.85	+4.50	4.00	-2.80	+4.50	7.00	-2.80	+7.00	10.00	-4.50	+12.00	15.00	-7.00
9.85	12.41	+5.00	5.00	-3.00	+5.00	8.00	-3.00	+8.00	12.00	-5.00	+12.00	18.00	-8.00
12.41	15.75	+6.00	6.00	-3.50	+6.00	10.00	-3.50	+9.00	14.00	-6.00	+14.00	22.00	-9.00

TABLE 17 ■ Running and Sliding Fits (Values in Thousandths of an Inch)

EXAMPLE: LC2 LOCATIONAL FIT FOR A Ø1.50 NOMINAL HOLE DIAMETER

LOCATIONAL CLEARANCE FITS

Nominal Size Range Inches		Class LC1			Class LC2			Class LC3			Class LC4			Class LC5			Class LC6		
		Hole Tol. GR6	Minimum Clearance	Shaft Tol. GR5	Hole Tol. GR8	Minimum Clearance	Shaft Tol. GR7	Hole Tol. GR10	Minimum Clearance	Shaft Tol. GR9	Hole Tol. GR7	Minimum Clearance	Shaft Tol. GR6	Hole Tol. GR9	Minimum Clearance	Shaft Tol. GR8	Hole Tol. GR9	Minimum Clearance	Shaft Tol. GR8
Over	To	-0		+0	-0		+0	-0		+0	-0		+0	-0		+0	-0		+0
0	.12	+0.25	0	-0.15	+0.4	0	-0.25	+0.6	0	-0.4	+1.6	0	-1.0	+0.4	0.10	-0.25	+1.0	0.3	-0.6
.12	.24	+0.30	0	-0.20	+0.5	0	-0.30	+0.7	0	-0.5	+1.8	0	-1.2	+0.5	0.15	-0.30	+1.2	0.4	-0.7
.24	.40	+0.40	0	-0.25	+0.6	0	-0.40	+0.9	0	-0.6	+2.2	0	-1.4	+0.6	0.20	-0.40	+1.4	0.5	-0.9
.40	.71	+0.40	0	-0.30	+0.7	0	-0.40	+1.0	0	-0.7	+2.8	0	-1.6	+0.7	0.25	-0.40	+1.6	0.6	-1.0
.71	1.19	+0.50	0	-0.40	+0.8	0	-0.50	+1.2	0	-0.8	+3.5	0	-2.0	+0.8	0.30	-0.50	+2.0	0.8	-1.2
1.19	1.97	+0.60	0	-0.40	+1.0	0	-0.60	+1.6	0	-1.0	+4.0	0	-2.5	+1.0	0.40	-0.60	+2.5	1.0	-1.6
1.97	3.15	+0.70	0	-0.50	+1.2	0	-0.70	+1.8	0	-1.2	+4.5	0	-3.0	+1.2	0.40	-0.70	+3.0	1.2	-1.8
3.15	4.73	+0.90	0	-0.60	+1.4	0	-0.90	+2.5	0	-1.4	+5.0	0	-3.5	+1.4	0.50	-0.90	+3.5	1.4	-2.2
4.73	7.09	+1.00	0	-0.70	+1.6	0	-1.00	+2.7	0	-1.6	+6.0	0	-4.0	+1.6	0.60	-1.00	+4.0	1.6	-2.5
7.09	9.85	+1.20	0	-0.80	+1.8	0	-1.20	+2.8	0	-1.8	+7.0	0	-4.5	+1.8	0.60	-1.20	+4.5	2.0	-2.8
9.85	12.41	+1.20	0	-0.90	+2.0	0	-1.20	+3.0	0	-2.0	+8.0	0	-5.0	+2.0	0.70	-1.20	+5.0	2.2	-3.0
12.41	15.75	+1.40	0	-1.00	+2.2	0	-1.40	+3.5	0	-2.2	+9.0	0	-6.0	+2.2	0.70	-1.40	+6.0	2.5	-3.5

Nominal Size Range Inches		Class LC7			Class LC8			Class LC9			Class LC10			Class LC11		
		Hole Tol. GR10	Minimum Clearance	Shaft Tol. GR9	Hole Tol. GR10	Minimum Clearance	Shaft Tol. GR9	Hole Tol. GR11	Minimum Clearance	Shaft Tol. GR10	Hole Tol. GR12	Minimum Clearance	Shaft Tol. GR11	Hole Tol. GR13	Minimum Clearance	Shaft Tol. GR12
Over	To	-0		+0	-0		+0	-0		+0	-0		+0	-0		+0
0	.12	+1.6	0.6	-1.0	+1.6	1.0	-1.0	+2.5	2.5	-1.6	+1.0	4.0	-2.5	+6.0	5.0	-4.0
.12	.24	+1.8	0.8	-1.2	+1.8	1.2	-1.2	+3.0	2.8	-1.8	+5.0	4.5	-3.0	+7.0	6.0	-5.0
.24	.40	+2.2	1.0	-1.4	+2.2	1.6	-1.4	+3.5	3.0	-2.2	+6.0	5.0	-3.5	+9.0	7.0	-6.0
.40	.71	+2.8	1.2	-1.6	+2.8	2.0	-1.6	+4.0	3.5	-2.8	+7.0	6.0	-4.0	+10.0	8.0	-7.0
.71	1.19	+3.5	1.6	-2.0	+3.5	2.5	-2.0	+5.0	4.5	-3.5	+8.0	7.0	-5.0	+12.0	10.0	-8.0
1.19	1.97	+4.0	2.0	-2.5	+4.0	3.6	-2.5	+6.0	5.0	-4.0	+10.0	8.0	-6.0	+16.0	12.0	-10.0
1.97	3.15	+4.5	2.5	-3.0	+4.5	4.0	-3.0	+7.0	6.0	-4.5	+12.0	10.0	-7.0	+18.0	14.0	-12.0
3.15	4.73	+5.0	3.0	-3.5	+5.0	5.0	-3.5	+9.0	7.0	-5.0	+14.0	11.0	-9.0	+22.0	16.0	-14.0
4.73	7.09	+6.0	3.5	-4.0	+6.0	6.0	-4.0	+10.0	8.0	-6.0	+16.0	12.0	-10.0	+25.0	18.0	-16.0
7.09	9.85	+7.0	4.0	-4.5	+7.0	7.0	-4.5	+12.0	10.0	-7.0	+18.0	16.0	-12.0	+28.0	22.0	-18.0
9.85	12.41	+8.0	4.5	-5.0	+8.0	7.0	-5.0	+12.0	12.0	-8.0	+20.0	20.0	-12.0	+30.0	28.0	-20.0
12.41	15.75	+9.0	5.0	-6.0	+9.0	8.0	-6.0	+14.0	14.0	-9.0	+22.0	22.0	-14.0	+35.0	30.0	-22.0

TABLE 18 ■ Locational Clearance Fits (Values in Thousandths of an Inch)

EXAMPLE: LT2 TRANSITION FIT FOR A Ø1.50 NOMINAL HOLE DIAMETER

TRANSITION FITS

Nominal Size Range Inches		Class LT1			Class LT2			Class LT3			Class LT4			Class LT5			Class LT6		
		Hole Tol. GR7	Maximum Interference	Shaft Tol. GR6	Hole Tol. GR8	Maximum Interference	Shaft Tol. GR7	Hole Tol. GR7	Maximum Interference	Shaft Tol. GR6	Hole Tol. GR8	Maximum Interference	Shaft Tol. GR7	Hole Tol. GR7	Maximum Interference	Shaft Tol. GR6	Hole Tol. GR8	Maximum Interference	Shaft Tol. GR7
Over	To	-0		+0	-0		+0	-0		+0	-0		+0	-0		+0	-0		+0
0	.12	+0.4	0.10	-0.25	+0.6	0.20	-0.4	+0.4	0.25	-0.25	+0.6	0.4	-0.4	+0.4	0.5	-0.25	+0.6	0.65	-0.4
.12	.24	+0.5	0.15	-0.30	+0.7	0.25	-0.5	+0.5	0.40	-0.30	+0.7	0.6	-0.5	+0.5	0.6	-0.30	+0.7	0.80	-0.5
.24	.40	+0.6	0.20	-0.40	+0.9	0.30	-0.6	+0.6	0.50	-0.40	+0.9	0.7	-0.6	+0.6	0.8	-0.40	+0.9	1.00	-0.6
.40	.71	+0.7	0.20	-0.40	+1.0	0.30	-0.7	+0.7	0.50	-0.40	+1.0	0.8	-0.7	+0.7	0.9	-0.40	+1.0	1.20	-0.7
.71	1.19	+0.8	0.25	-0.50	+1.2	0.40	-0.8	+0.8	0.60	-0.50	+1.2	0.9	-0.8	+0.8	1.1	-0.50	+1.2	1.40	-0.8
1.19	1.97	+1.0	0.30	-0.60	+1.6	0.50	-1.0	+1.0	0.70	-0.60	+1.6	1.1	-1.0	+1.0	1.3	-0.60	+1.6	1.70	-1.0
1.97	3.15	+1.2	0.30	-0.70	+1.8	0.60	-1.2	+1.2	0.80	-0.70	+1.8	1.3	-1.2	+1.2	1.5	-0.70	+1.8	2.00	-1.2
3.15	4.73	+1.4	0.40	-0.90	+2.2	0.70	-1.4	+1.4	1.00	-0.90	+2.2	1.5	-1.4	+1.4	1.9	-0.90	+2.2	2.40	-1.4
4.73	7.09	+1.6	0.50	-1.00	+2.5	0.80	-1.6	+1.6	1.10	-1.00	+2.5	1.7	-1.6	+1.6	2.2	-1.00	+2.5	2.80	-1.6
7.09	9.85	+1.8	0.60	-1.20	+2.8	0.90	-1.8	+1.8	1.40	-1.20	+2.8	2.0	-1.8	+1.8	2.6	-1.20	+2.8	3.20	-1.8
9.85	12.41	+2.0	0.60	-1.20	+3.0	1.00	-2.0	+2.0	1.40	-1.20	+3.0	2.2	-2.0	+2.0	2.6	-1.20	+3.0	3.40	-2.0
12.41	15.75	+2.2	0.70	-1.40	+3.5	1.00	-2.2	+2.2	1.60	-1.40	+3.5	2.4	-2.2	+2.2	3.0	-1.40	+3.5	3.80	-2.2

TABLE 19 ■ Locational Transition Fits (Values in Thousandths of an Inch)

EXAMPLE: LN2 LOCATIONAL INTERFERENCE FIT FOR A Ø1.50 NOMINAL HOLE DIAMETER

LOCATIONAL INTERFERENCE FITS

Nominal Size Range Inches		Class LN1 Light Press Fit			Class LN2 Medium Press Fit			Class LN3 Heavy Press Fit			Class LN4			Class LN5			Class LN6		
Over	To	Hole Tol. GR6 -0	Maximum Interference	Shaft Tol. GR5 +0	Hole Tol. GR7 -0	Maximum Interference	Shaft Tol. GR6 +0	Hole Tol. GR7 -0	Maximum Interference	Shaft Tol. GR6 +0	Hole Tol. GR8 -0	Maximum Interference	Shaft Tol. GR7 +0	Hole Tol. GR9 -0	Maximum Interference	Shaft Tol. GR8 +0	Hole Tol. GR10 -0	Maximum Interference	Shaft Tol. GR9 +0
0	.12	+0.25	0.40	-0.15	+0.4	0.65	-0.25	+0.4	0.75	-0.25	+0.6	1.2	-0.4	+1.0	1.8	-0.6	+1.6	3.0	-1.0
.12	.24	+0.30	0.50	-0.20	+0.5	0.80	-0.30	+0.5	0.90	-0.30	+0.7	1.5	-0.5	+1.2	2.3	-0.7	+1.8	3.6	-1.2
.24	.40	+0.40	0.65	-0.25	+0.6	1.00	-0.40	+0.6	1.20	-0.40	+0.9	1.8	-0.6	+1.4	2.8	-0.9	+2.2	4.4	-1.4
.40	.71	+0.40	0.70	-0.30	+0.7	1.10	-0.40	+0.7	1.40	-0.40	+1.0	2.2	-0.7	+1.6	3.4	-1.0	+2.8	5.6	-1.6
.71	1.19	+0.50	0.90	-0.40	+0.8	1.30	-0.50	+0.8	1.70	-0.50	+1.2	2.6	-0.8	+2.0	4.2	-1.2	+3.5	7.0	-2.0
1.19	1.97	+0.60	1.00	-0.40	+1.0	1.60	-0.60	+1.0	2.00	-0.60	+1.6	3.4	-1.0	+2.5	5.3	-1.6	+4.0	8.5	-2.5
1.97	3.15	+0.70	1.30	-0.50	+1.2	2.10	-0.70	+1.2	2.30	-0.70	+1.8	4.0	-1.2	+3.0	6.3	-1.8	+4.5	10.0	-3.0
3.15	4.73	+0.90	1.60	-0.60	+1.4	2.50	-0.90	+1.4	2.90	-0.90	+2.2	4.8	-1.4	+4.0	7.7	-2.2	+5.0	11.5	-3.5
4.73	7.09	+1.00	1.90	-0.70	+1.6	2.80	-1.00	+1.6	3.50	-1.00	+2.5	5.6	-1.6	+4.5	8.7	-2.5	+6.0	13.5	-4.0
7.09	9.85	+1.20	2.20	-0.80	+1.8	3.20	-1.20	+1.8	4.20	-1.20	+2.8	6.6	-1.8	+5.0	10.3	-2.8	+7.0	16.5	-4.5
9.85	12.41	+1.20	2.30	-0.90	+2.0	3.40	-1.20	+2.0	4.70	-1.20	+3.0	7.5	-2.0	+6.0	12.0	-3.0	+8.0	19.0	-5.0
12.41	15.75	+1.40	2.60	-1.00	+2.2	3.90	-1.40	+2.2	5.90	-1.40	+3.5	8.7	-2.2	+6.0	14.5	-3.5	+9.0	23.0	-6.0

TABLE 20 ■ Locational Interference Fits (Values in Thousandths of an Inch)

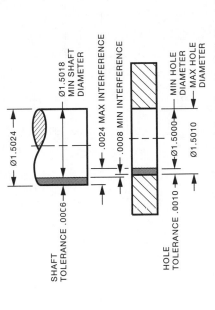

EXAMPLE: FN2 MEDIUM DRIVE FIT FOR A
Ø1.50 NOMINAL HOLE DIAMETER

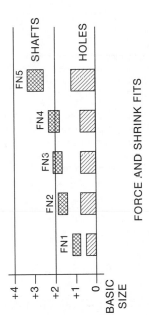

FORCE AND SHRINK FITS

VALUES IN THOUSANDTHS OF AN INCH

Nominal Size Range Inches		Class FN1 Light Drive Fit			Class FN2 Medium Drive Fit			Class FN3 Heavy Drive Fit			Class FN4 Shrink Fit			Class FN5 Heavy Shrink Fit		
Over	To	Hole Tol. GR6 −0	Maximum Interference	Shaft Tol. GR5 +0	Hole Tol. GR7 −0	Maximum Interference	Shaft Tol. GR6 +0	Hole Tol. GR7 −0	Maximum Interference	Shaft Tol. GR6 +0	Hole Tol. GR7 −0	Maximum Interference	Shaft Tol. GR6 +0	Hole Tol. GR8 −0	Maximum Interference	Shaft Tol. GR7 +0
0	.12	+0.25	0.50	−0.15	+0.40	0.85	−0.25				+0.40	0.95	−0.25	+0.60	1.30	−0.40
.12	.24	+0.30	0.60	−0.20	+0.50	1.00	−0.30				+0.50	1.20	−0.30	+0.70	1.70	−0.50
.24	.40	+0.40	0.75	−0.25	+0.60	1.40	−0.40				+0.60	1.60	−0.40	+0.90	2.00	−0.60
.40	.56	+0.40	0.80	−0.30	+0.70	1.60	−0.40				+0.70	1.80	−0.40	+1.00	2.30	−0.70
.56	.71	+0.40	0.90	−0.30	+0.70	1.60	−0.40				+0.70	1.80	−0.40	+1.00	2.50	−0.70
.71	.95	+0.50	1.10	−0.40	+0.80	1.90	−0.50				+0.80	2.10	−0.50	+1.20	3.00	−0.80
.95	1.19	+0.50	1.20	−0.40	+0.80	1.90	−0.50	+0.80	2.10	−0.50	+0.80	2.30	−0.50	+1.20	3.30	−0.80
1.19	1.58	+0.60	1.30	−0.40	+1.00	2.40	−0.60	+1.00	2.60	−0.60	+1.00	3.10	−0.60	+1.60	4.00	−1.00
1.58	1.97	+0.60	1.40	−0.40	+1.00	2.40	−0.60	+1.00	2.80	−0.60	+1.00	3.40	−0.60	+1.60	5.00	−1.00
1.97	2.56	+0.70	1.80	−0.50	+1.20	2.70	−0.70	+1.20	3.20	−0.70	+1.20	4.20	−0.70	+1.80	6.20	−1.20
2.56	3.15	+0.70	1.90	−0.50	+1.20	2.90	−0.70	+1.20	3.70	−0.70	+1.20	4.70	−0.70	+1.80	7.20	−1.20
3.15	3.94	+0.90	2.40	−0.60	+1.40	3.70	−0.90	+1.40	4.40	−0.70	+1.40	5.90	−0.90	+2.20	8.40	−1.40

TABLE 21 ■ Force and Shrink Fits (Values in Thousandths of an Inch)

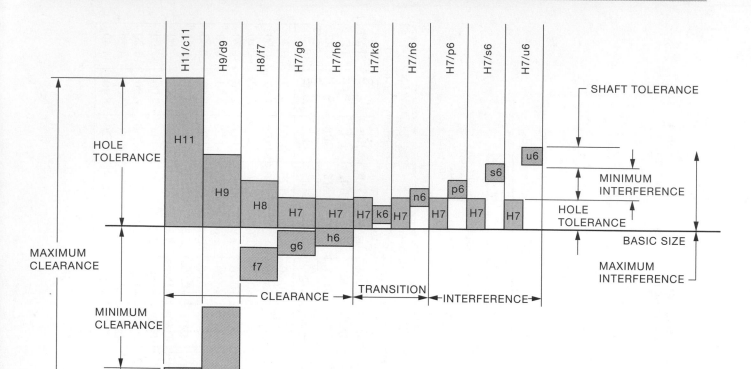

Hole Basis Symbol	Description
H11/c11 (RC9)	*Loose running* fit for wide commercial tolerances or allowances on external members.
H9/d9 (RC7)	*Free running* not fit for use where accuracy is essential, but good for large temperature variations, high running speeds, or heavy journal pressures.
H8/f7 (RC4)	*Close running* fit for running on accurate machines and for accurate location at moderate speeds and journal pressures.
H7/g6 (LC5)	*Sliding* fit not intended to run freely, but to move and turn freely and locate accurately.
H7/h6 (LC2)	*Locational clearance* fit provides snug fit for locating stationary parts, but can be freely assembled and disassembled.
H7/k6 (LT3)	*Locational transition* fit for accurate location, a compromise between clearance and interference.
H7/n6 (LT5)	*Locational transition* fit for more accurate location where greater interference is permissible.
H7/p6 (LN2)	*Locational interference* fit for parts requiring rigidity and alignment with prime accuracy of location but without special bore pressure requirements.
H7/s6 (FN2)	*Medium drive* fit for ordinary steel parts or shrink fits on light sections, the tightest fit usable with cast iron.
H7/u6 (FN4)	*Force* fit suitable for parts that can be highly stressed or for shrink fits where the heavy pressing forces required are impractical.

TABLE 22 ■ Preferred Hole Basis Metric Fits Description

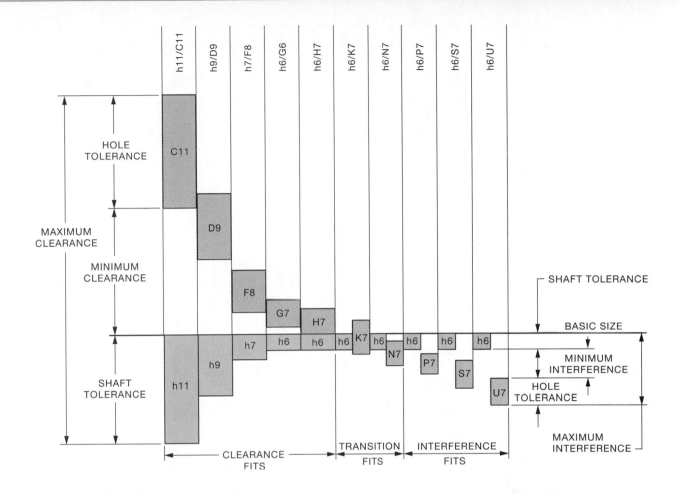

Shaft Basis Symbol	Description	Shaft Basis Symbol	Description
C11/h11	*Loose running* fit for wide commercial tolerances or allowances on external members.	K7/h6	*Locational transition* fit for accurate location, a compromise between clearance and interference.
D9/h9	*Free running* not fit for use where accuracy is essential, but good for large temperature variations, high running speeds, or heavy journal pressures.	N7/h6	*Locational transition* fit for more accurate location where greater interference is permissible.
F8/h7	*Close running* fit for running on accurate machines and for accurate location at moderate speeds and journal pressures.	P7/h6	*Locational interference* fit for parts requiring rigidity and alignment with prime accuracy of location but without special bore pressure requirements.
G7/h6	*Sliding* fit not intended to run freely, but to move and turn freely and locate accurately.	S7/h6	*Medium drive* fit for ordinary steel parts or shrink fits on light sections, the tightest fit usable with cast iron.
H7/h6	*Locational clearance* fit provides snug fit for locating stationary parts, but can be freely assembled and disassembled.	U7/h6	*Force* fit suitable for parts that can be highly stressed or for shrink fits where the heavy pressing forces required are impractical.

TABLE 23 ■ Preferred Shaft Basis Metric Fits Description

EXAMPLE: RC9 LOOSE RUNNING FIT FOR A Ø20 NOMINAL HOLE DIAMETER

Ø19.890 — MAX SHAFT DIAMETER
Ø19.760 — MIN SHAFT DIAMETER
0.370 MAX CLEARANCE
0.110 MIN CLEARANCE
Ø20.000 — MIN HOLE DIAMETER
Ø20.130 — MAX HOLE DIAMETER
SHAFT TOLERANCE 0.130
HOLE TOLERANCE 0.130

EXAMPLE: LC2 LOCATIONAL CLEARANCE FIT FOR A Ø40 NOMINAL HOLE DIAMETER

Ø40.000 — MAX SHAFT DIAMETER
Ø39.984 — MIN SHAFT DIAMETER
0.041 MAX CLEARANCE
0.000 MIN CLEARANCE
Ø40.000 — MIN HOLE DIAMETER
Ø40.025 — MAX HOLE DIAMETER
SHAFT TOLERANCE 0.016
HOLE TOLERANCE 0.025

Preferred Hole Basis Clearance Fits

Basic Size	Loose Running Hole H11	Loose Running Shaft c11	Loose Running Fit RC9	Free Running Hole H9	Free Running Shaft d9	Free Running Fit RC7	Close Running Hole H8	Close Running Shaft f7	Close Running Fit RC4	Sliding Hole H7	Sliding Shaft g6	Sliding Fit LC5	Locational Clearance Hole H7	Locational Clearance Shaft h6	Locational Clearance Fit LC2
5 MAX	5.075	4.930	0.220	5.030	4.970	0.090	5.018	4.990	0.040	5.012	4.996	0.024	5.012	5.000	0.020
MIN	5.000	4.855	0.070	5.000	4.940	0.030	5.000	4.978	0.010	5.000	4.988	0.004	5.000	4.992	0.000
6 MAX	6.075	5.930	0.220	6.030	5.970	0.090	6.018	5.990	0.040	6.012	5.996	0.024	6.012	6.000	0.020
MIN	6.000	5.855	0.070	6.000	5.940	0.030	6.000	5.978	0.010	6.000	5.988	0.004	6.000	5.992	0.000
8 MAX	8.090	7.920	0.260	8.036	7.960	0.112	8.022	7.987	0.050	8.015	7.995	0.029	8.015	8.000	0.024
MIN	8.000	7.830	0.080	8.000	7.924	0.040	8.000	7.972	0.013	8.000	7.986	0.006	8.000	7.991	0.000
10 MAX	10.090	9.920	0.260	10.036	9.960	0.112	10.022	9.987	0.050	10.015	9.995	0.029	10.015	10.000	0.024
MIN	10.000	9.830	0.080	10.000	9.924	0.040	10.000	9.972	0.013	10.000	9.986	0.005	10.000	9.991	0.000
12 MAX	12.110	11.905	0.315	12.043	11.950	0.136	12.027	11.984	0.061	12.018	11.994	0.035	12.018	12.000	0.029
MIN	12.000	11.795	0.095	12.000	11.907	0.050	12.000	11.966	0.016	12.000	11.983	0.006	12.000	11.989	0.000
16 MAX	16.110	15.905	0.315	16.043	15.950	0.136	16.027	15.984	0.061	16.018	15.994	0.035	16.018	16.000	0.029
MIN	16.000	15.795	0.095	16.000	15.907	0.050	16.000	15.966	0.016	16.000	15.983	0.006	16.000	15.989	0.000
20 MAX	20.130	19.890	0.370	20.052	19.935	0.169	20.033	19.980	0.074	20.021	19.993	0.041	20.021	20.000	0.034
MIN	20.000	19.760	0.110	20.000	19.883	0.065	20.000	19.959	0.020	20.000	19.980	0.007	20.000	19.987	0.000
25 MAX	25.130	24.890	0.370	25.052	24.935	0.169	25.033	24.980	0.074	25.021	24.993	0.042	25.021	25.000	0.034
MIN	25.000	24.760	0.110	25.000	24.883	0.065	25.000	24.959	0.020	25.000	24.980	0.007	25.000	24.987	0.000
30 MAX	30.130	29.890	0.370	30.052	29.935	0.169	30.033	29.980	0.074	30.021	29.993	0.041	30.021	30.000	0.034
MIN	30.000	29.760	0.110	30.000	29.883	0.065	30.000	29.959	0.020	30.000	29.980	0.007	30.000	29.987	0.000
40 MAX	40.160	39.880	0.440	40.062	39.920	0.204	40.039	39.975	0.089	40.025	39.991	0.050	40.025	40.000	0.041
MIN	40.000	39.720	0.120	40.000	39.858	0.080	40.000	39.950	0.025	40.000	39.975	0.009	40.000	39.984	0.000
50 MAX	50.160	49.870	0.450	50.062	49.920	0.204	50.039	49.975	0.089	50.025	49.991	0.050	50.025	50.000	0.041
MIN	50.000	49.710	0.130	50.000	49.858	0.080	50.000	49.950	0.025	50.000	49.975	0.009	50.000	49.984	0.000
60 MAX	60.190	59.860	0.520	60.074	59.900	0.248	60.046	59.970	0.106	60.030	59.990	0.059	60.030	60.000	0.049
MIN	60.000	59.670	0.140	60.000	59.826	0.100	60.000	59.940	0.030	60.000	59.971	0.010	60.000	59.981	0.000
80 MAX	80.190	79.850	0.530	80.074	79.900	0.248	80.046	79.970	0.106	80.030	79.990	0.059	80.030	80.000	0.049
MIN	80.000	79.660	0.150	80.000	79.826	0.100	80.000	79.940	0.030	80.000	79.971	0.010	80.000	79.981	0.000
100 MAX	100.220	99.830	0.610	100.087	99.880	0.294	100.054	99.964	0.125	100.035	99.988	0.069	100.035	100.000	0.057
MIN	100.000	99.610	0.170	100.000	99.793	0.120	100.000	99.929	0.036	100.000	99.966	0.012	100.000	99.978	0.000

TABLE 24 ■ Preferred Hole Basis Metric Fits (Values in Millimetres)

EXAMPLE: FN4 FORCE FIT FOR A Ø30 NOMINAL HOLE DIAMETER

EXAMPLE: LT5 LOCATIONAL TRANSITION FIT FOR A Ø50 NOMINAL HOLE DIAMETER

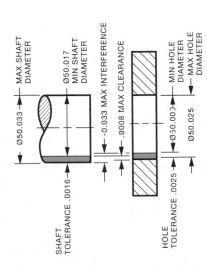

Basic Size	Locational Transn. Hole H7	Locational Transn. Shaft k6	Fit LT3	Locational Transn. Hole H7	Locational Transn. Shaft n6	Fit LT5	Locational Interf. Hole H7	Locational Interf. Shaft p6	Fit LN2	Medium Drive Hole H7	Medium Drive Shaft s6	Fit FN2	Force Hole H7	Force Shaft u6	Fit FN4
5 MAX	5.012	5.009	0.011	5.012	5.016	0.004	5.012	5.020	0.000	5.012	5.027	-0.007	5.012	5.031	-0.011
5 MIN	5.000	5.001	-0.009	5.000	5.008	0.016	5.000	5.012	-0.020	5.000	5.019	-0.027	5.000	5.023	-0.031
6 MAX	6.012	6.009	0.011	6.012	6.016	0.004	6.012	6.020	0.000	6.012	6.027	-0.007	6.012	6.031	-0.011
6 MIN	6.000	6.001	-0.009	6.000	6.008	0.016	6.000	6.012	-0.020	6.000	6.019	-0.027	6.000	6.023	-0.031
8 MAX	8.015	8.010	0.014	8.015	8.019	0.005	8.015	8.024	0.000	8.015	8.032	-0.008	8.015	8.037	-0.013
8 MIN	8.000	8.001	-0.010	8.000	8.010	0.019	8.000	8.015	-0.024	8.000	8.023	-0.032	8.000	8.028	-0.037
10 MAX	10.015	10.010	0.014	10.015	10.019	0.005	10.015	10.024	0.000	10.015	10.032	-0.008	10.015	10.037	-0.013
10 MIN	10.000	10.001	-0.010	10.000	10.010	0.019	10.000	10.015	-0.024	10.000	10.023	-0.032	10.000	10.028	-0.037
12 MAX	12.018	12.012	0.017	12.018	12.023	0.006	12.018	12.029	0.000	12.018	12.039	-0.010	12.018	12.044	-0.015
12 MIN	12.000	12.001	-0.012	12.000	12.012	0.023	12.000	12.018	-0.029	12.000	12.028	-0.039	12.000	12.033	-0.044
16 MAX	16.018	16.012	0.017	16.018	16.023	0.006	16.018	16.029	0.000	16.018	16.039	-0.010	16.018	16.044	-0.015
16 MIN	16.000	16.001	-0.012	16.000	16.012	0.023	16.000	16.018	-0.029	16.000	16.028	-0.039	16.000	16.033	-0.044
20 MAX	20.021	20.015	0.019	20.021	20.028	0.006	20.021	20.035	0.001	20.021	20.048	-0.014	20.021	20.054	-0.020
20 MIN	20.000	20.002	-0.015	20.000	20.015	0.028	20.000	20.022	-0.035	20.000	20.035	-0.048	20.000	20.041	-0.054
25 MAX	25.021	25.014	0.019	25.021	25.028	0.006	25.021	25.035	0.001	25.021	25.048	-0.014	25.021	25.061	-0.027
25 MIN	25.000	25.002	-0.015	25.000	25.015	0.028	25.000	25.022	-0.035	25.000	25.035	-0.048	25.000	25.048	-0.061
30 MAX	30.021	30.015	0.019	30.021	30.028	0.006	30.021	30.035	0.001	30.021	30.048	-0.014	30.021	30.061	-0.027
30 MIN	30.000	30.002	-0.015	30.000	30.015	0.023	30.000	30.022	-0.035	30.000	30.035	-0.048	30.000	30.048	-0.061
40 MAX	40.025	40.018	0.023	40.025	40.033	0.008	40.025	40.042	0.001	40.025	40.059	-0.018	40.025	40.076	-0.035
40 MIN	40.000	40.002	-0.018	40.000	40.017	0.033	40.000	40.026	-0.042	40.000	40.043	-0.059	40.000	40.060	-0.076
50 MAX	50.025	50.018	0.023	50.025	50.033	0.008	50.025	50.042	0.001	50.025	50.059	-0.018	50.025	50.086	-0.045
50 MIN	50.000	50.002	-0.018	50.000	50.017	0.033	50.000	50.026	-0.042	50.000	50.043	-0.059	50.000	50.070	-0.086
60 MAX	60.030	60.021	0.028	60.030	60.039	0.010	60.030	60.051	0.002	60.030	60.072	-0.023	60.030	60.106	-0.057
60 MIN	60.000	60.002	-0.021	60.000	60.020	0.039	60.000	60.032	-0.051	60.000	60.053	-0.072	60.000	60.087	-0.106
80 MAX	80.030	80.021	0.028	80.030	80.039	0.010	80.030	80.051	0.002	80.030	80.078	-0.029	80.030	80.121	-0.072
80 MIN	80.000	80.002	-0.021	80.000	80.020	0.039	80.000	80.032	-0.051	80.000	80.059	-0.078	80.000	80.102	-0.121
100 MAX	100.035	100.025	0.032	100.035	100.045	0.012	100.035	100.059	0.002	100.035	100.093	-0.036	100.035	100.146	-0.089
100 MIN	100.000	100.003	-0.025	100.000	100.023	0.045	100.000	100.037	-0.059	100.000	100.071	-0.093	100.000	100.124	-0.146

TABLE 24 (CONT'D) ■ Preferred Hole Basis Metric Fits (Values in Millimetres)

EXAMPLE: RC9 LOOSE RUNNING FIT FOR A Ø20 NOMINAL SHAFT DIAMETER

Ø20.000 MAX SHAFT DIAMETER
Ø19.870 MIN SHAFT DIAMETER
0.370 MAX CLEARANCE
0.110 MIN CLEARANCE
SHAFT TOLERANCE 0.130
Ø20.110 MIN HOLE DIAMETER
Ø20.240 MAX HOLE DIAMETER
HOLE TOLERANCE 0.130

EXAMPLE: LC2 LOCATIONAL CLEARANCE FIT FOR A Ø40 NOMINAL SHAFT DIAMETER

Ø40.000 MAX SHAFT DIAMETER
Ø39.984 MIN SHAFT DIAMETER
0.041 MAX CLEARANCE
0.000 MIN CLEARANCE
SHAFT TOLERANCE 0.016
Ø40.000 MIN HOLE DIAMETER
Ø40.025 MAX HOLE DIAMETER
HOLE TOLERANCE 0.025

Preferred Shaft Basis Transition and Interference Fits

Basic Size	Locational Transn. Hole K7	Shaft h6	Fit LT3	Locational Transn. Hole N7	Shaft h6	Fit LT5	Locational Interf. Hole P7	Shaft h6	Fit LN2	Medium Drive Hole S7	Shaft h6	Fit FN2	Force Hole U7	Shaft h6	Fit FN4
5 MAX	5.003	5.000	0.011	4.996	5.000	0.004	4.992	5.000	0.000	4.985	5.000	-0.007	4.981	5.000	-0.011
MIN	4.991	4.992	-0.009	4.984	4.992	-0.016	4.980	4.992	-0.020	4.973	4.992	-0.027	4.969	4.992	-0.031
6 MAX	6.003	6.000	0.011	5.996	6.000	0.004	5.992	6.000	0.000	5.985	6.000	-0.007	5.981	6.000	-0.011
MIN	5.991	5.992	-0.009	5.984	5.992	-0.016	5.980	5.992	-0.020	5.973	5.992	-0.027	5.969	5.992	-0.031
8 MAX	8.005	8.000	0.014	7.996	8.000	0.005	7.991	8.000	0.000	7.983	8.000	-0.008	7.978	8.000	-0.013
MIN	7.990	7.991	-0.010	7.981	7.991	-0.019	7.976	7.991	-0.024	7.968	7.991	-0.032	7.963	7.991	-0.037
10 MAX	10.005	10.000	0.014	9.996	10.000	0.005	9.991	10.000	0.000	9.983	10.000	-0.008	9.978	10.000	-0.013
MIN	9.990	9.991	-0.010	9.981	9.991	-0.019	9.976	9.991	-0.024	9.968	9.991	-0.032	9.963	9.991	-0.037
12 MAX	12.006	12.000	0.017	11.995	12.000	0.006	11.989	12.000	0.000	11.979	12.000	-0.010	11.974	12.000	-0.015
MIN	11.988	11.989	-0.012	11.977	11.989	-0.023	11.971	11.989	-0.029	11.961	11.989	-0.039	11.956	11.989	-0.044
16 MAX	16.006	16.000	0.017	15.995	16.000	0.006	15.989	16.000	0.000	15.979	16.000	-0.010	15.974	16.000	-0.015
MIN	15.988	15.989	-0.012	15.977	15.989	-0.023	15.971	15.989	-0.029	15.961	15.989	-0.039	15.956	15.989	-0.044
20 MAX	20.006	20.000	0.019	19.993	20.000	0.006	19.986	20.000	0.001	19.973	20.000	-0.014	19.967	20.000	-0.020
MIN	19.985	19.987	-0.015	19.972	19.987	-0.028	19.965	19.987	-0.035	19.952	19.987	-0.048	19.946	19.987	-0.054
25 MAX	25.006	25.000	0.019	24.993	25.000	0.006	24.986	25.000	0.001	24.973	25.000	-0.014	24.960	25.000	-0.027
MIN	24.985	24.987	-0.015	24.972	24.987	-0.028	24.965	24.987	-0.035	24.952	24.987	-0.048	24.939	24.987	-0.061
30 MAX	30.006	30.000	0.019	29.993	30.000	0.006	29.986	30.000	0.001	29.973	30.000	-0.014	29.960	30.000	-0.027
MIN	29.985	29.987	-0.015	29.972	29.987	-0.028	29.965	29.987	-0.035	29.952	29.987	-0.048	29.939	29.987	-0.061
40 MAX	40.007	40.000	0.023	39.992	40.000	0.008	39.983	40.000	0.001	39.966	40.000	-0.018	39.949	40.000	-0.035
MIN	39.982	39.984	-0.018	39.967	39.984	-0.033	39.958	39.984	-0.042	39.941	39.984	-0.059	39.924	39.984	-0.076
50 MAX	50.007	50.000	0.023	49.992	50.000	0.008	49.983	50.000	0.001	49.966	50.000	-0.018	49.939	50.000	-0.045
MIN	49.982	49.984	-0.018	49.967	49.984	-0.033	49.958	49.984	-0.042	49.941	49.984	-0.059	49.914	49.984	-0.086
60 MAX	60.009	60.000	0.028	59.991	60.000	0.010	59.979	60.000	0.002	59.958	60.000	-0.023	59.924	60.000	-0.057
MIN	59.979	59.981	-0.021	59.961	59.981	-0.039	59.949	59.981	-0.051	59.928	59.981	-0.072	59.894	59.981	-0.106
80 MAX	80.009	80.000	0.028	79.991	80.000	0.010	79.979	80.000	0.002	79.952	80.000	-0.029	79.909	80.000	-0.072
MIN	79.979	79.981	-0.021	79.961	79.981	-0.039	79.949	79.981	-0.051	79.922	79.981	-0.078	79.879	79.981	-0.121
100 MAX	100.010	100.000	0.032	99.990	100.000	0.012	99.976	100.000	0.002	99.942	100.000	-0.036	99.889	100.000	-0.089
MIN	99.975	99.978	-0.025	99.955	99.978	-0.045	99.941	99.978	-0.059	99.907	99.978	-0.093	99.854	99.978	-0.146

TABLE 25 ■ Preferred Shaft Basis Metric Fits (Values in Millimetres)

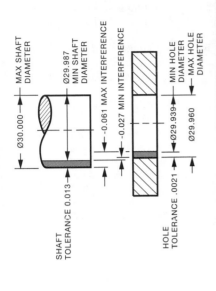

EXAMPLE: FN4 FORCE FIT FOR A Ø30 NOMINAL SHAFT DIAMETER

EXAMPLE: LT5 LOCATIONAL TRANSITION FIT FOR A Ø50 NOMINAL SHAFT DIAMETER

	Loose Running			Free Running			Close Running			Sliding			Locational Clearance		
Basic Size	Hole C11	Shaft h11	Fit RC3	Hole D9	Shaft h9	Fit RC7	Hole F8	Shaft h7	Fit RC4	Hole G7	Shaft h6	Fit LC5	Hole H7	Shaft h6	Fit LC2
5 MAX	5.145	5.000	0.220	5.060	5.000	0.090	5.028	5.000	0.040	5.016	5.000	0.024	5.012	5.000	0.020
MIN	5.070	4.925	0.070	5.030	4.970	0.030	5.010	4.988	0.010	5.004	4.992	0.004	5.000	4.992	0.000
6 MAX	6.145	6.000	0.220	6.060	6.000	0.090	6.028	6.000	0.040	6.016	6.000	0.024	6.012	6.000	0.020
MIN	6.070	5.925	0.070	6.030	5.970	0.030	6.010	5.988	0.010	6.004	5.992	0.004	6.000	5.992	0.000
8 MAX	8.170	8.000	0.260	8.076	8.000	0.112	8.035	8.000	0.050	8.020	8.000	0.029	8.015	8.000	0.024
MIN	8.080	7.910	0.080	8.040	7.964	0.040	8.013	7.985	0.013	8.005	7.991	0.005	8.000	7.991	0.000
10 MAX	10.170	10.000	0.260	10.076	10.000	0.112	10.035	10.000	0.050	10.020	10.000	0.029	10.015	10.000	0.024
MIN	10.080	9.910	0.080	10.040	9.964	0.040	10.013	9.985	0.013	10.005	9.991	0.005	10.000	9.991	0.000
12 MAX	12.205	12.000	0.315	12.093	12.000	0.136	12.043	12.000	0.061	12.024	12.000	0.035	12.018	12.000	0.029
MIN	12.095	11.890	0.095	12.050	11.957	0.050	12.016	11.982	0.016	12.006	11.989	0.006	12.000	11.989	0.000
16 MAX	16.205	16.000	0.315	16.093	16.000	0.136	16.043	16.000	0.061	16.024	16.000	0.035	16.018	16.000	0.029
MIN	16.095	15.890	0.095	16.050	15.957	0.050	16.016	15.982	0.016	16.006	15.989	0.006	16.000	15.989	0.000
20 MAX	20.240	20.000	0.370	20.117	20.000	0.169	20.053	20.000	0.074	20.028	20.000	0.041	20.021	20.000	0.034
MIN	20.110	19.870	0.110	20.065	19.948	0.065	20.020	19.979	0.020	20.007	19.987	0.007	20.000	19.987	0.000
25 MAX	25.240	25.000	0.370	25.117	25.000	0.169	25.053	25.000	0.074	25.028	25.000	0.041	25.021	25.000	0.034
MIN	25.110	24.870	0.110	25.065	24.948	0.065	25.020	24.979	0.020	25.007	24.987	0.007	25.000	24.987	0.000
30 MAX	30.240	30.000	0.370	30.117	30.000	0.169	30.053	30.000	0.074	30.028	30.000	0.041	30.021	30.000	0.034
MIN	30.110	29.870	0.110	30.065	29.948	0.065	30.020	29.979	0.020	30.007	29.987	0.007	30.000	29.987	0.000
40 MAX	40.280	40.000	0.440	40.142	40.000	0.204	40.064	40.000	0.089	40.034	40.000	0.050	40.025	40.000	0.041
MIN	40.120	39.840	0.120	40.080	39.938	0.080	40.025	39.975	0.025	40.009	39.984	0.009	40.000	39.984	0.000
50 MAX	50.290	50.000	0.450	50.142	50.000	0.204	50.064	50.000	0.089	50.034	50.000	0.050	50.025	50.000	0.041
MIN	50.130	49.840	0.130	50.080	49.938	0.080	50.025	49.975	0.025	50.009	49.984	0.009	50.000	49.984	0.000
60 MAX	60.330	60.000	0.510	60.174	60.000	0.248	60.076	60.000	0.106	60.040	60.000	0.059	60.030	60.000	0.049
MIN	60.140	59.810	0.140	60.100	59.926	0.100	60.030	59.970	0.030	60.010	59.981	0.010	60.000	59.981	0.000
80 MAX	80.340	80.000	0.530	80.174	80.000	0.248	80.076	80.000	0.106	80.040	80.000	0.059	80.030	80.000	0.049
MIN	80.150	79.810	0.150	80.100	79.926	0.100	80.030	79.970	0.030	80.010	79.981	0.010	80.000	79.981	0.000
100 MAX	100.390	100.000	0.610	100.207	100.000	0.294	100.090	100.000	0.125	100.047	100.000	0.069	100.035	100.000	0.057
MIN	100.170	99.780	0.170	100.120	99.913	0.120	100.036	99.965	0.036	100.012	99.978	0.012	100.000	99.987	0.000

Preferred Shaft Basis Transition and Interference Fits

TABLE 25 (CONT'D) ■ Preferred Shaft Basis Metric Fits (Values in Millimetres)

Quality	Metric Unit	Symbol	Metric to Inch-Pound Unit	Inch-Pound to Metric Unit
Length	millimetre	mm	1 mm = 0.0394 in.	1 in. = 25.4 mm
	centimetre	cm	1 cm = 0.394 in.	1 ft. = 30.5 cm
	metre	m	1 m = 39.37 in. = 3.28 ft.	1 yd. = 0.914 m = 914 mm
	kilometre	km	1 km = 0.62 mile	1 mile = 1.61 km
Area	square millimetre	mm^2	1 mm = 0.001 55 sq. in.	1 sq. in. = 6 452 mm^2
	square centimetre	cm^2	1 cm = 0.155 sq. in.	1 sq. ft. = 0.093 m^2
	square metre	m^2	1 m = 10.8 sq. ft.	1 sq. yd. = 0.836 m^2
			= 1.2 sq. yd.	
Mass	milligram	mg	1 g = 0.035 oz.	
	gram	g	1 kg = 2.205 lb.	1 oz. = 28.3 g
	kilogram	kg	1 tonne = 1.102 tons	1 lb. = 0.454 kg
	tonne	t		1 ton = 907.2 kg
				= 0.907 tonnes
Volume	cubic centimetre	cm^3	1 mm^3 = 0.000 061 cu. in.	1 fl. oz. = 28.4 cm^3
	cubic metre	m^3	1 cm^3 = 0.061 cu. in.	1 cu. in. = 16.387 cm^3
	millilitre	ml	1 m^3 = 35.3 cu. in.	1 cu. ft. = 0.028 m^3
			= 1.308 cu. yd.	1 cu. yd. = 0.756 m^3
			1 mL = 0.035 fl. oz.	
Capacity	litre	L	U.S. Measure	U.S. Measure
			1 pt. = 0.473 L	1 L = 2.113 pt.
			1 qt. = 0.946 L	= 1.057 qt.
			1 gal. = 3.785 L	= 0.264 gal.
			Imperial Measure	Imperial Measure
			1 pt. = 0.568 L	1 L = 1.76 pt.
			1 qt. = 1.137 L	= 0.88 qt.
			1 gal. = 4.546 L	= 0.22 gal.
Temperature	Celsius degree	°C	$°C = \frac{5}{9}(°F - 32)$	$°F = \frac{9}{5} \times °C + 32$
Force	newton	N	1 N = 0.225 lb (f)	1 lb (f) = 4.45 N
	kilonewton	kN	1 kN = 0.225 kip (f)	= 0.004 448 kN
			= 0.112 ton (f)	
Energy/Work	joule	J	1 J = 0.737 ft · lb	1 ft · lb = 1.355 J
	kilojoule	kJ	1 J = 0.948 BTU	1 BTU = 1.055 J
	megajoule	MJ	1 MJ = 0.278 kWh	1 kWh = 3.6 MJ
Power	kilowatt	kW	1 kW = 1.34 hp	1 hp (550 ft · lb/s) = 0.746 kW
			1 W = 0.0226 ft · lb/min.	1 ft · lb/min = 44.2537 W
Pressure	kilopascal	kPa	1 kPa = 0.145 psi	
			= 20.885 psf	1 psi = 6.895 kPa
			= 0.01 ton-force	1 lb-force/sq. ft. = 47.88 Pa
			per sq. ft.	1 ton-force/sq. ft. = 95.76 kPa
	*kilogram per square centimetre	kg/cm^2	1 kg/cm = 13.780 psi	
Torque	newton metre	N · m	1 N · m = 0.74 lb · ft.	1 lb · ft = 1.36 N · m
	*kilogram metre	kg/m	1 Kg/m = 7.24 lb · ft.	1 lb · ft = 0.14 kg/m
	*kilogram per centimetre	kg/cm	1 Kg/cm = 0.86 lb · in.	1 lb · in = 1.2 kg/cm
Speed/Velocity	metres per second	m/s	1 m/s = 3.28 ft/s	1 ft/s = 0.305 m/s
	kilometres per hour	km/h	1 km/h = 0.62 mph	1 mph = 1.61 km/h

*Not SI units, but included here because they are employed on some of the gauges and indicators in use in industry.

TABLE 26 ■ Metric Conversion Tables

INDEX

A

abbreviations, 3
  for spur gears, 325
  for structural steel shapes,
    289–290
absolute coordinate
  programming, 278
addendum (ADD) of gear tooth,
  324
Adjustable Shaft Support
  assignment, 217
alignment of parts and holes,
  250–251
  with dowel pins, 267
allowances
  defined, 116
  machining allowances, 68,
    69–70
alloys
  aluminum/aluminum alloy
    section lining symbol, 80
  section lining symbol, 80
  steel alloys, 221–223
Amazon New York, Industrial
  Press, 253
American Design and Drafting
  Association, 209
American Institute of Steel
  Construction, 291
American Iron and Steel
  Institute (AISI), 221, 291
American Society of Mechanical
  Engineers (ASME), 1, 73,
  83, 99, 157, 198, 243, 253,
  261, 271

drawing standards, 20–21
screw thread representation,
  141–143
American Standard Pipe
  Tread, 203
American Welding
  Society, 299, 317
Anchor Plate assignment, 19
angle dimensions, 47
  with decimal degrees, 49
angular contact ball
  bearings, 347
angularity tolerance, 403–404
  for feature of size, 409, 411
  and flatness tolerance, 406
  two directions, control in, 412
Animated Worksheets, 26, 62,
  224, 326, 332, 342, 357
ANSI/AWS A2.4-86, 307, 317
antifriction bearings, 215,
  347–355
  ball bearings, 347–348
  loads, types of, 348
  roller bearings, 349
arcs, 13. *See also* Circles
  assignments, 18–19
  with compasses, 14–15
  complete views, sketching,
    16–17
  dimensioning of, 38, 40
  freehand sketching, 16
  isometric sketches of, 58–59
  oblique sketching of,
    61, 62
  sketching, 13–17

arrangement of views, 21–22,
  196–201
  for Spider assignment, 270
arrowheads, application of, 36
ASME B1.1-2003, 149
ASME B1.13M-2001, 149
ASME B1.21M1997, 149
ASME B18.8.2-2000, 271
ASME B18.27-1998, 353
ASME B27.7-1977
  (R1999), 353
ASME Y14.6-2001, 149, 209
ASME Y14.7.1-1971
  (R2003), 326
ASME Y14.7.2-1978
  (R2004), 331
ASME Y14.24-1999
  (R2004), 285
ASME Y14.38-1999, 6
ASME Y32.2.3-1994
  (R1999), 209
ASME Y14.2M-1992
  (R2003), 6, 10, 17, 42
ASME Y14.3M-1994 (R2003),
  157, 175, 179, 184, 252, 261
ASME Y14.4M-1989
  (R2004), 62
ASME Y14.5M-1994 (R2004),
  42, 50, 119, 243, 368, 378,
  387, 399, 406, 417, 425,
  439, 445, 456, 463
ASME Y14.8M-1996
  (R2002), 236
ASME Y14.36M-1996
  (R2002), 71, 111

assemblies, threaded, 146
assembly drawings, 20–21,
  281–283
  bill of material for, 283
  break lines on, 71
  helical springs on, 283–285
  sectioned assembly
    drawings, 80
  typical drawing, 282
Auxiliary Pump Base assignment, 240–241
auxiliary views, 174. *See also*
  Primary auxiliary views
  arrangement of, 196–197
  secondary auxiliary views,
    183–189
axes. *See also* datums
  centre lines indicating, 13
axial assembly rings, 350
Axle assignment, 402

B

babbitt section lining
  symbol, 80
back or backing welds,
  305–307
ball thrust bearings, 348
barrel finishing, 109
barrel-type cams, 343
Base Assembly assignment,
  318–319
base line dimensioning, 243–245
Base Plate assignments, 18,
  51–53, 89
Base Skid assignment,
  308–309
"Basic Bearing Types"
  (De Hayt), 215, 353
basic hole system, 126
  assignments, 36, 127
  Inch Fits assignment, 127
basic shaft system, 126
basic size of dimension, 116
Bayer MaterialScience AG, 224
beaded dovetail joint, 192
Bearing Housing
  assignment, 427
bearings, 214–219. *See also*
  Antifriction bearings
  antifriction bearings, 215
  premounted bearings, 216

bedrock section lining
  symbol, 80
belt drives, 349
bevel gears, 322, 331–335
  formulas, 333
  nomenclature, 332
bevel groove welds, 303
bilateral tolerances,
  118–119, 452
bill of material, 283
bird's-eye view, pictorial
  sketch in, 57
blind holes, dimensioning
  of, 39
blocks
  for isometric sketches, 59, 60
  for oblique sketching, 62
Boiler Room assignment,
  212–213
bolts, 141
  sectioning of, 270
  types of, 141
boring, 39–40
  roughness range for, 109
boron steels, 223
bosses, 164
Boston Gear, 326, 332, 337
boundary lubrication, 103
Bracket assignment, 139, 195
brass
  pipe, 203
  section lining symbol, 80
break lines, 8–9, 71
  in primary auxiliary views, 178
  short/long breaks, 72
broaching, roughness
  range for, 109
broken-out sections, 80, 258
bronze
  for journal bearings, 215
  section lining symbol, 80
Brown and Sharpe taper, 91
bushings, 352, 353

C

cabinet oblique sketching, 59–60
  for sloped surfaces, 61, 62
CAD (computer-aided
  drawing), 3–4
  and numerical control
    (NC), 276

cams, 341–346
  displacement diagrams,
    341, 343
  types of, 341, 342
Canadian Institute of Steel
  Construction (CISC),
  221, 291
Canadian Standards Association
  (CSA), 1, 6
  drawing standards, 21
Canadian Welding Bureau, 299,
  307, 317
CAN3-B78.1-M83 (R1990), 6,
  17, 26, 62, 83, 97, 149, 157,
  175, 179, 184, 198, 252,
  261, 285, 291
CAN3-B95-1962 (R1996),
  71, 111
CAN3-B97.3-M1982 (R1992),
  132
CAN3-B232-75 (R1996), 166
CAN/CSA-B78.2-M91, 42, 92,
  97, 119, 126, 166, 243, 368,
  377, 387, 399, 406, 417,
  425, 439, 445, 456, 463
carbon steels, 221, 222
Carburetor Gasket
  assignment, 18
Caster Assembly
  assignment, 229
Caster Details assignment,
  112–113
casting processes, 230–242
  coping down, 234
cast iron
  pipe, 203
  section lining symbol, 80
  types of, 220
cavalier oblique sketching,
  59–60
cellular rubber, 223–224
Centering Connector Details
  assignment, 101
centre distance for gear trains,
  336–337
centre lines, 13
  for oblique sketching, 61
centre points, indication of, 13
chain dimensioning, 243–244
chamfers, 91, 92
change tables, 71

charts
  for coordinate tolerancing,
    431–432
    positional tolerancing, evalua-
      tion of, 440, 441
check valves, 204
chemical milling, 109
chilling process, 220
chordal addendum (ADDc) of
  gear tooth, 324
chordal thickness (Tc) of gear
  tooth, 324
chromium steels, 221, 222
chromium vanadium steels, 222
circles
  assignments, 18–19
  with compasses, 14–15
  complete views, sketching,
    16–17
  freehand sketching, 16
  isometric sketches of, 58–59
  oblique sketching of, 61, 62
  sketching, 13–17
  templates, 5, 14
circular features, 13
  orthographic sketching
    assignments, 31, 34
  pictorial sketching
    assignments, 66–67
  positional tolerances for,
    434–438
  runout tolerances, 458–460
  views of, 22–24
circularity tolerances, 385–386
  for noncylindrical parts, 386
  zones, 372
circular pitch (CP) of gear
  tooth, 324
circular thickness (T) of gear
  tooth, 324
classes of internal/external
  threads, 144
clearance (CL) of gear tooth,
  324
clearance fits, 122, 123
  locational clearance fits, 124
  metric fits, 133
clevis pins, 267
closed-cell sponge rubber,
  223–224
clutches, 349

coaxial datum features, 399
Coil Frame assignment,
  264–265
coil springs, 283–285
cold rolling, 109
commercial straight pins, 267
compasses, 14–15
complete views, sketching, 16–17
Completing Oblique Surfaces
  assignment, 177
composition section lining symbol, 80
Compound Rest Slide
  assignment, 53
computer numerical control
  (CNC), 276
concentric circles, 13
concrete in section lining, 79
cone point set screws, 163
conical tapers, 91, 93
conical washers, 291
connections for piping, 206
construction lines, 2
Contact Arm assignment, 248
Contactor assignment, 249
Control Block assignment,
  188–189
Control Bracket assignment,
  256–257
controlled profiles, 453–454
control surface texture requirements, 107–110
conventional drawing
  example, 99
coordinate tolerances, 428–432
  advantages/disadvantages
    of, 432
  charts for, 431–432
  maximum permissible error,
    429–431
  tolerance zones for, 429–431
coping down, 234
coplanar datum features, 399
copper
  pipe, 203
  section lining symbol, 80
  steels, 222
  tubing, 203
cored castings, 234–235
Corner Bracket assignment,
  218–219

corners on dovetails, 49
corrected addendum of gear
  tooth, 324
cotter pins, 266, 267, 268
counterbores, 82–83
countersinks, 82–83
Coupling assignment, 44
Cover Plate assignment, 18, 279
Crane Hook Assembly, 282
Crane Valve Group, 209
Crossbar assignment, 225
crossing of pipes,
  drawing, 206, 208
CSA-B97.3-1970 (R1992), 126
cup join, 192
Cut-Off Stand assignment, 408
cut point set screws, 163
cutting-plane lines, 78–79
cylindrical features
  break lines, 72
  circular runout for, 459–461
  dimensioning of, 39
  measuring principles,
    461–462
  orientation tolerances for, 410
  perpendicularity tolerance
    and, 415, 418, 419
  straightness of surfaces,
    366–368
cylindrical feeder cams, 341
  assignment, 344–345
cylindrical roller bearings, 348
cylindricity tolerances, 386–387

D

darkening lines
  for isometric sketches, 59, 60
  for oblique sketching, 62
datum features
  former ANSI symbol, 397
  geometric tolerances and,
    397–398
  size variations and, 395–396
  two features for one
    datum, 399
datums
  angularity tolerances and, 412
  assignments, 400–401
  defined, 243, 392
  dimensioning, 243–245
  establishing axis, 461–463

orientation tolerances
  and, 403
partial surfaces of, 398–399
primary datum, 392–393
secondary datum, 393
tertiary datum, 393
three-plane system,
  392–393, 394
for uneven surfaces, 393–395
datum symbols, 395–399
former ANSI symbol, 397
datum targets
  application of, 426
  areas, 423–424
  defined, 421
  on different planes, 424–425
  dimensioning for, 425, 426
  identification of, 422–424
  lines, 422–423
  points, 422
  symbol, 421–422
decimal-inch system
  circular features with, 34
  flat surfaces assignments,
    32–33
  machining allowances
    in, 69–70
  pictorial sketching of flat
    surfaces with, 64, 66
  sloped surface sketch with, 55
dedendum (DED) of gear tooth,
  324
degrees in angle dimensions, 47
De Hart, A. O., 216
Dehayt, A. O., 353
depth, 1
Design & Technology Online,
  194, 342, 353
detailed screw thread
  representation, 141
details, 20
  break lines on drawings, 71
  for isometric sketches, 59, 60
  multiple-detail drawings,
    96, 98
  for oblique sketching, 62
  in sectional views, 78
development drawings, 190
  with complete set of folding
    instructions, 191
diameter, 1
  abbreviation for, 3

limit dimensions for, 117
  symbol, 38
diametral dimensions, 38
diametral pitch (DP) of gears,
  324, 326
diamond knurls, 92, 93
die casting, 109
dimension lines, 35
  placement of, 38
  staggering of, 38
dimension origin symbol,
  164–165
dimensions/dimensioning, 35.
  See also Angle dimensions;
    Tolerances
  absolute coordinate
    programming, 278
  assignments, 45–46
  base line dimensioning,
    243–245
  basic elements, 36
  basic rules for, 38
  chain dimensioning, 243–244
  choice of, 38
  for cylindrical holes, 39
  for datum target locations,
    425, 426
  diameters, 40
  for dovetails, 49
  for fittings, 206
  holes, 146
  missing dimensions, 35
  not-to-scale dimensions, 70–71
  for numerical control
    (NC), 276
  for pipes, 206, 208
  placement of, 35, 39
  in primary auxiliary views, 178
  rectangular coordinate
    dimensioning, 165
  reference dimensions, 50
  relative coordinate
    programming, 278
  for repetitive features, 41
  similarly sized features, 41–42
  for spur gears, 325–326
  for two-axis coordinate
    system, 276–278
  in working drawings, 21
directions
  orientation tolerances
    and, 405

pictorial sketching, viewing,
  57–58
  straightness tolerances for, 368
double-line pipe drawings, 205
dovetails, 49
dowel pins, 266, 267
Drafting Zone, 27, 42, 50, 73,
  83, 92, 99, 111, 119, 126,
  149, 157, 166, 184, 194,
  368, 378, 388, 399, 406,
  418, 426, 440, 446, 456, 463
drawing callouts
  dimension origin symbol, 165
  for features of size, 372
  for fillet welds, 299
  for groove welds, 303, 307
  for sheet metal material, 193
  for straightness tolerance, 369
  for tolerances, 131, 134
drawings, 1, 2. *See also* Details;
    Development drawings;
    Welding drawings
  assembly drawings, 281–283
  multiple-detail drawings,
    96, 98
  for numerical control
    (NC), 276
  phantom outlines, 290–291
  piping drawings, 204–209
  revisions, 71
  to scale, 9–10
  for spur gears, 325–326
  standardization needs, 1–2
  working drawings, 20–21
drilling, 39
  roughness range for, 109
drill jigs, 266, 267
drills, 252
drive fits, 124
Drive Support Details assignment, 150–151
drum cams, 343
dry friction, 103
ductile iron, 220

E

eccentric plate cams, 343
edges, 190
  examples of, 192
Effective Training Inc., 368,
  378, 388, 399, 406, 418,
  426, 440, 446, 456

eFunda, 194, 216, 224, 236, 285, 299, 307, 317, 353, 368, 378, 388, 399, 406, 418, 426, 440, 446, 456, 463
elastomers, 223
electrical discharge machining, 109
electric cables, 80
electro-chemical process, 109
electrolytic grinding, 109
electron beam, 109
electro-polishing, 109
ellipse templates, 61–62
Emerson Power Transmission, 216
End Plate assignment, 468
end play rings, 350
Engineers Edge, 378, 388, 399, 406, 418, 426, 440, 446, 456, 463
Engine Starting Air System assignment, 210–211
enlarged scale, 9
erasers, 5
errors of circularity, 385
extension lines, 35
external threads, 142
inch threads, 144
metric threads, 145, 147
extra-fine thread series, 144
extruding, roughness range for, 109

F

Fastener Hut, 149
fastening devices. See Pin fasteners; Threaded fasteners
feature, defined, 243
feature control frames, 362–365
applications to, 364–365
features of size, 372
angularity tolerance for, 409, 411
assignments
orientation tolerances assignments, 420
straightness assignment, 382–383
definitions, 372–374
examples of, 374–375
orientation tolerances for, 409–412, 420

parallelism tolerances for, 411–412
perpendicularity tolerance, 409–411
straightness of, 376–377
Feed Hopper assignment, 43
ferritic iron grades, 220
field welds, 297
fillets, 74. See also Arcs
dimensioning, 41
intersections, filleted, 84
fillet welds, 294, 297–299
assignment, 301
fillister head screws, 141
finish. See Surface texture
FINISH ALL OVER (FA) notes, 69
first-angle views, 21–22
fits. See Inch fits; Metric fits
fittings, 203–204
dimensioning for, 206
flanged fittings, 204
screwed fittings, 203
welded fittings, 203
flame cutting, 109
flanged dovetail joint, 192
flanged fittings, 204
flanges
dimensioning for, 208
symbols for, 208, 209
flange welds, 316–317
assignment, 321
flare-bevel welds, 303
flare-groove welds, 303
flat back patterns for moulding, 232–233
flat head screws, 141
flat keys, 162–163
flatness tolerances, 384–385
angularity tolerance and, 406
flat point set screws, 163
flats, 164
flat surfaces
orientation tolerances for, 405
orthographic sketching assignments, 32, 33
pictorial sketching assignments, 64, 65
straightness of, 368
flat tapers, 91, 93
flaws, 105–106

Fluid Pressure Valve assignment, 286–287
Flying Pig, 326, 342
FN1 light drive fits, 125–126
FN2 medium drive fits, 126
FN3 heavy drive fits, 126
FN4/FN5 force fits, 126
foam rubber, 223–224
foot measurement abbreviation, 3
force fits, 124
types of, 125–126
foreshortened projections, 251
forging, 109
forming, 192–193
form tolerances, 365–368
assignments, 390–391
circularity tolerances, 385–386
cylindricity tolerances, 386–387
flatness tolerances, 384–385
Four-Wheel Trolley assignment, 293
fractional-inch system, 37
frames
for isometric sketches, 59, 60
for oblique sketching, 62
freehand sketching, 16
friction. See also Antifriction bearings
surface texture and, 103
front views, 22
full indicator movement (FIM), 458
full mould casting, 232
full scale, 9
full sections
side view in, 81
sketching assignment, 85–86
full thread studs, 141
functional dimensioning, 38
functional drafting, 97, 99

G

Garden Gate assignment, 11
Gasket assignment, 19
Gates Rubber Company, 224, 353
gate valves, 204
gauging in positional tolerancing, 444–445

Gear Box assignment, 181
gears, 322–330. *See also* specific
   types
   mechanical advantage of, 357
   sizes of teeth, 327
   terminology for, 323–324
gear trains, 336–340
   calculations assignment, 340
   ratio of gears, 336
general-use section
   lining, 79, 80
geometric tolerances, 360–365.
   See also Features of size;
   Maximum material
   condition (MMC)
   datum features and, 397–398
   datums for, 392
   feature control frame,
    362–365
   symbols, 363
Glaeser, W. A., 216
glass section lining symbol, 80
Globalspec, 357
globe valves, 204
Goldengrovehs, 17
Gothic lettering, 3
grey iron, 220
grid lines, sloped surfaces
   with, 54
grid sheets
   for isometric sketching, 58
   for oblique sketching, 60–61
grinding, roughness range
   for, 109
groove pins, 269
groove welds, 294, 303–310
   assignment, 311
   supplementary symbols,
    305–307
Guide Bar assignment, 75–76

H

half dog set screws, 163
half scale drawings, 9–10
half sections, 80, 81
   side view in, 81
   sketching assignment, 90
half views, 251, 252
Handle assignment, 94
Hanger Details assignment,
   114–115
heating system, piping for, 202

height, 1
helical springs, 283–285
hemmed (safe) edges, 192
Hexagon Bar Support
   assignment, 186–187
hexagon head bolts, 141
hexagon head screws, 141
hexagon washer head
   screws, 141
hidden lines, 8
   in oblique sketching, 59–60
   orthographic sketching, 30
   in secondary auxiliary
    views, 183
   in sectional views, 78
hole basis fits system, 130
holes. *See also* Inch fits;
   Metric fits
   alignment of parts and,
    250–251
   basic hole system, 126
   boring, 39–40
   drilling, 39–40
   internal dimensions, 131
   reaming, 39–40
   in sheet metal, 194
   threaded holes, 142
   true centre distance, showing
    in, 251
hollow spring pins, 269
honing, roughness range for,
   109
Hood assignment, 274–275
hot rolling, roughness range for,
   109
Housing assignment, 466–467
Housing Details assignment,
   152–153
Howstuffworks, 216, 224, 326,
   332, 337

I

Identifying Oblique Surfaces
   assignment, 176
IDS Development–Nebraska
   Education, 6, 10, 42, 83,
   291
inch fits, 122–128. *See also*
   specific types
   assignments, 127–128
   on basic shaft system, 126
   defined, 122

   locational fits, 122, 124
   RC4 close running fits, 124
   standard fits, 124–126
inch measurements, 37. *See also*
   Decimal-inch system
   abbreviation for, 3
   assignment problems in, 6
   for structural steel, 289
inch threads, 142, 144
   designation, 144–145
   metric threads compared, 148
   right/left-handed threads, 145
inch tolerances, 118, 120
Inclined Stop assignment, 182
inclined surfaces. *See* Sloped
   surfaces
Incompetech, 6
Index Pedestal
   assignment, 200–201
Indicator Rod assignment, 95
Industrial Coaters List, 236
Industrial Fasteners
   Institute, 149
Industrial Motion Control, 342
inner diameter (ID) of pipe, 202
Integrated Publishing, 10, 42,
   73, 92, 285
interference fits, 122, 123.
   See also Force fits
   drive fits, 124
   force fits, 124
   locational interference fits, 124
   metric fits, 133
   shrink fits, 125–126
Interlock Base assignment,
   246–247
Intermet, 236
internal cylindrical features
   orientation tolerances for,
    412–415
   parallelism tolerance for axis,
    413, 415
   perpendicularity tolerances,
    413–414
internal threads, 142
   inch threads, 144
   metric threads, 145, 147
International Organization for
   Standardization (ISO), 1, 368
   projection symbol, 22
   screw thread representation,
    141–142

International Tolerance (IT) Grade, 129–130
Internet resources, 6
intersection of unfinished surfaces, 83–84
investment casting, roughness range for, 109
isometric axes, 58
isometric ellipse templates, 58–59
isometric sketching, 2, 58–59. *See also* Pictorial drawings
basic steps for, 59, 60
of circular shapes, 58–59
paper for, 4, 58
for piping drawings, 206–207
of sloped surfaces, 58
items list, 283

J

j groove welds, 303
joints, 190, 192
journal bearings, 214–215

K

keys, 141, 162–163
common keys, 162
dimensioning of, 163
sectioning of, 80, 270
keyseats, 162, 163
knurls, 91–92, 93

L

lapping, roughness range for, 109
laser, roughness range for, 109
lay, 105, 106–107
LC1 to LC11 locational clearance fits, 125
leaders, 35, 36
lead section lining symbol, 80
least material condition (LMC), 373, 374
examples of, 374–375
positional tolerancing at, 428–429, 438, 439
symbol, 376
length, 1
of fillet welds, 298
of inclined surfaces, 47
in orthographic projections, 22

straightness on unit-length basis, 377, 381
lettering, 3
limit dimensioning method, 117
limits
concept, 116
definitions for, 116–117
MAX/MIN limits, 117
single limits, 117
two limits, 117
linear measurements, 49
lines. *See also* Mitre lines
break lines, 8–9
centre lines, 13
in geometric tolerancing, 362
hidden lines, 8
profile-of-a-line tolerances, 451–454
styles, 2–3
visible lines, 2, 8
Link assignment, 102
liquids section lining symbol, 80
LN1 to LN6 locational interference fits, 125
locational fits, 122, 124
clearance fits, 125
interference fits, 125
transition fits, 125
LT1 to LT6 locational transition fits, 125
lubricating
journal bearings, 214
surface texture and, 103
lugs, machining, 235

M

Machine Design, 149, 166, 216, 224, 236, 271, 278, 285, 299, 307, 317, 326, 332, 337, 353
Machine Design, 236, 271
machine pins, 266–268
machining allowances, 68, 69–70
machining lugs, 235
machining symbols, 68–70
outdated symbols, 69–70
removal of material prohibited, 70
magnesium section lining symbol, 80
malleable iron, 220

manganese steels, 221
manufacturing materials, 220–229. *See also* Steel
plastics, 223, 224
rubber, 223–224
marble section lining symbol, 80
Maryland Metrics, 132, 166
mass projection, 38
Matching Drawings assignment, 28–29
maximum material condition (MMC), 372–373, 374. *See also* zero MMC
application of, 375–376
examples of, 374–375
orientation tolerances and, 412
perpendicularity tolerance and, 413–415
positional tolerancing and, 428–429, 432–433, 434–436
straightness with, 377, 380
maximum permissible error, 429–431
maximum value, 376
perpendicularity tolerance and, 414
straightness of, 377
MAX/MIN limits, 117
The Mayline Company, 17
measuring principles, 461–462
mechanical advantage, 356–357
mechanical rubber, 223
Meehanite Metal Corp., 236
melt-through symbol, 307
metals. *See also* specific types
multiple-detail drawings, parts in, 96, 98
thicknesses, 190
Metrication.com, 10, 50, 149
metric fits, 129–140
assignments, 137–140
symbol, 130
types of, 133
metric system, 37
abbreviation for, 3
identification of metric drawing, 37
orthographic sketching assignments, 32–33

pictorial sketching
assignments, 64, 67
sloped surface sketch with, 56
metric threads, 145–148
designation, 145, 148
inch threads compared, 148
Michigan Tech, 175
microinches, 104
micrometres, 104
millimetres. *See also* Metric
system
assignment problems in, 6
machining allowances in,
69–70
for structural steel, 289
tolerances, 118–119, 121
milling, roughness range for, 109
minus tolerancing, 117–118
minutes, angle dimensions in, 47
mitre gears, 331
assignment, 334–335
mitre lines
for right-sided view, 24–26
for top views, 24–27
MMS Online, 278
module for gears, 326
molybdenum steels, 222
Morse taper, 91
motor drives, 337
assignment, 338–339
Mounting Plate
assignment, 199
multiple-detail drawings, 96, 98

N

naming of views, 252
nickel-chromium-molybdenum
steels, 222
nickel steels, 221, 222
nodular iron, 220
nonisometric lines, 58
notes
FINISH ALL OVER (FA)
notes, 69
part described by, 99
placement on drawing, 35
on roughness requirement, 107
for surface texture, 106
not-to-scale dimensions, 70–71
NPT (National Pipe Thread), 203
number of teeth (N) on gear,
323–324

numerical control (NC)
dimensioning for, 276
drawings for, 276
for two-axis coordinate
system, 276–278
nuts, 141
sectioning of, 270

O

object lines, 2
oblique sketching, 2, 59–62
basic steps for, 62, 63
paper, 4–5
receding axes in, 60
of sloped surfaces, 61, 62
oblique surfaces, 174–177
secondary auxiliary views
for, 184
Offset Bracket assignment,
75–76, 237
Offset Link assignment, 18
offset sections, 80, 81, 82
Oil Chute assignment,
226–227
1-inch grids, 5
one-view drawings, 96
open-cell sponge rubber,
223–224
orientation tolerances,
403–408
angularity tolerance, 403–404
for cylindrical features, 410
datum, reference to, 403
for features of size, 409–412,
420
for flat surfaces, 405
for internal cylindrical
features, 412–415
maximum material condition
(MMC) and, 412
parallelism tolerance, 403, 404
perpendicularity
tolerance, 403, 404
single feature, application
to, 405
two directions, control in, 405
origin (zero point), 277
o-ring seals, 349, 351
orthographic projections, 2. *See
also* Auxiliary views; Sloped
surfaces; Third-angle
projections

International Organization
for Standardization (ISO)
symbols, 22
in piping drawings, 206–207
sketching paper for, 4
views, 58
outer diameter (OD)
of cams, 341, 342
of gears, 324
of pipes, 202
oval head screws, 141
oval point set screws, 163
overrunning clutch, 349, 351

P

pads, 164
pan head screws, 141
paper. *See* Sketching paper
Parallel Clamp Assembly
assignment, 288
Parallel Clamp Details assignment, 228
parallelism tolerance, 403, 404
for features of size, 411–412
for internal cylindrical
features, 413, 415
partial datums, 398–399
partial sections, 80, 258
partial views, 251–252
parts and holes, alignment of,
250–251
pawls, 356
assignment, 19
pearlite iron grades, 220
pencils, 5
permanent mould casting, 109
perpendicularity tolerance,
403, 404
for external cylindrical features, 415, 418, 419
for feature of size, 409–411
perspective sketching, 2, 5
phantom outlines, 290–291
phosphorus steels, 221
pictorial drawings, 2, 24,
57–58. *See also* Isometric
sketching; Oblique sketching
assignments, 64–67
for circular features, 22–24
circular features assignments,
66–67

for shape description, 21
pillow blocks, 215
pin fasteners, 266–269
    machine pins, 266–268
    radial-locking pins, 268–269
pinion gears, 322
pins, 141
    sectioning of, 270
piping, 202–213
    dimensioning for, 206, 208
    drawings, 204–209
    fittings, 203–204
    isometric projections, 206–207
    kinds of pipe, 202–203
    orthographic projections,
        206–207
    projections, 206
    symbols for drawings, 205,
        206–209
    thread conventions, 203–204
    valves, 204
    wall thickness comparison, 203
pitch
    of coil spring, 285
    of fillet welds, 298
    of threads, 144–145
pitch circle for bevel gears, 331
pitch cone for bevel gears, 331
pitch diameter (PD) of
    gears, 323
Pittsburgh corner lock, 192
plain bearings, 214–216
"Plain Bearings" (Glaeser), 215
plain dovetail joint, 192
plain material, 289
planes. See also Datums; datums
    three-plane system, 294,
        392–393
planing, roughness range
    for, 109
plastics, 223
    common plastics list, 224
    pipe, 203
    section lining symbol, 80
Plate Cam assignment, 346
plating surface texture
    indications, 108
plug welds, 311–312, 320
plus tolerancing, 117–118
point of tangency, 13
points. See also datums
    in geometric tolerancing, 362

point-to-point
    programming, 278
polishing, roughness range
    for, 109
porcelain section lining symbol,
    80
positional tolerances, 428–443
    advantages of, 438–439
    angular hole, part with, 447
    assignments, 442–443
    datum selection
        assignment, 450
    charts for evaluating, 440, 441
    for circular features, 434–438,
        445–448
    datum features, selection of,
        444–450
    internal circular feature as
        datum, 447
    at least material condition
        (LMC), 428–429, 438, 439
    for long holes, 444–445
    on maximum material condi-
        tion (MMC) basis, 432–433,
        434–436
    methods, 428
    for multiple hole datums, 445
    regardless of feature size
        (RFS), 428
    on regardless of feature size
        (RFS) basis, 433, 438
    symbols, 432
    at zero MMC, 436–437
Power Drive assignment,
    354–355
Precision Devices, Inc., 111
Preferred Hole Basis Clearance
    Fit, 130
premounted bearings, 216
primary auxiliary views,
    178–182
    drawing examples, 180
primary datum, 392–393
profile-of-a-line tolerances,
    451–454
profile-of-a-surface tolerances,
    454–456
profile tolerances, 451–457
    all-around profile tolerancing,
        specifying, 452–453
    assignments, 457
    bilateral tolerances, 452

controlled profiles, 453–454
    extent, specifying, 453–454
    profile-of-a-line tolerances,
        451–454
    profile-of-a-surface tolerances,
        454–456
    symbols, 451
    unilateral tolerances, 452
punch and die components, 194
punch press, 193

Q

quick-release pin fasteners, 266

R

Rack Details assignment,
    170–171
radial-locking pins, 268–269
radial locking rings, 350
radius
    abbreviation for, 3
    dimension, 38, 40
    in freehand sketches, 16
    limit dimensions for, 117
Raise Block assignment,
    262–263
ratchet wheels, 356–359
RC1 precision sliding fits, 124
RC2 sliding fits, 124
RC3 precision running
    fits, 124
RC4 close running fits, 124
RC5 medium running fits, 124
RC6 medium running fits, 124
RC7 free running fits, 125
RC8 loose running fits, 125
reaming, 39–40
    roughness range for, 109
rectangles, break lines for, 72
rectangular coordinate
    dimensioning
    in tabular form, 165–166, 167
    without dimension lines, 165
REDRAWN AND REVISED, 71
red shortness, 221
reduced scale, 9
reducing fittings, 203
reference dimensions, 50
regardless of feature size (RFS),
    373–374, 376
    no maximum value
        specification for, 381

positional tolerancing and, 428, 433, 438
straightness and, 377
relative coordinate programming, 278
removable fasteners, 141
removal of material prohibited, 70
removed sections, 80, 157
of thread, 158
repetitive features, dimensioning, 41
retaining rings, 349–351
revisions, 100
tables, 71
revolved sections, 80, 155–156
ribs
alignment of parts and holes, 250–251
in sections, 80, 258–261
right-sided views, 24–26
rivets, 141
for joints, 190
sectioning of, 80, 270
rods, sectioning of, 270
roller bearings, 349
roller burnishing, roughness range for, 109
roller-element bearings. *See* Antifriction bearings
Roof Truss assignment, 12
root diameter (RD) of gear, 324
roughness
average, 104
defined, 104
description and application, 110
notes on, 107
sampling length ratings, 106–107
width/width cutoff, 104
roughness average ratings, 106–107
for common production methods, 109
recommendations for, 108
rounded intersections, 84
round head screws, 141
rounds, 74. *See also* Arcs
dimensioning, 41
rubber, 223–224
section lining symbol, 80

Rubber-Cal, 224
running fits, 122
RC3 precision running fits, 124
RC4 close running fits, 124
RC5 medium running fits, 124
RC6 medium running fits, 124
RC7 free running fits, 125
RC8 loose running fits, 125
RC9 loose running fits, 125
runout tolerances, 458–468
assignment, 464–465
circular runout, 458–460
for cylindrical features, 459–461
symbols, 459
total runout, 460–461
with two datum features, 462–463

S

Saltrie, 342
sand mould casting, 230–232, 235
coping down, 234
cored castings, 234–235
design of castings, 232–234
irregular castings, 232–233
preparation of moulds, 232
roughness range for, 109
sequence for, 231
set cores, 233–234
split patterns, 234
surface coatings, 235
sawing, roughness range for, 109
scale
drawing to, 9–10
not-to-scale dimensions, 70–71
1:1 scale, 9
scarf groove welds, 303
schematic screw thread representation, 141
screwed fittings, 203
screws, 141
set screws, 163–164
types of, 141
seamless brass/copper pipe, 203
seams, 190
examples of, 192
seam welds, 314–316
assignment, 321

secondary auxiliary views, 183–189
secondary datum, 393
seconds, angle dimensions in, 47
sectional views, 78–80
sectioned assembly drawings, 80
sections. *See also* Removed sections
broken-out sections, 80, 258
lining symbols, 79–80
partial sections, 80, 258
revolved sections, 80, 155–156
ribs in, 80, 258–261
spokes in, 80, 260, 261
types of, 80–82
webs in, 258–259
self-locking rings, 350
semi-permanent fasteners
pin fasteners, 266
threaded fasteners, 141
set cores, 233–234
set screws, 163–164
setup point for two-axis coordinate system, 277
Shaft Basis Fit System, 130–131
Shaft Intermediate Support assignment, 159
shafts. *See also* Gears
basic shaft system, 126
external dimension, 131
sectioning of, 80, 270
Shaft Basis Fit System, 130–131
Shaft Support assignments, 19, 160–161, 302
shapes in working drawings, 21
shaping, roughness range for, 109
shearing, 193
sheaves, 352, 353
sheet metal
punching hole in, 194
sizes, 190
stampings, 192–193
Sheetmetal Shop, 194
shimmed bearings, 215
shrinkage allowance, 232
shrink fits, 125–126
side outlet joints, 192
silicon steels, 221, 222

similarly sized features, identifying, 41–42
simple part, drawings for, 99
simplified drawing example, 99
simplified screw thread representation, 141
single-bevel groove welds, 307
single-groove welds, 304
single-line pipe drawings, 204, 205
   symbols/dimensioning, 208
single-row, deep groove ball bearings, 347
single v-groove welds, 307
sketching, 3–4
   arcs, 13–17
   circles, 13–17
   complete views, 16–17
   freehand sketching, 16
   full sections assignment, 85–86
   techniques, 5
   in third-angle projections, 24–26
sketching paper, 4–5. See also Grid sheets
   for isometric sketching, 58
SKF Co. Ltd., 353
slate section lining symbol, 80
sleeve bearings, 214–215
Slide Bracket assignment, 87–88
Slide Valve assignment, 242
sliding fits, 122
   RC1 precision sliding fits, 124
   RC2 sliding fits, 124
sloped surfaces, 47, 178
   decimal-inch system, sketches using, 55
   grid lines, sketches of objects with, 54
   isometric sketching of, 58
   metric system, sketches using, 56
   oblique sketching of, 61, 62
   simple objects with, 48
slots, 49–50
slot welds, 312–313
   assignment, 320
smooth finishes, 103
snagging, roughness range for, 109
Society of Automotive Engineers (SAE), 221

solder for joints, 190
solid centre lines, 13
solid insulation section lining symbol, 80
solid pins with grooved surfaces, 269
Spark Adjuster assignment, 254–255
   naming of views for, 252
specifications in working drawings, 21
spherical roller thrust bearings, 348
Spider assignment, 272–273
   arrangement of views for, 270
split journal bearings, 215
split patterns, 234
spokes in section, 80, 260, 261
spotfaces, 82–83
spot welds, 313–314, 320
spring drawings, 283–285
spur gears, 322, 323–326
   abbreviations, 325
   assignment, 328–329
   calculation examples, 326
   calculations, 330
   working drawings of, 325–326
square groove welds, 303, 307
square head bolts, 141
square keys, 162–163
squares, break lines for, 72
S-shaped beams, 290
stampings, 192–193
standardization, 1–2
Stand assignment, 407
steel, 221–223. See also Structural steel
   alloys, 221–223
   designations of, 222
   pipe, 202–203
   SAE/AISI systems of identification, 221
   section lining symbol, 80
straightedges, 5
straight knurls, 92, 93
straight line development, 190
straightness, 365–366
   assignments, 370–371, 382–383
   of feature of size, 376–377
   with maximum value, 377
   for several directions, 369

specifying, 378
   on unit-length basis, 377, 381
straight pins, 267
straight pipe threads, 203
structural steel
   conical washers, 291
   phantom outlines, 290–291
   shapes, 289–290
studs, 141
subassembly drawings, 281
sulfur steels, 221
superfinishing, roughness range for, 109
supplementary welding symbols, 296–297
Support Bracket assignment, 172–173
surfaces
   of feature control frame, 363–364
   in geometric tolerancing, 362
   oblique surfaces, 174–177
   profile-of-a-surface tolerances, 454–456
   total runout, application of, 460–461
surface texture, 103–106
   control requirements, 107–110
   definitions, 103–106
   ratings, 106–107
   for sand mould castings, 235
   symbols, 106
Swivel assignment, 140
symbols, 3. See also Machining symbols; Weld symbols
   datum symbols, 395–399
   for degrees, 47
   dimension origin symbol, 164–165
   for flanges, 208, 209
   geometric characteristic symbols, 363
   for lay, 105
   for least material condition (LMC), 376
   for maximum material condition (MMC), 374
   for minutes, 47
   for not-to-scale dimensions, 70–71
   piping drawing symbols, 205, 206–209

for positional tolerances, 432
for profile tolerances, 451
for runout tolerances, 459
for seconds, 47
for surface texture, 106
symmetry symbol, 47–49
for valves, 208, 209
symmetrical features, 47–49

T

tabular form, rectangular coordinate dimensioning in, 165–166, 167
tangency, point of, 13
tapered pipe threads, 203
taper pins, 266, 267, 268
tapers, 91, 93
TechSourcer.com, 216
TechStudent.Com, 6, 27, 224, 236, 278, 285, 326, 332, 337, 342, 353, 357
teeth terms for gears, 323–324
templates, 5
    circle templates, 5, 14
    ellipse templates, 61–62
    isometric ellipse templates, 58–59
Terminal Block assignment, 168
Terminal Board assignment, 280
Terminal Stud assignment, 169
tertiary datum, 393
texture. *See* Surface texture
thermoplastics, 223, 224
thermosetting plastics, 223, 224
thickness, 1
    chordal thickness (Tc) of gear tooth, 324
    circular thickness (T) of gear tooth, 324
    of metals, 190
    piping wall thickness comparison, 203
third-angle projections, 21–22, 22–24
    assignments, 30–34, 45–46
    for circular features, 22–24
    sketching views in, 24–26
threaded both ends studs, 141
threaded fasteners, 141–154.
    See also Inch threads
    common threaded fasteners, 146

multiple start threads, 144–145
removed section of thread, 158
right/left-handed threads, 145
standards, 142
threaded holes, 142
three-dimensional sketching paper, 4
three-plane system, 392–393, 394
thrust bearings, 215
The Timken Company, 353
title blocks, 9
title strips, 9
tolerances, 116. *See also* specific types
    circular tolerance zones, 372
    definitions for, 116–117
    deviations permitted by, 361, 362
    drawing callout, 131, 134
    flatness tolerances, 384–385
    International Tolerance (IT) Grade, 129–130
    methods, 117–119
    metric tolerance symbols, 129–131, 135
    minus tolerancing, 117–118
    plus tolerancing, 117–118
    straightness tolerances, 365–368
top views, mitre lines for, 24–27
total runout, 460–461
transition fits, 122, 123
    locational transition fits, 124
    metric fits, 133
trigonometry set, 5
Trip Box assignment, 238–239
Truarc Retaining Rings, 353
truss head screws, 141
T slots, 49–50
turning, roughness range for, 109
twist drills, 252
two-axis coordinate system, 276–278
two-dimensional sketching paper, 4
two-view drawings, 96, 97

U

u groove welds, 303
undercut oval head screws, 141

undercuts, 91, 92
unfinished surfaces, intersection of, 83–84
Unified Engineering, Inc., 299, 307, 317
Unified Thread system, 144
unilateral tolerances, 118–119, 452
unit production, 38

V

valves, 204, 208
vanadium steels, 222, 223
V-belt drives, 349, 352
V-Block Assembly assignment, 154
v groove welds, 303
views. *See also* Arrangement of views
    mitre lines for constructing, 24–26
    partial views, 251–252
    phantom outlines, 290–291
    selection of, 96
    spark adjuster, naming of views for, 252
virtual condition, 373
visible lines, 2, 8
    orthographic sketching, 30
Vogelsang Corporation, 271

W

wastebasket construction example, 193
waviness, 106
wear control, 103
webs in sections, 258–259
welded fittings, 203
welding drawings, 294
    all-around symbols, 300
    basic joints, 295
    multiple reference lines for, 297–298
    supplementary welding symbols, 296–297
welds
    flange welds, 316–317
    plug welds, 311–312, 320
    seam welds, 314–316
    slot welds, 312–313
    spot welds, 313–314, 320

weld symbols, 294–297
  for groove welds, 305–307
  melt-through symbol, 307
  tail of, 296–297
white iron, 220
white metal section lining
    symbol, 80
whole depth (WD) of gear
    tooth, 324
width, 1
Wikipedia, the Free
    Encyclopedia, 6, 27, 179, 368

Winch assignment, 358–359
wood
  break lines, 72
  in section lining, 79, 80
woodruff keys, 162–163
working drawings, 20–21
worm gears, 322
worm's-eye view, pictorial
    sketch in, 57
wrought-iron pipe, 202–203

Z

zero MMC, 377
  perpendicularity tolerance
    and, 414
  positional tolerancing at,
    436–437 •
  straightness and, 379

CONTINUING

LIBRARY
NSCC WATERFRONT CAMPUS
80 MAWIO'MI PLACE
DARTMOUTH, NS B2Y 0A5 CANADA

DISCARDED